PERGAMON INTERNATIONAL
of Science, Technology, Engineering an
The 1000-volume original paperback library
industrial training and the enjoyme

Publisher: Robert Maxwell, M.C.

Geophysics in the Affairs of Man

A PERSONALIZED HISTORY OF EXPLORATION
GEOPHYSICS AND ITS ALLIED SCIENCES OF
SEISMOLOGY AND OCEANOGRAPHY

Jan. 21, 1991

To my friend, Doug Knox
With my complements
and best wishes
Robert Rice

Related Pergamon Titles of Interest

Books

*ANGEL
A Voyage of Discovery

*ANNALS OF THE INTERNATIONAL GEOPHYSICAL YEAR
(in 48 volumes)

GRIFFITHS AND KING
Applied Geophysics for Geologists and Engineers (The Elements of
Geophysical Prospecting)

*MELCHIOR
The Tides of the Planet Earth

RHODES
Language of the Earth

*WHITELEY
Geophysical Case Study of the Woodlawn Orebody

Journals

Computers & Geosciences

Continental Shelf Research

Deep-Sea Research

Journal of African Earth Sciences

Ocean Engineering

Physics and Chemistry of the Earth

Progress in Oceanography

Full details of all the above publications/free specimen copy of any
Pergamon journal available on request from your nearest Pergamon
office.

*Not available under the terms of the Pergamon textbook inspection
copy service.

Geophysics in the Affairs of Man

A PERSONALIZED HISTORY OF EXPLORATION GEOPHYSICS AND ITS ALLIED SCIENCES OF SEISMOLOGY AND OCEANOGRAPHY

CHARLES C. BATES

Honorary Member, Society of Exploration Geophysicists, USA

THOMAS F. GASKELL

Formerly with the Oil Companies International Exploration and Production Forum, London, UK

and

ROBERT B. RICE

Past President, Society of Exploration Geophysicists, USA

PERGAMON PRESS

OXFORD · NEW YORK · TORONTO · SYDNEY · PARIS · FRANKFURT

UK
Pergamon Press Ltd., Headington Hill Hall, Oxford OX3 0BW, England

USA
Pergamon Press Inc., Maxwell House, Fairview Park, Elmsford, New York 10523, USA

CANADA
Pergamon Press Canada Ltd., Suite 104, 150 Consumers Road, Willowdale, Ontario M2J 1P9, Canada

AUSTRALIA
Pergamon Press (Aust.) Pty. Ltd., PO Box 544, Potts Point, NSW 2011, Australia

FRANCE
Pergamon Press SARL, 24 rue des Ecoles, 75240 Paris, Cedex 05, France

FEDERAL REPUBLIC OF GERMANY
Pergamon Press GmbH, 6242 Kronberg-Taunus, Hammerweg 6, Federal Republic of Germany

First edition 1982

Library of Congress Cataloging in Publication Data
Bates, Charles C. (Charles Carpenter), 1918–
Geophysics in the affairs of man.
(Pergamon international library of science, technology, engineering, and social studies)
Bibliography: p.
Includes indexes.
1. Geophysics. 2. Prospecting — Geophysical methods.
I. Gaskell, Thomas Frohock. II. Rice, Robert B. III. Title.
IV. Series.
QC807.B37   1982   551   82-7605

British Library Cataloguing in Publication Data
Bates, Charles C.
Geophysics in the affairs of man.
(Pergamon international library)
1. Physical geology
I. Title II. Gaskell, Thomas F.
III. Rice, Robert B.
551 QE28.2
ISBN 0-08-024026-7 (Hardcover)
ISBN 0-08-024025-9 (Flexicover)

Printed in Great Britain by A. Wheaton & Co. Ltd., Exeter

Acknowledgments

First and foremost, we thank our wives — Bartie, Joyce and Elaine — whose willingness to put up with the long absences and many pressures of geophysical endeavors make them true partners in every sense of the word.

Second, we wish to thank those geophysicists who took the time not only to provide background information but also to provide thoughts regarding the most satisfying facets of their careers. These helpful individuals were Robert Abel, Milo Backus, Anthony Barringer, F. Gilman Blake, Bruce Bolt, Wayne Burt, Albert Crary, C. Hewitt Dix, William Ford, Anton Hales, James Heirtzler, Wilmot Hess, Ralph C. Holmer, J. A. Jacobs, Sir Harold Jeffreys, Sidney Kaufman, Helmut E. Landsberg, Dale F. Leipper, Gordon G. Lill, Roy Lindseth, Waldo Lyons, Arthur Maxwell, Patrick McTaggart-Cowan, H. William Menard, Jack Oliver, Louis Pakiser, Frank Press, David Potter, Roger Revelle, S. Keith Runcorn, Robert Sheriff, Rear Admiral Edward Stephan, USN (Retired), Harris B. Stewart, H. I. S. Thirlaway, James Wakelin, Edward Wenk, Jr., J. Tuzo Wilson and J. Lamar Worzel.

Third, we are particularly indebted to six individuals who provided the vignettes that so vividly and poignantly describe how the profession of geophysics really was a quarter of a century or more ago. These contributors are the late Dr. Wilbert M. Chapman and Professor Maurice Ewing, plus Cecil H. Green, John Holmes, W. Harry Mayne and Ned A. Ostenso.

Fourth, the concepts and patience expressed during personal interviews are deeply appreciated. The persons who underwent such interrogation were Richard Arnett, Bettye Athanasiou, Joseph Berg, Ralph Bennett, Robert Bernstein, Howard Breck, Tate Clark, Charles Darden, Decker Dawson, Richard H. Fleming, William Green, Sigmund Hammer, Eugene Herrin, John Hyden, Carl Kisslinger, Paul L. Lyons, Robert H. Mansfield, Donal Mullineaux, Walter Munk, Louis Nettleton, William Nierenberg, John Northwood, Hugh Odishaw, Boyd Olson, Wallace Pratt, Waverly Person, N. Allen Riley, Sandra and Carl

Savit, Harold Sears, John Sherwood, Father William Stauder, SJ, John Stephens, Robert Van Nostrand, Captain Donald Walsh, USN (Retired), Edward Wenk, Jr., Joseph Whalen and Harry Whitcomb. Oral histories of Dale Leipper, Gordon G. Lill, Walter Munk, William Nierenberg, Boyd Olson, George Shor and Athelstan Spilhaus assembled by Professor Robert A. Calvert and filed in the Texas A&M University Library also provided highly useful background material.

Fifth, we acknowledge with thanks the sizeable group of individuals who dug deep into personal or office files to provide background information. These persons include the Honorable Bruce Babbitt, Peter Badgley, Willard Bascom, Peter W. Basham, Rhonda Boone, Lloyd Breslau, John Borchert, Hazel Bright, Reid Bryson, R. Frank Busby, William Clayton, Arthur E. Collin, B. King Couper, George Cressman, John Crowell, Vicky Cullen, Elizabeth Dickson, Colonel Joseph Fletcher, USAF (Retired), Joan V. Frisch, Wilburt Geddes, Richard Geyer, Alexander R. Gordon, Jr., Jerry Henry, William Heroy, Mrs. Harry H. Hess, George D. Hobson, John H. Hodgson, Sue Hood, M. King Hubbert, William Jobst, David Johnson, Dianne Johnson, Barbara Jones, Jean Keen, John Knauss, Eugene La Fond, Anthony Laughton, K. M. Lawrence, Father Daniel Linehan, SJ, Paul D. Lowman, Jr., Eric B. Manchee, Geoffrey May, Colonel Arthur Merewether, USAF (Retired), J. Murray Mitchell, Richard Perdue, Jon Peterson, Willard Pierson, Marc Pinsel, Andreas Rechnitzer, Edward Ridley, Captain Elliott Roberts, USC&GS (Retired), Henry Rowe, Carol Hastings Sanders, Murray Schefer, A. Schnapf, Mary Sears, Mark Settle, Betty Shor, Howard Simons, S. Fred Singer, Brigadier General Kenneth Spengler, USAFR, Athelstan Spilhaus, A. F. Spilhaus, Jr., Richard Stephens, John Stringer, George Sutton, Marie Tharp, Morley K. Thomas, Warren Thompson, Robert M. White and David Yowell.

Sixth, we are particularly grateful to the learned societies (particularly the Society of Exploration Geophysicists and the American Meteorological Society), and various publishers and authors, all acknowledged in the text, for permission to reproduce copyrighted material in the form of quotes, tables and diagrams.

Finally, several persons reviewed portions of the text to ensure that the facts were properly presented. These helpful persons included Samuel J. Allen, Decker Dawson, Craig Ferris, Valerie Godley, Sigmund Hammer, Roy Lindseth, Paul Lyons, Louis Nettleton, Carl Savit, H. I. S. Thirlaway and Owen Thomas. William B. Shedd, Esquire, of Leonia, New Jersey, kindly edited much of the text to ensure that the layman would find it readable, while Peter A. Henn, Senior

Publishing Manager at Pergamon Press, provided useful advice concerning the book's organization. However, any shortcomings in the book, whether they be of omission or commission, are strictly responsibilities of the authors.

Contents

Approximate Conversion Factors

Monetary Units (8 October 1981)

One British pound (£) equals 1.87 U.S. dollars ($) or 2.24 Canadian dollars (C$). One Canadian dollar (C$) equals 0.83 U.S. dollar ($) or 0.44 British pound (£). One U.S. dollar ($) equals 0.53 British pound (£) or 1.20 Canadian dollars (C$).

Rate of Inflation

Year	U.S. Consumer Price Index	British Wholesale Price Index
1600	—	54
1700	—	83
1800	51	130
1900	25	75
1940	42	100
1950	72	350
1960	89	420
1970	116	530
1980	260	1200 plus
(1981)	(275)	(1350 plus)

References: U.S. Bureau of the Census, *Historical Statistics of the United States*, 1975, page 211.
Jastram, R., *The Golden Constant*, John Wiley & Sons, 1981.

Approximate Conversion Factors

Weights, Measures and Speed

From Metric (SI) to British units

Symbol	When you know	Multiply by	To find	Symbol
kg	kilograms	2.2	pounds	lb
t	tonnes (1000 kg)	1.1	short tons	
cm	centimeters	0.4	inches	in
m	meters	3.3	feet	ft
m	meters	0.55	fathoms	fm
km	kilometers	0.6	miles (statute)	mi
km	kilometers	0.54	miles (nautical)	nm
km^2	square kilometers	0.4	square miles	mi^2
m/s	meters/second	1.9	knots	kt
m/s	meters/second	2.2	miles per hour	mph

Note: 1 metric tonne of crude oil is roughly equivalent to 7.6 barrels or 308.8 gallons (U.S.) depending upon the type of crude oil.

From British to Metric (SI) units

lb	pounds	0.45	kilograms	kg
in	inches	2.54	centimeters	cm
ft	feet	0.30	meters	m
fm	fathoms	1.8	meters	m
mi	miles (statute)	1.6	kilometers	km
nm	miles (nautical)	1.85	kilometers	km
mi^2	square miles	2.6	square kilometers	km^2
mph	miles per hour	0.45	meters/second	m/s
kt	knots	0.5	meters/second	m/s

Frequency

cps	cycle per second	1.0	hertz	hz

Prologue

Every aspect of human life is directly affected by the earth on which we live. The air we breathe, the water we drink, the food we eat, the shelter we construct, the energy we consume, and the elements through which we travel are all directly related to man's optimal use and improved understanding of the planet Earth. "Geophysics" is that part of observational and experimental physics which pertains to this same planet, particularly its atmosphere, hydrosphere, crust, mantle, and core. As a consequence, geophysics is an intriguing scientific and technical field of endeavor full of sharp contrasts, and it provides an unusual blend of the theoretical and the practical, the laboratory and the field, the non-profit effort and the profit-making venture, the keeping of peace and the conduct of war.

This book describes many of the key and intriguing developments which took place within several major fields of geophysics. The authors' main emphasis is placed on describing the geophysical enterprise as an interplay of technical, social, and economic factors, for even the purest science cannot prosper when divorced from the human world around it. Yet this is not a book written by science historians. Instead, it is written by members of the profession who were reasonably well acquainted with many of the key actions and players during the time that the major events took place. All three authors have engaged in the commercial search for oil and natural gas, and all have actively carried out and fostered basic research. Bates and Gaskell first met, however, while engaged in a different mode of geophysics — that of geophysical warfare. This took place on a cold February day in 1944 in a frigid, sunny office just off London's Whitehall where Gaskell was serving as Deputy Science Advisor to General Laycock of Combined Operations, and Bates, as a U.S. Army weather officer posted to the British Admiralty, was looking for assistance in the acquisition of reliable wave data from the English Channel.

Our training was acquired when neither the individual nor his community believed that "London", "Ottawa" or "Washington" could —

xvii

or should — solve all types of local problems, whether they be schooling, physical or mental health, energy supply and pricing, or optimal land use. Instead, our youthful goals were to wrest an ever better standard of living from the earth's resources for the betterment of mankind. As a result, we view with dismay the opinion held by many of our younger political leaders that the objectives of business and government are basically inimical, that a prime role of "big government" is to protect its constituents from the "spoilers of the national patrimony", and that more emphasis should be placed on the redistribution of wealth than on its creation.

Upon reviewing the literature describing the evolution of geophysics over the past six decades, it appeared to us that none of the existing books adequately described the extensive and valuable interplay between academia, industry and government that has so characterized this particular area of science and technology. For example, excellent textbooks by such authors as Sverdrup, Johnson, and Fleming (1942), Byerly (1942), Dobrin (1976), Richter (1958), and Sheriff (1980) paid scant attention to the human and economic sides of geophysics. Intriguing books by wives and daughters of sea-going scientists, such as those by Raitt and Moulton (1967), Schlee (1973), Shor (1978), and Deacon (1971) also tended to emphasize activities of classical, rather than applied, geophysicists. To be sure, recent books by practicing geophysicists and engineers, such as Wenk (1972), Bolt (1976), Petty (1976), and Sweet (1978), filled in parts of this gap in the literature but still addressed only certain aspects of the profession. A fourth approach has been that by skilled science writers, such as in *Continents In Motion* by Walter Sullivan (1974) of the *New York Times,* in *The Politics of Pure Science* by Daniel Greenberg (1967), publisher, *Washington Science News Report,* and in the ongoing historical monograph series of the American Meteorological Society. The fifth approach to documenting the actions and hopes of geophysicists has also become quite common — that of the "special pleading" study generated by the non-profit community and targeted at the need to spend more taxpayer's money on a specific field of interest. Excellent products of this school of thought can be found in the National Research Council's reports such as *Polar Research — A Survey* (1970) and *Trends and Opportunities in Seismology* (1977), the National Academy of Engineering's *Toward Fulfillment of a National Ocean Commitment* (1972), and the National Academy of Science's *Directions for Naval Oceanography* (1976).

But if none of these approaches totally portray the broad scope of how geophysics has affected the affairs of man, what is it that we hope to achieve? Our approach is to address, in a personalized history format, the discipline of exploration geophysics and its closely allied sciences of seismology and oceanography. This means the intentional omission of other important facts of modern geophysics— meteorology, hydrology, glaciology, volcanology, aeronomy, and magnetospheric physics — even though these, too, are incorporated within the International Union of Geodesy and Geophysics's realm of interest. Space limitations have also required the saga to be primarily limited to those activities undertaken by the citizens of Canada, Great Britain, and the United States of America, with emphasis on the latter, for that is where the largest number of people have found it appropriate to call themselves "geophysicists".

Because this chronicle stresses challenge and change, the overview is bracketed by two major flex points in Western civilization — the initial waging of deadly global war (1914—1918) and the onset of skyrocketing energy costs that leaves the world's economies and politicians in disarray (1973 on). To be sure, there have been about 100 successive lifetimes since the Sumerians first invented writing. Yet it has been our particular fortune to spend nearly one cent of civilized time within a period of enormous and catastrophic change. Thus, our generation has lived simultaneously in the "Nuclear Age", the "Age of Velocities (or Jet Engines)", the "Age of Electronics and Computers", the "Age of Mass Communications", and the "Age of Population Explosion". Perhaps the most worrisome facet of this change, besides the rapid growth in population and the potential for nuclear war, is the reappearance of appalling monetary inflation rates and the associated turmoil and uncertainty in human and property values. Before 1940, prices had been so stable over the previous century that a "nickel beer" was a "nickel beer" from the times of Presidents Thomas Jefferson to Woodrow Wilson, and a "three-penny beer" the same from the times of the Duke of Marlborough to David Lloyd George.[1] Today, unfortunately, the consumer price index stands at more than six times that existing at the start of World War II. The reader is cautioned, therefore, to make appropriate price adjustments as he goes along because monetary values expressed here are in the values of the moment (see listing of conversion factors).

The narrative style is semi-technical in order that the text may prove interesting to others besides practicing geophysicists. Much that is contained herein should actually appeal to those family associ-

---

[1.] In fact, English wholesale prices remained relatively stable for two centuries beginning in 1717 when the physicist, Sir Isaac Newton, then Master of the Mint, placed the pound on a gold standard (Jastram, 1981). Because gold is too scarce to be debased, this worked well until it became necessary to finance World War I during 1914. Even today, the annual production rate for gold is only about two percent of estimated above-ground supplies.

ates who wondered why their favorite geophysicist found it necessary to be away for weeks to months at a time. Certain events, if not well described elsewhere and personally known to us, are narrated in extra detail. The overall and quite complex story is portrayed in a discursive manner, for there are no short-cuts within the "time machine". To accomplish this, the book treats seven geophysical epochs, roughly broken into decades. This time sequence is then supplemented by a major chapter describing the business side of geophysics, for it is one of the most rapidly growing industries on the international scene. Finally, because there is such a strong intermingling of science, technology, and human endeavor within geophysics, the closing chapter is written by the profession itself. This feat is accomplished by having nearly two score noted geophysicists delineate what they consider the most outstanding actions they were ever involved in, and then supplementing that material by six vignettes written in the first person that clearly portray the uniqueness of the profession.

After having carefully reviewed and contemplated the accomplishments of geophysicists during the past century, we continue in the firm belief that the best process for understanding and optimally using the physics of the earth is the continued fostering of a strong partnership between academia, industry, labor, and government in which all readily participate and none forcibly dominates the others. Today's trend towards dealing with social and economic problems of high geophysical content by manipulating public opinion via slanted "public hearings", noisy public demonstrations and biased media presentations, rather than through careful consideration of finite physical facts, distresses us. Even the art of "grantsmanship" appears to lean too far towards use of the scary "Chicken Little sky-is-falling" syndrome or towards that trusty alternative, the claim that "The Russians are coming!" To you of the coming generation who will apply the tools and techniques of geophysics in solving many of the world's key problems, we cite Herman Wouk's dedication (in Hebrew) of the book, *The Winds of War*, to his two young sons. His thought therein was but one simple word of guidance — REMEMBER!

CHAPTER 1

Some Antecedents to the Modern-day Profession of Geophysics Through World War I

The roads you travel so briskly lead out of dim antiquity, and you study the past chiefly because of its bearing on the living present and its promise for the future.

Lieutenant General James G. Harbord, U.S. Army (Retired) (1866–1947)

Diffusing Geophysical Knowledge

Scientists have long been trained to build on the successes or failures of their predecessors, their teachers, and their fellows largely through scientific associations and their publications. Such societies range from small, local ones to huge organizations with membership drawn from over 100 countries. The oldest and most prestigious for geophysicists is the Royal Society, given both its name and charter by Britain's King Charles back in 1660. The Royal Astronomical Society chartered in 1820 has also had a marked interest in geophysical matters, even to the extent of publishing a *Geophysical Journal,* because the earth is very much a part of the planetary system. Within the United States, the prestigious *N*ational *A*cademy of *S*ciences (NAS) was started as an ally of government at the initiative of President Abraham Lincoln who asked the scientific community in 1863 for technical assistance with the war effort.[1] Geophysical societies *per se* did not appear until the early 1900s. As a result of the great San Francisco earthquake of that year, the *S*eismological *S*ociety of *A*merica (SSA) was formed in 1906. The *I*nternational *U*nion of *G*eodesy and *G*eophysics (IUGG) came into being during 1911, while its U.S. interface, the *A*merican *G*eophysical *U*nion (AGU), was finally

---

[1.] In 1916, the NAS further expanded its capabilities for meeting the wartime and civil needs of government for scientific and engineering advice by creating the *N*ational *R*esearch *C*ouncil (NRC) which could draw on far more than the 150 members of the Academy. By 1979, the NRC had 800 boards, committees, panels, and *ad hoc* study groups.

1

organized in 1919.[2,3] The field of exploration geophysics lagged even further, with the *Society* of *Exploration* *Geophysicists* not being incorporated until 1930. In Canada, achieving a critical mass for independent societies outside of such major groupings as the Royal Society of Canada was difficult to achieve, and Canadian geophysicists tended to orient themselves toward the appropriate "American" grouping based in the United States.

Some Geophysical Forbearers Prior to the 19th Century

However, long before the advent of scientific societies, perceptive men had been contending with the physical forces of nature. In fact, Aristotle (384–322 B.C.) compiled the first known geophysical treatise, the *Meteorologica*, less than half of which pertained to weather matters, thereby allowing the remainder to deal with oceanography, astronomy, and "meteors" (also called shooting stars). Formal seismic instrumentation appeared as early as A.D. 132 when Chang Heng set up a seismoscope in China that not only indicated that an earthquake had occurred but also the direction of the first motion. However, man's formal knowledge of the physics of the earth did not change much from the time of Aristotle until late in the European Renaissance, when the fertile mind of Leonardo da Vinci (1452–1519) initiated new thinking on this subject, as he did in so many others. Early in the 16th century, he studied, for example, the tides of the Euxine (Black) and Caspian Seas, as well as the mechanics and inherent dangers of rock slippage along a geological fault near Florence, Italy. He also deduced that Alpine rocks were at one time submerged for he found embedded sea shell fossils. He observed that the speed of a falling body increases with time, but did not derive the exact relation. This amazing man also had some practical knowledge of hydrodynamics, including wave propagation on the water's surface. Studying light and sound, he recognized that reflection of an optical image is like a sound echo, and that the angle of reflection is equal to the angle of incidence. Thus, five centuries ago, Leonardo demon-

2. As of 1981, the IUGG was funded at the rate of $236,000 annually. The United States provided the largest annual donation ($24,000), followed by Great Britain, the Federal Republic of Germany, and France ($16,000). The next lower category included the Soviet Union and Japan ($12,000). Thirty-three countries, alphabetically ranging from Algeria to Zimbabwe, contributed $800 annually. The IUGG's President was Dr. George D. Garland of the University of Toronto, while its Secrétaire Général was Dr. P. Melchior based at the Royal Observatory, Brussels, Belgium.

3. The AGU, as a standing committee of the NRC, had as its objectives: ". . . to promote the study of problems concerned with the figure and physics of the earth, to initiate and coordinate researches which depend upon international and national cooperation, and to provide for their scientific discussion and publication".

strated an elementary knowledge of geology, the earth's gravity field, and wave propagation and reflection, all of which are an integral part of modern geophysics.

Modern-day exploration geophysics is based on four important earth properties: (1) density, typically measured as the local force of gravity; (2) magnetization, expressed in terms of local magnetic force; (3) electrochemical, variously measured by surface or down-hole electrodes, Geiger-type counters, aerial antennae, or gas-samplers; and (4) acoustical response, measured in terms of voltages derived from seismometers or hydrophones. How scientists and engineers learned to measure these properties from the time of Leonardo to the present has been described in extensive detail in the various editions of the *Encyclopaedia Britannica* and in numerous technical publications issued over the years. Suffice it to say that many of the early "geophysicists", whose notable actions are encapsulated in Table 1.1, remain world famous and their basic concepts are utilized even today.[4]

A Brief Overview of Earth Physics During the Period 1800–1919

Although today's geophysical forbearers were able to stake out and explain specific physical phenomena of the earth, they also faced the difficult problem of having their particular segment of the total system, as was every other one of the component parts, subject and responding to a complex set of forces, each of which involved a large number of variables. Even more confusing to these early workers was that many of the observable variables changed with time and physical position. To make sense out of such confusion, the researcher is heavily dependent on four allied and supporting technologies — instrumentation, transportation, communication, and computation. Thus, despite the unique skills and great talent of the early "geophysicists", major advances in this field that would truly benefit mankind on a large scale were not possible until after the invention of the telegraph in 1840, the radio in 1900, and the vacuum tube amplifier in 1905.

As a consequence, the development of the geophysical profession during the 1800s was a very slow one and largely academic in nature.

[4.] The term, "geophysics", first appeared in 1853 when a German lexicon used it as a substitute for the term "earth physics".

TABLE 1.1

Some Notable Geophysical Findings in the 16th through the 18th Centuries

Scientist	Accomplishments of geophysical interest
William Gilbert (1540–1603)	While Court Physician to Queen Elizabeth, founded the sciences of magnetism and electricity; observed that north-pointing tip of magnet dipped at angle dependent on latitude.
Viscount Francis Bacon (1561–1626)	Besides serving as Lord Chancellor (1618), expounded on nature of gravitational and magnetic forces and stressed need for physical experiments. Suggested southern tips of South America and Africa would prove to be homologous.
Galileo Galilei (1564–1642)	Developed correct formulae for motion of a pendulum (basic term for force of gravity now termed a "gal"); invented primitive thermometer (1593).
René Descartes (1596–1650)	Founded analytic geometry; published 10-part discourse, *Meteorology*, on winds, clouds, and precipitation (1637); measured index of refraction to explain rainbow.
Unknown	Used magnetic compass to search for Swedish iron ore deposits (*circa* 1640).
Evangelista Torricelli (1608–1647)	Invented mercury barometer to demonstrate presence of atmospheric pressure (1643).
Christian Huygens (1629–1695)	*Traité de la Lumière* explains refraction and behavior of wave impinging on an interface, as well as diffraction (1693).
Sir Isaac Newton (1642–1727)	Discovered (independently from Baron von Leibniz) concept of the calculus (1669); stated basic laws of motion and explained true cause of ocean tides in his *Principia*.
Pierre Bouguer (1698–1758)	Compared regional mass of earth with local mass of mountain via pendulum measurements (1740). (Correction of gravity measurements for elevation above a datum is now termed the "Bouguer correction".)
Benjamin Franklin (1706–1790)	Studied lightning and recommended use of lightning rods (1749); elected member of Royal Society from America (1750); published comprehensive chart of Gulf Stream based on current and temperature measurements (1770); postulated continental drift (1780); attempted to track storms over North America.
Reverend John Michell (1724–1793)	As Woodwardian Professor of Geology at Cambridge University, published important first paper on cause of earthquakes (1760); described method for measuring gravitation field by a "torsion balance" (1777). (Charles Coulomb also built similar device to study magnetic and electrical attraction in 1777 as well.)
Captain James Cook, RN (1728–1779)	First global navigator to apply scientific method to geographic exploration, including initial extensive use of chronometer to determine longitude rapidly during three long cruises to the Pacific Ocean region between 1768 and 1779. (His chief scientist was Sir Joseph Banks.)
Count Pierre Laplace (1749–1827)	Paper on tidal theory written in 1778 uses equations similar to those in 1835 when Coriolis explained why gases and fluids, such as air and water, tend to flow along, rather than across, lines of equal pressure (isobars) on a rotating sphere, such as earth.

To be sure, in 1845 the Irish engineer, Robert Mallet (1810–1881), began some pioneering work on the use of "artificial earthquakes" to measure the velocity of seismic waves through differing surface materials. His approach consisted of detonating buried gunpowder charges and timing, by an electrical chronometer, the disturbance of a spot of light falling on the surface of a bowl of mercury. This effort was well conceived, but the glitter method still failed because it did not show the early weak arrivals of the all-important "compressional wave" typically used for purposes of seismic exploration.

The second major geophysical advance of the 1800s was that of mapping and interpreting the earth's gravity field. Much of this advance was due to the efforts of Baron Roland von Eötvös (1848–1919), a Professor in Experimental Physics at the University of Budapest. By 1890, von Eötvös had completed his first single-beam torsion balance as a sizeable improvement over those built by Coulomb and Michell more than a century before. Then, after making extensive field surveys in Hungary with his single-beam device, von Eötvös invented during 1902 the double-beam torsion balance, essentially the instrument of today, and used it to delineate the subsurface extension of the Jura Mountains in France. The chief competitive method for measuring gravity – using a "gravity meter" to measure the displacement of a weight on a spring – had been suggested back in 1833 by Sir John Herschel (1792–1871), but never developed further during that century to provide accuracies comparable to those obtained by measuring the beats of a pendulum.

The most significant global initiative of the period was the establishing and interpretation of earth physics data from observational networks. At the urging of the famed German geographer and climatologist, Baron Friedrich von Humboldt (1769–1859), magnetic stations were set up in many places on the globe during the 1830s. One of the key contributors was Professor Karl F. Gauss (1777–1855) who, as director of the University of Göttingen's observatory between 1807 and 1855, invented the bi-filar magnetometer and began the Magnetischer Verein (Society for Magnetic Research) which started making accurate, simultaneous magnetic measurements throughout Europe between 1836 and 1841.[5]

Seismic networks lagged magnetic networks by more than a half century. The first accurate earth displacement meter (or seismometer) was not constructed until 1880 by a team of British workers, Gray, Ewing, and Milne, at the Imperial College of Engineering in Tokyo.

[5.] The standard unit of magnetic intensity is termed the "gauss", while the curve generated by plotting observational error is known as a "gaussian curve".

Their approach was to support a heavy mass within a horizontal bracket pendulum such that the mass had a free restoring period of 12 seconds, or about six times that of the vertical pendulum in a "grandfather clock". With great inertia and a relatively independent suspension, the seismometer's undamped, slow-moving mass would lag sharply behind the earth's movement during an earthquake and, when linked to a stylus, trace out the differential motion between the pendulum and the earth during a seismic event. Mechanical linkages for multiplying this motion — true "Rube Goldberg" devices — were used even long after the Russian, Prince B. B. Galitizin, developed electromagnetic seismographs as early as 1906 that drove mirrored galvanometers which deflected beams of light over photographic paper. Thus, as late as 1922, de Querrain and Piccard built the "Universal seismograph" at Zurich, Switzerland, which used a mass of 21 tons. In this instance, the vertical sensor had a one-second period and a static magnification of 1,500, while the two horizontal components had three-second periods and a static magnification of 1,300.

The first seismographs in California were installed at Berkeley and the Lick Observatory during 1887 by the University of California's Department of Astronomy. In 1895, Professor John Milne (1850–1913) finally returned to London's Royal School of Mines and by 1897 was able to have the British Association for the Advancement of Science propose that a world-wide network of seismic stations be established. The concept was well received by directors of meteorological and astronomical observatories, and in 1899, a coordinated network of 27 stations was operating on all continents (Herbert-Gustar and Nott, 1980). The Association then began publishing a list of global seismic events, and the International Seismological Association was formed in 1903, with ultimate headquarters at the Institut de Globe, Strasbourg, France, to facilitate data exchange.[6] With so much raw seismic data available, it was necessary that the results be interpreted, and in 1901 the first, of many, "Institute of Geophysics" was started by Professor Emil Wiechert at the University of Göttingen.

Almost immediately, analysis indicated that earthquakes recorded from the same distant area looked amazingly alike on seismograms obtained at a given site. As the detectors improved, it was also apparent that the large waves in the most conspicuous part of the record would be preceded by as much as a half hour by smaller, shorter-period wavelets which could be further split into primary and secondary

6. The British listing of seismic events continues even today at the International Seismological Centre near Newbury, Berkshire.

tremors (see Fig. 1.1). To sort this all out, seismologists agreed on an international nomenclature based on Latin. The initial wave, which was compressional-rarefactive, was labeled "P", the immediately following transverse (or shear) waves, "S", and the later, large, long-period wave train was labeled as consisting of "L", "M", and "C" waves. Study soon showed that the "P" and "S" waves traveled directly through the earth and thereby arrived at distant stations from below, while the larger waves were primarily surface waves, including some of the type originally predicted by John William Strutt (Lord Rayleigh) and expanded upon in 1911 by Augustus E. H. Love.[7]

Because of the multiplicity of wave paths between the seismic event and the seismic recorder, sorting out which phase was arriving at a given time could become a complex endeavor. In fact, a common saying soon arose that "two (seismic) shocks are the last refuge of a seismologist", as two near-simultaneous shocks from different epicenters could explain almost any series of recorded phases.

In 1909, the Jugoslav, A. Mohorovičić, in studying seismograms from an earthquake which occurred not far from Zagreb, observed

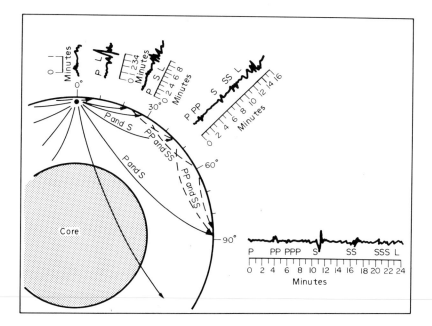

FIG. 1.1. Schematic diagram showing increase in complexity and duration of seismic wave train with distance from epicenter.

[7] Two of the dominant surface phases are now called "Rayleigh" and "Love" waves.

that if the records were taken within 150 km of the source, the initial waves were sharp, followed by a lesser motion, and possessed a velocity of about 5 km/sec. Farther out, however, initial arrivals started with a comparatively small, long-period motion (now called P_n waves), followed by at least one larger, sharper wave train of shorter period, and possessed a velocity of about 8.0 km/sec. His calculations soon showed that the faster waves came in first at the greater distances because they traveled deeper and through rock with much higher velocities than that which existed near the surface. This interface between slower and faster rocks was explained by there being a marked difference between "crustal" and "mantle" rocks. As a consequence, the interface was labeled the "Mohorovičić discontinuity" (or "Moho" for short) and has since been found to exist at depths of as little as 5 km below the ocean floor and to more than 60 km under certain mountain roots. In 1912, 23-year-old Dr. Beno Gutenberg (soon to be a meteorologist in the German Army) found, while working at the Strasbourg center, that something was blocking reception of P_n wavelets at distances of $108°$ to $143°$ of latitude from earthquake epicenters.[8] Such a blockage, he calculated, could be created by an "earth core" slightly larger than the Planet Mars starting at a depth of 2,900 km. This important finding, of course, verified the earlier claims of Emil Wiechert (1896) and Richard D. Oldham (1897) that the earth consisted of two major components, rather than just one vast homogeneous sphere (Brush, 1980). Thus, just 13 years after the start-up of the first global seismological network, geophysicists were able to point out that the earth consisted of three important elements — a core, a mantle, and a crust.

Despite the frequent occurrence of damaging earthquakes within the United States, local chambers of commerce and state legislators were not anxious to see the issue publicized. Hence, when the Seismological Society of America caused a bill to be introduced into the Congress during 1910 which would establish a "Bureau of Seismology" within the Smithsonian Institution, the concept was rejected even though it would have cost but \$20,000 per year. As a consequence, the U.S. Weather Bureau, as part of the Department of Agriculture, operated just a few seismographs at selected weather stations until 1925, when the function was finally transferred to the *C*oast and *G*eodetic *S*urvey (C&GS) within the Department of Com-

---

[8.] P_n waves are compressional-rarefactive in nature and analogous to sound waves. Associated with them, but traveling at about 0.6 their velocity, are S_n waves, which are transverse (or shear-like) to the direction of wavefront advance, as in the transverse vibration of a string.

merce.[9] To fill this gap, Professor Willis Moore, Chief of the Weather Bureau, encouraged the highly educated and organized Society of Jesus (SJ) to enter the field.

In a letter of 2 February 1909 addressed to all Jesuit colleges in North America, Father F. L. Odenbach, SJ, of St. Ignatius College, Cleveland, Ohio, pointed out that the Society was in a favorable position to establish a truly standard seismic network not only domestically but abroad. Such standardization was much needed for the stations then in existence used widely varying instruments, thereby causing an analyst to spend as much time puzzling over instrument response as over earth motion. Moreover, establishing such a standard network would be relatively cheap for it could use the Wiechert pendulum manufactured by Spindler and Hoyer of Göttingen and costing but $112 including timing clock. His letter concluded:

> Now, dear Father, if you think you can join in this movement to the great benefit of your institution, the glory of our Society, and the honor of that Church which has been proclaimed an enemy to all modern progress and enlightenment, please drop me a line of encouragement so that I may proceed in this important work and send you more detailed information.

The response to Odenbach's plea was excellent. By early 1911, 15 Jesuit colleges in the United States and one in Canada (St. Boniface College, Saint Boniface, Manitoba) had seismic observatories either operating or planned, with the central data collection point being St. Ignatius College, which then forwarded consolidated information on to the center in Strasbourg.

In Canada, the geophysical situation was somewhat the same. The initial Milne seismographs had been given to the Meteorological Service of Canada to be operated at Toronto and Victoria. In 1905, the central government then established the Dominion Observatory at Ottawa, and, eight years later, permitted the creation of three geophysical subdivisions (magnetics, gravity, and seismology) led by Charles French, A. H. Miller and Ernest A. Hodgson under the general direction of the lead geophysicist, Dr. Otto Klotz. Although Hodgson had been educated as an astronomer, Klotz taught him to read seismograms and determine epicenters; as a result, Hodgson did not receive formal training in seismology until 1930, when he took his Ph.D. under Father Macelwane at St. Louis University. In the meantime, he had become Chief of a separate Seismological Division in 1919, a position he held until retiring 32 years later. Much of this career was actually spent in gradually expanding the Canadian Seismic

---

9. As late as 1918, the U.S. Weather Bureau's Chief, Professor C. F. Marvin, began a two-year stint as President of the Seismological Society of America.

Network, which included picking up the stations of the Meteorological Service in 1936, even though the annual budget for new equipment was seldom more than C$200 and his staff grew at the rate of about one per decade.

Artillery Location by Acoustic and Seismic Means

Experience gained in using early seismic networks to locate the origin of distant earthquakes soon led to locating heavy artillery positions of the enemy once the "Western Front" stabilized during October, 1914. Using carbon-grain microphones instead of seismometers as energy detectors, the French led in introducing "reparage par le son". Signal recording was a problem; after trying three methods, they eventually used the "Telegraphic militaire" system of mechanical levers in order to avoid the complexities of photographic recording under battle conditions (Sweet, 1978). But even when their acoustic units located "Big Bertha", the 100-km range gun which regularly fired on Paris, the French still chose to direct their counter-battery fire at a camouflaged dummy gun spotted by aerial photography.

After familiarization with the French methods, Captain Lawrence Bragg, the 1915 Nobel Prize winner (physics), was sent to the British front during September, 1915 with another officer and six men to set up an experimental sound ranging system of their own. Bragg's unit typically spread a number of microphones, usually six, along a 4-km base parallel to the front line but about 1.5 km behind it, and then wired the microphones back to a central recording station. Once three microphones had recorded the same gun, it was then a simple matter of solving the hyperbolic equation indicating where the source had to be in order to create the observed arrival times along the microphone array. Unfortunately, the British microphones were more sensitive to the higher frequencies, such as caused by rifle and machine-gun fire, than they were to the low-frequency "booms" of heavy guns. Within a year, Lance Corporal Tucker had developed a "hot-wire" microphone, however, that worked on the "cooling principle", and soon excellent fixes were being obtained whenever meteorological conditions were favorable, as when the wind blew from the enemy lines or, better still, there was quiet, foggy weather. The British Sound Ranging Division grew rapidly in size, and by the end of hostilities there were over 40 units in the field employing 500 men.

Across the lines, German geophysicists had been pursuing a rather similar line of attack. For two years after the outbreak of war, Dr. Ludger Mintrop, a product of Göttingen's Institute of Geophysics, had tried to convince his General Headquarters of the desirability of using portable mechanical seismographs to locate enemy guns. In 1917, General Ludendorff finally did order "100 seismic troops" into the field, but never with full implementation. Mintrop, in fact, ended up playing only a minor part in this field deployment.

This sharply contrasts with the American approach in 1917. Upon entering the war, General Pershing, Commander of the American Expeditionary Force, immediately ordered Dr. Charles B. Bazzoni, a University of Pennsylvania physicist, to determine which was the best approach — French or British. While a first lieutenant, Bazzoni chose the British system and stayed on to direct U.S. Sound Ranging Section #1 when it went into action along the St. Mihiel sector on 11 March 1918. Eventually nearly 700 Americans were involved in this facet of applied geophysics before Armistice Day later that year. Most of this expansion used American equipment patterned after the British gear. Such American instrumentation was largely developed by the Western Electric Company in cooperation with the Sound Section of the U.S. Bureau of Standards. This bureau support effort is of special interest historically for it involved four men who later would become famous names in petroleum geophysics — Dr. Engelhardt A. Eckhardt, Technical Chief of Section; Burton McCollum, Administrative Chief of Section; Dr. William P. Haseman, on leave from the University of Oklahoma, and John C. Karcher, a brilliant student of Haseman's who had moved on to the University of Pennsylvania's graduate school where he studied physics in the same department that employed Bazzoni and Eckhardt as faculty members. Although none knew it at the time, their involvement in artillery location would soon blossom within the coming decade into a totally new and highly profitable profession — that of seismic exploration for oil and natural gas. Moreover, with a suitable touch of irony, their initial major competitor would again be the German seismologist, Dr. Ludger Mintrop.

Oceanography Simultaneously Becomes a Science and a Technology

One of mankind's earliest and most persistent technologies has been that of using a weighted (lead) line for determining water depth. There is actually a diagram of a sounding lead made by an Egyptian artist of the 12th Dynasty (approximately 1790 B.C.), and the technique did not go out of use until introduction of electronic echo-sounding in the early 1920s. As a discipline, however, marine science can probably be said to date back only to the 1830s. Between 1831 and 1836, the British vessel, HMS *Beagle*, with Charles Darwin on board as chief naturalist, made its epochal cruise during which he formed his concept of evolution and also his theory that the formation of coral atolls was related to the sinking of volcanoes.

In the same year that the *Beagle* returned, the U.S. Congress also authorized that fledgling nation's first major scientific cruise, the U.S. Exploring Expedition, a forerunner to the great federally-funded oceanographic cruises of more than a century later. In naval orders handed to the expedition's leader, Lt. Charles Wilkes, USN, expedition goals were stated as including ". . . the promotion of the great interests of commerce and navigation, and the extension of scientific knowledge". The six-ship expedition, which continued from 1838 to 1842, covered much of the same general area as had Captain Cook's explorations; in addition, Wilkes' party landed on the actual Antarctic land mass. His observational program was particularly extensive and thorough, for it included magnetic dip, intensity and declination, tides, solar radiation, surface and sub-surface currents, ocean transparency, wave conditions, deep-sea soundings, bottom samples, and marine biology. End-products were impressive for, when printed, they resulted in a series of 17 volumes and 106 new nautical charts (Stanton, 1975).

The biggest "nautical operator" of the mid-1800s, however, was Lieutenant Matthew Fontaine Maury once he became Superintendent, Navy Depot of Charts and Instruments (the forerunner of today's Naval Oceanographic Office) on 11 July 1842. By combining sound judgment, a far-ranging mind, a talent for promotion, and tremendous energy despite an improperly healed broken leg which kept him from sea duty, Maury soon compiled a remarkable series of "Pilot Charts" and "Sailing Directions" that promptly earned him the sobriquet of "Pathfinder of the Seas".[10] Based on an early version of "pressure pattern navigation", his work enabled sailing ships to cut transit times by as much as 50 percent (Lyman, 1948). Convinced of the possibility of laying telegraph cables between Europe and North America, by 1852 he had three ships engaged in deep-sea research, and in 1854, reported to the Secretary of the Navy that the sea floor between Ireland and Newfoundland was "a plateau which seems to have been placed there especially for the purpose of holding the wires of a submarine telegraph, and of keeping them out of harm's way". In the meanwhile, he was writing one of the world's all-time "scientific

---

10. By 1858, the Depot had issued 20,000 copies of "Sailing Directions" and 200,000 copies of "Wind and Current Charts" without charge in return for cooperating mariners submitting complete and accurate daily marine observations.

best-sellers" — his *Physical Geography of the Sea.* First published in 1855, the book went through 22 editions in England alone and had as its objective what sounds very modern even today:

> A philosophical account of the winds and currents of the sea; of the circulation of the atmosphere and ocean, of the temperature and depth of the sea; of the wonders that lie hidden in its depths; and of the phenomena that display themselves at its surface. In short . . . of the economy of the sea and its adaptations — its salts, its waters, its climates, and its inhabitants, and of whatever there may be of general interest in its commercial uses or industrial pursuits.

But Maury also had his detractors. The most prominent was Alexander Bache, the great-grandson of Benjamin Franklin, and then Superintendent of the United States Coast Survey. Their battle echoes down through the years even until today as policy-makers continue to debate the proper role and scope of effort for their two agencies — the military directed Naval Oceanography Command and the civilian-run National Oceanic and Atmospheric Administration. Maury's theories on the cause of ocean currents also seemed so implausible to a Tennessee school-teacher, William Ferrel, that he put forward his own more scientific ideas as early as 1856. Eventually, Ferrel built one of the earliest tide-prediction machines and became, at least in the opinion of some, the world's foremost physical oceanographer in the 1870s and 1880s (Burstyn and Schlee, 1979).

As Ferrel faded from the scene, the dominance in marine science shifted back to Europe. In fact, the most noted scientific cruise of all time — that of HMS *Challenger* — took place between 1872 and 1876, an effort so comprehensive that oceanographic papers based on its biological and bottom sampling were appearing as late as the 1920s. In a physical sense, however, the dominance shifted to the Institute of Geophysics at Bergen, Norway. A decline in North Sea fisheries had led to forming the International Council for Exploration of the Sea in Copenhagen, Denmark, during 1903. The famed Norwegian polar explorer, Dr. Fridtjof Nansen, then 42 years old, took a deep personal interest in this joint effort which not only addressed the issue of biological productivity of the sea but also the associated physical oceanography. Much needed was a measurement technique which would show the nature of current-flow patterns within the fishery region. Nansen interested the distinguished Norwegian hydrodynamicist, Professor Vilhelm Bjerknes, in this problem, and by 1905, two of Bjerknes' students, J. W. Sändström and Bjørn Helland-Hansen, developed a method displaying the field of geostrophic flow at any surface above one of "no motion", such as might be found at a depth of a thousand meters.[11] One of the major

---

11. This type of flow field was determined by taking extremely precise salinity and

users of this technique came to be the International Ice Patrol operating within the Grand Banks and Labrador Sea region. Formed by the U.S. Coast Guard just two years after the disastrous sinking of the "unsinkable" *Titanic* in 1912, the Patrol, at the suggestion of Dr. Henry B. Bigelow, Curator of Oceanography at Harvard University, attempted to use the Norwegian "hydrographic station" technique during their very first year on the Grand Banks. However, it never did succeed in getting useful results until a young Coast Guard officer, Lt. Edward H. Smith, went to Bergen in 1924 with several years of back data to find out what they were doing wrong.[12]

Although the Americans had lost the lead in marine science, they did eventually create a new marine technology — that of underwater acoustics. As early as 1901, a Boston-based group of scientists and inventors, convinced that sea water was the most reliable medium for transmitting signals to ships, had formed the Submarine Signal Company (Fay, 1963). In 1910, Professor Reginald Fessenden joined the firm; second only to Thomas Edison in inventive ability, his assignment was to develop an underwater communication system for military use.[13] By 1912, "Sub Sig's" underwater hazard signaling system, based on the use of submarine bells (or clappers) and ship-mounted port-and-starboard underwater microphones, was installed on hundreds of ships, while submerged signaling devices were operated at dangerous coastal points by such countries as the United States, Canada, England, France, Portugal, Spain, Italy, Brazil, Chile, and China. By steadily improving on an underwater oscillator he invented in 1911, within three years Fessenden had a device that could send up to 20 "Morse Code" words a minute for up to 80 km. He also had learned early on that such an oscillator would detect incoming underwater pulses as well as transmit them. Accordingly, he finally took his heavy transceiver (it weighed about 250 kg) to sea aboard the U.S. Coast Guard cutter, *Miami*, and was able to demonstrate, while terribly seasick on 27 April 1914, that the device could detect

---

temperature samples at various levels via a string of vertically spaced reversing water bottles (known as Nansen bottles), and then converting those data to a density (and pressure) field above the layer of no motion.

12. A protégé of Bigelow's, Smith obtained the first American Ph.D. degree in physical oceanography (Harvard, 1925), became commander (with rank of Rear Admiral) of the Greenland Patrol during World War II, and served as the third director of the Woods Hole Oceanographic Institution (1951–1957).

13. A one-time teacher of electrical engineering at Purdue University (1892–1893) and the Western University of Pennsylvania (now University of Pittsburgh) (1893–1900), he received the first patent for using "acoustic signals" to detect underground ore bodies in 1917. As a result, the Society of Exploration Geophysics established the Reginald Fessenden Award in his honor as ". . . the originator of the concept of reflection and refraction surveying . . .", for the courts eventually held that "acoustic" also meant "seismic".

icebergs up to 20 km away. In this instance, the outgoing signal took about 30 seconds to reach the iceberg and return; however, he and his co-workers, Robert F. Blake and William Gunn, also noted one persistent echo that continually appeared about two seconds after the outgoing signal. This proved to be the bottom return, for the sea-floor was about 1.5 km below the ship. Thus, on just one cruise on a strange ship during adverse weather, talkative, persistent Fessenden demonstrated that both horizontal and vertical echoes could be generated within the sea, a finding with tremendous implications for the fields of hydrographic charting, undersea warfare, and seismic exploration. He even dreamed of the possibility of using long, sub-audible acoustic waves to signal across the Atlantic Ocean, a concept whose proof would have to wait for three decades when Professor Maurice Ewing and J. L. Worzel were finally able to detect a small TNT charge detonated in the deep-sound channel over 10,000 kilometers away (Ewing and Worzel, 1948).[14]

Fessenden's echo-ranging technique was immediately called to the U.S. Navy's attention, and soon "Sub Sig" submarine-detection devices were being installed on warships. The device, however, had its problems for local noises from propellers, waves slapping on the ship's sides, and hydrophone movement in the water, tended to drown out the weak echo return. To overcome this, one practice was to stop the ship dead in the water, shut all the machinery down, and listen — not a very pleasant thing to do in combat or a heavy seaway. In addition, a single detector could not indicate which direction the echo was coming from. However, the British Navy utilized a much higher frequency (10 to 20 kHz) sound source in order to achieve some directionality for the acoustic beam. The resulting ASDIC system with its quartz, rather than mechanical, oscillators and receivers was elaborated on considerably after 1919 and went on to form much of the basis for present-day sonar detection and bottom echo-sounding.

Shoreside applications of underwater sound detection were also quickly proposed. On 28 February 1917, or about six weeks before the United States entered World War I, "Sub Sig's" second vice president, H. J. W. Fay, wrote the chairman of the Navy Consulting Board and proposed the development and production of "efficient devices for detecting submarine boats from fixed stations near shore"

---

14. The "deep sound channel" is an acoustic duct that exists in much of the ocean at depths ranging from near the surface (in polar regions) to about one kilometer in tropical regions. Its presence permits the use of the "*Sound Fixing and Ranging*" (SOFAR) technique for tracking underwater buoys, fixing explosions dropped by surface and aerial platforms, monitoring the T-phase of underwater earthquakes, and even recording the catastrophic sounds of imploding submarines that have gone below their safe depth.

(Fay, 1963). Although slow in coming, this application was in place in the seaward approaches to such military installations as the Panama Canal during World War II. After that war, the Western Electric Company began installing for the Navy the massive CAESAR system which eventually extended from Iceland to Barbados. Even as late as Fiscal Year 1981 (FY 81), the U.S. Navy was budgeting $105 million for "oceanographic research" in support of CAESAR as part of the overall *Sound Surveillance Subsystem* (SOSUS) effort.

Despite the increased tempo in marine technology during the early 1900s, America, outside fishery and hydrographic charting areas, held little interest in acquiring additional oceanographic information. There were no formal college courses in the subject, and what little private field work was being done centered at marine biological laboratories at Woods Hole, Massachusetts (1888), and San Diego, California (1903), both patterned after those started at Naples, Italy (1872), and Plymouth, England (1888). But even these laboratories were primarily summer facilities where biological professors could bring their best students and work in tidal and coastal waters only. In Canada, the situation was even worse, for there were no non-government marine science facilities at all.

CHAPTER 2

Geophysics Comes of Age—The Roaring Twenties and the Depressing Thirties

He who struggles with joy in his heart struggles the more keenly because of that joy. Gloom dulls, and blunts the attack. We are not the first to face problems; as we face them we can hold our heads high.

Vannevar Bush, *Pieces of the Action,* 1970 (Courtesy of William Morrow and Company, New York)

The Golden Days of Exploration Geophysics

Most stories about the early days of petroleum geophysics start with the formation of the Geological Engineering Company.[1] We mentioned in the prior chapter that four talented men — W. P. Haseman, J. C. Karcher, E. A. Eckhardt, and Burton McCollum — had ended up working together at the U.S. Bureau of Standards in 1917 to develop improved detectors of blast waves generated by enemy artillery. While working there, Haseman discussed with Karcher his belief that reflected sound waves could detect potential petroleum-bearing structures. After resuming teaching at the University of Oklahoma, Haseman wrote Karcher, then finishing his Ph.D. degree in physics at the University of Pennsylvania, to ask whether he would like to form a seismic exploration firm upon graduation. Karcher was enthusiastic and immediately began working on a patent application for the concept. They also invited Eckhardt and McCollum to join them. Because McCollum already had eleven patents, he and Karcher jointly filed in January, 1919 a patent application for "determining the contour of subsurface strata ... ". Actually, they prepared four separate applications, two on reflection and two on refraction methods. To stay solvent, Dr. Eckhardt and

---

1. For additional details, including patent citations and copies of early seismic records, see Sheriff and Geldart (1982).

17

the new Dr. Karcher continued working at the Bureau of Standards and, during off hours, built a three-trace recorder from an oscillograph and converted some radiotelephone receivers into electrodynamic geophones. With this crude gear, they then obtained the world's first intentional seismic reflections on 12 April 1919 within a Maryland rock quarry.[2] The Geological Engineering Company, the first seismic contracting organization was then incorporated in Oklahoma during April, 1920. By early 1921, Haseman, having resigned from the university, had raised $28,000 to get the company under way. On 4 June 1921, the firm conducted the first fully valid field test of the technique on what was then the outskirts of Oklahoma City at a site marked today by a monument erected by the Oklahoma City Geophysical Society. In mid-July, the crew moved to the Arbuckle Mountains of southern Oklahoma to measure velocities of specific outcropping formations by refraction, the same method as used by Robert Mallet 80 years before. This time, accurate velocities were obtained for three different formations. Next, it was surmised that it should be possible to obtain a reflection from the interface between the Sylvan shale and the underlying Viola limestone at a place where the Viola formed a dome. After adding a second detector, by late July they had obtained the first reflection seismograph cross-section showing depth to the Viola limestone over a geologic dome occurring near Sulfur, Oklahoma (see Fig. 2.1).

With proof in hand, the partners interested a friend, E. W. Marland of the Marland Refining Company, in underwriting bare crew costs for two months of reflection work in the Ponca City, Oklahoma, area.[3] Unfortunately, that area is a very difficult one in which to do reflection prospecting, and their results were extremely poor. Eventually, they were hopelessly in debt and without a work contract, so McCollum, being more optimistic, agreed to clear all debts in exchange for the equipment and sole ownership of the Karcher–McCollum patents. After adding two more patents in 1928, McCollum sold the entire seismic group to Texaco Development Corporation for a lump sum plus 20 percent of any royalties collected by Texaco. In 1933, Texaco "invited" all oil companies and seismic contractors to begin paying royalties, causing much litigation but

---

2. The technique used consisted of dynamite shot in holes for an energy source, electrodynamic geophones built of radiotelephone receivers, and a three-trace recorder converted from an oscillograph.

3. Marland became a major player on the petroleum scene, becoming Governor of Oklahoma in 1934 after losing control of his firm to the financier, J. P. Morgan, in 1928. The firm became the giant Continental Oil Company (Conoco) which also came to be bought out by the even larger E. I. DuPont de Nemours and Company during the summer of 1981.

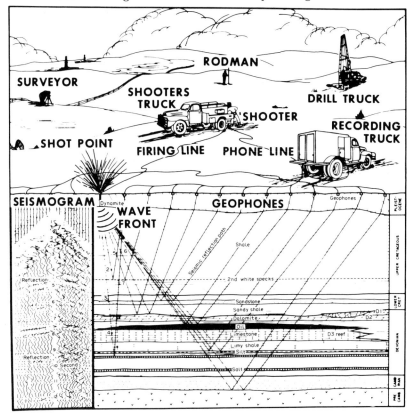

FIG. 2.1. Schematic diagram of reflection seismograph crew at work. (Courtesy of Texas Instruments)

eventually earning an out-of-court settlement and subsequent royalty payments, quickly making McCollum rich. He never lost interest in seismic exploration and later developed seismic recording and "stacking" on analog magnetic tape, as well as a weight-dropping technique as an alternative seismic energy source.

With the war over, Dr. Ludger Mintrop modified some of his military gear and filed in December, 1919 for a German patent entitled, "Method for Determination of Rock Structures".[4] He,

---

4. John W. Evans, a Fellow of the Royal Society and on the staff of the Imperial College of Science and Technology, and Willis B. Whitney, also applied during July, 1920 for a British patent on "Improvements in and relating to Means for Investigating the Interior of the Earth's Crust". Their proposed method involved generating earth vibrations and detecting reflection and refraction paths. Nothing major came out of their initiative, however.

too, set up a seismic prospecting firm called Seismos, Limited in 1921, and, via Marland's Chief Geologist, W. A. J. M. van der Gracht, obtained a contract to conduct seismic surveys for Marland starting in 1923.[5] Refraction shooting was tried in northern Oklahoma and in East Texas in search of faults, and along the Gulf Coast of Texas and Louisiana for salt domes. However, the Mintrop crew used mechanical seismographs, and signal amplification was so low that it did not indicate any shallow salt domes (there were none, in fact) nor deeper salt domes (which there were). In spite of this limitation, another Seismos crew working for Gulf Production Company is credited with the *first seismic discovery* of a salt dome — the Orchard dome in coastal Texas — during 1924 which found commercial amounts of oil.

While Marland was ineffectively sponsoring geophysical surveys for petroleum prospects, another Oklahoma-based organization, Amerada (a combination of *Amer*ica and Can*ada*), funded by Britain's Lord Cowdray, was also looking into this approach. Amerada's Vice President and General Manager was Everette De Golyer, a geologist with a special knack for finding oil, and one of the key founders of the American Association of Petroleum Geologists in 1916. With the war over, DeGolyer contacted members of the physics faculty at Cambridge University to review the possibility of using artillery sound-ranging methods and gravity surveys to locate possible oil-bearing structures. He obtained an Eötvös torsion balance, which were available by 1922, and surveyed the famous Spindletop oil field near Beaumont, Texas. The survey indicated a pronounced positive gravity anomaly, thereby making it the first *salt dome* and American oil field to be mapped by geophysical techniques. Amerada then discovered in 1924 the Nash salt dome in Brazoria County, Texas, with the torsion balance and completed the first producing well in the structure during early 1926, thereby making this the *first discovery of an oilfield* by any method of geophysics.

Through his Cambridge University contacts, DeGolyer had heard of Reginald Fessenden and subsequently had many discussions with him in Boston during the early 1920s about using seismic waves for locating ore bodies and oil fields. As a consequence, "De" organized the Geophysical Research Corporation (GRC) in 1925 as an Amerada subsidiary and hired Karcher, as GRC's new vice president, to acquire a license from Fessenden for Amerada and GRC to use the funda-

---

5. Seismos Party #1 also started working in the Golden Lane region of Mexico during 1923 for Mexican Eagle (Shell) Petroleum Company. The monthly charge for three men and necessary equipment was $2,050 when working geologically unknown areas.

mental Fessenden patent for ore body location issued eight years before. Once the license was in hand, Karcher ordered the construction of a set of seismic instruments under the direction of Eugene McDermott, a young M.A. (physics) from Columbia University with special training in vacuum tubes and radio. When this equipment arrived in Houston, Texas, during January, 1926, it was extremely simplistic by today's standards. The entire unit consisted of an electrical seismometer weighing about 20 kg, a vacuum tube amplifier, and a single trace recording camera modified from an old hand-cranked 35-mm movie camera in such a way that timing lines were projected onto the film by shining a light through slits attached directly to prongs of a large 50-cycle tuning fork. Shot time-breaks were shown by an interruption in a radio signal transmission from the shotpoint, while distance between shotpoint and geophone was derived from arrival time of the air wave at a "blast-phone" similar to those used at the front in World War I.

GRC's first field demonstration for the Gulf Production Company was running a line of stations across the Spindletop dome. However, on developing the first day's film back in a hotel room that night, the data proved worthless because of a light-leak in the camera. The leak was soon fixed, and the next day's work showed the "fast" salt refraction signal arrival everyone had hoped to see. Gulf asked to hire *two* GRC crews, and they were ready to go in April, 1926. Within three months, one of the crews had discovered two salt domes on the featureless coastal plain, partly because the sensitivity of electrical instruments allowed them to work with shotpoint-to-detector distances up to nine km, thereby permitting some of the recorded signal to penetrate so deeply as to pass through salt in deeply buried domes, whereas the Seismos gear was limited to working at maximum ranges of five km. L. P. Garrett, Gulf's Chief Geologist, also suggested that the detectors should be successively placed along an arc encircling the shotpoint, thereby introducing highly effective "fan-shooting". GRC began obtaining exceptional results for other clients as well. During 1927 and 1928, GRC was able to locate eleven salt domes in nine months for the Louisiana Land and Exploration Company by working an area suggested on the basis of stratigraphic evidence. As a result, the average cost of finding these domes was less than *$50,000 apiece,* whereas the last previous discovery in the region without the use of geophysical techniques had cost some $20 million four years before (Rosaire, 1940).

By late 1926, Amerada operated three regular GRC crews, plus a small experimental crew, for its own account. The latter crew became the first to map reflections from a salt dome (the Nash dome) and also was able to obtain strong reflections from the Viola limestone in Oklahoma's Seminole Plateau.[6] The other three crews were immediately brought in but were plagued in some localities by excessive low-frequency, high-amplitude surface waves, known as "ground roll", which obliterated the desired higher frequency reflections. The GRC laboratory solved this by providing a new amplifier which rejected all low frequencies, including the ground roll. To speed up field production, by early 1928 two recording trucks were used per shotpoint with two cameras per truck, thereby giving four traces per shot. Only somewhat later did four-trace cameras come along to simplify greatly the interpretation effort. A new geophone also made for much better record quality. As a result, GRC had a total monopoly on the use of the reflection seismograph in commerce and was employing seventy percent of all the world's seismic exploration specialists. Because of Amerada's successful "in-house" discovery of new oil fields by the reflection seismograph, the decision was then made that the use of GRC's reflection crews should be limited exclusively to itself and to its subsidiary, the Rycade Oil Company.[7]

Amerada's General Manager, De Golyer, so strongly disagreed with this restrictive policy that during February, 1930, he secretly financed the formation of Geophysical Service Inc. (GSI) to do reflection work independently. Karcher and McDermott shifted over to become GSI's President and Vice-President, respectively, and in a whirlwind sales campaign, Karcher lined up, in just a two-month period, contracts for ten field crews (see Table 2.1). McDermott simultaneously was building seismic detection and recording equipment, employing a method different enough to avoid infringing the Fessenden patents, and the first crew was in the field within four months of the firm's incorporation.

This action broke GRC's monopoly and started GSI on the path to becoming the world's largest geophysical contractor during the 1950s as well as spawning what is today a multi-national conglomerate, Texas Instruments, Inc. Within just eight years, GSI had 34 seismic crews working in the United States, Canada, South America, Saudi Arabia, India, and the Far East. In view of GSI's rapid success, other seismic companies were quickly formed, generally by entrepreneurs

---

[6.] By skilful geological mapping, Cities Service brought in a well near Seminole during March, 1926 that opened up the Greater Seminole series of fields. Within a year, this one area was providing ten percent of all the oil being produced within the United States.

[7.] For an interesting detailed account of life within the GRC set-up between 1926 and 1930, read Mr. Henry Salvatori's letter of 9 September 1948 to C. T. Jones (Sweet, 1978).

TABLE 2.1
Geophysical Services' Reflection Crews Fielded in 1930

Crew number	First Party Chief	Client	Remarks
301	Roland F. Beers	Pure	Founded Geotech Inc. in 1936
302	Kenneth E. Burg	Gypsy	Directed GSI's digital research effort in the 1950s
303	Henry Salvatori	Indian Territory Illuminating and Oil Company	Founded Western Geophysical Company in 1933
304	H. Bates Peacock	Carter	GSI president, 1948–1951
305	L. E. Randall	Humble	
306	Chalmers V. A. Pittman	Carter	Founded Geochemical Surveys, Inc. in 1942
307	R. P. P. Thompson	Empire Gas and Fuel	
308	John P. Lukens	Shell	
309	Earl W. Johnson	Atlantic	Founded General Geophysical Company (1935)
310	Cecil H. Green	Sun (Twin State)	GSI president, 1951–1955 (see vignette)
311	Ben S. Melton	Cities Service (Empire)	Senior geophysicist with Air Force Technical Applications Center, 1950s–1970s

existing within the GRC or the GSI organizations. As a result, over 30 U.S. seismic contracting firms came into existence during the 1930s, including the Western Geophysical Company in 1933 which would exceed even GSI in size by the 1970s. Most of these new firms built instruments similar to the new GSI design, but, since the GSI equipment had some inherent problems, it took several years before their equipment was producing results as good as those obtained with the latest GRC equipment.

In a few instances, contracting firms developed seismic equipment of their own design. For example, the Petty Geophysical Engineering Company, a family corporation founded by two Petty brothers, Dabney and Scott, in September, 1925, made a prototype seismic detector and tested it in the basement of the family home against vibrations caused by street-cars passing two blocks away. Thus, starting with $10,000 and a vacuum tube "ultra-micrometer" invented by the University of Dublin's Professor John J. Dowling, their detectors were capable of measuring all three components of wave motion and were so sensitive that they could obtain good salt dome reflection records with as little as 20 pounds of dynamite instead of the several hundred-pound charges that Seismos Ltd. had to use. They also could obtain seismic reflections and actually mapped one structure with a geologic relief of only 150 feet which 20 years later proved productive when drilled to a depth of three km by Pan American Production Company. Between 1927 and 1931, Petty operated three refraction crews in Venezuela and then, in 1932, began in north-western Oklahoma for the Sinclair Oil Company some of the first continuous profile shooting ever undertaken.

Sinclair proved to be an excellent client and kept Petty crews under contract for the next 33 years despite the infamous accident on 4 June 1934 when safety practices were disregarded and all seven men, including the crew's Sinclair representative, gathered at a shotpoint were inadvertently killed. In this case, although normal practice called for only three men at a shotpoint (the shooter and two helpers) and not making up charges in advance, the shooter apparently fired a capped charge lying on the surface rather than the one deep in the shot hole. Two hundred pounds of dynamite blew and dazed even the remaining member of the crew, Wendell Crawford, seated in the recording truck 100 m distant (Sweet, 1978). By 1940, Petty had 24 crews operating, seven of which were abroad, and had become nearly as large as its two main competitors, GSI and the United Geophysical Company. Technically, Petty stayed in the forefront and introduced during the 1950s one of the two great advances in reflection seismology — the common-depth-point technique of Harry Mayne's (see vignette by Mayne).

Another genius active in the early days of reflection geophysics was Frank Rieber, who had a small research laboratory of his own as early as 1915 although offered a research position by the great Thomas Edison. After doing some radio research with Lee De Forest, inventor of the vacuum tube amplifier, Rieber set up a geophysical laboratory in Los Angeles during 1924 and applied for his first seismic patent a year later. This refraction equipment did not work too well, and his firm went bankrupt during the early depression days of 1930. However, three years later he returned with a set of reflection seismograph equipment, and in 1938, filed for a patent on a radical new approach to seismic recording called the "Sonograph" (Johnson, 1938). The Sonograph performed two novel functions. The first was recording ten geophone traces simultaneously in variable density form on film, and the second was the ability to add optically these traces with any desired increasing or decreasing time offset between them.[8] Although three major oil companies

---

8. In this instance, ground motion is expressed in varying degrees of opaqueness from

acquired rights to the Sonograph patents, Rieber's ideas and equipment were too advanced for the time, and it was not until two decades later, with the advent of magnetic tape and disc recording, that the variable density form of plotting seismic cross-sections came into its own for interpretive purposes.

While major oil companies made extensive use of contract geophysical crews, particularly for "shooting" areas in which they wished to hide their presence, almost all large oil firms also created their own geophysical divisions early on for both field survey and research work. Thus, in 1924, the Humble Oil and Refining Company (then over 50 percent owned by Standard Oil of New Jersey) organized its own geophysical group under Dr. Norman H. Ricker at the urging of its Chief Geologist, Dr. Wallace Pratt, one of the nation's great oil finders.[9] Humble started with mechanical seismographs and made three structural discoveries in East Texas during 1927, but shifted over to electrically-magnified refraction instruments the following year. By 1929, its geophysical department had 119 employees. In 1934, it discontinued the four refraction crews, but by 1937, was operating eight domestic and eleven foreign reflection crews. Because of Pratt's aggressive leasing campaign, Humble also controlled well over two billion barrels of oil reserves in 1937, about half of which had been found by geophysical techniques.

Humble did not restrict itself just to exploration seismology. Between 1927 and 1931, its staff built the first truly field-worthy gravity meter, bringing its weight down from 60 to 40 kg, followed by a new model in 1939 with a weight of but 16 kg. During 1940, this device turned up 15 prospects in southern Mississippi alone. Actually, as far back as 1925, Humble had started making torsion balance surveys, and by 1927 had a maximum of seven such crews in the field. During the next two years, these crews found five Gulf Coast salt domes, a record for that type of surveying; however, because of its tedious nature, this type of survey was dropped in 1937.[10] Humble started using the magnetometer as a geophysical survey tool in 1926 and eventually had as many as nine crews,

---

light to dark within a narrow band, rather than by indicating signal amplitude via a "wiggly" line.

9. As of late 1981, Dr. Pratt, now in his nineties, continued to carry a professional card in the *Bulletin* of the AAPG. Upon being queried about his memories of Ludger Mintrop, Pratt advised that Mintrop was to be noted far more for his sales ability than for his oil-finding capability.

10. Humble learned early on that the Nash dome was truly anomalous, for the higher-than-normal gravity (positive) readings obtained there were caused by high-density cap rock and not by the salt itself. Many other domes, with poorly developed "cap rock", therefore showed up as negative gravity anomalies but proved just as attractive commercially.

primarily in west Texas, where the magnetic prospects consistently yielded dry holes. However, in 1932, such surveys did turn up an intrusive volcanic plug in central Texas that eventually proved to have a commercial oil field associated with it. In 1937, Humble also began experimenting with direct detection of oil- and gas-bearing strata by contracting with the early GRC-GSI geophysicist, Dr. E. Eugene Rosaire, for two "Eltran" (*electrical tran*sient) crews. Although these efforts continued for four years, Humble never publicly reported whether the results were useful or not. By late 1938, Humble's Geophysical Department had 336 employees, a very large group by even today's standards, but with the advent of World War II, staff members assigned to geophysical projects quickly dwindled to less than 90 by late 1942.

Although the Gulf Oil Company was Pittsburgh-based, its Texas subsidiary, the Gulf Production Company, was among the first of the oil companies to use geophysics on a large scale. Their effort began in 1924 with a Seismos refraction crew, which, along with a second Seismos crew soon released by Marland, found five salt domes during that and the following year. In 1926, Gulf added two GRC crews which then discovered a dozen salt domes by 1928. In using this aggressive new approach, Gulf fielded, during that year alone, six seismic, fourteen torsion balance, fifteen magnetometer, and six electrical transient crews; however, the electrical crews were dropped two years later.

To provide adequate laboratory support, Gulf hired Dr. Paul D. Foote away from the Bureau of Standards in 1927, and he, in turn, hired Dr. E. A. Eckhardt away from Marland to head up Gulf's Geophysical Research Division, a post Eckhardt held until 1946. Because of the research group's size — it numbered 221 employees by 1934 — a research facility site was purchased at Harmarville, Pennsylvania which in time proved to contain both a rich coal vein and several million dollars of crucible clay. Eckhardt's early years at Gulf were extremely productive, for he introduced new reflection seismograph equipment in 1931, as well as a field pendulum more suitable for measuring gravity than the torsion balance then in extensive use. In 1933, work started on a gravity meter and soon produced a device that could be operated even by remote control while underwater. Such a meter could occupy up to 400 stations per month, and by 1940 some 200,000 stations had been measured. During the first half of 1937, just one gravimeter survey using two instruments turned in salt dome discoveries at an average rate of one per week for a string of over twenty findings (Eckhardt, 1940). In fact, the cost of that particular gravity meter campaign was less than the average cost of discovering just one salt dome using seismic surveys.

Outside of the United States, the Anglo-Persian Oil Company had been using geological studies for spotting potential petroleum-bearing structures as far back as 1901.[11] In the 1920s, it, too, sought Mintrop's advice and experimented with both torsion balance surveys and with seismic refraction measurements using portable earthquake-type seismographs. The oil fields being sought were within 300 m-thick sections of hard limestone folded into structures several km in both width and length, and thereby proved ideal targets for this type of prospecting. During the 1930s, Anglo-Persian elaborated on the fan-shooting technique and used charges of dynamite weighing several tons and placed in hand-dug pits to find many new oil and gas fields not only in that decade but during the war years as well.

Despite the tremendous success obtained with geophysical exploration techniques, there were few academic courses and no suitable textbooks at all during the 1920s and the 1930s (see Fig. 2.2). Consequently, petroleum geologists had to take it largely upon themselves to resolve the validity and utility of the methods being offered. In 1939, the Houston Geological Society therefore formed its own study group and reported results shown in Table 2.2.

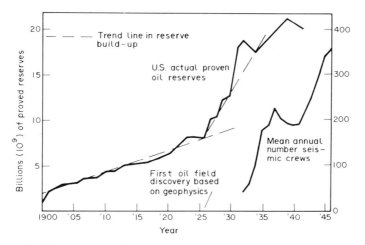

FIG. 2.2. Increase in oil reserves and seismic exploration crews within the United States.

[11.] As First Sea Lord of the Admiralty in 1914, Winston Churchill persuaded the British Government to purchase 48.9 percent of Anglo-Persian at the cost of £2,000,000. In 1981, its successor firm, British Petroleum, Ltd., held net assets of the order of £20,000 million, and the British Government still held a major portion of the common stock.

TABLE 2.2

*Comparative Values of Geophysical Survey Methods for
Locating Subsurface Structural Uplift as of 1939*

Method	Useful in rapid reconnaissance	Suitable for detail work prior to drilling well	Remarks
Magnetometer	X		
Gravity meter	X		
Torsion balance	X	X	Largely replaced by gravity meter after 1934
Electrical prospecting (Eltran method)	X		
Chemical soil analysis (Subterrex method)	X		Results clouded by biogenic and man-introduced hydrocarbons
Refraction seismograph	X	X	Very expensive
Reflection seismograph	X	X	Far superior to the other methods, particularly in lessening or correcting for asymmetry in deep-seated anomalies*

Source: Billings (1940).

*Further improvements anticipated, particularly in identifying deep-seated fault displacements.

Costs of discovering new fields on the Gulf Coast with the reflection seismograph during the 1930s were unusually modest. By assuming an average lag of two years between the time of locating a geophysical anomaly and actual discovery of oil associated with the anomaly, Rosaire arrived at the statistics shown in Table 2.3. These data demonstrate that the reflection seismograph, during its heyday on the Gulf Coast, had found 131 fields, many of which would produce more than 100 million barrels of oil each, at an average geophysical cost of only $164,000 per discovery. Such a record, of course, was not typical of all petroleum prospecting provinces. For example, although Turner Valley had come in as a major field in western Alberta as early as 1914, none of the petroleum-prospecting techniques worked in that area until finally the reflection seismograph turned up an anticlinal anomaly near Leduc, Alberta, which, when drilled in 1947, proved to be a prolific reef-producing structure, and thereby opened up western Canada as a major oil-producing province.

Because geophysical survey crews had to move incessantly by

the very nature of their work — and still do — crew members were rather like "dry-land sailors" and, if assigned to a seismic crew, called "doodlebuggers" because of their strange equipment.[12,13] A seismic crew for a major oil company was typically staffed during the late 1930s in the manner shown in Table 2.4.

Such a crew normally shot eight to ten holes a day, although as many as 14 holes could be obtained should the record quality be good and the shot holes stand up well between successive explosions in them. Explosive charge size was normally two pounds of dynamite, but could range between 0.5 and 20 pounds depending on local conditions. Shot holes, typically spaced a mile apart during reconnaissance surveys, were normally 13 to 20 m deep, although depths of 50 m or more were sometimes used in areas of loose sand and low water-table. Detection was normally by a six-unit spread of two-kg geophones buried just below ground level at 50-foot intervals between 250 and 500 feet from the shotpoint, although there were instances when three geophones might be connected to a single detector "take-out". Recording was at one instrument setting (although automatic gain control was incorporated) and directly on to fast-moving photographic paper, so typically it took five shots

TABLE 2.3
Seismic Exploration Costs versus Discovery Rates on the Gulf Coast, 1932—1939

Year of discovery by drill	Field discoveries	Year of seismic discovery	Reflection crew years that year	Crew years per discovery	Estimated cost of a crew year	Exploration cost per discovery
					(in thousands of $)	
1932	2	1930	1.0	0.5	80	40
1933	2	1931	1.6	0.8	80	64
1934	9	1932	10.1	1.1	80	88
1935	12	1933	26.6	2.2	85	187
1936	20	1934	32.6	1.6	90	144
1937	34	1935	42.3	1.2	99	104
1938	29	1936	56.9	2.0	100	200
1939	23	1937	59.6	2.5	105	262

Source: Rosaire, 1940.

[12.] While with the Carter Oil Company between 1939 and 1941, Charles Bates moved 14 times in 22 months, served on three different crews, worked in four different states, and helped locate one commercial oil field (the Stuttgart Field in northern Kansas).

[13.] With oil being "where you find it", which could just as likely be a swamp, desert, muskeg, tundra, or other roadless area, as in a mountainous region, seismic gear and its supporting drilling equipment underwent rapid modification as the case warranted, ranging from portable equipment fitting on the back of humans to unique installations going on air-boats and special all-purpose terrain vehicles.

TABLE 2.4
*Staffing and Salary Level of Oil Company
Seismic Crew (1939)*

Specialty	Approximate monthly pay
Party chief	$350 to $400*
Computer	$275 to $300*
Instrument observer	$275 to $300*
Surveyor	$250*
Shooter	$200 to $225*
Student computer	$180*
Student observer	$180*
Attached helpers (3)	$125 to $150
Driller	On contract
Driller's helpers (2)	On contract

*Also allocated a tax-free $85 per month expense account.

or more per hole to obtain two satisfactory records, one for the home office and one for the party's own resident computer (a man, not a machine).[14]

Because the profession was young, the people in it were young. "Observers" were typically electrical engineering or engineering physics graduates, while "computers" were geologists or mathematicians. There were no company manuals or training courses, either centralized or by correspondence, and the first decent textbooks did not appear until 1940.[15,16] Even comprehensive journals in exploration geophysics were hard to come by. The first magazines of this nature were published in Germany — the *Zeitschrift für angewandte Geophysik* (Magazine for Applied Geophysics) in 1924 and the *Zeitschrift für Geophysik* (Magazine for Geophysics) in 1925. The first American equivalent appeared in 1929 as a collection of papers, *Geophysical Prospecting,* published by the American Institute of Mining and Metallurgical Engineers (AIME), followed in 1932 by the *Transactions* of the Society of Petroleum Geophysicists

[14.] By introducing a double row of galvanometer lights and dual strips of photographic paper within the recording box, the Carter Oil Company speeded up its field work during 1940 by 10 percent through just this simple modification.

[15.] Three books of this type finally appeared in 1940: C. A. Heiland's *Geophysical Prospecting,* J. J. Jakosky's *Exploration Geophysics,* and L. L. Nettleton's *Geophysical Prospecting for Oil.* Reviews were highly favorable, although somewhat critical of excessive use of mathematical developments, for, as one reviewer pointed out, nature involves much heterogeneity and very, very many layers (Blau, 1941). In no case was seismology given over 30 percent of the space within any of the three books.

[16.] The first workbook providing *Lessons in Seismic Computing* was written by Dr. Morris M. Slotnick in 1949 and, after editing by Dr. Richard Geyer, published by the SEG in 1959. By mid-1981, over 9,300 copies had been sold.

published in two locations — in Volume XV of the *Bulletin,* American Association of Petroleum Geologists, and in the journal, *Physics,* of the American Physical Society. The first issue of the new *Journal* of the Society of Petroleum Geophysicists then appeared in July, 1935, followed six months later by its successor, *Geophysics,* which has continued until today. In the early years, because of the competitive nature of the oil industry, top quality papers were extremely hard to obtain, particularly those dealing with field techniques and case histories of field discoveries. On 1 January 1937, the sponsoring society relabeled itself as the "Society of Exploration Geophysicists" (SEG) in order not to limit its interests as to geographic realm or type of application. Even so, the SEG's first meeting independent of the larger American Association of Petroleum Geologists did not occur until October, 1955.

Acquiring a formal degree in exploration geophysics was largely unheard of in North America, for only two schools granted such degrees (see Table 2.5).

The question at the time was whether an aspiring geophysicist fresh out of college should go to work for a contractor or for a major oil company. The contractor had no oil wells of its own in order to be free of any taint of surreptitiously leasing promising prospects on the side, so such a firm did not pay well, and it was also noted for laying off whole crews without notice.[17] On the other hand, this "feast or famine" type operation gave promising young geophysicists a much better chance to move up than in an oil company where seniority stifled mobility. To complicate matters further, oil company management made it standard practice to keep geologists and geophysicists apart so that one would not taint the other's interpretations of a given prospect, just as if Mother Nature had ordained that one approach was totally exclusive of the other.[18] Dr. Donald Barton, a key member of Humble's Exploration Department and the SEG's first president, described it rather nicely (Barton, 1938):

> The rank and file, composing perhaps two-thirds of the personnel of exploratory geophysics, is not, and should not be, composed of highly trained men. The routine, stereotyped character of the work and the impossibility of promotion of more than a few from these lower ranks even under the best circumstances justify companies in filling the bulk of these rank and file positions with men having only a bachelor's degree at the most.

---

[17]. As soon as exploration budgets were cut back, major oil companies were notorious for canceling crew contracts on 24-hours notice, thereby exhibiting, as one noted geophysicist expressed it, "all of the loyalty expected from a French whore".

[18]. The situation is almost reversed in the early 1980s, for geologists are now required to review their prospects with the geophysical department before recommending exploratory drilling.

TABLE 2.5

Status of University Geophysical Institutes as of 1939

Country	Institute location	Remarks
Canada	Toronto McGill Dominion Observatory at Ottawa	University institutes were subordinated to physics departments; governmental institute provided both research and instruction
Great Britain	Cambridge*	Stressed fundamental problems concerning physics of the earth, seismology, earth structure, and meteorology
	London	Stressed geophysical prospecting under H. Shaw and Lancaster-Jones
France	Strasbourg	Founded 1919 with curriculum leading to diploma of geophysical engineer
	Paris	Founded 1921; "Certificate of Geophysics" covered constitution of earth, seismology, terrestrial magnetism and electricity, transmission of electromagnetic waves in atmosphere and solid earth, meteorology, aerology, radiation, and ionization of the atmosphere
	Clermont-Ferrand Toulouse Algiers }	Instruction a minor activity at these institutes
Germany	Göttingen	Founded as Observatory of Terrestrial Magnetism in 1832 by Gauss; Wiechert became director in 1898
	Leipzig Königsberg Potsdam (Berlin)	No organized curricula for geophysical engineers appeared to exist within German academia
	National Institute of Earthquake Research	Located at Jena
	Prussian Geological Survey, Berlin	Well-staffed Institute of Applied Geophysics
Italy	Trieste	Reale Istituto di Geofisica under Vercelli
	Rome	Istituto Nazionale di Geofisica which provides central administrative function
	Other Institutes at various cities	Established for observation and research rather than for instruction
United States of America	St. Louis University	Founded 1925 as first full-fledged Department of Geophysics as outgrowth of Geophysical Observatory started in 1909
	Colorado School of Mines, Golden	Founded 1927 to train men for practical geophysical work and to inspire picked men to carry on further research in industry and educational institutions
Japan	Tokyo	Seismological Institute at University of Tokyo Earthquake Research Institute at the Imperial University Central Meteorological Observatory conducted some seismological re-research

Thus, geophysicists with geological backgrounds were often frustrated by their Chief Geophysicists, particularly if the latter had been a university physics professor until a few years before. For example, Party Three of the Carter Oil Company was led by the University of Oklahoma geologist, Sam Zimerman. During August, 1936, Zimerman's crew found the sizeable Louden structure in central Illinois. He soon knew it to be a major structure for one day his seismic party would be ahead of the leasemen, and the next day the situation would be the reverse.[19] Once the structure was delineated, "Sam" started the crew working to the southwest and soon found a reversal in dip suggesting that a second anticlinal probably lay in that direction. But before he could prove this out, "Tulsa" called and ordered the crew to stop and immediately go to southern Arkansas. Zimerman protested to no avail. As a result, a year later Texaco sent a seismic crew into the area that Party #3 had almost worked and found the Salem Field, the largest ever located in Illinois. Looking back, perhaps there is poetic justice in the subsequent conversion of the Carter Oil Company into Exxon's coal company! Zimerman, in the meantime, moved on to become ARAMCO's chief geophysicist during that firm's fabulous oil-finding days in the 1950s and early 1960s.

The Appearance of Electrical Well Logging

Aside from the seismic method, the most important technical advance in petroleum exploration and production was introducing the capability to measure the various physical properties of earth strata via a borehole. Today, such measurements are, for geologists, an indispensable tool in constructing subsurface cross sections and structural maps of an area's stratigraphy (type and number of rock

---

[19.] The structure proved to be 6 by 24 km in productive extent, and to have a closure of 60 m. By the end of 1979, it had produced 369 million barrels of oil. In comparison, the Salem field produced 370 million barrels in the same time frame.

Source: Macelwane (1940).

*Professor Edward Bullard emphasized "hands-on" surveys. Working first with Kerr Grant and then with Gaskell as his research assistant during 1937–1938, Bullard mapped the nature of the Paleozoic basement rocks underlying East Anglia by using seismic techniques both on and offshore. In this effort, they also obtained extensive assistance from Dr. W. B. Harland of the University's Department of Geology. Bullard was equally anxious to check other methods of geophysical surveying and soon had Ben Browne at sea in Royal Navy submarines to measure gravity with Vening Meinesz type pendulums. Bullard also started making heat-flow measurements in Africa and Great Britain and, in 1939, even carefully evaluated a well-known water-diviner. As "Teddy" remarked, "If it works, it will be the cheapest method of prospecting known!"

layers) and determining the lithology (rock composition, pore characteristics, and fluid content). In addition, petroleum engineers find such measurements invaluable in evaluating, developing, and producing reservoirs of petroleum and natural gas. Geophysicists also need such data to check actual subsurface conditions against those indicated by their surface surveys, and mining engineers likewise make extensive use of electrical logs to assess and delineate mineral deposits.

Surface electrical prospecting can be said to have started with two French professors, Conrad and Marcel Schlumberger (Allaud and Martin, 1977). They made the world's first geophysical discovery of a non-magnetic ore deposit in 1913, the first tectonic study by geophysical means (of a French sedimentary basin using man-made electrical sources) between 1912 and 1914, and the first large-scale electrical survey for petroleum in 1922 which led to discovery and mapping a petroliferous dome in Romania. In 1927, while talking to the manager of a drilling company about surface electrical prospecting, Conrad Schlumberger heard instead of drilling problems, such as the inability to know exactly what the drill had cut unless one engaged in the slow, costly process of cutting cores (Johnson, 1962). As a consequence, Schlumberger wondered whether one could log such holes electrically after they had been drilled, and in September of the same year, the two brothers ran the first electrical resistivity log in France's Pechelbronn oil field using a four-electrode arrangement now known as a "lateral" configuration. The technique gained rapid acceptance, and logs were being made in California as early as 1929.

During the next decade, Schlumberger Ltd. offered new techniques at a steady pace, not only to show the presence of oil or gas within a given geological horizon, but also to show variations in borehole diameter, drilling mud characteristics, and thickness of geologic strata. The initial use of hand recording via an electrical nulling device was replaced by photographic recording in 1936 and by digital recording in 1961. Although Schlumberger dominated the world market for well-logging services from the first, competitors began to appear in the late 1930s.[20] Thus, the Seismograph Service Corporation and its subsidiary, Well Surveys, Inc., developed a

---

20. Although headquartered in New York City, Schlumberger retains such a French flavor that there is a saying among its executives, "No matter how high you get, there's always a Frenchman ahead of you!" Schlumberger still provides over half of the world's wireline services and, in 1980, grossed 5.1×10^9, had net profits of $994 million, and employed 45,000 persons. The Schlumberger family still owns about 25 percent of the common stock.

gamma-ray log in 1939 that could operate through well pipe in place and with any kind of borehole fluid. In this case, the technique was licensed to another logging firm, Lane-Wells, for commercial use. In 1941, Well Surveys then developed the neutron gamma-ray log, which bombards the rock surrounding the hole with neutrons and measures the amount of secondary gamma-ray radiation resulting therefrom. Continued improvements even four decades later now make it possible to obtain basic analyses of individual rock constituents, including the ability to identify petroleum-rich shaly sands that previous logging techniques could not indicate. In fact, the use of modern logging devices is so sophisticated and highly specialized that it is impossible here to give even a cursory rundown of the state of the art, and it is suggested that the reader, if interested, should delve further in books addressing this single subject such as those by Guyod and Shane (1969), Pirson (1978), Hilchie (1979), and Asquith (1980).

Classical Seismology Strikes Out on its Own

Once World War I was past, there came an expanded interest in the exact nature and structure of the earth. In Great Britain, Harold Jeffreys, after returning to Cambridge University, was able to publish his landmark treatise, *The Earth,* in 1924.[21] In Germany, Beno Gutenberg and his associates also began work on the monumental series, *Handbuch für Geophysik,* publishing the first volume in 1929. Interest in seismology likewise quickened in the United States. Membership in the Seismological Society of America had held relatively constant at about 400 between 1911 and 1920, but membership then started growing and reached a peak of 980 in 1928, a number not exceeded for the next quarter of a century.

The Jesuit Order also revitalized its seismic network which had fallen into disarray.[22] To do this, it sent a young assistant professor in physics at St. Louis University, Father James Macelwane, SJ, in 1921 to the University of California (Berkeley) to earn his doctorate in physics with minors in seismology and mathematics. A year later, his work was so impressive that Professor Andrew C. Lawson, the fiery, heavily bearded chairman of the geology depart-

---

21. Jeffreys was eventually knighted in recognition of his outstanding work concerning the constitution of the earth. He was a co-winner of the Vetlesen Award during 1962, sharing it with the Dutch gravity specialist, Professor F. A. Vening Meinesz.

22. The only American Jesuit seismic stations that operated continuously during the 1910s were: Spring Hill at Mobile, Alabama; Georgetown in the District of Columbia; Sacred Heart in Denver, Colorado; and St. Louis University in Missouri.

ment, asked him to join the department and operate the university's seismic network as his life's work. Graduating in 1923, Father Macelwane did become an assistant professor of geology there, but soon received a letter from Father M. Neumann, SJ, director of the seismic station at Granada, Spain, which said in part: "I believe it would be very much AMDG if our chain of seismographic stations in the United States would rise once and for all out of its present state which you will pardon me if I dare call 'wretched'."[23]

Father Macelwane had the courage and foresight to send this letter to all four American Provincials (the regional directors) of his Order. By July, 1925, he was told to report to St. Louis University and develop and chair a Department of Geophysics there. In addition, the Missouri Provincial and the university rector authorized him to offer St. Louis University as the "central station" should a nation-wide meeting of Jesuit delegates on 24 August 1925 decide that their seismological network should be resurrected. This time around, the Jesuit Seismological Network turned into a highly productive entity, not only providing telegraphic data regarding the intensity and arrival times of large earthquakes to Science Service (which paid for the messages) but also to the U.S. Coast and Geodetic Survey which had just taken over the Seismic function from the U.S. Weather Bureau. The Network also published its own *Preliminary Bulletin* which was circulated throughout the globe, causing Macelwane to note later:

> It is particularly consoling to read, in published articles and official bulletins, the epicenters accredited to the Jesuit Seismological Association, issued even by countries such as Switzerland and Norway which will not permit Jesuits within their borders.

In Washington, D.C., once the Coast and Geodetic Survey acquired the seismological mission from the Weather Bureau in 1926, Com-mander (later Captain) Nicholas H. Heck was placed in charge of the new Division of Geomagnetism and Seismology.[24] Depending pri-marily on cooperative stations independent of the Jesuit network, the Survey's prime interest was better defining seismic zones within the United States and learning more about strong, close-in motion generated by earthquakes, mining activities, and quarry explosions for use by architects, builders, and home-owners. Within certain

---

[23.] "AMDG" is the Jesuit abbreviation for their motto, "Ad majorem Dei gloriam", or, when expressed in English, "For the greater glory of God". Because of their extensive intellectual training, tight discipline, and a high degree of initiative, they have been referred to as the Pope's "marines".

[24.] An incisive person known as a "human IBM machine" within the C&GS, Heck eventually became president of both the Seismological Society of America and the Inter-national Seismological Association.

states, particularly California, such information became particularly important as building codes began to require earthquake-resistant buildings.

The small but highly accomplished Carnegie Institution of Washington, which had started a Department of Terrestrial Magnetism in 1904, formed its own Advisory Committee on Seismology in 1921.[25] One of the group's recommendations was that a seismological laboratory and associated seismic network be started in Southern California much like the one operated by the University of California in the central part of the state. In 1927, with the assistance of the *Cali*fornia Institute of *Tech*nology (CalTech), Carnegie was able to open a Seismological Laboratory in western Pasadena under Harry Wood and a staff of four who also operated three auxiliary seismic stations at Riverside, Santa Barbara, and La Jolla. Once two seismic laboratories were operating in the same state, competition set in. After the Lompoc earthquake of 4 November 1927 some 200 km west-northwest of Pasadena, Father Macelwane's successor at the University of California, the 30-year-old Dr. Perry Byerly, immediately traveled the 400 km to the site, studied the area, and published a report. This was a social error in Wood's eyes (he, too, had worked at Berkeley betweeen 1904 and 1912), and he wrote Byerly suggesting that they agree on an informal jurisdictional line across the state roughly midway between Pasadena and Berkeley (Bolt, 1979). After some appropriate adjustments, the "Byerly—Wood line" came into existence and was generally respected by the two groups until well into the 1970s.[26] Wood also succeeded in bringing another strong-willed man on board during 1930 — Dr. Beno Gutenberg. As a consequence, Byerly and Gutenberg, being both rather "prickly" by nature, were able to carry on some famed disagreements during society meetings, particularly with regard to the applicability of the fault plane solution technique.

In 1931, both Harvard University and the *M*assachusetts *I*nstitute of *T*echnology (MIT), although located in the same community of Cambridge, also saw fit to launch formal educational programs in

---

[25.] One of the initial projects of this Department was to conduct a worldwide magnetic survey, particularly of the oceans which had not been magnetically mapped since the demise of the wooden sailing ships. To do this, the Institution had the totally non-magnetic vessel *Carnegie* built. The ship then made seven extensive cruises between 1909 and 1929 before blowing up in Apia Harbor, Samoa. In addition to taking magnetic data, she also took extensive meteorological and oceanographic data, particularly on the last ill-fated voyage. Scientists of note who worked on the latter type of data included Andrew Thomson of Canada, Harald Sverdrup of Norway, and Richard Seiwell and Woodrow C. Jacobs of the United States.

[26.] CalTech took over operation of the Pasadena Seismological Laboratory in late 1936.

geophysics. Dr Don L. Leet took responsibility for Harvard's long-time seismic station at its Blue Hill Observatory on the outskirts of Boston and then published, in 1938, one of the pioneering American texts entitled *Practical Seismology and Seismic Prospecting,* although it was quite weak in the latter facet. MIT's new geophysical instructor was Dr. Louis B. Slichter, who already had spent two years with the Submarine Signal Corporation and seven years as a partner in a geophysical firm specializing in the use of electrical methods to locate ore bodies. One of his particularly imaginative efforts at MIT was persuading the local public utility firm to let him interconnect 50 km of electrical power circuits in a massive electromagnetic induction experiment, the results of which indicated the earth's conductivity profile down to a depth of eight km. He also developed three-component short-period seismographs for crustal studies, which he performed in New England and Wisconsin using explosive charges, and was among the first to urge that refraction seismology be made a practical tool for exploring the earth's crust and upper mantle (Knopoff *et al.,* 1979).

During the mid-1930s, Professor Jeffreys at Cambridge University, with the help of a very able Australian graduate student, Keith Bullen, developed a truly consistent set of travel times for all major seismic wave trains passing through the earth. These monumental Jeffreys—Bullen Seismological Tables were finally published in 1940 and remained "gospel" for the next quarter of a century until travel times based on the known times of underground nuclear explosions permitted regional refinements as Jeffreys had long urged. Jeffreys was also a key contributor to the U.S. National Research Council's monograph, *The Internal Constitution of the Earth,* edited by Dr. Beno Gutenberg and published in 1939.[27] In critiquing the book for the *Bulletin* of the American Association of Petroleum Geologists, the reviewer noted (Bates, 1941):

> Questions of the earth's interior can be answered, if at all, only through the medium of geophysics, the term being used in a broad sense to refer to that "boundary science" which overlaps the domains of astronomy, geology, physics, chemistry, and seismology . . . (thus) trained scientists of considerable intellectual ability and imagination are qualified to reconstruct rationally the history of the earth's origin and the nature of its interior. . . .

27. The collected works of Sir Harold Jeffreys total six volumes and over 300 papers. Within the fields of seismology and gravitation, it is difficult to find a research topic that Sir Harold has not already written about. However, when geophysicists from British Petroleum, Ltd. approached him with the hope of learning some new survey techniques, they first explained at length the difficulties in finding new oil-bearing structures in Iran. He replied, "I'm glad it's your problem, not mine!"

The Rebirth of Extensive Oceanographic Endeavors

The onset of a strong yearning for peace led to sharp reductions in the size of major battle fleets during the early 1920s. Thus, extensive study of the sea and application of this knowledge to real-life problems became rather a "scientific frill" for the next decade. In Great Britain, the principal oceanographic effort was a continuing study of the Southern Ocean encircling Antarctica utilizing the Royal Research Vessels (RRV) *William Scoresby* and *Discovery II* in support of the large British whaling fleet then working those waters.[28] In the United States, there were, in 1920, eight marine laboratories and perhaps 125 persons who could call themselves "oceanographers" if you included marine biologists within the group. Consequently, almost all new knowledge about the high seas was acquired by Federal facilities in the normal course of their work.[29]

The first oceanographic breakthrough of the 1920s was Dr. Harvey Hayes' development at the Naval Marine Engineering Station, Annapolis, Maryland, of an improved Fessenden underwater oscillator which generated extremely short signals (of about 2 m wavelength) in sea water. By using this oscillator vertically aboard the USS *Stewart,* it became possible to obtain the first ocean-wide line of acoustic soundings between Newport, Rhode Island and Gibraltar in June, 1922, finally freeing sailors from the age-old, tedious use of inaccurate lead lines. Two years later, Dr. H. G. Dorsey of the Submarine Signal Company further refined the Navy's "echo-sounder" into a "Fathometer"® which provided depth-values by a rotating arm tipped with a tiny red neon light that flashed when the acoustic signal was outgoing at the zero mark and flashed again when the reflected bottom signal returned. Water depth was then read from a circular dial behind the arm calibrated directly in fathoms via the assumption that sound had a constant speed of 800 fathoms per second through sea water.[30] The device was steadily improved, until eventually it had the ability to observe at as many as four different ranges of depth and to burn a permanent trace of depth

---

[28.] Key participants in this difficult effort included H. F. P. Herdman and George Deacon. The latter became the Director of the National Institute of Oceanography at Wormley, Surrey, when it was founded during 1949.

[29.] These institutions were: U.S. Naval Hydrographic Office (ocean currents and hydrography), U.S. Coast Guard (International Ice Patrol), U.S. Coast and Geodetic Survey (coastal tides, hydrography, and currents), U.S. Bureau of Fisheries (marine geography of fishing areas), and the U.S. Weather Bureau (marine climatology and marine weather phenomena).
® is a registered trademark of the Raytheon Company.

[30.] In practice, sound velocity in water can vary about two percent depending upon salinity and temperature.

values by having the arm transmit a spark to special recording paper at the times of signal transmission and signal reception.

On the Pacific Coast, the Scripps Institution for Biological Research, located on a barren tract of coast about 20 km north of San Diego, California, began a significant phase of its existence on 1 February 1924 when an imperious, rather stuffy but dapper, member of the U.S. Geological Survey, Dr. T. Wayland Vaughan, took over the laboratory's directorship. Guidance from the outgoing director, Dr. William E. Ritter, was quite explicit:

> . . . the Institution we have developed on the coast of Southern California ought to be treated as a nucleus for an oceanographic institution worthy of the magnitude of the oceanographic problems of the largest ocean now on earth and one of the richest countries on earth.

However, Vaughan's institution had an annual budget of only $50,000 and did not even own a research vessel. But the vision persisted, and on 14 October 1925, the name was changed to the "Scripps Institution of Oceanography", and early reports noted that the efforts of the six-man professional staff were divided among four areas — biological, chemical, physical, and geological oceanography.[31] Industrial funding was also obtained wherever possible, particularly in support of Professor George McEwen's theory, advanced as early as 1918, that negative ocean water temperature departures from the normal were followed by positive departures in seasonal rainfall.[32] Despite limited financing, Vaughan slowly but steadily built up Scripps, introducing during 1930 the first formal degree-granting program in oceanography within the hemisphere, as well as doubling the annual budget to $100,000 before he retired in 1936 (Raitt and Moulton, 1967).[33]

The U.S. Navy was not as fortunate as Scripps in setting a new oceanographic course, even though on 2 June 1924, the Acting Secretary of the Navy, Theodore Roosevelt, Jr. (son of the 26th President), wrote to several government departments:

> The Navy Department has had under consideration for some time a research expedition in oceanography, proposed to be conducted in a vessel or vessels of the

---

[31.] Despite the name change, biologists so dominated the laboratory that one person wrote University of California management as follows: "Calling a mongrel by some high sounding name does not make him a pedigreed dog. With its mongrel characteristics, it is hard to explain . . . why this institution is called an Institution of Oceanography" (Allen, 1929).

[32.] McEwen's outside funding started at the $500 level in 1925 and grew to $12,000 a year by 1932, the donations eventually provided by a dozen power companies.

[33.] However, the first doctoral degree in physical oceanography had already been awarded in 1925 by Harvard University to Lt. Edward H. Smith, U.S. Coast Guard, for research performed in conjunction with the International Ice Patrol.

Navy. . . . The desirability of concentrating and intensifying the research and experimental work carried out by the Navy in connection with the various problems presented under the general heading of oceanography has been recommended for a long time, and it is believed that a scientific investigation along these lines should no longer be delayed. . . .

The subsequent meeting on 1 July provided some very positive statements. Dr. David White, the U.S. Geological Survey's Senior Geologist, proposed that there be a ten-year research program, and it was definitely agreed that "research in oceanography will take a permanent place among the activities of the Navy". But despite all this, nothing tangible happened.

The next attempt to get something major going in oceanography came from academia. Funded by a grant of $75,000 from the Rockefeller-financed General Education Board, the U.S. National Academy of Sciences appointed a Committee on Oceanography on 27 April 1927 under Professor F. R. Lillie, Chairman of the Department of Zoology, University of Chicago, and President (in 1928) of the famed Woods Hole Marine Biological Laboratory, then, and to some extent today, a summer operation. The "blue-ribbon" nature of the Committee is clear, for its chairman moved up to be President of the prestigious parent Academy on 1 July 1935, allowing the group's secretary and instigator, Dr. Henry B. Bigelow of Harvard University, to become Chairman.[34] The Lillie Committee was charged with determining ". . . the share of the United States of America in a world-wide program of oceanographic research", and began writing reports on these topics: physical and chemical properties of the sea, submarine geology, the interrelationship between oceanography and meteorology, life in the sea, economic value of the sea, including fisheries, navigation, shoreline and harbor currents; and submarine cable laying, weather forecasting for oceanic areas, and the unity of the physical–chemical–geological–biologic aspects of the ocean.

Although the Committee took three years to prepare its basic 165-page mimeographed report to the Academy, its recommendations were to have both rapid and long-term results for they provided the ground work for much of the training and employment of U.S. oceanographers for decades to come.[35] The report's major thrust was that while existing marine institutions could stand some financial

---

[34.] From the very first, Dr. T. Wayland Vaughan was not satisfied with the group's composition. On 26 May 1927, he wrote Chairman Lillie: ". . . at present, the committee is composed of three biologists, one geologist and paleontologist, and myself, a cross between a geologist and an oceanographer." Only Vaughan was from the Pacific Coast.

[35.] The panel later issued two regularly printed books, *Oceanography: Its Scope, Problems and Economic Importance* (Bigelow, 1931) and *International Aspects of Oceanography* (Vaughan, 1937), neither of which had any lasting impact.

help, the big need was for a single well-equipped oceanographic institution on the Atlantic coast "to supply necessary facilities for research and education, hitherto lacking, and to encourage the establishment of oceanography as a university subject" (National Academy of Sciences, 1930). Despite the onset of the great economic depression, the Rockefeller Foundation moved quickly to provide over $3 million during 1930 to carry out these recommendations. Table 2.6 shows how the allocations were made.[36] A rundown of the donations indicates why the Scripps staff felt short-changed, because Scripps received just one percent of the total amount even though Scripps was the only operationally viable, year-round non-federally operated marine research laboratory in the Western Hemisphere. Dominated by Easterners, the Lillie Committee rationalized that the State of California and the rich E. W. Scripps family could provide any additional funds that might be needed.

Dr. Bigelow also approached Secretary of the Navy Curtis D. Wilbur and Captain C. S. Freeman, USN, Superintendent of the

TABLE 2.6
Allocation of Rockefeller Foundation Funding to Foster Enhanced Oceanographic Research (1930–1932)

Recipient	Amount	Purpose
Woods Hole Oceanographic Institution, Woods Hole, Massachusetts	$3,000,000 (1930)	To construct an oceanographic laboratory, to build an oceanographic vessel (the 142-foot ketch, *Atlantis*), and to provide a $2 million operating fund endowment
	$1,000,000 (1932)	Additional operating endowment funds
Bermuda Biological Station St. Georges, Bermuda	$50,000	Expand Station and stabilize its research funds
University of Washington, Seattle, Washington	$45,000	Purchase 23-m-long research vessel (*Catalyst*)
	$25,000	Operating funds for research vessel
	$200,000	Construct oceanographic laboratory on campus (*Note*: State was to provide $50,000 for equipment and other construction costs)
University of California Scripps Institution of Oceanography	$40,000	Additional facilities to be placed in Ritter Hall, then under construction
TOTAL	$4,360,000	

36. To place this donation of Rockefeller funds in proper perspective, note that the entire annual budget for the U.S. Weather Bureau in 1932 was but $4,497,720!

Naval Observatory who operated informally as a scientific spokesman for the Navy, to explore how much oceanographic cooperation academia could expect from the Navy. Freeman knew something had to be done along these lines, for he had not the least bit of doubt that the Coast and Geodetic Survey was anxious to expand its coastal operations to the high seas, an action that could easily "spell the forfeiture of naval cognizance over the Hydrographic Office". To meet such a peacetime challenge, the Navy formed a new board under Rear Admiral Frank H. Schofield, USN, to study the situation. This board then reported back to the Secretary of the Navy on 22 March 1929 and suggested two possible approaches, one, the assignment of a vessel devoted exclusively to oceanographic work, and the other (and much less costly); to conduct oceanographic studies from naval vessels on their regular voyages. The Navy chose the "poor-boy" route, bought some oceanographic gear that could be shifted from ship to ship, routed vessels through unsurveyed ocean areas whenever possible, and sought closer association with the private oceanographically oriented institutions.[37] For example, arrangements were made for the famed Dutch geophysicist, Dr. F. A. Vening Meinesz, working in cooperation with the Carnegie Institution of Washington and Princeton University, to conduct gravity surveys in the Gulf of Mexico and the Caribbean region aboard navy submarines during 1928, 1932 and 1936. Over 100 gravity stations were made in this manner and interpreted by Princeton's 30-year-old Dr. Harry Hess, who became a naval reserve officer to facilitate his presence at sea.[38] The Scripps Institution of Oceanography was also asked to provide a consulting oceanographer during 1933 to direct a supplementary oceanographic observation program as the USS *Hannibal* surveyed the approaches to the Panama Canal.[39]

On Cape Cod, the quick availability of the Rockefeller millions during 1930 caused much to happen. The "*W*oods *H*ole *O*ceanographic *I*nstitution" (WHOI) came into being with no less than Dr. Bigelow as its Director and Dr. Lillie as Chairman of its Board of Directors. As soon as a laboratory building was under construction,

---

[37.] Full-time oceanographic survey vessels were not assigned to the Hydrographic Office until 18 years later.

[38.] As Commanding Officer of the attack transport, USS *Cape Johnson,* during 1944–1945, Hess modified the echo-sounder on his ship and was able to detect the deepest known spot of the ocean as of that time, as well as to discover many "drowned" flat-topped seamounts which he called "guyots".

[39.] The consultant was Richard H. Fleming, who would be the first (in 1935) to earn a doctoral degree in "pure oceanography" from the University of California (Berkeley), Scripps not being authorized to grant degrees on its own.

Bigelow had a Danish shipyard start building a large double-ended ketch, the *Atlantis,* to serve as the laboratory's research vessel. Bigelow picked a favorite student, the 26-year-old Columbus O'Donnell Iselin II, son of a New York banker and member of a prominent yachting family, to be the *Atlantis*'s master. Despite the Rockefeller endowment funds, WHOI's annual income as late as 1936 was but $102,000, and the practice was maintained wherein the shore facility was a "summer laboratory" and the *Atlantis* kept at sea the year around (Schlee, 1978).[40] As a result, during mid-summer, visiting investigators and associated students might be in residence, but in winter the professional staff would drop to six or even less. In addition, no formal teaching took place in order to keep costs to a minimum, and it was not until 1970 that WHOI began granting graduate degrees in its own right.

Because neither his laboratory nor ship were operational in 1930, Bigelow committed some of the initial Rockefeller funds to support Captain Sir Hubert Wilkins, a noted Australian polar explorer, in his attempt to cruise under the Polar Pack of the Arctic Ocean in a submarine and perhaps even make a submerged run to the North Pole. The U.S. Navy also cooperated by providing an over-age submarine which Wilkins renamed the *Nautilus.* Dr. Harald Sverdrup, who had served as Chief Scientist of the *Maud*'s polar drift north of Siberia between 1917 and 1925, was named to a similar post on the Wilkins' expedition. Once the *Nautilus* was modified with an ice drill, an echo-sounder, and a special pressurized forward hold permitting lowering Nansen water bottles, reversing thermometers, plankton nets, and bottom samplers while submerged, she sailed for the ice pack north of Spitzbergen in August, 1931. Upon reaching the edge of the permanent ice pack, it was learned that she had lost a diving rudder and could only submerge by shoving herself via a top-side rail under small ice floes. However, this action did permit her to operate the oceanographic kit of instruments from the pressurized compartment for a 10-day period before returning to civilization. Sverdrup later confided that it was a godsend that the vessel could not dive normally as she was so decrepit that they would have never returned.

Despite her inherent limitations, the *Atlantis* provided the all-important work platform for many young scientists anxious to learn

---

40. To scientists who worked aboard her, the *Atlantis* was the "Hell Ship" for she was crowded, wet, smelly with diesel fumes, and often at sea in winter storms. In 1936, WHOI's academic roster consisted of H. B. Bigelow, C. O'D. Iselin, Alfred E. Parr, Alfred C. Redfield, Charles E. Renn, Carl-Gustav Rossby, Harry R. Seiwell, Norris Rakestraw, Henry G. Stetson, Selman Waksman, E. E. Watson, and Captain Sir Hubert Wilkins.

more about the sea. Iselin, for example, accomplished some excellent three-dimensional surveys of the currents and chemistry of the Gulf Stream system off North America. Maurice Ewing of Lehigh University, assisted by such top students as J. Lamar Worzel, Allyn Vine, and Albert Crary, began a series of marine experiments using seismic techniques to provide some inkling of the geological structure beneath the continental shelf and the North Atlantic Basin.[41] WHOI's meteorologist, Dr. Carl-Gustav Rossby, who held a joint appointed with the *Massachusetts Institute of Technology* (MIT), started work on an "oceanograph" that could trace out the continuous relationship between water temperature and pressure (hence depth) down to 270 m below the surface. However, it proved to be quite unwieldy, and it was not until 1938 that Rossby's imaginative student from South Africa, Athelstan Spilhaus, was able to file for a patent on a much improved device that "Spilly" called a *bathythermograph* (BT). Even this was hard to use under way, and it took some re-engineering by Ewing and Vine during the summer of 1940 before it became the streamlined, well-balanced device used extensively ever since.[42]

The availability of the BT could not have been more timely. In late 1936, Iselin had been advised by Lt. William Pryor, USN, of the destroyer USS *Semmes,* that the ship's highly-touted "searchlight sonar" (Asdic gear to the British) was experiencing severe losses in submarine detection ranges while off southern Cuba during hot, sunny afternoons. Pryor's initial thinking was that the diurnal release of oxygen by phytoplankton might be causing this "afternoon effect", but Iselin, after sailing the *Atlantis* down to Cuba to work with the *Semmes,* soon showed that afternoon surface heating was causing an upward bending sound refraction pattern which made detection impossible under the warm surface layer. As Iselin noted later (Iselin, 1959): ". . . thus U.S. oceanography had its first introduction to the ASW (anti-submarine warfare) problem!" Fortunately for the Allies, the German submarine force never did become aware of this diurnal limiting effect on sonar ranges because their sound

---

41. Bullard and Gaskell were doing much the same by making seismic refraction measurements of the British continental shelf starting in 1938 from the survey ship HMS *Jason* (Bullard and Gaskell, 1941). During this effort, the important discovery was made that the hydrophone used for timing underwater shots also recorded the incoming seismic signals just as well as did the water-proofed land geophones resting on the sea floor, thereby negating the continued use of the latter.

42. BT's were used and lost by the hundreds during World War II. In fact, Major Spilhaus once advised Captain Bates during 1945 that he was making more money from the BT patent than he was from his Air Force pay.

gear was tested in the shallow Baltic Sea where acoustic reflections off the sandy bottom kept the temperature effect from controlling.

In contrast to WHOI, Scripps remained largely land-bound during the early 1930s for their vessel, the *Scripps,* was so small and so old that it could not venture more than 15 km offshore. Even so, Vaughan did inveigle Sverdrup, who was by then considered to be the world's leading physical oceanographer under age 50, to come to Scripps as his replacement in September, 1936.[43] Upon reporting in, Sverdrup found a modest-sized staff who rarely went to sea, a problem compounded further when the *Scripps* blew up while at anchor during the following November.[44] However, Sverdrup was a master of public relations, and on 17 December 1937, "Bob" Scripps of the family that ran the Scripps—Howard newspaper chain presented the institution with the *E. W. Scripps,* a sleek 31-m luxury yacht formerly named the *Serena* when owned by the actor, Lewis Stone. After that, seagoing research took priority at Scripps as Sverdrup laid down a basic dictum: "Never sail without a hypothesis to test! To take data for data's sake is the sign of both an amateur and a wastrel!"[45] As in the case of exploration geophysics, the field of oceanography lacked a good text for teaching purposes. Sverdrup set out to remedy this situation in late 1938 with two of his assistant professors, Martin Johnson and Richard Fleming. After four years, they produced the monumental 1,087-page *The Oceans — Their Physics, Chemistry, and General Biology,* expecting total sales of about a thousand copies. In practice, the total count came to over 20,000 copies!

---

43. Sverdrup's personality combined an unusual ". . . harmony between . . . the explorer and the scholar, the naturalist and the theorist, the teacher and the administrator . . ." (Revelle and Munk, 1948).

44. The professional staff totaled eight, of which two were instructors. Five graduate students, including Roger Revelle, were in residence. Instructors' pay was $125 per month, but they were allowed to live in on-campus cottages renting for but $7 per month. However, Sverdrup's wife, Gudrun (a dentist by training), blurted when she saw this housing, ". . . how can men do good scientific work when they live so sordidly? Why, little Norway's peasants live better than these people!" (Sargent, 1979).

45. This practice was not always followed at WHOI. In the late 1940s, Iselin advised Bates that WHOI was taking far more oceanographic data than it could analyze because ". . . you never knew when the Federal money might run out!"

Entrance to International Seismological Centre near Reading, England. Left to right: Anthony Hughes (Director), T. Usami (Earthquake Research Institute, Tokyo) and David McGregor (Deputy Director).

Dr. Ernest Hodgson, the first Canadian to earn a doctoral degree in seismology (1930). (Courtesy J. Hodgson.)

Diagram of binaural equipment aboard a naval vessel to perform sub-sea direction finding during World War I. (Courtesy Raytheon Corporation.)

Advanced binaural headset for underwater direction finding during World War I. (Courtesy Raytheon Corporation.)

Professor Reginald Fessenden. (Courtesy Raytheon Corporation.)

Burton McCollum, one of the founders of reflection seismology, by a torsion balance. (Courtesy Bettye Athanasiou.)

Monument on outskirts of Oklahoma City, marking site of first comprehensive testing of the reflection seismograph. (Courtesy GSI.)

Sir Edward Bullard, one of the most eminent British geophysicists active from the 1930s to the 1970s. (Courtesy SIO.)

Dr. John C. Karcher, another of the founders of reflection seismology. (Courtesy GSI.)

O. S. Petty beside a tent for developing seismograph film in the Louisiana swamps *circa* 1925. (Courtesy Geosource Inc.)

Elephants providing motive power for a Petty Geophysical Company recording van in India *circa* 1939. (Courtesy Geosource Inc.)

Multiple explosions at a seismic shotpoint. Note the falling rocks. (Courtesy Bettye Athanasiou.)

Sir Hubert Wilkins' exploratory submarine, *Nautilus*, surfaced in the ice pack north of Spitzbergen during 1931. (Courtesy H. Sverdrup.)

Dr. Henry B. Bigelow, first Director of the Woods Hole Oceanographic Institution. (Courtesy of WHOI.)

Drs. Columbus O'D. Iselin and Edward H. (Iceberg Eddy) Smith, second and third Directors of the Woods Hole Oceanographic Institution. (Courtesy of WHOI.)

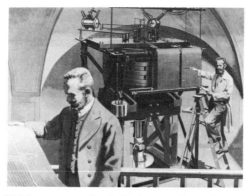

Weichert-type mechanical seismograph as of about 1900. The central mass could range up to 21 tons. (Courtesy Texas Instruments.)

Dr. Harry Hess and the Vening—Meinesz gravity meter he operated aboard submarines of the U.S. Navy during the early 1930s. (Courtesy NAVOCEANO.)

Research vessel *Atlantis*. (Copyrighted and Reprinted with permission of National Geographic Society.)

Athelstan F. Spilhaus working with an early version of his bathythermograph. (Courtesy of A. F. Spilhaus.)

Technical Sergeant Groverman and Private Samuel Loundy operating the sound recording equipment of Battery A, 2nd Field Artillery Observation Battalion, during mid-1941.

Dr. William F. Ford making a "BT" lowering from CSS *Hudson* during 1965. (Courtesy of W. F. Ford.)

Dr. Maurice Ewing rigging a deep-sea ocean-bottom corer. (Courtesy Lamont-Doherty Geological Observatory.)

The Scripps Institution of Oceanography during the mid-1930s. (Copyrighted with all rights reserved; reprinted courtesy of Dr. Eugene C. La Fond.)

The Scripps Institution of Oceanography during the mid-1970s. (Courtesy of the Archives, Scripps Institution of Oceanography, La Jolla, California.)

Retirement gathering in Assistant Secretary of the Navy Frosch's office honoring Butler King Couper in January 1972. Left to right: Richard Vetter, Hugh McClellan, Gordon Lill, Robert Abel, J. Lamar Worzel, Arthur Maxwell, the Honorable Robert Frosch, Gordon Hamilton, John Ewing, B. King Couper, Eugene La Fond, Maurice Ewing, J. B. Hersey, Rear Admiral R. C. Gooding, USN (Vice Commander, NAVSHIPS), Donald Martineau, Feenan Jennings, Captain Donald Walsh, USN, and Captain Donald Keach, USN.

Army Air Force officers assigned to Scripps Institution of Oceanography in mid-1943. Left to right: Lieutenants H. G. *Venn*, John C. *Crowell*, *Timpson*, Redfield, Charles C. *Bates*, R. S. *Klopper*, Professor Harald U. Sverdrup, Cletus J. *Burke*, J. Armstrong, G. S. Holliday, E. Yale Dawson, Dale F. *Leipper*, and Boyd E. *Olson*. (Courtesy of SIO.) (Names in italics are wave forecasters.)

CHAPTER 3

Geophysicists at War—1939-1945

War must be for the sake of peace, business for the sake of leisure, things necessary and useful for the sake of things noble.

Aristotle (*Politics* VI)

Exploration Geophysics During a Time of Global War

Despite the successful conclusion of the "War to end all wars" and the subsequent formation of the League of Nations, the implementing Treaty of Versailles proved unworkable because its basic premises were not based on a practical understanding of human nature and the need for adequate natural resources at the national level. With a revitalized Germany behind him by the late 1930s, Hitler was busy expanding his "Lebensraum" and being imitated by the leaders of Japan and Italy. Such actions rapidly upset the stability of the "Versailles World Order" concept. In view of this drift toward a new geo-war, Professor Richard M. Field noted in his presidential address to the American Geophysical Union on 30 April 1941:[1]

> From the dawn of history this method of conquest and colonization has led to the rise and fall of "master races" and imperial governments, equally aided and abetted by organized science and organized religion and organized trade.... Much as we may wish it otherwise, the true history of the rise of civilization is the history of organized science in which, until quite recently, the most important facts were either unmentioned or misinterpreted by historians. That is why "we learn from history that we do not learn from history".

This lack of historical understanding and a diplomatic inability to resolve the associated geopolitical problems then brought on a series of conflicts that would involve practically all of the world's geophysicists.

[1.] During the 1930s, Field was a major force in encouraging the use of seismic surveys to solve geological problems, particularly those associated with the nature of the sub-oceanic crust in the region of the continental shelf and continental slope. Edward Bullard of Cambridge University, Maurice Ewing of Lehigh University, and Harry Hess of Princeton University became particularly intrigued with the area because of Field's writing and speeches.

47

Looking back to 1940, one finds that this was a time when the United States was producing nearly two-thirds of the world's petroleum supply. In contrast, the Axis Nations were producing only the minuscule amount of 43,000 barrels per day within their home territory (see Table 3.1). Within the United States, it was still very much business as usual. The price of crude oil was holding steady at $1.00 a barrel after having risen to as much as $1.18 in 1937 following the price collapse of 1931–1932 that had taken the price to as low as $0.10 a barrel. The basic reason for such conservative pricing was that the United States had built up its proven reserves of petroleum in the ground from eight billion (10^9) barrels in 1926, when exploration geophysics first came on the exploration scene in a big way, to 20 billion barrels in 1940, thereby providing a known 14-year reserve of oil (see Fig. 2.2).

Even as late as April, 1941, Robert E. Wilson, president of the Pan American Petroleum and Transportation Company, felt confident enough to observe that ". . . no industry of comparable importance to defense has so few even potential bottlenecks as has the petroleum industry" (Wilson, 1941). Yet geophysicists were even then worrying about the seismic exploration technique becoming more and more deficient as an oil-finding tool once the salt domes and large anticlinal structures had been mapped.[2] Thus, Dr. W. T. Born of the Geophysical

TABLE 3.1
Crude Oil Production in 1940 by Geographic Region*

Geographic region	Barrels produced per day	Percent of total
United States of America	3,692,000	63.0
Other Western Hemisphere (Mexico, Canada, etc.)	866,000	14.0
Soviet Union	593,000	10.1
Middle East (Persian Gulf)	335,000	5.7
Netherlands East Indies	166,000	2.8
Romania	118,000	2.0
Germany, Poland, Albania, Hungary, France, and Japan	43,000**	0.7
Rest of World	49,000	0.8
	5,862,000	100.0

*Source: Wilson, 1941.
**Includes synthetic products manufactured in Germany. These products had a very high wax content and made German war vehicles difficult to operate on the critical Russian front during winter months.

2. The importance of geophysical work by this time is indicated by the succinct observation of one of the all-time great oil finders, Everette De Golyer, ". . . geophysicists (are) the gentlemen who do for geologists that which they are unable to do for themselves" (DeGolyer, 1942).

Research Corporation (the "grand-daddy" firm of the seismic business) found it necessary to state in his presidential address to the Society of Exploration Geophysicists on 3 April 1941:

> It seems certain that much of the seismic work on the future will be directly concerned with the location of stratigraphic traps . . . I believe that both (the geologist and the geophysicist) would profit also by studying together the subsurface picture of known stratigraphic traps. . . . In fact, I believe that it would be quite worth the cost to conduct a few seismic surveys over a number of known stratigraphic traps in the hope that the opportunity of comparing the results obtained with the known conditions might enable similar conditions in the same areas to be recognised. Certainly, some experimental work of this sort should be done before concluding that the seismic method is not applicable.

As it turned out, Born's prediction would not become truly valid until three decades later when the use of additional detector traces (up to 1,024 channels in lieu of six channels), longer detector spreads (up to four km *vice* 0.2 km), multiple signal sources richer in higher frequencies above 100 Hz, and vastly enhanced signal processing (using electronic computers) would finally make "stratigraphic seismology" a commonly used term. In fact, the continued military build-up and actual combat requirements caused the type of research he called for to be largely shelved for the duration. Within the United States, "oil-patch" employers did not normally attempt to shelter their staff from call-ups by the local draft board, this being considered an unpatriotic action.

Because of the lack of military deferments, a shortage of drill and production pipe, and low national priority, by 1942 the United States was discovering only 260 million barrels of oil per year. This was one-third the discovery rate of just five years before and but 19 percent of the actual consumption rate for it had been decided that the United States should "float to victory on a sea of oil". Gradually, the nation's entire petroleum exploratory effort comprised less than 25,000 persons: geophysical staff — 4,500 (40 percent technical), geological staff — 5,500, land-men — 1,500, and "wildcat drillers" — 12,000 (Heroy, 1943). Manpower turnover was also excessive, reaching as high as 75 percent in 1942 when, out of a geophysical payroll of 4,370, 1,428 persons joined the direct war effort and another 1,974 went with other industries (Goldstone, 1943). Even so, the geophysical industry mounted 225 seismic crews and 50 gravity meter parties during 1942, using women where feasible and extending the regular work-week to 50 hours and even, in some cases, to a steady 60 hours per week. As a result, exploration costs more than tripled by the end of the war, even though the price of crude oil stayed nearly fixed. Consequently, the financial incentive for finding new oil had not only disappeared but became a negative value (see Table 3.2).

Once their personnel loss rates approached 50 percent, the commercial geophysical laboratories admitted that "survival is the first rule in war" and became actively engaged in direct war work, either in manufacturing or in developing equipment. As Herbert Hoover, Jr., President of United Geophysical Company, noted at the annual meeting of the Society of Exploration Geophysicists in 1942, there was no profit in such work, but it did provide a lot of valuable experience in allied fields. His firm specialized in developing and constructing equipment for measuring and analyzing static and dynamic strains, vibration, and flutter in high performance aircraft. Humble's large geophysical instrumentation shop also went after war work and, by late 1942, fully 75 percent of its effort was used to design and manufacture special items of radar test equipment. Gulf Research and Development, on the other hand, had interested the Navy in Gulf's embryonic airborne magnetometer and became deeply involved in readying that device for field testing under combat conditions. Once the sensor was ready for routine manufacture, GSI turned to and built nearly half of these units used by the U.S. Navy during World War II.

During the war years, intensive geophysical exploration work also took place in many European countries (Vajk, 1949). In Great Britain, geochemical and magnetometer surveys were undertaken in the Midlothian region of Scotland in 1939, as were reflection seismograph surveys in northeastern Yorkshire and gravity surveys in the eastern Midlands. Refraction seismograph crews were then deployed from 1940 through 1943 and again in 1945 in central England, as were seismic reflection crews in 1943 and 1944. Using data acquired by Anglo-Iranian Petroleum, Ltd., Mobil promptly drilled the best of

TABLE 3.2

*Average Cost Per Barrel of Finding New Oil in the United States Between 1936 and 1944 (in dollars)**

Year	Exploration cost	Field development cost	Lease operating cost	Total cost of new oil	Average field price for crude oil
1936	0.18	0.21	0.26	0.65	1.06
1941	0.26	0.20	0.29	0.75	1.14
1942	0.42	0.22	0.31	0.95	1.19
1943	0.43	0.30	0.31	1.04	1.19
1944	0.91	0.38	0.31	1.60	1.25

**Source*: Jakosky (1947).

the gravity anomalies and discovered the 40-well Eakring Field within Sherwood Forest of Robin Hood fame.

German geophysicists also searched for additional oil and gas throughout Europe between 1939 and 1945. Although Prakla and Seismos were competent geophysical firms, they made relatively limited use of the reflection seismic method because of obsolete seismometers.[3] Their work areas also lacked good reflecting horizons and often had very complex geological conditions as well. Still, seismic surveys were made in the sedimentary basins of Austria, the Netherlands, Hungary, Czechoslovakia, Yugoslavia, Romania, and Poland, as were gravity and magnetic surveys. In the summers of 1942–1943, three refraction and four reflection seismograph parties, plus five gravity meter and seven torsion balance parties, worked in the conquered Ukraine region, mostly in the Rommy area. A magnetometer party also made a regional survey of much of this vast basin. Italian geophysicists were active, too, in the 1940–1944 period, occupying 2,600 shotpoints for seismic reflection work in north central Italy as well as 12,800 gravity meter stations in Italy and in a concession within Hungary. Exploration results were poor, and only 12 oil fields, *all* in Germany, were found during the war years. In view of these results, as well as those of the Allies, during the war years, it is quite legitimate to state that exploration geophysics played only a modest role during World War II. Yet, in another sense, one can argue that the skills and techniques of geophysics played a major role during the war. Let us now look at some of these applications.

Classical Seismology during the War Years

With the fall of France in 1940, the Central Seismological Bureau of the International Association of Seismology under Dr. J. P. Rothé in Strasbourg could no longer service the non-Axis nations with its monthly bulletin of important earthquake epicenters. Consequently, the Science Service Epicenter Project in Washington, D.C., with the help of the Jesuit Seismological Association and the Coast and Geodetic Survey, took on this task for English-speaking nations. In the United States, there was also one major attempt to use classical seismology in the war effort. In mid-1943, the Hurricane Microseismic Research Project was approved for implementation by the Meteorological Committee, Joint Chiefs of Staff. The goal of the effort was to obtain a dependable way of obtaining advanced warning of the paths and intensities of hurricanes and typhoons as they moved over

---

3. Mintrop's European patents for seismometer design were inclusive enough to prohibit the importation of better American geophones until 1937.

remote ocean areas. Lt. Commander Marion H. Gilmore, USNR, served as project leader. Reverend Macelwane, SJ, served as lead technical consultant during 1943 and 1944, followed by Professor Gutenberg in 1945 (Gilmore, 1947). The U.S. Navy was the action agency, with technical assistance by the U.S. Weather Bureau and the U.S. Coast and Geodetic Survey. The first tripartite seismic station became operational at Guantanamo Bay, Cuba, in 1944, followed later by stations at Roosevelt Roads, Puerto Rico, and Richmond, Florida, in early 1945, and at Corpus Christi, Texas, and on the islands of Trinidad, Antigua, Swan, and Guam during 1946 (Macelwane, 1951).[4] Eventually, the seismic detection approach for tracking tropical storms was shut down, however, once reconnaissance aircraft developed an all-weather capability and could penetrate such storms during hours of darkness as well as during daylight. Project findings did include: (1) most well-formed, dominant microseisms apparently come from a point under a deep low-pressure area moving over deep water, and (2) a tripartite microseismograph station can determine the direction from which microseisms approach from storms up to 600 miles distant in the Caribbean Sea and up to at least 1,000 miles distant in the Pacific Ocean.

A month before the war ended, the seismic networks of CalTech and the University of California (Berkeley) observed on 16 July 1945 a major train of seismic waves coming from the east associated with an air-blast so strong that the microbarograph's light-spot moved off its recording paper at Pasadena. This air wave affected the strain seismograph deep in the instrument tunnel, thereby suggesting that even the hill over the tunnel was measurably compressed during the wave's passage. Using data from five Coast and Geodetic Survey seismographs in Arizona and Nevada plus those from the six-station CalTech network, Benioff determined that this strange event had taken place at 11:29:21 GCT 1,000 km away in the desert near Alamogordo, New Mexico.[4a] The event, of course, later proved to be the world's first and highly secret atomic explosion, Trinity. Later on, the time announced for the event was never given to closer than 15 sec, causing Benioff's origin time to be as good as any (Gutenberg, 1946).[5]

4. A typical tripartite seismic station was laid out as a 2.4-km equilateral triangle using electromagnetic, horizontal component Sprengnether seismometers of six-second period.

4a. Professor Don Leet of Harvard University was the only seismologist invited to New Mexico to monitor this totally unique blast.

5. Gutenberg calculated that the air wave had moved at a velocity of 299.2 m/sec, a rather low velocity suggesting that the blast had traveled to Pasadena high up in the atmosphere where the temperature was about $-30°$ C. Seismic velocities through the earth's upper mantle (P_n) worked out to be 7.99 km/sec between Tucson, Arizona, and Riverside, California, and 8.04 km/sec over the near-by Tucson—Mount Palomar, California, path.

Geophysical Detection of Enemy Artillery

British and American armies made good use of sound and flash ranging during World War II just as they did during World War I. Even so, as of early 1941, the sound-ranging equipment used by the 2nd *Field Artillery Observation Battalion* (2nd FA Obsn Bn) at Fort Sill, Oklahoma, was little different from the sound equipment on hand at the end of the prior war. To be sure, the unit's Commanding Officer, Major Herbert Kruger, USA, held a master's degree in physics from the University of Pennsylvania, the same school that had provided doctoral degrees to such early exploration seismologists as Karcher, Eckhardt, and Haseman. Moreover, the two Sound Ranging Sections of Kruger's battalion were largely manned by draftees plucked from seismograph exploration crews. As it worked out, a party chief rated a sergeant's billet, an experienced computer a corporal's billet, and college-trained junior geophysicists the rank of private first class. It was well that this was so, for even as the "Second" at Fort Sill, Oklahoma, had cloned off the "First" at Fort Bragg, North Carolina, the summer before, the "Second", too, spun off the 8th Field Artillery Observation Battalion even before the attack on Pearl Harbor. In fact, by the end of the war, the Army had over a dozen such battalions in action.[6]

Records of the U.S. Army Field Artillery School indicate that the "Second", with its motto of "We Find Them", finally went into action, still as part of the 18th Field Artillery Brigade, to the southwest of Mount Camino and to the north of Naples, Italy, on 27 November 1943. Although they made their first battle target locations that evening, the Sound Platoon of Battery "A" (Bates' old outfit during 1941–1942) soon learned that their position was untenable because of intense enemy shell fire. As a result, the Platoon was unable to spot any enemy targets during their first day of combat because wire communications between the microphones and the plotting center were being continually shot out.[7] Yet two months later, the battalion turned in its most productive month of the war. In this case, the Germans had artillery well dug in along the line, San Giorgio–Esperia–Pontecorvo–Mount Cassino–Piedmonte–Aquino, and intended to hold it. Using four different types of sound

---

[6.] Battalion manning was approximately 24 officers, two warrant officers, and 429 enlisted men.

[7.] Radio was so unreliable on an active battle-front that the battalion could have out 380 km of communications wire when fully established. As much as possible of this wire would have to be retrieved before moving on to the next deployment area, a most difficult task, for this wire would be mixed in with communications wire of other units.

bases but primarily the five microphone, four-second base type, plus four-post flash bases either 5,000 or 6,000 m in length, 943 useful fixes were obtained during the 29-day month of February, 1944 (see Table 3.3). Despite mountainous terrain and German use of mobile artillery, accuracies achieved were normally surprisingly good. Later surveys indicated that only about 18 percent of the sound locations were false or unprofitable. Also, a careful examination during July, 1944 of 443 hostile gun positions found that the sound locations had an average error of 84 m and the flash locations 59 m.

While winding up operations on the Cassino Front in March, 1944, the "Second" cooperated with the British 8th Survey Regiment, which provided similar services on the Minturno Front, as well as with the British 3rd Survey Regiment. By the end of the month, in fact, the "Second" turned over an eight-microphone base to the rear of Mount Trocchio for the latter unit to operate, as well as a four-post, 7,000-m flash base along the line, Trocchio—Cedro—Maggiore. By 31 March 1944, the "Second" had each day for 100 consecutive days located one or more hostile gun positions, a most satisfying

TABLE 3.3
Major Events of 2nd FA Observation Battalion
(February, 1944)

Type of event	Frequency of occurrence
Hostile guns located	740 by sound ranging and 243 by flash ranging
Artillery fire registration	4 by sound ranging and 68 by flash ranging
Adjustment of friendly batteries	23 by flash ranging
Other targets located	23
Meteorological messages*	219
Surveillance of bombing runs	4
Survey assistance	19 days (support provided to 13 artillery units, 2 anti-aircraft regiments, and 36th New Zealand Survey Regiment, which also had an enemy gun location capability)

*Meteorological messages were actually ballistic wind values provided to Corps Artillery and the corporate Sound Ranging Platoon on a three-hourly basis based on pilot balloon ascents and radiosonde data obtained from the Air Force weather unit assigned to the nearest Army Corps Headquarters. When winds blew from the enemy lines toward the battalion, one had "good" sound conditions, and vice versa.

situation.[8] By then, however, it was time to move on to other battles. It soon participated in the Anzio Beachhead near Rome (April—May, 1944), the invasion of southern France (15 August 1944), and the forcing of the Rhine River at Ludwigshaven (30 March 1945). All told, the unit made some 3,252 fixes of enemy gun locations (2,438 by sound and 814 by flash) in Italy, 1,048 similar fixes (934 by sound and 114 by flash) in eastern France, and 30 locations (29 by sound and one by flash) in southwestern Germany.[9] One of the "Second's" more exciting moments towards the end of the campaign came near the French Maginot defense line at Reimerswiller when, during a German counter-offensive, Battery A's sound base had German troops on both sides. Enemy gun positions thus had to be plotted lying 180° from each other!

Each of the four corps comprising the U.S. First Army in the Normandy invasion had its own observation battalion: Vth Corps — 17th FA Obsn Bn; XIXth Corps — 8th FA Obsn Bn; VIIth Corps — 13th FA Obsn Bn, and VIIIth Corps — 12th FA Obsn Bn. Observation battalion units traditionally began to arrive on the beachhead on either D-Day or D+1 and had to stay as close to the front as possible no matter how fluid the battle situation. Finally a major price had to be paid for this necessary practice. On 17 December 1944, Kampfgruppe Peiper overran Battery "B" of the 285th FA Obsn Bn during Germany's last-gasp offensive through the Ardennes Forest.[10] All 125 men of the battery were taken from their trucks, led to a field, and massacred to the last man — the infamous "Malmédy Massacre". In general, however, casualty rates were low for front-line units. Lt. Colonel Wells once reported: "The battalion suffered nine casualties during the month of December (1944). Of these, two were killed, six wounded, and one shocked; all resulted from enemy shelling. Morale was excellent". Observation battalion services were in so much demand that once into the battle line, the unit seldom came off. Thus, the 13th FA Obsn Bn did not receive a rest period after reaching Normandy until 24 January 1945, eight months later.

Building on the experience and traditions of their predecessors in

---

[8.] By January, 1945, the battalion's members had won: Legion of Merit — one; Silver Star — seven; Bronze Star — 21, and Purple Hearts — 30.

[9.] Many of the unit's original officers stayed with it for the duration. In May, 1945, Lt. Colonel James Wells, a former Coast and Geodetic Survey Officer, became Battalion Commander; as of then, Major Jack Foster was Operations Officer, Major Harold Tucker, Executive Officer, and Captain William Contole, Assistant Operations Officer. The irrepressible, wise-cracking Jewish comedian, Samuel H. Loundy, had also risen from lowly draftee to the rank of 1st Lieutenant with duties as Assistant Sound Officer.

[10.] The 285th had been brought up to fill a gap between the areas covered by the 12th and 13th FA Observation Battalions.

World War I, the geophysicists of World War II turned in a highly commendable target-fixing service under the worst of combat and field conditions. As a by-product, they also provided corps and divisional artillery, plus many other units, with accurate geographic coordinates. Over and over again they proved, "It can always be done!" Unfortunately, the exigencies of battle did not permit using any of the electronic miracles in communications and signal processing even though radio links were actually field tested, but with poor results, in Italy.

Influence Mines in Naval Warfare — A New Application of Geophysical Sensors

For centuries, mines have played an effective role in warfare both on land and at sea. As far back as 1585, Gianibelli, an Italian engineer working for the hard-pressed Dutch, used a time bomb to blow up a Spanish floating bridge closing off Antwerp, throwing the Spanish Army into considerable confusion and terror as well as causing a thousand casualties. The Confederate Navy used mines in the War between the States and ran up a score of 32 Union ships mined (of which 27 sank), whereas only nine Union ships were sunk by gunfire.[11] The Russians at the close of the 19th century started their continuing interest in the extensive use of naval mines. In the Russo-Japanese War of 1904—1905, after decisive naval defeats at sea, they were still able to sink two Japanese battleships and five attack cruisers by mines but none by gunfire or torpedoes.

In World War I, all major powers used contact mines as a major naval defensive weapon. In the abortive Dardanelles campaign of 1915, Churchill credits unswept mine fields as indirectly leading to abandoning the invasion of Gallipoli. In return, the Allies (and particularly Great Britain) used mines anchored at depths which allowed surface vessels to transit safely but created a major threat to submarines running submerged. The magnitude of this effort is indicated by the British planting 128,650 and the Americans 56,000 mines just to create the "North Sea Barrage". On the other side, the Germans used over 43,000 naval mines and were able to sink 44 British

---

11. Rear Admiral David Farragut, USN, while in the rigging of his flagship, USS *Hartford*, on 5 August 1862, called down the immortal words, "Damn the torpedoes (mines) — full steam ahead!", as his task force began to penetrate the Confederate mine field protecting the entrance to Mobile Bay. Even so, he soon lost the ironclad USS *Tecumseh* to a mine off Fort Morgan.

12. Lord Kitchener, the British Minister of War, was lost, as were 585 more of the 600 men aboard the cruiser HMS *Hampshire* in June, 1916 when she hit a mine in the North Sea *en route* to Russia.

war vessels, 225 British auxiliaries and 586 allied merchant men by such devices.[12]

The big need in mine warfare was the obvious one to extend the influence of a passing ship on a mine detonator beyond pure physical impact. During World War I, the idea of using a magnetically detonated mine triggered by a ship going by had occurred almost simultaneously in Germany, Great Britain and the United States.[13] Of the three, the Germans had the best instrumentation and largest amount of skill and experience. Within the United States, most of the exploratory development was done by scientists of the Department of Terrestrial Magnetism in the Carnegie Institution of Washington. Their first efforts were simple — determine a ship's magnetic field by floating in a battleship or submarine into a dry-dock, measure the surrounding field, and then re-do the measurements once the ship has left the dry-dock.[14] Professor E. L. Nichols, on leave from Cornell University, soon conceived of a mine firing mechanism capable of being triggered by such a magnetic field.[15] The war ended before he had perfected such a trigger, but being of retirement age, he left Cornell in 1919 and began working for the Navy's Bureau of Ordnance, primarily at the Standard Magnetic Observatory of the Department of Terrestrial Magnetism. Ultimately, he achieved a trigger setting that would detonate the "M-1" mine with a change of less than two percent in the earth's magnetic field. Although a thousand M-1s were then built, they were never used in the field for their earthenware cases easily broke during shipment, nor could they be adjusted for placement at differing geographic latitudes.[16]

British mine inventors overcame problems similar to those faced by the Americans and actually had a magnetic mine built by the end of World War I. Work at the Admiralty's Mine Design Department continued during peacetime, and magnetic triggers were in stock for use when needed. Such an event occurred in September, 1939. German aircraft were observed parachuting long cylinders into British channels and harbors, and even the Home Fleet's flagship,

---

13. The steel and iron in a ship's hull acquires an appreciable magnetic field because of all the hammering conducted during construction, as well as the presence of strong electrical currents in the vessel during welding and while under way.

14. One of the key workers was Dr. John A. Fleming, Chief Geomagnetician at the Carnegie from 1904 to 1946, General Secretary of the American Geophysical Union from 1925 to 1947, and Honorary President of the Union during 1947.

15. Nichols was also the founder of the esteemed journal *Physical Review* and edited it from 1893 to 1912.

16. The handling of the problem of converting an advanced laboratory device that only the inventor can operate and maintain into a field device routinely and reliably operated and maintained by enlisted men at sea is, in large part, what makes military research, development, test and evaluation so expensive and time consuming in the naval sphere.

HMS *Nelson*, received mine damage while in Loch Ewe. On 23 November 1939, a fisherman reported an unusual object on the tidal flat near Shoeburyness. Admiralty mine specialists, upon viewing it, decided that here was the new German mine. The Germans at first dropped their "secret weapon" into British waters in small numbers, but ship casualties were high in proportion to the number of mines employed. In one weekend, 45,000 tons of shipping were lost from mines in the Bristol Channel of western England. Fortunately, heavy laying of these magnetic mines, for that is what they proved to be, did not start until April, 1940.

Once the nature of the mine was known, Professor Edward Bullard of Cambridge University's Geodesy and Geophysics Department was called from the Admiralty Anti-Submarine Research Establishment to work on the problem with the help of Commander Charles Goodeve and other members of the Mine Design Department as well as staff of HMS *Vernon*, the Navy's torpedo and mining establishment at Portsmouth. The overall group made excellent progress and soon had not only learned how to sweep magnetic mines but also to cancel enough of the magnetic field of a ship so that it was not so likely to trigger the magnetic sensor. Sweeping a large area could be done by pulsing a large electric current through parallel cables streamed behind two ships moving in line abreast. Design of such a cable proved a problem at first. However, the designers were told on good authority that such floating cables had been used pre-war by Canadians on the Great Lakes of North America. This information gave the necessary confidence to ensure rapid development of a streamlined cable buoyed up by foam rubber, although, in fact, in the cited Great Lakes project, standard underwater electric cable had been suspended by wooden beams. Once "Double-L" sweeps were in action, German magnetic mines were harmlessly exploded in large numbers, even if a given area had to be swept many times because some of the mines were fitted with mechanisms which could be pre-set to fire on the first actuation or on any subsequent triggering up to a dozen or so. In the case of reducing the ship's magnetic field, winding permanent electric coils around the whole ship and passing suitable currents through them did the trick. However, each ship had its own unique magnetic field; hence, after installation, the "degaussing belts" had to be calibrated and adjusted while the ship steamed over a magnetic range composed of a line of magnetometers placed on the sea bed.

The German mine designers faced a problem that, even today, is a difficult one. While devices were fitted which made mines detonate when they hit land rather than at sea, such devices had trouble in differentiating land impact from that of striking shallow water. To add a further complication, much of the English coast is subject to a large rise and fall of tide, so a mine laid at the proper water depth might end up high and dry a few hours later. In addition, the parachute of an air-dropped mine often caught in telephone wires, trees, bridges, or other obstacles, making it the rule, rather than the exception, that a mine was recovered soon after a new mechanism was introduced by the enemy. The Germans realized this and tried to deter the brave naval officers and scientists who took enemy mines apart to extract the working mechanisms by fitting booby traps. On one occasion as a new type of mine was introduced, specimens were dropped well inland at Portsmouth and Portland. The Portsmouth mine was then defused and taken into the HMS *Vernon* mining shed. But on taking off the back-plate, a 4.5-kg charge detonated, killing some of the expert mine-removal team. Ways were soon worked out to circumvent booby traps, and the skills subsequently proved to be a good back-up for de-fusing mines dropped as bombs during the "blitzes".[17] Providing a totally tamper-proof mine should it fall into enemy hands was always at variance with providing safety to the crew of the mine-laying aircraft should it have to make a forced landing with mines on board.

Throughout any period of warfare, decisions must be made on how best to introduce improved or new weapons. For example, should the Germans have used "secret weapons" in massive saturation raids or should they have used the course they did, namely, try them out in small numbers at first? The Allies subscribed to the first technique and believe that the Germans erred when using the second approach, which was the same they used to introduce poison gas to the battlefield during World War I. Even Eisenhower observed that it was German practice to commit reinforcements piecemeal as they arrived to hold back an Allied breakthrough, rather than keeping such troops intact and rested until a concentrated counter-attack could carry the day.

The effectiveness of the German magnetic mine during 1940 was duly noted by the U.S. Navy.[18] Several of its brighter young officers were accordingly placed on liaison duty with HMS *Vernon* to observe developments, and the British also supplied several German triggering

---

[17.] The British Broadcasting System has an excellent TV series, "Danger-UXB", which vividly portrays the adventures of bomb-disposal squads of the Royal Engineers.

[18.] Enemy loss rates from aerial mining during World War II were reported as: German mines — 1940, 1 ship loss per 65 mines, and 1941, 1 ship loss per 112 mines; British mines — 1 ship loss per 52 mines, and United States mines in the massive mine laying campaign against Japan in 1945 — 1 ship loss per 23 mines. Operations analysis studies indicate air-dropped influence mines were about ten times as economical as submarine attacks in sinking a ton of enemy shipping during this particular war.

devices for detailed study.[19, 20] However, developing an effective chain-of-command to contend with the threat of influence mines took the U.S. Navy almost the entire duration of World War II. This problem arose in large part because of a type of "class war". Key policy components of the Office of the Chief of Naval Operations (CNO) were largely staffed by regular officers who had been through the "trade-school" (the Naval Academy at Annapolis), while the technical support bureaus, instead of reporting directly to the CNO, reported to the civilian-oriented Office of the Secretary of the Navy. In addition, these bureaus (with the exception of the Bureau of Ordnance) were staffed primarily by "special duty only" regulars with their own promotion boards and by reservists. In May, 1940 the Secretary of the Navy added a further complication when he ruled that the Bureau of Ordnance would have jurisdiction over the physical recovery and rendering safe of enemy mines and over degaussing, the Bureau of Construction and Repair (assisted by the Bureau of Engineering) over magnetic sweeping by ships, and the Bureau of Aeronautics over magnetic sweeping by airships and aircraft.[21] Such an organizational nightmare was just the reverse of British practice. There, to conserve scarce manpower and materials and simultaneously to accelerate appropriate technical response to Fleet needs, the closely related activities of degaussing, mine sweeping, mine recovery, procurement, and supporting research were melded into one tight organization whose officers were also assigned to field commands as needed.

In early 1940, the U.S. Naval Bureau of Ordnance finally directed its Naval Ordnance Laboratory (NOL) to begin an aggressive research programme for magnetic and acoustic mines and their counter-measures.[22] Under prodding from the Bureau Chief and Dr. Vannevar Bush, President of the Carnegie Institution of Washington, Commander

---

[19.] Lt. Odale D. Waters, USN, was one of the first Americans assigned to HMS *Vernon* to learn how to render safe and then recover German ground mines. Upon returning State-side, he started the Navy's Mine Recovery School during June, 1941 and remained there as its Officer-in-Charge until January, 1943. Waters became Commander, Mine Warfare Forces, Pacific in 1964, followed by serving as the third "Oceanographer of the Navy" between 1965 and 1971.

[20.] To place magnetic mines in the field quickly, the U.S. Naval Ordnance Laboratory duplicated the German magnetic firing mechanism for both submarine and air launched mines and had 1,500 mines ready to be laid out of Cavite Naval Base in Manila Bay when the Japanese struck on 7 December 1941.

[21.] The Bureaus of Engineering and of Construction and Repair were combined into a "Bureau of Ships" two months later.

[22.] Even as late as July, 1940, the NOL technical staff consisted of two scientists and three engineers, although a sizeable number of skilled technicians and craftsmen were also on board.

J. B. Glennon, USN, the Laboratory's Commander, met with staff members of the Institution's Department of Terrestrial Magnetism only to learn that the Department had been well along on solving the magnetic problem 22 years before but now the all-important records were hopelessly lost.[23, 24] Actually the magnetic mine defense problem was already solved because in March, 1940 the unfinished passenger vessel, *Queen Elizabeth*, steamed into New York Harbor with a great degaussing girdle encircling her. A month later, the New York Naval District sent in a piece of degaussing cable for all to see, plus the information that the British Navy was having over 30 km of the cable manufactured in the United States each day! As a consequence, by June, 1940, U.S. naval vessels began to receive degaussing coils and in mid-1941, half of the nation's electrical cable manufacturing capacity was assigned to making degaussing cable.

To get a competent research program going immediately, Commander Glennon leaned heavily on two reservists, Professor Louis W. McKeehan, Director of Yale University's Sloane Physics Laboratory, and Professor Ralph D. Bennett of MIT's Electrical Engineering Department. Both reported in as Lieutenant Commanders in mid-1940, and both would eventually reach the rank of Captain by the war's end. For the first year, Bennett specialized in bringing on board component technical staff, and by late 1941, nearly a thousand scientists were working at NOL.[25] Of the 66 members of the Society of Exploration Geophysicists known to be civilians involved in the war effort as of 1945, thirteen were at NOL.[26] One of the first efforts of the expanded NOL staff was to look at the entire realm of detect-

---

[23.] During World War I, Bush worked on the submarine detection problem. He then became a professor (1918—1932) at MIT, followed by becoming that school's Dean of Engineering and Vice President (1932—1938). In 1941, he was named Chairman, National Defense Research Committee, reporting directly to President Roosevelt and thereby controlling the nation's academic research program on behalf of the war effort.

[24.] After activating a mine during the Normandy shore bombardment on 8 June 1944, the USS *Glennon*, a destroyer named after Commander Glennon's father, eventually sank with a loss of 25 crew members.

[25.] Recruiting duties over, Bennett became Chief, Mine and Depth Charge Division, at the Laboratory from 1941 to 1944, after which he became the Laboratory's Technical Director, first as a captain from 1944 to 1947, and then as a civilian (1947—1954). He then moved into the civilian field of nuclear energy for General Electric, followed by becoming Vice President, Research and Engineering, the Martin Company (1961—1966). Bennett is considered to have been one of the Navy's two best laboratory directors (the other was Dr. William McClean); Bennett also became the first engineer to be designated a "supergrade scientist" under the provisions of Public Law 313 (1950).

[26.] These included Milton B. Dobrin (ultimately Vice-President of United Geophysical Company and author of *Geophysical Prospecting*), Lewis Mott-Smith (inventor of the Mott-Smith gravity meter), and Noyes D. Smith (eventually Chief Geophysicist of The California Company, a major subsidiary of the Standard Oil Company of California).

able physical fields that might be caused by a passing ship.[27] Magnetics, of course, was the best known influence field, but the second type — developed almost simultaneously by the Germans, British and Americans — was the acoustic field intimately related to vibrations set up by propulsion machinery and by the action of a ship's propeller on the water.

When, in September, 1940, the Germans played their next secret card, the acoustically-activated mine, the British "Sweeping Section" ("S-Section"), had been operational for six months under Professor Edward Bullard.[28] Part of HMS *Vernon*, the Section was close to ship facilities, plus cable and stores, in the vast Portsmouth dockyard. Experimental magnetic and acoustic ranges were set up at Clarence Pier on the Southsea water front, with laboratory facilities adjacent in the "Assembly" or "A" and "B" rooms converted from a pre-war dance hall. German acoustic mine behavior was determined here at Clarence Pier, as well as the research and design of standard hydrophones needed to test the noise of ships and mine sweepers. After six months, the Pier burned during an air raid, and "S" Section moved to Edinburgh to ensure that British sweepers would be technically one step ahead of any new mines from the enemy. It was by coincidence, rather than German spite, however, that Clarence Pier and all its elegant gear tuned to perfection for the occasion was destroyed the evening before the First Lord of the Admiralty was due to visit his successful acoustic mine-sweeping department. Needless to say, it was quickly agreed that the charred remains of recording huts and cable ends did not need an official blessing.

After moving to Edinburgh, the Sweeping Section of the Mine Design Department concentrated on the full frequency range of underwater sound from fractions of a cycle per second (hertz) through low frequency (one to 50 hertz) to high frequency (1,000 to over 100,000 hertz). Appropriately for a nation that has produced much of the world's classical music, the German acoustic mine had a response peaking sharply at "Middle C". The work at Portsmouth,

---

[27.] Other physical fields known to exist but difficult to exploit technologically included the ship's gravitational field and the change in the cosmic ray field as a ship passes by. The Germans also looked at both effects and actually constructed experimental devices using them. Target detection ranges of 15, 30 and 60 m gave enough flexibility for most operational situations. The effect of "non-sweepability" was mitigated by intentionally including arming delay and sterilizing devices set for various time patterns.

[28.] Bullard brought from Cambridge University his inventive laboratory technician, Leslie Flavill, and one of his research aides, Thomas Gaskell. As a consequence, the experiments at Portsmouth were almost a continuation of work within the "G and G Department" back at Cambridge. Bullard was knighted in 1953 for his many research contributions to the war effort.

besides determining the behavior of the German mine, involved making a broad comparison of countervailing noise-makers proposed to the navy by many budding inventors.

One of the early useful sound sources proved to be the old Fessenden oscillator produced for underwater signalling (see Chapter 2). By good chance, it operated at 512 hertz, twice that of the mine's principal resonance, and it was not at all difficult to adapt the Fessenden device to give a good signal at 256 hertz. Underwater sirens, steam-driven oscillators, and many other peculiar devices were tried, but, in the end, the simplest, most robust noise-maker was the "hammer-box", a conical housing fitted with a diaphragm hit about 20 times a second with a KANGO® riveting hammer.[29] The diaphragm's diameter and thickness were chosen to have a resonant frequency of 256 hertz in water. The physics textbook consulted to derive these parameters was none other than Lamb's 19th-century *Treatise on Sound* (Volume II). The old professor would definitely have been delighted to see his mathematical reasoning so nicely confirmed, even though it took a war to do so. When time allowed, other elegant experiments to verify Lamb's deduction were carried out, including recording the line spectra of sound output from hammer boxes and the propagation of sound waves in water layers with bottom boundaries of differing reflectabilities. Before effective mine sweeps could be devised, it was also found possible to render an acoustic mine temporarily inoperative by producing a loud, sharp sound underwater. This particular feature had been intentionally incorporated into the mine to prevent the explosion of one mine within a field causing all its neighbors to fire as well. However, the Portsmouth group found it was possible to go through an acoustic minefield by merely firing machine gun bullets into the water ahead of the ship at intervals determined by the time that the mine lay dormant after each explosion.

Bullard's team also developed a highly useful standard hydrophone using quartz as the active sensor for measuring the level of underwater sound.[30] This device was then used to measure the noise of many hundreds of ships passing over the "acoustic range" in the River Clyde, and also to find the change of noise type and intensity with varying speed of ships moving over the experimental range off Port Edgar in the Firth of Forth. In addition, special trials were carried out to alert

---

[29.] The Americans had good results by merely letting two pieces of pipe collide with each other as they streamed in a ship's wake.

[30.] The U.S. standard hydrophones had been calibrated against a secondary level, i.e. a standard noise source in air, rather than in water. Thus, when the nation's two standard hydrophones were compared, a 10 percent discrepancy showed up, traced to an error in the U.S. air standard. Discovering this fact was extremely important if one were to measure correctly the sensitivity of mines and noise-maker outputs in various parts of the world.

submarine commanders to the degree of quiet they must maintain when lying on the sea bed within sound range of the enemy. At one time, the Minesweeping Section was called in to listen and determine whether the Germans might be tunnelling under the English Channel, for photographic interpreters had mistaken some defence preparations on the French coast as spoil from nefarious tunnelling operations. In this instance, the Section's geophysicists were delighted to get back to measuring noise in the ground. A quartz accelerometer was quickly constructed and used to listen for digging activity of the type that might be expected in the experimental five km of Channel tunnel dug in the 1880s. Field tests soon indicated that a series of seismic monitoring posts could detect enemy operations to a distance of about 0.8 km, enough to provide adequate warning should suitable countermeasures need to be employed.

Supersonic noise produced by ship's propellers was also measured in good time to produce noise-makers to confuse the German acoustic torpedo. Two noise-makers, code-named Foxers, were towed behind the ship and the wretched torpedo was confused by the double sound source into following one and then the other source alternately, so that it wore itself out doing figure-of-eight courses around the noise sources. The U.S. Naval Ordnance Laboratory also made a great advance in producing, in a very short time, an airborne acoustic homing torpedo working on the audible sound range. By this time (1943), a good exchange of both people and results existed between Britain and the United States. Gaskell went, in fact, to a six-week course at the Bell Telephone Laboratories at Summit, New Jersey, to learn the intricacies of the Mark-24 torpedo in case unforeseen operational troubles should occur when the new weapon was used by Coastal Command or from aircraft carriers against submarines in the eastern North Atlantic. The British team of scientists and naval technicians crossed the dangerous North Atlantic via the speedy, unescorted, independently routed *Queen Elizabeth*. They did have to earn their passage to some extent, for the ship's master asked them to mend the radar, a very new device in those days used to give advance warning of icebergs. Although radar was new to Gaskell as well, he applied such time-tested geophysical fault-correction techniques as hitting the apparatus and smelling for burned condensers with good success. The end result was particularly favorable, for the Captain then offered drinks in what was, for the passengers, essentially a "dry" ship.

By late 1941, there was no doubt that the British were ahead of the Germans on the mine-sweeping front.[31] However, the "ding-dong"

[31.] This conclusion is based on a report by Lt. Danckwerts, RN, who carried out a post-war interview with the head of the German Mine Department.

struggle between mine-maker and mine-sweeper was much the same on each side. For example, the British combined magnetic—acoustic mine had an initial success in sinking the large enemy liner, *Gneisenau*, but was soon controlled by updated German minesweepers. Reversely, the British had expected to see a combination acoustic—magnetic mine in their own waters, and appropriate field trials and sweeping instructions had been implemented before the new German mine — Sammy — appeared, having been, as usual, kindly dropped on land by mistake. British minesweepers operated in the new sweeping formation the very next morning, and one of the East Coast mine-sweeping bases then sent in their daily telegram to "S" Section as follows:

"ROLLS ROYCE AND BEN MEDIE (two of their vessels) SWEPT TWO FOUR MINES BEFORE BREAKFAST".

By the spring of 1942, NOL had finished initial development of a mechanism to convert the Mark-25 2,000-pound aerial magnetic mine into pressure and subsonic acoustic mines; however, follow-up work was given such low priority that field-worthy triggers for this mine were not ready until three years later.[32] This was a deliberate delay because high-level decision-makers, backed by the Operations Research Group, held to the belief that new types of mines should not be deployed until adequate in-house sweeping counter-measures for such a new mine were available. Inasmuch as the Bureau of Ships had not developed the necessary counter-measure to the potential modification, a strong apprehension existed that if the Germans learned of the development, both British and U.S. naval forces would be at grave risk. In practice, the German scientific war effort came up with the same approach on its own and, as noted earlier in this chapter, introduced it as the Oyster mine five days after the Normandy invasion started with deadly effect, particularly in the Cherbourg area. Had the Germans introduced this "unsweepable" mine earlier when its air force could more freely range over British coastal waters or had their submarines laid such mines in constricted water close-in to such harbors as New York City, the results could have been extremely damaging. Ironically, Germany's key naval planners did not allow such an early introduction for the identical fear that the Allies would acquire, analyze, and deploy such an unsweepable mine against their own navy!

[32.] NOL made extensive use of a mine-test station established with the help of the Royal Canadian Navy in the Bay of Fundy for the purpose of testing mine survivability after high-speed water entry. Bombers would lay the mines at high tide in 13 m of water, and technical crews then dashed out with jeeps onto the tidal flat at low tide to examine and test fire the mine mechanisms prior to the onrush of the next tidal bore.

As a result of the fetish for "tight military security", huge inefficiencies were generated because of the ponderous U.S. Navy communication channels by which operating field commands were informed of what would shortly become available from the massive military research and development effort. Vice versa, the military laboratories had no simple, direct way of learning via the written word of unforeseen operational needs of the true nature of enemy threats and reaction capabilities, and of the empirical but useful facts developed on the battlefield at considerable expense in lost lives and material. Even when Dr. David Katcher tried to set up an "Operational Research Group" at NOL early in 1942, his concept was shot down because "coordination of operational information with technical development was considered to be too secret to study by such a low echelon, and by civilians to boot!" (Johnson and Katcher, 1973). As a direct consequence of working in a "near vacuum", NOL's highly competent research staff spread itself over too many projects. Some 65 different mine devices were given "Mark" or "Mod" designations by the U.S. Naval Bureau of Ordnance during World War II. Thirty-four of these were released to service, but only seven were used in combat. In fact, the U.S. Navy's biggest production run — the Mark 16 mine — never saw combat because of field unsuitability.

Technical and military liaison was far better handled in British circles. Here the "regulars" and civilian scientists worked side-by-side taking mines apart and on the field trials of sweep apparatus and the determination of ship properties relevant to triggering influence mines. Casualty data caused by submarine attack and mine-laying were also routinely supplied to the scientific side of the establishment as well as to the operators. Scientists were taken as a matter of course to meetings where operational matters and force requirements for weapons were threshed out.[33, 34] In looking back, it is quite obvious that effective minesweeping, aided in considerable measure by technical leadership and knowledge from university and oil company geophysicists, played a necessary and significant role in keeping the

---

[33]. Professor J. D. Bernal, as Scientific Adviser (x-SA1) to Combined Operations Headquarters, attended some of the highest-powered meetings on Whitehall. In fact, his assistant, Dr. Gaskell (x-SA2), was also brought in occasionally.

[34]. As far back as the Revolutionary War, that esteemed student of lightning and the Gulf Stream, Benjamin Franklin, engaged in operational research with the British by writing Joseph Priestley: "Britain, at the expense of three millions, has killed 150 Yankees in this campaign at £20,000 per head. And at Bunker Hill she gained a mile of ground, half of which she lost by our taking post on Ploughed Hill. During the same time 60,000 children have been born in America. From these data any mathematical head will easily calculate the time and expense necessary to kill us all and conquer our entire territory."

nation's coastal approaches and harbors operations at a particularly critical time.[35] In general, the British system worked well even when using two services for laying offensive mines in enemy waters. Such mines were designed, constructed and tested by the Royal Navy. The Navy also specified where they should be laid, but the actual laying was the responsibility of the Royal Air Force. In the instance of using special mines to harass German submarine movements proceeding in and out of the French seaports facing the Atlantic Ocean, these mines, at first, were activated by the passage of the minesweeper preceding the submarine and then firing as the submarine went by. The Germans soon responded by running two sweepers in tandem before the submarine, but a fast-moving team of Admiralty's Mine Design Department was apprised of this and thereafter delivered several hundred mines with altered properties every few weeks to fool the enemy. But the major glitch came relative to the lack of mining canals on the Continent. The Mine Design Department had gone ahead and produced a small mine suitable for canal placement by fast, low-flying Mosquito aircraft, and its use would have been quite complementary to the bombing of railway lines then being vigorously pursued by the Royal Air Force. But upper naval echelons dragged their feet regarding this tremendous opportunity, for certain captains thought that canal warfare was not a legitimate naval responsibility.

Despite the complicated organizational structure by which the U.S. Navy pursued mine warfare, offensive aerial mining did come to play a major, although somewhat unsung, role in waging war in the Far Pacific from 1943 on. To do so took a lot of perseverance, salesmanship, and, in special measure, outstanding bravery by flight crews.[36] These weapons were used, however, in impressive numbers by the Royal Air Force, the Royal Australian Air Force, the 14th and 20th U.S. Air Forces, the U.S. Navy's Fleet Air Arm and Commander, Task Force 58, to make operations of the Japanese Navy and merchant marine costly all the way from Rangoon, Burma to the waters of occupied China and even to the major harbors and waterways of Japan itself. All told, some 21,389 successful launchings of aerial mines were made in the Pacific Theater, 93 percent of which

---

35. The petroleum geophysicists were drawn largely from the staff of British Petroleum, Ltd., then known as Anglo-Iranian, Ltd.

36. When there is a plethora of manpower, as there was in Washington, D.C. during World War II, headquarters organizations tend to expand, as noted by C. D. Parkinson, into complex structures wherein it is far easier to get "no" for an answer than it is a "yes". Highly motivated persons at the working level still go ahead and get the job done as best they can, even if it takes "truthful deceit" from time to time.

were on target, with a loss of only 55 aircraft.[37, 38] Experience indicated that the B-29, with its heavy cargo-carrying capacity and long range, was the most effective of the aerial mine layers.[39] Most types of aircraft did their mining during daylight or moonlight, although the B-29s preferred moonless nights with cloudy or bad weather and altitudes of 1.5 to 3.3 km as they used radar for navigation. In fact, the B-29s were able to drop some 12,000 mines into every important Japanese harbor between March–August, 1945. On one night alone, just before the invasion of Okinawa, these "Superfortresses" were able to plant nearly a thousand magnetic and acoustic mines in the Shimonoseki Straits and the Inland Sea. As a result, by August, a virtual stoppage existed of all Japanese shipping wishing to transit the Straits to southern Japan where an invasion appeared imminent.[40] This aerial minelaying effort, supplemented by mines laid by the U.S. Submarine Force, sunk or damaged 1,075 Japanese ships, including 109 combat vessels. The total tonnage lost was more than 2.25 million tons, or approximately one-fourth of the pre-war strength of the Japanese merchant marine. Prince Konoye of the Imperial Household and one of the Empire's principal leaders estimated that the mining attack proved just as effective as had the bombing attack on hampering Japanese industry (Johnson and Katcher, 1973).

Use of a tireless, passive unmanned weapon that could lay in wait for the appropriate geophysical signature from a passing enemy ship was extremely cost-effective in terms of U.S. lives saved and cost per ton of enemy ship casualty within Japanese waters. In studying the situation afterwards, the delivery vehicle investment per enemy ton casualty worked out to $16 when mine delivery was by aircraft and $100 when delivery was by submarine mine and torpedo (Johnson and

---

[37.] Aerial mines expended during the overall campaign were as follows:

Magnetic:	12,053	Low frequency:	768
Acoustic:	4,448	Contact (drifters):	268
Pressure-magnetic:	2,992	Magnetic–acoustic:	124
		Dummies:	106

[38.] Mining sorties flown by the several services were: U.S. Army Air Force: 2,078; Royal Australian Air Force: 1,218; Royal Air Force: 631; U.S. Naval Air Arm: 486.

[39.] Much of the credit for persuading tough-talking, aggressive Major General Curtis H. Lemay, USA, Commander 20th Air Force, to utilize his unique B-29s in aerial minelaying belongs to Lt. Commander Ellis A. Johnson, USNR, a Carnegie Institution geomagnetician on the staff of NOL who visited Lemay in the Marianas during late 1944 for just this purpose. Going so far out of regular channels almost brought Johnson a court martial, however.

[40.] In March, 1945, 520,000 tons of Japanese shipping moved through the Shimonoseki Straits; by May, the monthly tonnage was 150,000 tons, and August but 10,000 tons.

Katcher, 1973).[41] Similarly, the cost of lost U.S. delivery vehicles per enemy ton casualty works out at $6 for aerial mining and at $55 for submarine attack. In this overview, the same authors concluded:

> The weaponsmakers of World War II, who were mostly military amateurs (scientists and engineers) from the universities and industry, *were* enormously productive but, through no fault of their own, they were late. . . . For every useful weapon there were dozens or hundreds that might just as well not have been born.
>
> Mines were largely developed by amateurs and their final use was guided almost entirely by Reserve Naval Officers who were representatives of these same amateurs. . . . The Scientist Naval Reserve Officers were amazed by the fact that wars are won principally by military skill, rather than by new weapons, and that in the actual running of the war in combat areas, civilian scientists, except as expert technicians or observers, are about as useful in operations as are professional military men in running a research and development programme. The wrong shoe pinches on either foot . . . neither group can claim honestly to possess proven competence in the practice of the other's profession.

Geophysical Aspects of Undersea Warfare

When World War II opened in September, 1939, those who controlled naval tactics and design believed that submarines should be hybrid vessels, capable of operating and fighting on the surface as well as below. Submarines of all major navies carried deck guns almost as heavy as those on destroyers. Although Grand Admiral Erich Raeder of Germany had planned to build up his peacetime fleet until it had 250 U-boats, Hitler moved up the date for starting World War II so much that Raeder had only 64 submarines available when war was declared.[42] Nevertheless, they and Raeder's successor, Admiral Karl Dönitz, believed that they might be able to win command of the sea via submarine warfare as they had come so close to doing in 1917—1918. Hence, the bulk of Germany's ship construction capability went into mass production of submarines during the 1939—1945 period. Typical of the German submarines was the Type VII-C class (one 8.9-cm gun and five torpedo tubes) that could run at 18 knots on the surface and seven knots submerged. At the height of the German

41. Between March—August, 1945, aerial mining by American B-29s produced approximately 60 percent of the attrition of enemy shipping. During that period, some 17,791 members of the U.S. 20th Air Force participated in 1,424 mining sorties with a loss of 15 aircraft and 103 aircrew killed or missing. The average distance flown on a non-stop mission carrying twelve 1,000-pound mines or seven 2,000-pounders was 5,280 km (the Marianas—Shimonoseki Straits and return).

42. As of the end of 1938, submarines in commission and being built by the major world powers were: Axis countries — 218/74 (Germany — 57/50, Italy — 104/11, and Japan — 57/13); Allied nations — 232/51 (United Kingdom — 59/14, France — 78/27, and United States — 95/10). It is an interesting but forgotten fact that Italy had the largest submarine fleet when war broke out. (*Note*: the number following the slash mark is the number of submarines under construction at that time.)

submarine campaign in 1942, about 65 U-boats were assigned to the Mediterranean Sea, some 40 to the South Atlantic Ocean, approximately 30 to the Indian Ocean and the remaining 250 craft to the western approaches of Europe and the varied shipping routes of the North Atlantic Ocean, the Gulf of Mexico and the Caribbean Sea. Before the war was over, these undersea vessels had sunk nearly 3,000 Allied vessels with a gross weight of over 14 million tons and would require an armada of thousands of naval vessels, aircraft and blimps to subdue them. Obviously, the price the German Navy paid was high for it eventually lost 809 U-boats out of a total of 1,228 and 27,491 officers and men out of a total seagoing corps of 55,000.

Turning the tide against the German submarine force was not easy and took many and diverse actions and technologies. Typical of the brute-force suppression tactics were extensive air patrols to seaward in order to keep submarines submerged during daylight hours and the mounting of frequent bombing raids over construction yards and refitting bases. In turn, the Germans could counter-measure by recharging batteries on the surface at night and by using anti-aircraft defenses and reinforced-concrete sheltering "pens". The Allies also made extensive use of three electronic devices: (1) ship-based and shore-based "High Frequency Direction Finding" sets that could obtain a bearing (and even an estimate of distance) from the frequent radio transmissions German U-boats were required to make so that their headquarters were kept posted, (2) the rapid development of ship-mounted radar, followed soon by airborne radar, and (3) interception and deciphering of German radio messages by the British Secret Intelligence Service.[43, 44] However, the Allied action which led to winning the "Battle of the Atlantic" was the introduction in late 1942 of well-handled, heavily defended convoys after practically every lesson learned in the prior submarine war had been relearned at great cost in terms of men, supplies, and ships.

Once Dönitz was able to use the French ports of Brest and Lorient in mid-1940, he began to overcome his deficiency in aerial reconnaissance by introducing "wolf-pack" tactics. Initially, submarines would be deployed broadside along favored convoy routes. Once a detection

[43.] By 1939, British patrol aircraft were equipped with radar operating at a wavelength of one meter, thereby providing a device good enough to see surface submarines at night out to a distance of about eight km. In 1942, the operational length had been brought down to centimeters, doubling the detection range. By 1944, the Germans were using "snorkels" (breathing tubes) in order that the entire submarine did not have to emerge to charge batteries. On the other hand, British Coastal Command aircraft radar could now detect such metal protuberances in a seaway as much as sea state four.

[44.] Such sensitive decoded messages were classified "Ultra" and could only be acted on if confirmed by some other intelligence source known to the Germans.

was made, the word was passed to the other near-by submarines, and a massed submarine attack would start after a proper concentration had been achieved which, in 1941, was about six, but in 1943 had grown to as many as 20. Such attacks normally came at night because once an U-boat fired its first load of torpedoes, it had to surface and run ahead of the convoy while reloading its torpedo tubes. With even the slowest convoy making seven knots (about the speed of a submerged U-boat) and the fast ones between 10 and 14 knots, such surface running had to be done if a second attack were to be made. For the Allies, the crux of the game was to keep the attacking submarines submerged the entire time, whether by air attack from shore or carrier-based aircraft or by surface craft detecting their quarry with underwater sound and then attacking with depth charges. The detonation of such depth charges was always spectacular but not necessarily deadly, for the tough pressure hulls of the German submarines could withstand the shock from conventional 200-pound depth charges if the explosion took place more than seven meters away. Further, the underwater acoustic conditions at the battle scene could be quite complicated. There could be sharp distortion and bending of the acoustic "pings" released by the surface vessel of the type found by Iselin in 1937, causing "shadow zones" and vertical bearing inaccuracies. The turbulent wakes of the involved vessels also caused natural barriers through which coherent echoes would not pass; at the same time, the wake could even provide strong echoes which might be mistaken for the submarine.

As more and more destroyers, destroyer escorts, and corvettes came on the line, it was typical for an escort, after having gained acoustic contact, to call in another escort and work together, one vessel holding contact on the target and the other making runs guided by radio to ensure dropping her depth charges on the most likely target location. The problem of how to optimize the single escort attack still remained. The British had ended World War I with an "Asdic" capable of detecting submarines at ranges of as much as a thousand meters. However, the kill weapon was then dropped from the stern or propelled sideways only after the attacker moved at full speed over the apparent location of the acoustically detected target to release the depth-charge pattern. During such an approach and the subsequent time in which the depth charges were sinking, valuable seconds elapsed during which a cunning submarine commander would take evasive action and be well away from the danger point when the depth charges finally exploded. After some advance thinking at the British Anti-Submarine School during 1919 and 1920, it was obvious it would be much better to take necessary pot-shots at the target submarine during the initial run-up while it was still in the sights of the "Asdic" equipment. In the 1920s, a forward thrown weapon was produced by one of the British armament manufacturers, but it showed up badly on its trials and the project was relegated to low priority until the Second World War loomed ahead. Another unfortunate mis-judgment of this period was the decision to abandon the tilting mechanism of the early "Asdic" gear which made it possible to turn the acoustic beam downwards in the event the submarine went deep. In fact, one of the attributes of the new German submarine of the late 1930s was an extra depth capability, and the omitted "Asdic" depth control could have been most helpful. Some rapid war-time development work (with the help of B. C. Browne of Cambridge's "G and G Department") then produced a tidy sonar device which could be fitted to the bow of a destroyer and follow the submarine to any depth within range. With this in hand, it was only logical to develop the "hedge-hog", a mechanism looking somewhat like a bank of "roman candles" that would fire out a flock of light-weight projectiles in an area pattern ahead of the escort. Should any of the flock directly contact the submarine's hull, the resulting explosion, though small, would cause mortal damage because of its immediate proximity and the magnifying effect of the "water tamping" pheno-menon.

To solve technical problems of this type, the British Anti-Submarine Experimental Establishment had the ocean-oriented geophysicists at Cambridge University — Edward Bullard, B. C. Browne, and Thomas Gaskell — immediately appointed to their staff once war broke out in 1939. As mentioned earlier, all three had been making geophysical measurements from Navy ships, in large part because the Hydro-grapher of the Navy was a member of the Management Board for Cambridge's Department of Geodesy and Geophysics. Just prior to the war, Bullard and Gaskell had badly needed a signal recorder cap-able of working from a pitching, rolling ship, and they had been most pleased to receive for free a six-string galvanometer designed for artil-lery sound ranging back in World War I. But with another war on, what was now called the "Cambridge device" was being used by Browne and Gaskell in Portland Harbor to determine the underwater trajectory of forward-thrown anti-submarine projectiles. Such a projectile would be fitted out by taking the steel nose of a three-inch shell, inserting a series of delayed detonators firing at intervals of one-quarter and one-half second, and then firing it out of an airgun for a distance of a hundred meters or so. Four hydrophones located on the sea floor would then record the succession of "bangs" as the projectile traveled down through the water. For each "bang", there would be three time differences recorded on the old multiple-string galvanometer; from this, one could compute a position fix for each of the detonations and map the underwater path of the projectile being tested. It was soon found that spherical noses caused the projectile to career downwards in a zig-zag course because the projectile's movement caused a cavity to form into which the projectile repeatedly flopped. In contrast, flat-nosed projectiles stood firm within such a cavity and followed the desired trajectory. With experimentation such as this, plus the lifting of restrictions on staff and money, the Royal Navy began to have the detection devices, the weapons and the tactics needed to impact sharply on the burgeoning German submarine fleet. In fact, by 1943, Admiral Dönitz found it necessary to state:

> For some months the enemy has rendered the U-boat war ineffective. He has achieved this, not through superior tactics or strategy, but through his superiority in the field of science; this finds its expression in the modern battle weapon — detection. By this means he has torn from our hands our sole offensive weapon in the war against the Anglo-Saxons.

To be sure, the sharp improvement in submarine detection was not just a British invention. Back in 1940, the Naval Advisory Committee of the U.S. National Academy of Science (NAS) had started an analysis of America's capability to wage successful anti-submarine warfare (ASW).[45] Concerning itself only with the detection problem, the Committee reported back that the "sonar" developed by the Naval Research Laboratory and the Submarine Signal Company gave good performance. However, the training and quality of the Navy's sonar-men was below standard, and there were environmental irregularities in the gear's performance that could only be remedied by starting a broad academic research program into the fundamentals of underwater acoustics. On 10 April 1941, Rear Admiral S. M. Robinson, USN, of the Navy wrote Dr. Vannevar Bush, by then Chairman of the National Defense Research Council (NDRC), and asked the NDRC to study the problem. The NDRC knew what it wanted and immediately assigned the task to its Division "C" (Communication and Transportation) chaired by Dr. Frank B. Jewett, President of both the NAS and the equally famed *Bell* Telephone *Lab*oratories (Bell Labs). By the end of the month, Jewett was able to announce that there should be two contract laboratories — one on the east coast at New London, Connecticut, under the sponsorship of Columbia University

---

[45.] Chairman of the study group was Dr. E. H. Colpitts. Other members were: W. D. Coolidge, Vice President, General Electric Company; Professor V. O. Knudsen, an outstanding acoustician at UCLA; Dr. H. G. Knox; and Dr. Louis B. Slichter, Professor of Geophysics at MIT.

and one on the west coast at San Diego, California, under the sponsorship of the University of California.[46] In fact, Jewett was even able to specify the proposed laboratory directors — Mr. Timothy E. Shea and Dean Vern O. Knudsen, respectively (Tate, 1946).

In addition to having the ability to move quickly, the NDRC also had money to spend.[47] By 11 June 1941, at the first formal meeting of NDRC's Section C-4 (Sub-Surface Warfare), the Section's Chairman, Dean John T. Tate, had his group arrive at an initial budget of $740,000. From this amount, the Western Electric Company (a sister company to Bell Labs) and the General Electric Company were each awarded $5,000 to start developing magnetic detection methods. Today, of course, such an amount would not even cover the cost of preparing such a proposal. Equally of interest is that Tate's group did not make a similar award to the Gulf Research and Development Company even though that firm's Dr. Victor Vacquier had just developed a magnetic (flux) gate magnetometer capable of being flown from an aircraft. However, Vacquier's superior at Gulf, Dr. Paul D. Foote, soon joined the steering group and Vacquier had his support. Because the development of such a sensor was extremely important to both ASW and geophysical exploration, its further evolution is described here briefly in the words of Gulf's research and development chief, R. D. Wyckoff (who served as President of the SEG in 1943—1944) (Wyckoff, 1948):

> ... the problem of orientation stabilisation was attacked by the then most expedient means in order to establish the practicability of airbone operations with the least possible delay. The first relatively crude orienting scheme successfully demonstrated the practicality of the scheme. In fact, the experiment was rather quickly followed by the successful detection and disposal of an enemy submarine in an operational test flight of this early experimental model in December 1942. (Author's note: Would all operational test flights be so kind!) ... It resulted in an immediate and vastly expanded programme of development on the part of NDRC. Thus arose the Airborne Instruments Laboratory at Columbia University under NDRC where a very creditable piece of work was done in further development and design and from which arose the final service model of the MAD....
>
> Concurrent with the NDRC activity, work underway at the Naval Research Laboratory and associated Bell Telephone Laboratories resulted in a basically similar

---

[46.] Most of the Scripps Institution of Oceanography's staff, including even some of the biologists, abandoned the La Jolla campus in favor of working at what was known as the "University of California Division of War Research" at Point Loma. Dr. Harald Sverdrup initially headed the environmental component until he ran into security problems because he had family members in Norway. After that, the team was headed by the junior author of *The Oceans*, Dr. Richard H. Fleming.

[47.] Dr. Columbus O'Donnell Iselin, as Associate Director of the Woods Hole Oceanographic Institution, was quick to obtain $100,000 from the NDRC to study the adverse "afternoon effect" on "searchlight" sonar. By the end of World War II, he had brought in sufficient federal funding to have 600 full-time people on the institution's pay-roll investigating a wide variety of military problems.

apparatus. This unit, also designed for anomaly detection rather than magnetometric mapping, was suitably modified during the war and flown in cooperation with the U.S. Geological Survey on geophysical surveys, evidently with the view to expedite exploration of certain Naval petroleum reserves.

Despite what they claimed at the time, war work was relatively idyllic for the academicians who flocked from the student-deserted campuses to the university-operated "Divisions of War Research" and "Applied Physics Laboratories".[48] The pay was better, the local Draft Board gave deferments, and now, after the long, penniless Depression years, there were adequate funds, plus ship and aircraft time, to work on problems of joint national and personal interest. However, physicists and engineers used to working in a homogeneous medium or in a near-vacuum soon found how messy and uncomfortable the ocean was to work in. The ocean was not only layered, but it contained bountiful marine life, was corrosive, acted as a good electrical conductor, and caused apparatus to leak from a surrounding pressure field that increased rapidly with depth.

There were also two major initiatives within the United States's military establishment to provide improved knowledge concerning the physical properties and behavior of the sea to those directly engaged in combat. The first catalyst of this type was Lt. Commander Rawson Bennett, USN, a Naval Academy graduate (1927) and an experimental engineer posted to the Bureau of Ships (and its predecessor, the Bureau of Engineering) between 1937 and 1941.[49] Bennett had been kept up to date by Lt. Roger Revelle, USNR, at the Navy's Sound and Radio Laboratory in San Diego and by Dr. Columbus Iselin at Woods Hole on the new work which showed a close relationship between the temperature-depth trace of a BT and ship sonar performance. Accordingly, ten new reserve ensigns were posted in August, 1941 to a three-months course to learn to be "Bathy-thermograph Pilot Instructors" at Woods Hole. In addition to Drs. Iselin and Ewing, class instructors included Allyn Vine, Fritz Fuglister, and Henry Stommel, all of whom ultimately earned high repute in American oceanography.[50]

48. The number of science and technology students who, but for the war, would have received B.A. degrees was estimated at 150,000 and the comparable Ph.D. degree deficit at 3,375 (Bush, 1945). The only way an able-bodied student could stay on campus was to be enrolled in engineering, aviation and medical programs such as the Navy V-12 and the Army Special Training Program.

49. An academician at heart, Bennett served as Director for Electronics Design in the Bureau during 1942–1946, as Commanding Officer and Director, U.S. Naval Electronics Laboratory (the old Radio and Sound Laboratory) between 1946 and 1950, and as Chief, Office of Naval Research, during 1955–1961.

50. In the 1970s, both the American Geophysical Union and the Society of Exploration Geophysicists gave "Maurice Ewing" awards. In addition, the work-house of America's

Eventually three classes involving 30 ensigns were trained to deploy the BT without bashing it against the side of a fast-moving ship. Soon after the Pearl Harbor attack, these fledgling instructors scattered throughout the operating fleet and did their best to convince convoy commanders how to space their escorts in an optimal fashion depending upon the water condition existing below. One of these instructors won the Navy Cross for heroism, and several were war casualties.[51] Ultimately, the most eager users of BT equipment and data were the submariners.[52] By having such data in advance, they could plan diving trim adjustments before penetrating the thermocline, the sharp interface between the lighter, warm surface water and the dense, colder water existing below. Similarly, they also knew exactly how thick the surface layer was when they wished to make a hurried escape running below the layer once they fired a torpedo at periscope depth. In other words, a good knowledge of real-time oceanography helped carry out that old Arabic adage, "He who runs away lives to fight another day!"

However, the first entity specifically labeled as an "Oceanographic Unit" within the military was created not by the Navy but rather by the Weather Directorate, Headquarters, Army Air Forces, early in 1942. This action was largely due to some aggressive efforts by the reservist, Captain Richard Seiwell, USAR, one of the earliest staff members of the Woods Hole Oceanographic Institution.[53] The Unit's major efforts included sponsoring research by Dr. Harald Sverdrup and his assistant, Walter Munk, on the quantitative forecasting of sea, swell, and surf, the charting of ocean currents in the Pacific theater of war, and establishing a course in military oceanography at

---

deep-diving research submersible fleet is called the Alvin in honor of Vine's work on her behalf. And during 1981, the American Geophysical Union issued the self-explanatory volume, *Evolution of Physical Oceanography — Scientific Surveys in Honor of Henry Stommel*, edited by B. A. Warren and C. Wunsch.

51. The outstanding "90-day" wonder from this course proved to be Ensign Gordon Lill, a geologist from the University of Kansas, who received his first naval training under Lt. John F. Kennedy, USNR (ultimately the 35th President), at the University of Chicago between December, 1940 and March, 1941. Lill ended his federal career as Deputy Director, National Ocean Survey, in 1980.

52. The implementor of the BT hardware program during World War II was a softspoken, extremely conscientious Carolinian, Lt. (j.g.) Butler King Couper, USNR. After the war, "King" eventually became the senior civilian oceanographer with the U.S. Navy Bureau of Ships (ultimately the Ship Systems Command). The BT also became a favorite tool of civilian oceanographers throughout the world in the post-war period and was further modified into expendable ($20 a copy) and air-dropped versions.

53. Although Seiwell was considered to be a very promising chemical oceanographer, his personal traits were quite bizarre. Aboard the *Atlantis*, he had the reputation for being able to make "booze" out of anything. His personal ash-tray, built out of a human skull, always took one's attention, as did the nude painting of his wife, a graphic work of art prominently displayed in the family living room.

the Scripps Institution of Oceanography during mid-1943 (Sverdrup and Munk, 1944; Bates, 1949).

The Navy's first formal "Oceanographic Unit" was a bit slower in coming. To be sure, as early as June, 1941, Scripps's brightest young marine geologist, Lt. Roger Revelle, USNR, had been called up for active duty. By mid-1942, Revelle had been assigned to the Navy's Hydrographic Office at Suitland, Maryland, but with collateral duty to the Bureau of Ships in nearby downtown Washington, D.C. Revelle quickly learned that the Bureau would issue research contracts while "Hydro" would not, so that was the end of his interest in duty at Suitland. Finally, in March 1943, "Hydro" had a chance to inherit the Air Force's Oceanographic Unit (less Major Seiwell). Under the command of a salty marine biologist, Lt. (j.g.) Mary Sears, USNR(W), the Unit began turning out a series of very useful end-products.[54] These included the oceanographic chapter of the *Joint Army—Navy Intelligence Study* series, a submarine supplement to the *Sailing Directions* series, and survival drift charts based on Sverdrup's personal research and printed on silk cloth so that they could be used, in the worst case, as bandannas to ward off sunburn.

To close out this chapter on geophysicists at war, we observe that most science historians label World War II as the "War of the Physicists". And yet we wonder whether the desire and enthusiasm of the physicists to create nuclear weapons was all that necessary, in view of the steadily increasing capability of influence mines based on geophysical principles, to close down the Japanese economy. Be as it may, perhaps a thousand young men and women learned the excitement of applied geophysics during the war years in the applications just described. However, a much larger use of technically-oriented manpower was in the allied field of military meteorology where, in the United States alone, the weather service of the Army Air Force trained and utilized over 4,000 meteorological officers and 15,000 enlisted personnel between 1940 and 1945, while the U.S. Navy did the same for approximately 1,300 aerological officers and 5,000 aerographer mates. As a result, once the war was over, there was a unique pool of manpower existing within the geophysical profession. Many elected to stay in the field during the rest of their careers. How this occurred and what they accomplished is a major theme to be touched on repeatedly in the remainder of this book (see Table 3.4).

[54]. After the war, Dr. Sears returned to a very active scientific career at the Woods Hole Oceanographic Institution. She kept in close contact with Dr. Revelle, however, and the two of them were prime organizers of the first International Oceanographic Congress held at the United Nations Building, New York City, during August—September, 1959.

TABLE 3.4

Notable Graduates from America's University Training Program for Military Meteorologists and Oceanographers

Class and date	School	Graduate	Future role
I: 9/40 to 6/41	MIT	Kenneth C. Spengler	Executive Director, American Meteorological Society, 1946–present
II: 6/41 to 2/42	Chicago	Norman Allen Riley	President, Chevron Oil Field Research Corporation, 1959–1980
	New York	George P. Cressman	Director, National Weather Service, 1965–1979
III: 3/42 to 11/42	MIT	Robert M. White	Administrator, National Oceanic and Atmospheric Administration, 1970–1977
	Chicago	Reid A. Bryson	Director, Institute for Environmental Studies, University of Wisconsin, 1970–present
	MIT	John Borchert	President, American Association of Geographers, 1968–1969
IV: 7/42 to 5/43	UCLA/Chicago (joint course)	Charles C. Bates	Science Advisor to the Commandant, U.S. Coast Guard, 1968–1979
		John C. Crowell	Chairman, Division of Earth Sciences, National Research Council, 1980–1982
		Dale F. Leipper	Founding Chairman, Department of Oceanography, Texas A & M University, 1949–1964
		Boyd E. Olson	Scientific and Technical Director, U.S. Naval Oceanographic Office, 1969–1980
V: 3/43 to 12/43	UCLA	Donald W. Pritchard	Founding Director, Chesapeake Bay Institute, Johns Hopkins University, 1949–1978
	UCLA	Robert O. Reid	Chairman, Department of Oceanography, Texas A & M University, 1981–present
	Chicago	Roscoe R. Braham, Jr.	Chairman, Department of Geophysical Sciences, University of Chicago (during the 1970s)
VII:	USN Post-graduate School	Wayne V. Burt*	Founding Chairman, Department of Oceanography, Oregon State University, 1959–1978
	CalTech	Carl F. Romney*	Deputy Director, Defense Advanced Research Projects Agency, 1979–present

*Burt and Romney were naval ensigns during training; the others were aviation cadets in the Army Air Force.

CHAPTER 4

Reversion to Peacetime, 1945-1950

New frontiers of the mind are before us, and if they are pioneered with the same vision, boldness, and drive with which we have waged this war we can create a fuller and more fruitful employment and a fuller and more fruitful life.

President Franklin D. Roosevelt
(In a letter of 17 November, 1944 to Dr. Vannevar Bush, Chairman, Office of Scientific Research and Development)

A Time of Transition

Once the Axis Nations' hopes for world empire had been frustrated in 1945 at a cost of some 50 million lives, thousands of geophysicists — and particularly those in uniform — had one basic question — what next?[1] Within the United States, there were three easy answers: (1) return to your former job (your employer was required to provide an assignment equivalent to the one you left), (2) stay in the military service, or (3) go back to school as a student under the "GI Bill of Rights". If one did return to his employer, he found that management tended to hold the opinion that you had learned very little, if anything, during the half decade you had been away.[2] Moreover, in the world of petroleum exploration, the policy was still enforced that there must be a sharp distinction between the use of geological and geophysical approaches in finding oil.[3] However, the situation

[1.] The Allied Nations mobilized 62 million personnel (USA: 14 million, British Empire: 12 million) and the Axis Nations 30 million personnel (Germany: 17 million).

[2.] When Bates inquired of the Carter Oil Company whether he could try being the firm's "environmentalist" and take the weather factor out of exploration, construction and refining, the answer was: "If Gulf Oil had one, we'd have one — but they don't!"

[3.] Upon requesting a six-month leave of absence from Carter Oil to participate in the geological survey of Bikini Atoll associated with the atomic bomb tests of mid-1946, Bates was advised by Carter's Chief Geologist, Mr. Melvin Fuller, "Who ever found oil in a coral reef? Hell no!" A year later, Imperial Oil Ltd., after a quarter of a century searching, opened up the prolific reef oil field at Leduc, Alberta. The irony of it all was that the wildcat well was spotted using seismic data acquired by a Carter crew under Malcolm Reiss. Paul L.

was quite different in the military, where command policy was to retain as many as possible of the bright young reserve officers. As one Air Force general put it, "In the next war, we're going to have our own damned Ph.D.'s already in uniform!", thereby omitting the need to use academics in high-ranking consultancy posts. To this end, Colonel Don Yates, USAAF, now Chief of the Air Weather Service, wrote to promising but demobilized weather officers and requested that they apply for permanent military commissions, particularly since "plans for the post-war Weather Service provide for the annual assignment of outstanding officers to leading universities for further education in meteorology (including) . . . the opportunity for acquiring Master and Doctor degrees". But it was just as attractive to many young officers to return to graduate school as civilians and live off the provisions of the "GI Bill of Rights", a route taken by two reserve captains who eventually succeeded Commander Francis Reichelderfer as Chief, United States Weather Bureau, namely, Robert M. White (1963–1965) and George P. Cressman (1965–1979).[4]

Exploration Geophysics Bounces Back

In his presidential address to the Society of Exploration Geophysicists on 21 March 1944, Dr. R. D. Wyckoff of the Gulf Research and Development Company predicted that war developments would have little to offer to geophysical exploration during the immediate future. While his prediction was reasonably correct, his own laboratory soon broke it, for Gulf was quick to make a workable exploration tool out of its now militarized magnetic anomaly detector (MAD). During 1946, the equipment was licensed to two air surveying companies, the Aero Service Corporation and Fairchild Aerial Surveys. Full-scale commercial use came a year later, with most of the aeromagnetic work conducted outside the United States; domestically, the device was used primarily in the Rocky Mountain region

---

Lyons worked on the records back in Tulsa and found a bulbous-nose type of structure with a curious ridge open to the east so that the closure was not complete. Nevertheless, the decision was made to drill to the Viking sandstone. When this horizon was finally hit on a Sunday, it proved dry. However, the drilling superintendent was out playing golf. In the absence of shut-down orders, the drillers kept making hole and hit the productive Devonian reef rock later in the day and thereby opened up an oil province that totally rejuvenated the Canadian petroleum-exploration industry.

4. In 1965, Dr. White moved up to direct the Environmental Sciences Service Administration (ESSA), and Dr. Cressman took over what was now called the "National Weather Service". White received his Sc.D. from MIT (1950), while Cressman earned his at the University of Chicago (1949).

by both petroleum and mining interests. Experience soon indicated that "five aircraft engaged in aeromagnetic surveying could run as much profile as did all the seismograph parties operating in the United States . . ." during a given year (Eckhardt, 1948). Also, this type of surveying could be conducted quickly over such difficult terrain as jungle, swamp, tundra or desert. In addition, the quality of the aeromagnetic data was actually better than that obtained by land magnetometers because measurements at flight altitude averaged out minor surface and near-surface variations of no interest to petroleum exploration. Furthermore, the aerial technique permitted flying the same survey at different elevations in order to delineate better the depth of any anomalies that might be present.

As time went on, the principal use of aeromagnetic surveying for petroleum exploration was found to be the mapping of basement structure, a feature which could then be reflected in the structure of overlying sedimentary rocks. Because aeromagnetic surveying usually cost about five percent of the cost of seismic surveying, the airborne magnetometer offered a rapid, cheap method of obtaining an initial assessment of the petroleum potential of large, unexplored areas. In Libya, for example, all of the large structurally-controlled oil fields discovered in the first years of exploration were reflected in basement structure and could have been found by the initial aeromagnetic survey alone.[5] Unfortunately, a great deal of costly seismic work had already been done before this was realized. However, the seismic work would have likely been undertaken anyway for further delineation of the producing structures. The importance of the airborne magnetometer in petroleum exploration shows clearly in the report, "Geophysical Activity", published in *Geophysics* each year. By 1960, aeromagnetic surveying exceeded 500,000 line miles per year, a figure that has frequently been exceeded since. As recently as 1975, there were four individual surveys that amounted to more than 100,000 line-miles each, and one was over 200,000 line-miles.

The number of line-miles surveyed per year shows that aeromagnetometry became more important in the exploration for ore bodies than in petroleum exploration. In fact, the number of ore bodies discovered by the method has been very substantial, and one large steel firm has reported satisfying within a decade of aeromagnetic effort its need for delineating a 75-year iron-ore reserve requirement (Reford and Sumner, 1964). Mineral exploration

---

[5.] This situation replicated itself in the neighboring Sudan, where Chevron Overseas Petroleum used aeromagnetic data supplemented by information from "LANDSAT" photographs taken in space and found enough oil during the late 1970s to justify laying a pipeline to the Red Sea (Paul and Mascarenhas, 1981).

TABLE 4.1
Geophysical Activity During 1948 (expressed in crew months)*

Country	Seismic	Gravity	Magnetic	Electrical
Alaska	21	—	—	—
Canada	381	109	8	—
United States	5,520	1,320	200	?
Mexico	160	47	—	12
Guatemala	—	13	—	—
Nicaragua	4	—	—	—
Bahamas	7	5	—	—
Cuba	6.5	8	1	—
Trinidad	?	24	—	—
Venezuela	285	102	1	—
Colombia	100	56	37 (a)	—
Peru	12	12	—	—
Chile	14	12	—	—
Argentina	?	?		
Paraguay	34	—	—	—
Brazil	24	15	—	—
Great Britain	12	12	12	—
Netherlands	31	12		
France	18	16	—	10 (b)
Denmark	12	6	—	—
Germany	157 (c)	26 (d)	—	—
Hungary	?	35?		?
Russia	?	?	?	?
Italy	30	8	—	6 (b)
French Morocco	12	12	12	19 (c)
Algeria	—	1	—	—
Tunisia	—	11	—	2 (b)
Nigeria	8	8	—	—
French Equat. Africa	—	8	8	—
Mozambique	—	—	3 (f)	—
Madagascar	—	16	8	—
Egypt	24	17	11	—
Turkey	18	9	—	—
Cyprus	4	—	—	—
Transjordan	—	4	4	—
Syria	3	12	8	—
Iraq	14	12	8	—
Qatar	3	—	—	—
Trucial Coast	2	6	3	—
Arabia	12	27	—	—
Bahrein	2	—	—	—
Kuwait	1	—	—	—
Iran	34 (g)	—	—	—
Pakistan	18	4	—	—
Philippines	—	?	—	—
Netherlands East Indies	25	42	—	—
British Borneo	8	—	—	—
Papua & E. New Guinea	2	12	8	—
Portuguese Timor	—	4	—	—
Australia	—	8	8	—

surveys are more costly per line-mile than for petroleum because of the nature of the anomalies sought. However, the cost is still quite reasonable. In 1962, the average line-mile cost was only $6.07, and in 1974, it was still only $13.70. In both years, almost a million line-miles were flown for mineral exploration, and in other years, the mileage has been considerably higher. Thus, the aeromagneto-meter has been an invaluable addition to our petroleum and mineral exploration tools, particularly after two completely new and dif-ferent types of sensors were developed since 1955 to improve upon the flux-gate instrument. These are the nuclear magnetic resonance (proton) magnetometer available in land, aerial, and ship-towed models, and the optical pumping magnetometers (Dobrin, 1976).[6]

During the late 1940s, geophysical techniques for oil exploration did not undergo dramatic change except for the just mentioned introduction of the aerial magnetometer. By 1948, the industry was spending about $125 million annually for geophysical surveys through the world as shown in Table 4.1, 95 percent of which was accomplished by U.S.-based firms.

Equipment-wise, the late 1940s saw the development of much smaller, lighter, more sensitive geophones, the use of arrays of geophones per single trace to reduce seismic noise, and the expan-sion of recording instruments to a standard of 24 simultaneous channels. Figure 4.1 is an example of a good-quality, 24-trace seismic record of the period. The technique used here is a "split-spread" recording with 12 geophones on each side of the shotpoint and results in the first refraction arrivals (usually those travelling just beneath the bottom of the weathered surface layer) appearing

---

[6.] The National Aeronautics and Space Administration has also flown an intriguing series of magnetometer-equipped, earth-oriented satellites. The first flights of this type were aboard the Polar Orbiting Geophysical Observatories (POGO-2, -4 and -6) between 1965 and 1971, followed by the MAGSAT satellite during 1979—1980.

---

*A dash (—) indicates no activity while a question mark indicates activity, or probable activity, but of unknown amount.

Meaning of other symbols:
(a) Does not include aerial magnetic survey by Gulf in Llanos Basin east of Andes.
(b) Telluric current method.
(c) Comprising 146 reflection and 11 refraction crew-months.
(d) Includes 5 torsion-balance crew months.
(e) Includes 12 crew months by telluric current method.
(f) Aerial magnetometer.
(g) Divided between 24 refraction and 10 reflection crew-months.

Source: Weeks, L.G., "Highlights of 1948 Developments in Foreign Petroleum Fields", *Bull. Amer. Assoc. Petrol. Geol.* 33, 1029—1124. (Reprinted courtesy of the AAPG; all rights reserved.)

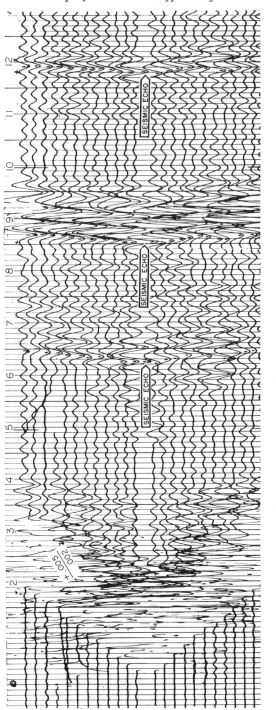

FIG. 4.1. Twenty-four trace split-spread reflection seismic record typical of the 1950s.

symmetrically on the two center traces generated by the geophone clusters closest to the energy source. Three bands of reflections labeled "seismic echo" are marked by arrows and demonstrate that the automatic gain control (AGC) has maintained a signal level at a fairly uniform value throughout the entire recording. The hand-marked two-way travel times are shown at the top of the record, with "10" meaning 1.000 sec. Thus, the vertical timing lines are spaced 0.1 sec or 100 milliseconds apart. The 0.022 (sec) time written on the upper left-hand part of the record is the time from the shot location at the bottom of the shothole to the surface and was obtained from the first break on the 6th trace as recorded by a geophone very close to the shothole. The +.005 (sec) denotes the time by which the shot time-break on the 4th trace comes in ahead of the zero timing line; hence, this is the correction to be added to any time picked on the record. The two center traces represent essentially vertical travel paths to the reflecting horizons, thereby requiring no correction for path geometry. Hence, these were the traces primarily used to determine reflection times to various horizons — and actual horizon depths if average velocities were known from a nearby well survey. The chief purpose of the outer traces was to provide information about the tilt, or "dip", of strata in the plane of profile. Or if records were shot using "cross-spreads" of geophones laid out at right angles to each other, the true dip of the strata in three dimensions could also be determined (Rice, 1953 and 1955).

Seismic exploration for the first 25 years consisted principally of reconnaissance work to locate large subsurface structures. Hence, it was only necessary to shoot into a single spread or cross-spread of geophones at intervals governed by the size of structures sought, say every mile. This was called "correlation shooting" because reflections from a given horizon had to be correlated from point to point on the basis of character alone. However, by the 1950s, more detailed exploration in most areas required the use of "continuous profiling". This referred to any field procedure which gave a continuous set of reflection points along a profile line. In the case of the split-spread method, this was accomplished by shooting at both ends of each half geophone spread. Thus, ignoring minor corrections, reflections on the upper trace of the record in Fig. 4.1 would have the same arrival times as the corresponding reflections on the lower trace of the next split-spread record in that direction. This permitted much more accurate mapping of individual horizons and of faults than the correlation technique.

By the late 1940s, a considerable amount of seismic work was being done in regions of swamps and muskeg, thereby requiring the drilling and recording equipment to be carried on boats, barges, or specially-wheeled or tracked vehicles such as "swamp buggies" and "weasels". Commercial "shooting" also regularly took place in open water areas, such as the Gulf of Mexico and the Persian Gulf, following ideas and concepts tried out during the pre-war period by Professor Maurice Ewing and his student Allyn Vine, J. L. Worzel and Albert Crary off the Delaware–New Jersey coast in 1937, by Shell Oil off Texas in the same year, by Professor Edward Bullard and Thomas Gaskell off the United Kingdom in 1938, and by GSI in the Persian Gulf during 1939.[7] However, such offshore work did not acquire specific recognition until 1953 when the annual "Geophysical Activity" report in *Geophysics* reported 102 offshore crew-months for the year. Even so, the recording equipment was still essentially land equipment, and the seismometers were trailed along a cable towed behind the recording boat which stopped each time it was necessary to detonate a charge of dynamite or ammonium nitrate near the water surface. The tow cable itself was usually suspended from surface floats and had another boat at the far end in order to keep the recording line straight and taut.

Accurately positioning the aeromagnetic and offshore seismic survey lines became an ever more important issue during the late 1940s. At first, offshore positioning used visual methods, usually land surveyors on tall towers sighting on boat-streamed targets such as mirrors hanging from kite-like balloons. However, visual methods were badly hampered by visibility and line-of-sight considerations, and radar was not accurate enough to provide a substitute method. Fortunately, once wartime security measures were lifted, several useful electronic positioning schemes became available. For the Normandy invasion, the Royal Navy had utilized a precise-positioning scheme using hyperbolic lines of position to fix control boats. With the war over, the Decca organization was able to commercialize this technique under the leadership of the system's inventors, Harvey F. Schwarz and W. J. O'Brien, by building and operating navigational transmitters and renting out the receivers. Later on, portable "Decca" transmitters were developed especially for the offshore geophysical industry. Within the United States, two other electronic positioning schemes came on the market — one an Air Force development called

---

7. As a young party chief for the Shell Oil Company, Dr. Sidney Kaufman had been directed to map an onshore structure along the South Texas coast. When the structure appeared trending seaward, Kaufman had his crew shoot a six-km long line offshore using "marsh" geophones linked to an eight-channel recorder. His positive action caused raised eyebrows at headquarters for he had wasted money and crew time on an area that couldn't possibly be drilled!

SHORAN (*Short Range Navigation*) and the other a National Advisory Committee for Aeronautics development eventually named "RAYDIST" (*Radio Distance*) by its inventor, Charles Hastings. Another system, also based on the same heterodyne phase comparison principle as RAYDIST, was offered as LORAC by the Seismograph Service Corporation in 1949, thereby giving rise to a long-running patent infringement suit that was not resolved until 11 years later. Typically, however, when one had a positioning problem, one would turn to a commercial positioning service, such as that by Offshore Navigation Incorporated (ONI) which would use whatever positioning technique best met the client's particular need.[8]

Classical Seismology Stays Much the Same

Of all the geophysical sciences reviewed here, classical seismology was the least affected by World War II. Thus, transition back to a sleepy science dominated by academia took place with little public notice. Even Father Macelwane's valiant attempt to induce the U.S. Navy to track hurricanes by operating seismic arrays at air stations scattered throughout the Gulf Coast, the Caribbean, and the North Pacific Ocean was phased out during the early 1950s as special Navy and Air Force aircraft became regular hurricane trackers. However, in early 1948, the Atomic Energy Office (AF-OAT-1) of Headquarters, U.S. Air Force contracted with Beers and Heroy, Inc. of Troy, New York, to study various ways in which geophysics might be used in nuclear test detection. Based on Gutenberg's earlier demonstration that seismographs in California and elsewhere had clearly detected the Trinity nuclear explosion near Alamogordo, New Mexico, in July, 1945 and to a lesser degree the underwater shot, Baker, at Bikini Atoll in July, 1946, the Beers–Heroy report suggested that seismic detection be given special attention. To be sure, as long as the nuclear explosion was in the atmosphere, there was a strong airblast, a pronounced electro-magnetic pulse, and large amounts of radioactive fall-out.[9] Even so, the resulting seismic wave train would,

---

8. ONI was formed in 1946 by the demobilized Colonel Robert Suggs, USAAF (Reserve), who had become familiar with SHORAN positioning while in the course of making bombing runs. Returning to the Gulf Coast, he spotted the industrial need for better positioning and, after scraping up $30,000, bought an unused SHORAN set from RCA and went into business for himself. Today, ONI still dominates the industry and operates throughout the world. Suggs also spotted another industrial need — that of offshore helicopter support — and formed in 1948 the equally successful firm of Petroleum Helicopters, Incorporated. In 1981, ONI still owned 14.5 percent ($7.7 million worth) and Colonel Suggs 38.6 percent ($20.5 million worth) of the latter firm's stock.

9. The first Soviet nuclear explosion did all these things on 29 August 1949.

it appeared, be useful in refining the position and magnitude of the explosion. So, in 1949, Beers and Heroy were placed under contract by the *A*ir *F*orce *T*echnical *A*pplications *C*enter (AFTAC) to begin development of a workable seismic detection system for nuclear events.[10] Two of their first employees for this effort were excellent selections, for the 26-year-old electrical engineer, Richard Arnett, would rise to the post of president with the Geotechnical Corporation (1966–1970), while the 25-year ex-Navy aerologist, Carl Romney, would eventually become the nation's top specialist in this arcane art. To supplement this "out-house" effort, the newly-founded Air Force Cambridge Research Laboratories also created a "Wave Propagation Group" in 1948 headed by sage Dr. Norman Haskell, who had taken his doctoral degree in geology–geophysics under Professor Don Leet at Harvard University back in 1936.

The late 1940s also saw one major academic initiative in classical seismology when Professor Maurice Ewing persuaded Columbia University to let him use the manorial Thomas J. Lamont estate on the Hudson River 30 km northwest of the Morningside in-town campus in order that his geophysical equipment would be housed in a quiet environment. Approval for the move came through in 1949 but carried the proviso that Ewing would have to raise almost all of his own operating funds, thereby starting a Herculean task that never went away until Ewing finally departed in 1972 for the more financially hospitable Galveston campus of the University of Texas (see vignette by Ewing).

The Opening of the Canadian Arctic

Until World War II came along, the American Arctic, whether on the Canadian or the USA side of the border, was largely untouched by civilization, and native settlements were pretty much left to themselves. For example, at Barrow Village near Point Barrow, Alaska, the northernmost point in contiguous North America, the residents received supplies but once a year when the U.S. Department of Interior's vessel, *North Star,* arrived late each summer. However, the war did cause additional attention to be paid to the area because of aircraft ferrying needs, the development of uranium deposits on Great Bear Lake and oil at Norman Wells in Canada's Northwest Territories, and the mounting of a major exploration program

---

10. An excellent review of how seismic instrumentation developed to meet this need within the United States between 1948 and 1976 has recently been published in the *Transactions* (EOS) of the American Geophysical Union for 1981 (pp. 505–10 and 545–48) by B. Melton.

within the U.S. Navy's Petroleum Reserve Number Four using a new base facility built six km from Point Barrow. Yet it was primarily aviation interests and meteorologists who sold the two governments on the need to open up the far north by arguing that weather forecasting could not be notably improved until there was a high-quality observational network in place throughout that region.

At the repeated urging of Colonel Charles J. Hubbard, USAAF (Reserve), and others, the U.S. Congress passed in February, 1946 Public Law 276 (79th Congress) that proposed establishing a joint Canadian–United States weather network in the Canadian Arctic. The Canadian Cabinet moved more slowly, and it was not until January, 1947 that the Cabinet approved a "Joint Arctic Programme" calling for nine arctic weather stations to be built between 1947 and 1949 according to these guidelines:

1. Canadian Government would provide officer-in-charge, pay and subsistence of half of all the staff, and all permanent installations, including the adjacent air strip.
2. United States would bear all other costs including equipment, transportation, fuel, and Arctic supplies.
3. All persons assigned to these weather stations would be subject to applicable laws of Canada and the Northwest Territories, including game laws and the "Scientists and Explorers Ordinance", which would require supernumerary scientists, such as geophysicists, to obtain a license from the Commissioner of the Northwest Territories.

Finding suitable sites in this remote area the size of France and lying 800 to 1,600 km north of the Arctic Circle was far from simple for four key criteria had to be met: (1) the presence of a smooth strip of natural ice permitting initial aircraft landing and delivery of supplies, (2) a suitable environmental exposure permitting unbiased sampling of weather conditions, (3) an adequate water supply, and (4) a suitable permanent camp site within five miles of a shoreside landing strip. However, five sites were finally chosen (see Fig. 4.2). The first of these, Eureka, went in on 7 April 1947 and the last, Alert, on Easter Sunday, 1950, by military airlift in each case.[11] Once a station became operational, its mission was not only to accumulate meteorological data for better defining the interaction between polar and mid-latitude weather conditions, but also to

---

[11.] As of October, 1980, Station Alert, which operates the world's most northern post-office at 84° North Latitude, was manned by 21 women and 180 men, most of them on the payroll of the Canadian Department of National Defence, although four were still employees of Canada's Atmospheric Environment Service.

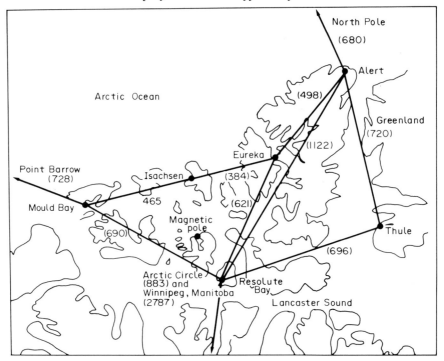

FIG. 4.2. Location of Canadian weather outposts installed during the late 1940s
(distances are in km).

determine the feasibility of regularly scheduling commercial flights within the Canadian Archipelago and along great-circle routes between North America and Europe or Asia.

Looking back, these early Arctic weather stations must be rated a major technological success. They not only supplied important geophysical research data for studies of polar meteorology, seismology, geomagnetism, aeronomy, and ice behavior but also provided the landing strips, radio stations, and habitation points from which a vast geological and geophysical exploration effort could be launched from the early 1950s on, of the Canadian Arctic Islands and the Polar Continental Shelf. Among the key players in this massive effort were Dr. W. E. van Steenburgh, Director General of Scientific Services in the Canadian Department of Mines and Technical Surveys, and Dr. Fred Roots, the initial Director (until 1972) of the skilfully executed Polar Continental Shelf Project which started working out of the Isachsen Station in 1959 and out of Mould Bay in 1964. In addition, such notable entrepreneurs as the late consulting geologist, Dr. J. C. Sproule of Calgary, Alberta, John P. Gallagher and others of Dome Petroleum Ltd., and Charles R. Heatherington and his staff of Panarctic Oils Ltd. utilized these Arctic bases to demonstrate that the Canadian Arctic, both onshore and offshore, might be capable of being made into a "mini-Middle East" with respect to oil and gas production well before the year 2000, particularly in the area, Resolute Bay–Mould Bay–Eureka, and in the region further to the southwest off the mouth of the Mackenzie River.

Oceanography Comes of Age

Eight months before World War II came to an end, Dr. Iselin of Woods Hole and Dr. Sverdrup of Scripps met at La Jolla, California, with key officials of the National Defense Research Committee and the University of California's Division of War Research to chart a peacetime continuation of oceanographic research that would be of interest to the Navy. One of the first tangible results of this meeting was Secretary of the Navy James Forrestal formally approving on 29 January 1946 the establishment of an "Oceanographic Division" within the U.S. Navy *Hydro*graphic Office (Hydro). Three days later, Rear Admiral George S. Bryan, USN (Retired), issued Order Number 27 which established such a component within his command and specified:

> This Division shall be responsible . . . for the collecting, codifying, coordinating and implementing basic oceanographic research required by various bureaus of the Navy Department, the U.S. Coast Guard, the War Department, and other government agencies, both by its own staff and through contracts with oceanographic institutions.[12]

But despite these fine statements, by mid-1946 the Hydrographic Office had lost three-fourths of its wartime oceanographic personnel, and the Office's Chief Engineer, Guillermo Medina, remained of the firm opinion that "Hydro's" main mission in life was to develop total global chart coverage, even though this meant duplicating the British Admiralty's extensive chart series. In addition, the Office of Naval Research (ONR) came into existence on 1 August 1946 with the specific mission of funding basic research. As a consequence, when Dr. Richard H. Fleming, co-author of *The Oceans,* arrived in April, 1946 to head up the new Division of Oceanography, he was granted a minimal budget and a staff of 15 out of a total "Hydro"

---

[12]. As it worked out, Hydro never was given a line item in the Navy's "Research, Development, Test and Evaluation" budget until 1960 when it acquired the responsibility of developing the Navy's "Anti-Submarine Warfare Environmental Prediction System" (ASWEPS). By Fiscal Year 1968, Hydro's research program utilized 255 man-years of effort costing $10.8 million on an annual basis.

complement of 625 persons. Moreover, all of his professional oceanographers had been seconded by Commander Roger Revelle, USNR,
for conducting oceanographic surveys needed in support of the
Bikini Atoll atomic bomb tests.

These tests — labeled Operation Crossroads — had been authorized
by President Harry Truman and were designed to: (1) determine the
effect of aerial and underwater detonations on modern (but surplus)
naval vessels, (2) provide the first thoroughly instrumented nuclear
detonations in the atmosphere (Shot Able) and underwater (Shot
Baker), and (3) demonstrate to the general public and world political
leaders the urgent and immediate need for rigid controls on nuclear
weapons.[13] The assigned task force (JTF-1) involved 42,000 men,
over 200 ships (including such famous target vessels as the Japanese
battleship *Nagano,* the German cruiser *Prinz Eugen,* the U.S. aircraft
carrier *Saratoga,* and the U.S. battleships *Arkansas, Nevada, New
York,* and *Pennsylvania*), and some 150 aircraft. Key to the operation were "before and after environmental surveys", plus a reliable
prediction as to when mixing and drift of the radioactivity from
the underwater shot would permit inspection of any sinking ships.
Even before the task force sailed, Vice Admiral W. H. P. Blandy,
USN, as its commander, sought to reassure the public by noting:

> ... the bomb will not kill half the fish in the sea and poison the other half so they
> will kill all the people who eat fish hereafter. (Moreover), it will not start a chain
> reaction in the water converting it all to gas and letting all ships drop to the bottom.
> It will not blow out the bottom of the sea, letting all the water run down a hole.
> It will not cause an earthquake or push up new mountain ranges. It will not cause a
> tidal wave. It will not destroy gravity.

In his role as "Officer-in-charge, Oceanographic Section, Technical
Staff (013B.1b)", Revelle had virtually an unlimited budget and first
call on the services of JTF-1's survey unit, which included the large
survey ship, USS *Bowditch* (AGS-4), two smaller survey ships (USS
Blish and USS *Gillis*), and six yard mine-sweepers (YMSs). By pointing
out that the Bikini study would be the greatest oceanographic investigation of the Pacific Ocean since that of HMS *Challenger* in the
1870s, Revelle quickly recruited a scientific support staff of many
of America's most promising young marine scientists. After two
months of hectic field work, the team turned in its initial predictions
during early May, 1946 and indicated that, under normal trade-
wind conditions, contaminated water would flush out of the lagoon,
particularly through the wide channel in the southeast corner of the

---

13. By early 1946, Dr. J. Robert Oppenheimer, the technical director of the secret Los
Alamos, New Mexico, laboratory that built these nuclear devices, was warning government
leaders about the "greater danger" inherent in keeping nuclear bomb technology secret.

atoll, at a rate of about three percent per day. The tests came off more or less on schedule, although the carefully drilled bombing crew of the B-29, Dave's Dream, dropped the initial nuclear bomb on 1 July 1946 450 to 600 m off target above the 93 ship array, partially messing up the experimental design but still sinking five ships and badly damaging three others. The underwater shot then followed on 25 July 1946 (Bikini Atoll date) creating a tremendous mass of radioactive water. Within two days, lagoonal mixing permitted radiological monitors, washdown crews, and damage inspectors to penetrate the outer part of the target area, although the central portion remained "too hot" for doing any useful work even after another two days had passed.

In the Baker test with an estimated yield of 20 kilotons of high explosive equivalent, a central core of radioactive water 675 m across rose to a height of 1,800 m before falling back into the lagoon to form waves 25 to 30 m high within about 300 m of ground zero (JTF-1 Historian, 1946). The phenomenological question then was — how high would these waves be when they hit the beach at Bikini Island about four km away, the island itself being but three m high? To find out, various techniques were deployed, including stereographic cameras, but none were simpler than John Isaacs's technique of nailing a series of protected beer cans one above the other on selected coconut trees. However, the deadly base surge which washed over the close-in ships quickly died out during transit through the 40-m-deep lagoon, and the final uprush at Bikini Island's beach was but two m high.[14]

Once these two detonations were over, Blandy's meteorological staff officers, Colonel Benjamin Holzman, USAAF, and Captain A. A. Cumberledge, USN, studied the surrounding meteorological situation and concluded that the nuclear releases had so quickly injected heat and moisture into the atmosphere that ". . . no significant meteorological influence other than purely local cloud effects

---

14. This finding verified some wartime work of Bernal and Gaskell regarding the possibility that large charges (one kiloton or so) of high explosive detonated a few km off an enemy held beach might set up a pre-invasion tidal wave that would sweep away both the defenders and their shore defenses. Their calculations indicated that the explosive volcanic outburst which destroyed Krakatau Island in Indonesia during 1883 (leaving a five-km-diameter crater) was in the 100-megaton range. They compared the size of this crater with those of craters formed by exploding charges on beaches and by aerial bombs and developed a rule relating crater size to charge size. They also knew that the Krakatau blast produced a tidal wave about 30 m high when it passed a lighthouse a known distance from the volcano. By comparing the two sets of data, it was apparent that any wave created by an explosive charge in the low kiloton yield range would be about a meter in height when striking a beach one km or so away. In other words, the proposal to clean off enemy beaches by large chemical explosions would not work.

resulted from the Bikini tests 'Able' and 'Baker' " (Holzman and Cumberledge, 1946). Because of the continuing intense interest in learning whether the ecology of Bikini Atoll and surrounding environs had been damaged, in 1947 Revelle directed an elaborate re-survey of the area patterned somewhat after the grand naturalistic surveys of the 19th century. By so doing, he was delivering on his promise of the year before:

> to carry out an integrated investigation of all aspects of the natural environment within and around the atoll; the currents and other properties of the ocean and lagoon waters, the surface geology, the identity, distribution, and abundance of living creatures, and the equilibrium relationships among all these.

Their key findings included the good news that normal marine life was quickly repopulating the underwater detonation area, while there appeared to be no adverse effects on the commercial tuna population existing thoughout this island archipelago.

By 1947, research funds were trickling into U.S. oceanographic research institutions from sources as the new Office of Naval Research, the Navy's Bureau of Ships and its Hydrographic Office (for a special cruise by the *Atlantis* to the Mediterranean and Aegean Seas), from the oil companies beginning to go offshore, and even from the State of California and its troubled sardine industry.[15] When the National Academy of Sciences in 1949 appointed its second Committee on Oceanography (NASCO-II), this group found some $2.3 million per year going into oceanographic research (three times the pre-war rate). But rather than being happy about this, the Committee warned:

> . . . the effects of this great outpouring of money from state and federal sources is not entirely healthy. . . . Oceanography does not at present need, and probably could not effectively use, very much larger sums than now available . . . the demand for trained people will remain small so that few universities will be willing to maintain an adequate department of oceanography. . . . (In fact), the demand for trained personnel is likely to decline so that perhaps half a dozen good recruits per year will suffice for the openings now envisaged.

The sour viewpoint was further reiterated when four of the nation's top marine scientists — Dean V. Knudsen of UCLA, Dr. Alfred Redfield of WHOI, Dr. Roger Revelle of Scripps, and Dr. Robert Shrock of MIT — stated in the 23 June 1950 issue of *Science* that they, too, believed the large demand would level off and only ten or

---

15. The disappearance of sardine schools during the 1946—1947 fishing season left the fate of a hundred sardine canneries, including those of Steinbeck's "Cannery Row" in Monterey, California, so much in doubt that the fishing industry was willing to be specially taxed in order to provide research funding. The sardines never did come back, and the industry had to switch to fishing for anchovies and tuna.

so newly trained oceanographers would be needed each year "to take care of normal replacements and expected growth".[16]

Never was an academic group more off target. Six additional universities had already gone ahead during the late 1940s and started teaching oceanography to sizeable classes of demobilized veterans.[17] The reasons for such start-ups varied. For example, in early 1947, the Texas A&M University Research Foundation, using oil company funding, began an extensive three-year study of oyster mortality in the Louisiana marshes.[18] However, the problem proved so complex that it was apparent that this research-oriented school needed a "Department of Oceanography" even if it were located 225 km from the sea. On 9 January 1949, the Board of Trustees approved a "Department of Oceanography and Meteorology" and later agreed to hiring of Dr. Dale F. Leipper, one of Bates' fellow wartime classmates at Scripps, as departmental chairman. When Leipper opened his first class — one in "Introductory Oceanography" — in February, 1950, 54 students were on hand. Starting with five instructors, by 1963 the department was large enough to split into separate departments of oceanography and meteorology for its technical staff now numbered 154 and had an annual budget of two million dollars. By then, the school was also turning out more professional meteorologists and oceanographers — about 30 a year — than any other American academic institution.[19]

On the Washington, D.C. scene, Commander Revelle found early

---

16. In 1939, Dr. Sverdrup advised Walter Munk to think carefully before going into oceanography for he (Sverdrup) could not think of a single job that would open up during the next ten years.

17. The schools were: Johns Hopkins University, 1948; Brown University, University of Rhode Island, Cornell University, and Texas A&M University, in 1949, and Lamont Geological Observatory of Columbia University, 1950.

18. The Louisiana oyster harvest fell from about 3.5 million bushels in 1943 to approximately 2.1 million bushels in 1948. Based on favorable court rulings in 1944, the oystermen filed some 100 damage suits totaling over $30 million against oil companies operating in the Barataria Bay area during the mid-1940s, thereby giving rise to the request that the Texas A&M Research Foundation find out once and for all what was the cause of excess oyster mortality. On the east side of the delta, Gulf Oil Corporation faced over 60 similar lawsuits with claims in excess of $20 million. The upshot of all this was three years of extensive field work costing over two million dollars and involving over a hundred scientists and technicians, including Professor Sewell Hopkins, Dr. J. Mackin and Sammy M. Ray for Texas A&M, plus Dr. Albert Collier and A. H. Glenn and Associates (specifically C. Bates) as consultants for Gulf Oil. Eventually enough data were at hand to convince the judge trying the multiple lawsuits simultaneously that the increasing number of levee and river outlet cut-offs, plus the presence of a new oyster parasite, *Dermocystidium marinum,* was causing the increased oyster mortality. Consequently, the case was finally dismissed in 1954.

19. Scripps remained notoriously slow in granting advanced degrees. As Munk once muttered, "I starved six years to get my doctor's (from 1941 to 1947), and my students will, too!"

on that ONR would permit him to write simple, open-ended and easily renewable research contracts with his "chosen" oceanographic institutions — Scripps, Woods Hole, Columbia University, and the University of Washington.[20] This done, he returned to the academic life at Scripps in 1948 with his slot as Chief, Geophysics Branch taken by an University of California (Berkeley) seismologist, Dr. John Adkins. Of greater importance was that Gordon G. Lill, one of the initial BT officers, became Head, Earth Science Section, within the Branch. At essentially the same time, ONR had also opened the *N*aval *A*rctic *R*esearch *L*aboratory (NARL) at Point Barrow, Alaska.[21] Among Lill's first initiatives was encouraging oceanographers and other types of geophysicists to work in the Arctic, particularly offshore. He noted:

> If the Arctic Ocean is to be changed from an useless ocean into an useful one, large scale efforts of research teams will be required.... The aid, suggestions and interest of all geophysicists are solicited in order that the study of one of the world's most abundant minerals, ice, might receive the attention it merits.

The major response to this standing offer came from two iron-willed oceanographers at Woods Hole, John Holmes and Gerald Metcalf.[22] Their initial hope was to go offshore with the Norwegian polar explorer and rescue expert, Colonel Bernt Balchen, USAF, who planned, in his capacity as Commander, 10th Rescue Squadron, to do polar ice landings similar to the ones the Russians had been routinely doing as far back as 1937. The piggy-back operation did not work out, so Lill and Bates then explored the possibility of the naval air arm providing light aircraft or helicopters. The Navy brass was violently opposed to this offshore landing concept, fearing that an offshore crash would create a difficult search and rescue effort. As a consequence, neither the U.S. Air Force nor the U.S. Navy moved into the Polar Basin during 1949 or 1950.

However, in June, 1949 ONR and Hydro were able to sponsor Operation Cabot, the first multi-ship survey of the Gulf Stream co-ordinated on scene.[23] The project had been jointly urged by Gulf

---

20. A typical multi-million dollar work statement would read about as follows: "... to conduct a continuing series of research investigations into the physical behavior of the ... Ocean, its chemistry, its geology, and its interaction with the atmosphere."

21. NARL led a very checkered history. Operated by a series of universities under contract to the Navy, its first three directors were physiologists or biologists. As time wore on, the laboratory became used more by the Departments of Commerce and Interior than by the Navy; as a consequence, it was placed in a stand-by basis during 1980 as a money-saving action.

22. Holmes was an aviator in his own right. When Metcalf's infant son died suddenly of viral pneumonia, "Gerry" continued working at NARL. In those days, flying out of Point Barrow during winter darkness and cold was far from simple.

23. At Bates' suggestion, the survey was named in honor of the British explorer, John

Stream enthusiasts at Woods Hole — C. O'D. Iselin, Fritz Fuglister, Val Worthington, and "Rocky" Miller — and by Dr. William Ford, by then the Canadian Navy's senior oceanographer. Their case rested heavily on the ability of Dr. William von Arx's new *geo-magnetic electro-k*inetograph (GEK) to measure the set and speed of the surface currents while the research ships were under way. Operating under the technical command of Dr. Richard Fleming aboard the USS *Rehoboth* (AGS-30), the scientific task force also included the USS *San Pablo* (AGS-50), the Royal Canadian Navy's *New Liskeard,* and the Woods Hole research vessels, *Atlantis* and *Caryn.* Each day a new ship was introduced into the Gulf Stream off Norfolk, Virginia, and followed the Stream's "cold-wall" northeast until off Nova Scotia.[24] This edge was also tracked by William Kielhorn of ONR's Geophysics Branch flying daily along the Gulf Stream in a Navy patrol aircraft.[25] As the raw data poured into the *Rehoboth* and were analyzed by Walter Munk and Fritz Fuglister, it was obvious that the Gulf Stream was "snakey" and in the form of two great north–south looping meanders, one to the south of Cape Cod and the other off Nova Scotia.[26] Munk labeled these Edgar and Neddy. But the Stream became "sneaky". On the return leg going back up the Stream, the spaced-out ships found that the western downstream loop had broken completely away from the main current and was moving southwards towards the Sargasso Sea with a cyclonic circulation and a cool central core. Munk immediately renamed the feature Eddy in lieu of Edgar. This finding was an exciting event, for until then oceanographers had thought that a change in a major oceanographic phenomenon took months, not days (Von Arx, 1962).[27]

Cabot, who had first noted the existence of the Gulf Stream in 1498 on the basis that the beer in the ship's hold seemed unduly warm for being that far north (the first crude thermo-meters did not come until a century later).

[24.] After many years of hoping, the USS *Rehoboth* and the USS *San Pablo* were assigned to the U.S. Naval Hydrographic Office in 1947 for use in an elaborate series of "*A*coustic–*M*eteorological–*O*ceanographic–*S*onar" (AMOS) cruises co-sponsored by the Naval Under-water Sound Laboratory at New London, Connecticut.

[25.] By the late 1970s, Kielhorn was operating a small aerial remote-sensing business based in Groton, Connecticut. Surveys of this type are especially useful in mapping the effluent patterns off power plants, sewage outfalls, and river mouths.

[26.] In order to re-cross the "Cold Wall" promptly from the warm side, the ships some-times had to sail 45 degrees west of north, a heading never conceived of whatsoever during the planning stage.

[27.] With the advent of meteorological satellites and oceanographic research aircraft in the 1960s, oceanographers in both the U.S. Navy and the national weather service began rou-tinely plotting the position of the Gulf Stream. Tanker operators, such as Exxon, found that by taking the position of this strong current into account, their ships could make several knots better speed over the ground, just as Benjamin Franklin had found while serving as

Despite the sizeable advances in oceanographic manpower, facilities, funding, and education made in the United States during the immediate post-war period, the same did not hold true in Canada. There oceanography had existed traditionally only to support the fisheries. After the war, however, the Canadian Navy took the lead role. It was natural, then, that Dr. William Ford, upon leaving Woods Hole in 1948, should join the Naval Research Establishment. As Head, Underwater Physics Section (although basically he was a chemist), Ford had the challenging task of developing the Canadian variable-depth sonar system, a device much needed because of the rough water in which Canadian ASW vessels often operated.[28] Even so, during this post-war transition period, not a single Canadian was taking graduate training in oceanography at any Canadian university. The number of Canadian practitioners in this general field could be counted on the fingers of one's hands if he excluded the marine biologists and hydrographers; including them would have taken only a few more sets of hands.[29]

Although no key Canadian official was interested in correcting this situation, a group of maritime enthusiasts took it upon themselves to do just this. The group consisted of Drs. Andrew Thomson and Patrick McTaggart-Cowan of the Meteorological Service, Drs. William Ford and William Cameron of the Defence Research Establishment, Dr. John Tully of Fisheries, and a few others in the Ministry of Mines and Technical Surveys. They soon induced Dr. W. E. Van Steenburgh, the highly perceptive, personable Director General for Research in the latter agency, to chair a small group called the *Joint Committee on Oceanography* (JCO). Membership was limited to government officials who could provide tangible support to oceanography via their respective agency. They agreed to be "honour bound" to maintain their allotted contribution to the overall plan designed to make Canada self-sufficient in maritime studies and surveys. The JCO was soon able to persuade the National Research Council to allocate $60,000 a year to the University of British Columbia and to Dalhousie University at Halifax, Nova Scotia for establishment and operation of oceanographic institutes. As the years went on, these behind-the-scenes "shakers and movers" also obtained funding for

---

Deputy Postmaster General for the American colonies two centuries before.

28. This four-year program developed advanced techniques in high-speed towing, noise suppression, cable reliability, and cable fairing, all later adapted (with the collaboration of a parallel U.S. effort) to many non-military applications in marine exploration geophysics, including the latest generation of sub-bottom reflection profilers.

29. A generation later, each of these three groups number in the hundreds.

the Great Lakes Institute at the University of Toronto and, by bringing in university representatives, converted the JCO into a Combined Committee on Oceanography (CCO). Later, oceanography was so well established at such major federal facilities as the Bedford Institute of Oceanography in Dartmouth, Nova Scotia (1965), and the Pacific Institute of Ocean Sciences at Sidney, British Columbia (1978), that the informal steering group was eventually disbanded.[30]

In Great Britain, oceanography fared rather well during the immediate post-war period. Dr. Edward Bullard returned to the Geodesy and Geophysics (G and G) Department at Cambridge University, as did B. C. Browne. Although the Department never used the term "oceanography" in its title, the two were able to build it into one of the world's leading centers of marine geophysics.[31] The group not only received conventional fiscal support from the university and the British government but also from the U.S. Navy's Office of Naval Research and from the international oil industry which, at long last, had become fully awake concerning the value of offshore scientific research. As a result, Browne was able to expand his research in the area of making gravity measurements at sea, while Dr. Maurice Hill and other demobilized members of the Admiralty's Mine Design Department were able to turn their efforts towards developing improved sonobuoys for deep-water seismic measurements.

Within the government establishment, the Beach Reconnaissance Committee (with Gaskell as its Secretary) vigorously pointed up the nation's continuing lack of adequate knowledge relative to such fundamental properties of the sea as waves and changes in beach profiles due to storms. Soon two timely research institutes were formed — the Hydraulics Research Institute at Wallingford under the internationally recognized research engineer, Brigadier R. A. Bagnold, Royal Engineers, and the *National Institute of Oceanography* (NIO) at Wormley, near Godalming, Surrey. Housed in a surplus radar research building and initially funded by the Royal Navy, the fledgling NIO began during 1949 with a staff of approxi-

---

[30.] In a letter dated 30 March 1981, McTaggart-Cowan notes: "Dr. Van Steenburgh (now dead) deserves the major credit but each of the members, including the Met Service, made substantial contributions in kind and in influence. (Thus), from this humble and unorthodox beginning, post-war oceanography has grown quite well to its present stature."

[31.] However, Bullard liked to travel. During 1948–1949, he served as a physics professor at the University of Toronto, followed by returning to England to be Director, National Physical Laboratory, before joining the Cambridge University faculty for a third time in 1956. One of the first Bullard–Browne post-war students to make his ultimate mark was H. I. S. Thirlaway, who became Director of the Pakistani seismological network after graduation. In 1961, Thirlaway returned permanently to England to become one of the pioneers in using seismic arrays to improve the state-of-the-art in underground nuclear explosion detection and identification.

mately 30 drawn from "W" Division at ARL and biologists assigned to assessing RRV *Discovery* field data. The Director, amiable George E. R. Deacon, remained in that position for the next 22 years, and had much to do with setting the tone of high-quality research that typifies the work conducted at that federal institution.[32] Much of this work continued to emphasize improving man's knowledge of the formation and decay of ocean waves, including the important analytical work by Dr. Michael S. Longuet-Higgins and others based in large part upon wave-measuring devices that Norman Barber and M. J. Tucker were able to install on British weather ships operating in the stormy North Atlantic Ocean.

[32.] A graduate of King's College, London, Deacon made frequent oceanographic surveys in the Antarctic region between 1927 and 1939 aboard RRVs *William Scoresby* and *Discovery*. In addition to his NIO role, Deacon served as Chairman, National Committee for Geodesy and Geophysics of the United Kingdom (1955–1960), President of the International Association of Physical Oceanography (1960–1963), and Chairman of the British National Committee for Oceanic Research. For these and many other professional services, he was knighted in 1971 and also designated a Commander, Order of the British Empire.

Aerial view of the U.S. Navy Hydrographic Office, Suitland, Maryland (on the right), and the U.S. Census Building (on the left) during 1946. "Hydro's" Division of Oceanography occupied a basement wing of the latter building from 1953 to 1978.

Oceanographic team *en route* to Bikini Atoll aboard the USS *Bowditch* during February, 1946. Left to right, front row: Walter Munk, William von Arx, Gordon Riley, William Wilson (fisherman), William Ford and H. Turner; rear row: Vernon Brock, Melvin Trailor, Mark Antonich (fisherman), Martin Johnson, Marston Sargent, Jack Marr, Leonard Schultz, (unknown fisherman), Joshua Tracey, Randolph Taylor, Clifford Barnes, Thomas Austin and Kenneth Emery. (*Note*: Missing are Roger Revelle, Charles Bates, John Lyman, and John Isaacs.)

Aftermath of detonating the Baker device underwater at Bikini Atoll on 25 July 1946. Water depth was approximately 40 meters.

Lt. Commander Mary Sears, USNR(W), Commander, Oceanographic Unit, U.S. Navy Hydrographic Office (1943–1945). (Courtesy of WHOI.)

Professor George P. Woollard, founder of Geophysical and Polar Research Center, University of Wisconsin, and later Hawaii Institute of Geophysics. (Courtesy University of Hawaii.)

Dr. William E. van Steenburgh, Director-General for Research, Canadian Ministry of Mines and Technical Surveys during the 1950s and 1960s.

The world's largest oceanographic vessels of their time — the USS *San Pablo* and the USS *Rehoboth* — alongside the quay at Monaco *circa* 1952. (Courtesy NAVOCEANO.)

Ned Ostenso, one of Woollard's students, aboard the ice-breaker, *Glacier*. (U.S. Coast Guard photograph.)

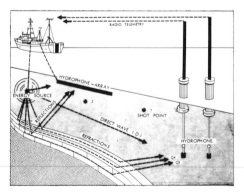

Schematic diagram demonstrating seismic reflection and refraction measurements at sea while under way. (After Ludwig and Houtz, 1979.)

Hydro's first full-time ice observer, Murray Schefer, before his reconnaissance aircraft, a RCAF Lancaster. (Courtesy M. Schefer.)

Fred Bucy and Cecil H. Green of Geophysical Service, Inc. scrutinize a "magnedisc" during September, 1954. (Courtesy of *Dallas Morning News*.)

Seismic detector cable used at sea; some strings can be as much as 3.5 km in length. (Courtesy of Western Geophysical Company.)

Dr. Gordon G. Lill (Deputy Director, National Ocean Survey, 1970–1979). (Courtesy G. Lill.)

Monument and crosses commemorating the crew of the RCAF Lancaster which crashed at Alert in northern Canada during July, 1950. (Courtesy Canadian Atmospheric Environment Service.)

The "Pineapple Special" — first aircraft, flight crew and geomagnetic observing team for Project Magnet (1953). Front row, left to right: R. L. Robbins, L. D. Burke, B. C. Byrnes (Project Director), M. A. Roskowski, and R. Fussell. Rear row: R. J. Touch (Navigator), B. T. Smith (Pilot), and G. F. Hinton, W. H. Geddes, G. R. Lorentzen, V. J. Covello and R. W. Seaton (Geomagnetic Observers). (Courtesy NAVOCEANO.)

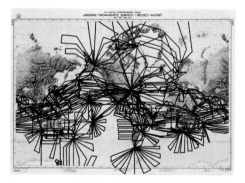

Tracks of Project Magnet's aerial magnetic surveys, 1959–1970. (Courtesy NAVOCEANO.)

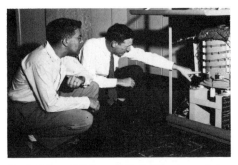

Dr. J. Edward White and R. B. Rice inspect during 1956 the 4,000-word magnetic storage drum of an ElectroData Datatron computer; this device was the first of its type used for seismic research at Marathon Oil's Denver Research Center.

Operator at the console of Phillips Petroleum Company magnetic tape playback center for seismic data enhancement. (Courtesy H. Mendenhall.)

Miniature geophones used by the dozen to improve signal-to-noise ratios. (Courtesy Bettye Athanasiou.)

W. Harry Mayne, inventor of the CDP technique, and Paul L. Lyons, compiler of the first seismic map that showed the Prudhoe Bay geological anomaly. (Courtesy SEG.)

Mechanical vibrator technique for acquiring seismic data. (Courtesy GSI.)

Dr. Sydney Chapman offering after-luncheon comments on 18 July 1968. To his left, respectively, are Dr. D. H. Sadler, President of the Royal Astronomical Society, and Sir Harold Jeffreys, the noted British geophysicist.

Delegation of the United States of America to the Geneva Nuclear Test Ban Talks in action during 25 November 1959. Left to right, front row: W. Panofsky, James Fisk and Carl Romney; second row: Frank Press, Anthony Turkevich, Hans Bethe and John Tukey. (Courtesy Carl Romney.)

Soviet delegation to the same test ban talks. Left to right: Igor P. Passechnik, K. E. Boubkin, M. A. Sudovsky, A. I. Ustyuhenko, E. K. Federov, Y. V. Riznichenko, V. Vishustov (Diplomatic Corps), and V. I. Keilis Borok. (Courtesy Carl Romney.)

CHAPTER 5

The 1950s—A Burgeoning Era of Geophysics

"The search for Truth is in one way hard and in another easy. For it is evident that no one can master it fully or miss it wholly. But each adds a little to our knowledge of Nature, and from all the facts assembled there arises a certain grandeur."

> Aristotle
> (Translation of the Greek transcription on the façade of the U.S. National Academy of Sciences Building, Washington, D.C.)

A Short Overview of the Decade

Of all the decades during the past half century, the 1950s may have come the closest to providing a steady-state condition for the pursuit of geophysics. Economic conditions stabilized throughout the world. The European Economic Community was formed in 1957, Japan aggressively rebuilt its domestic economy, the Soviet Union underwent a period of "de-Stalinization", and the United States, under the benign leadership of General Dwight Eisenhower as President, had an average inflation rate of about one percent per year between 1953 and 1960. Crude oil, whose price had jumped from $1.81 a barrel in 1947 to $2.76 a barrel later that year once price controls came off, was priced at $3.11 a barrel in 1954, a charge that stayed roughly the same for the remainder of the decade.[1] At this price level, the United States had enough surplus oil-producing capacity to fill Europe's basic needs during a boycott by Middle East producers in 1956—1957 following attacks by Great Britain, France, and Israel on Egypt in October, 1956. The exploration geophysical industry opened the decade doing about $200 million a year in business and built up to an all-time peak of 1,040 seismograph crews working throughout the world during 1952. The British introduction of the jet-powered airliner in 1953 made it necessary for meteorologists to develop and provide terminal and upper-air

[1.] Price cited is for 34° gravity U.S. Gulf Coast crude.

101

weather forecasts of greatly increased scope and accuracy along the world's commercial airways. The new potential and feasibility of weather modification intrigued not only the meteorological community but also world leaders in the fields of agriculture, power generation, and ranching. But the geophysical highlight of the decade was the International Geophysical Year (1957—1958), man's most comprehensive international scientific undertaking, which featured the Soviet Union launching the first two space satellites, SPUTNIKS I and II, in the latter part of 1957.

Despite the good feelings rampant within the international scientific community, East—West relationships were still not good. North Korea invaded South Korea on 25 June 1950, setting off a hard-fought conflict that placed United Nations troops on one side and North Korean and Chinese soldiers on the other. Behind the Iron Curtain, the Soviet Union saw fit to suppress revolts in East Germany (1953), Poland (1956), and Hungary (1956). The United States (1952), the Soviet Union (1953), and Great Britain (1957) also detonated their first hydrogen fusion (H-bombs) during the decade, causing a major jump in the amount of nuclear fall-out. In response to the consequent public outcry, these nations — and many others — started a long-running series of nuclear test-ban talks at Geneva, Switzerland, during 1958. So important was the science of geophysics in designing test-ban monitoring systems for these treaty talks that both Great Britain and the United States found it necessary to launch a "New Seismology" heavily based on the use of arrays and digital computers.

Correcting the Manpower Shortage in Exploration Geophysics

In contrast to meteorology and oceanography, exploration geophysics entered the 1950s with a severe manpower shortage. Although the industry had doubled the number of seismic crews between 1946 and 1952, the military draft and call-up of reserve forces threatened such severe cutbacks that Dr. E. Eckhardt observed at the SEG's annual meeting in 1951 that something had to be done to keep skilled geophysicists from being called to military duty (Eckhardt, 1951).[2] The fledgling *A*merican *G*eological *I*nstitute (AGI), located in Washington and representing 17,000 members of twelve national

---

2. Between 1950 and 1953, the U.S., with 5,764,000 military personnel involved in the Korean War, suffered 157,000 dead or wounded. The British Commonwealth (including Australia, Canada, and New Zealand) suffered 7,268 casualties (including missing).

geological and geophysical societies, tried to do this by issuing a statement which said in part (American Geological Institute, 1951):[3]

1. The United States can neither sustain an expanding defense economy nor meet a war emergency without accelerating the rate of discovery of essential minerals such as oil, gas, metals and strategic materials. . . .
4. The loss of geologists and geophysicists from the search for minerals to the Armed Services will slow down the rate of mineral discoveries and imperil the nation's ability to prepare adequate defense or wage successful war.

Despite this initiative, nothing much changed relative to the trained manpower situation for the next three years.[4] At the SEG's annual meeting in 1954, Father Macelwane, SJ, found it necessary to report (Macelwane, 1954):

> The shortage of professionally trained geophysical engineers, as of all types of engineers, continues to be acute. In spite of the encouraging growth of offerings in the field of geophysical education, the needs of industry, government and the universities are such as to induce a degree of competition for available manpower that is far from healthy.
> There has been no increase in the number of departments of geophyscis and geophysical engineering . . . A total registration of 3,492 in geophysical courses (including meteorology) was reported by 65 educational institutions. The number of undergraduate students whose major field is geophysics, geophysical engineering or meteorology is 364, and the number of graduate student majors is 338. The total number of bachelor's degrees in geophysics given between July, 1952 and July, 1953 was 113. The number of masters' degrees conferred in the same period was 47. Doctor's degrees in geophysics conferred in this same period were (16). . . .[5]

This being a period of "God helps those who help themselves", the perspicacious Cecil H. Green of Geophysical Services, Inc.

[3.] As late as 1955, the Tulsa, Oklahoma-based earth science societies (AAPG and SEG) were unhappy about the AGI being cited in Washington, D.C.; Longwell notes (1955): "The aversion to Washington as a home for AGI is astonishing. Even if we should admit a dislike for some aspects of the federal government, can we avoid dealing with it? Shall we go hide our heads in the sand instead of keeping ourselves informed about government actions that affect us closely? . . . And since governments are human affairs, we may, with proper contacts, exert some influence in fashioning important decisions."

[4.] Having spent the World War II years interfacing with the Federal Government in development of the airborne magnetometer, Eckhardt held strong opinions about the skilled manpower policy of that time (Eckhardt, 1952): "Much has been heard in recent years from Gobble-de-gook Center about equality of sacrifice or the equal distribution of the burden. It is an idea to which all of us can readily agree in principle. . . . In truth, our greatest need is to win the struggle in which we are engaged. In sheer weight of manpower we are outnumbered. . . . Should we take technologists out of technology in order that equality of sacrifice may be more completely achieved? A complementary question is, 'Should we permit anything that will make the winning of the struggle more uncertain, lengthen its duration and increase its cost?' An intelligent answer to the second question automatically provides an answer to the first. This answer may not be socially palatable, but we had better heed it and thereby improve the chances of saving our skins."

[5.] The schools granting doctoral degrees: St. Louis University (3); University of Toronto (3); CalTech (2); Texas A&M University (2); University of California (Berkeley — 1) (Los Angeles — 1); University of Oklahoma (1); Columbia University (1), Johns Hopkins University (1); and MIT (1).

joined with Professor Robert R. Shrock of MIT's Department of Geology in 1951 to create an elaborate "University–GSI Cooperative Plan in Student Education" (Green, 1953; Shrock, 1966). During its 14-year life-span, the program involved 268 students (both undergraduates and graduates) in 78 colleges, universities, and institutes throughout North America.[6] During orientation week, a group of approximately 18 students would hear 40 presentations from some of the highest powered geophysicists in existence before leaving for their field crew. The Green–Shrock "Plan" was so broadly oriented that eventually 300 speakers participated over the years, including earth scientists from 33 oil companies, 41 educational institutions, and 11 federal agencies. Once the summer field effort was over, the selected students had no further responsibilities to the program except turning in a report to GSI and their sponsoring university as to what they had learned.

Several universities also operated rough-and-ready field programs of their own for budding geophysicists. Professor Maurice Ewing, once he had acquired a former yacht, the *Vema,* in 1953, saw to it that he spent six months at sea each year where he served as much as a teacher as a research investigator.[7] An equally effective teacher of geophysicists was Professor George P. Woollard at the University of Wisconsin, who had claimed as far back as 1938 that teaching geophysics within a conventional department of geology was almost impossible for such students had inadequate training in mathematics, physics and chemistry (Hubbert, 1955). But a decade later, Woollard found he could work within such an academic structure by utilizing the medieval concept of master and apprentice (Woollard, 1955).

Much of Woollard's success came when he found that the research agencies of the Navy and Air Force were willing to issue "invitational travel orders" to their academic contractors that were essentially blank checks for going anywhere in the world at no cost via the Military Air Transport Service, the Naval Air Transport Command, and the ships of the U.S. Navy and Coast Guard. Even though Woollard had only 23 graduate students between 1949 and 1955, they provided the manpower for two global projects and one major regional project. After coordinating with the Woods Hole Oceanographic Institution

---

[6.] Universities supplying the largest number of students were: MIT (60); Colorado School of Mines (24); University of Texas (13); University of Massachusetts (12); Iowa State University (11), and Pennsylvania State University (10).

[7.] During 1953, Ewing was washed overboard from the *Vema* during a howling gale. Although groggy from a blow to the head, he clung to a nearby fuel drum until the ship could be headed back into the wind and accomplish the rescue.

and the geodesists at Ohio State University, Woollard set out to establish a "World Gravity Base Network" and to accomplish the "Integration of International Gravity Standard Values". The goal of the first project was to establish 3,000 gravity bases in 85 countries using special Worden gravimeters of large enough range to permit measuring the entire gravity field of the earth without resetting. This effort permitted tying together isolated oil-company gravity base networks existing in many countries, as well as the creation of benchmarks from which new gravity surveys could take off should the area still be virgin relative to gravity surveys. His second world-wide project hoped to learn whether earlier gravity networks had different datums from the one now being determined by gravimeter, the fear being that gravity data obtained with early survey pendulums had been tainted by the susceptibility of such pendulums to recognize changes in the intensity and orientation of the earth's magnetic field.[8] The third simultaneous effort was portrayal of the regional geophysics of the United States for the scientific record. To do this, he had his students run a network of gravity and magnetic traverses across the country along major highways (with stations at 15- to 20-km intervals) external to areas already surveyed by oil companies.[9] Then the University of Wisconsin data were supplemented by equivalent data supplied by the oil companies to the special Committee for the Geophysical and Geological Study of the Continents operated by the American Geophysical Union.

The Geophysical Industry Grapples with Managerial and Technical Problems

At first glance, it would appear that the geophysical industry was doing extremely well within the United States during the early 1950s.

[8.] This required running a line of triplicate measurements between Paso Cortez, Mexico, and Fairbanks, Alaska, at roughly 150-milligal intervals over a range of about 4,850 milligals using three compound quartz pendulums of the Gulf Research and Development Company. These were then carefully compared with data read off Cambridge University's compound invar pendulum and the single invar pendulum system of the U.S. Coast and Geodetic Survey. His fear proved correct, for the original world gravity datum at Potsdam, Germany, proved to be wrong and eventually had to be adjusted by -14.0 milligal. As a result, in October, 1980 he published posthumously with co-author Valerie Godley (he had died during April, 1979) the report, "Changes in International Gravity Base Values and Anomaly Values". This corrected by approximately 14.7 milligals the 773 international gravity values published by John C. Rose and himself in an SEG volume issued back in 1953.

[9.] In 1951, the United States had only three complete magnetic observatories — those of the Coast and Geodetic Survey at Cheltenham, Maryland, Tucson, Arizona, and Houston, Texas. About this time, Ruska Instrument Corporation of Houston, Texas, built four more units for installation at Point Barrow, Alaska, Honolulu, Hawaii, the Philippine Islands, and Argentina, plus another four for stations in Europe.

By 1951, the nation was producing an all-time high of 2.24×10^9 barrels of crude oil per year, exceeding by over ten percent the previous peak year of 1948. In addition, enough new oil had been found to add 1.4×10^9 barrels to the country's estimated oil reserves. Leaders of the profession did, however, view with concern the heavy concentration of exploration effort within the already well-worked "oil patch" of Texas, Louisiana, and Oklahoma.[10] Yet much of the United States and the outside world remained unamenable to geophysical exploration because of complex geology, seismic noise problems, and elusiveness of stratigraphic traps.[11] The overall picture of seismic reflection quality for an entire sub-continent is shown in Fig. 5.1. Excluding mountainous and other regions considered to be non-prospective (about 25 percent of the total area), greatly improved exploration techniques were still required for nearly half of the tentatively prospective area remaining within the United States (Lyons, 1951).

One of the better geophysical frontiers still lay in the public domain, particularly that lying within the 11 states comprising America's arid west. As of 1953, 827 million acres of land remained in this category, with some 365 million acres in Alaska alone. Federal policy called for the management of this land to be oriented towards multiple usage in order to provide the greatest good to the greatest number of citizens. Hence, exploration of this vast tract was largely accomplished on an honor system without tight federal regulations. Because such lands were largely roadless, the lead vehicle in a seismic party might well be a bulldozer. Unfortunately, it was quickly learned that five miles of new "seismograph trail" could destroy enough grass to carry a cow and calf for one month. Still some 16,000 km of these trails were cut into the Big Horn Basin in Wyoming between 1949 and 1952. Similar trails across the tundra of the North Slope of Alaska were even slower to heal and often led to extensive erosion. Thus in 1953, fully two decades before environmentalists made exploration of the public domain costly and time-consuming, an official of the U.S. Bureau of Land Management was warning the industry (Killough, 1953):

[10.] As of early 1953, 300 seismic crews (38 percent of the world's total crews) were working in Texas and Louisiana.

[11.] In 1930, one U.S. seismic crew could expect to find 14 million barrels of oil per year on the average; by 1954, this figure had become less than two million barrels per year. One factor in this decrease was that the size of fields being discovered was also falling off – from 26 million barrels per field in 1938 to about three million barrels per field in 1954 (Lyons, 1956).

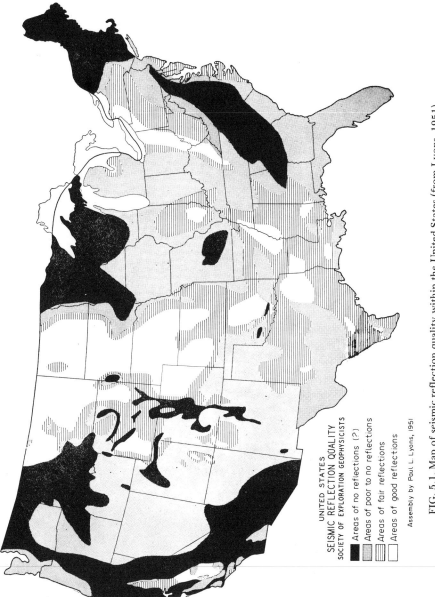

UNITED STATES
SEISMIC REFLECTION QUALITY
SOCIETY OF EXPLORATION GEOPHYSICISTS

Areas of no reflections (?)
Areas of poor to no reflections
Areas of fair reflections
Areas of good reflections

Assembly by Paul L. Lyons, 1951

FIG. 5.1. Map of seismic reflection quality within the United States (from Lyons, 1951).

Disregard for the public interest and abuse of our national resources is tipping the balance, and the petroleum industry must now decide by its own actions its future place on our public domain. Will it choose conservation and voluntary cooperation or legislation and enforced regulation?[12]

In the early 1950s, the economics of oil production tipped sharply from favoring United States production to favoring foreign production. By this time, because of pro-rationing and production capability, the average American oil well was producing about twelve barrels per day, while in Venezuela the typical well produced a thousand barrels a day and, in Saudi Arabia, five to ten times the South American rate. This tilt meant a boom in foreign geophysical work, and by 1958, 45 percent of all seismic work took place outside the United States. But the very productiveness of this new exploration effort brought on its own downfall in 1959 when the total number of crew months decreased to 9,700 as compared to 12,000 crew months three years before. As a result, many geophysicists and allied technicians were laid off, and a general sense of "gloom and doom" pervaded the working level (Weatherby, 1959).

The Advent of Magnetic Recording and Electronic Data Processing

If a geophysical firm could come up with a data-processing method that could use more of the valid reflection information in a seismic record than could be extracted by simple visual examination backed up by crude travel-time corrections made with the help of a desk calculator, that organization would pick up, rather than continue to lose, business. The best route to go was development of a mechanism by which field recordings could be repeatedly played back for analysis and refinement until an optimal set of traces had been obtained. To this end, transparent film was tried as the recording medium in lieu of photographic paper but with little success. Implanting signal recordings magnetically on wire for later playback was tried during World War II, and by the early 1950s, the technology was so improved that similar recordings could now be easily placed on magnetic discs and tape. By 1954, in fact, the Field Research Laboratories of the Magnolia Petroleum Company were able to report that it had developed a 13-channel magnetic tape recording and playback system (Loper and Pittman, 1954). Their work clearly showed a decided improvement in record quality achievable by

---

12. By 1981, the cost of land permits, whether on private lands or public domain, had undergone such escalation that it could amount to 15 percent or more of the cost of operating a seismic crew in the field.

making a broad-band field recording through recording signals over a wide range of frequencies and then selecting the optimum filter setting(s) through experimentation on playback equipment built for use in a field office. These authors also foresaw the time when such permanent recordings might be fed into digital computers "many years later in the event that new analytical techniques are developed".

The big break-through, however, came from technology created in-house by the old-line geophysical contractor, GSI, and its technically oriented foster parent, Texas Instruments, which had been formed in 1952 to play a more aggressive role in the entire field of applied electronics. By 1954, GSI announced a magnetic-disc field recording system tied into an advanced amplifier that would be the first in a series of standard-setting systems. The following year, they unveiled a magnetic playback system for these discs that permitted the making of corrections for near-surface weathering, elevation, and the geometrical effects of various recording configurations. They also led the industry by introducing a field unit capable of combining or "stacking" the magnetic disc recordings from as many as 17 separate shots in order to build up reflection signals and to cancel random noise.[13] By 1955, a technology review paper in *Geophysics* reported that seven different manufacturers had delivered 145 magnetic recording units, and 36 more were on order (Dobrin and Van Nostrand, 1956). This number accounted for 20 percent of all operational seismic crews in the world. Thus, the magnetic recording revolution was well under way, even though plagued by the insistence of each manufacturer that it use its own tape format and recording head arrangement, thereby making the equipment and tapes of various manufacturers and contractors incompatible.

A revolutionary method of making best use of the total magnetic recording and processing system was conceived in 1950 by soft-spoken W. Harry Mayne of Petty Geophysical Engineering Co. (see vignette by Mayne; also Mayne, 1956 and 1962).[14] A U.S. patent

13. Signal-detection theory calls for the ratio of signal to *random* noise to improve at the rate of the square root of the number of shots or as the number of detectors increases in an array feeding a single recording channel. Thus, a nine-detector array should give a signal-to-noise ratio twice as good as that obtained with a three-detector array. This improvement does not take into account additional betterment that may occur through cancellation of non-random noise, such as horizontally travelling surface waves.

14. In 1965, the SEG presented Mayne with the Reginald Fessenden Award for inventing the CDP method which, by then, had become recognized as one of the great innovations in exploration seismology. The Geological Society of London subsequently gave Mayne the prestigious William Smith Medal during May, 1979 for development of "Common Reflection Data Stacking Techniques". Smith (1769–1839) had been the first person to utilize fossils in a systematic way for determining the age of rocks.

was issued in 1955 on this method, which is called the "*common-depth-point*" (CDP) or "*common-reflection-point*" (CRP) shooting and processing technique. The incentive for the development of this method came from the fact that, as seismic geophone and shot-hole arrays were increased more and more in a brute-force attempt to improve the signal-to-noise ratio, the subsurface area from which the reflections were being detected increased correspondingly in size. This caused an increasing amount of smearing and distortion of subsurface details being sought. To improve the signal-to-noise ratio and still preserve subsurface detail, Mayne's CDP method introduces many times more individual shots into a given geophone spread. In the case of reflection horizons that are horizontal, all subsurface reflection points lie vertically beneath the midpoint between the shotpoint and detector locations. In such a case, a given reflection point can be mapped as many times as desired by using many different shotpoint—detector combinations having the same mid-point.[15] As shown in Fig. 5.2, if a 24-channel geophone spread is used, the degree of multiplicity of subsurface coverage is governed by whether an energy source is triggered at every geophone location in the spread (24-fold coverage), at every other location (12-fold coverage), at every fourth location (six-fold coverage), and so forth. Then, by correcting all the CDP traces for path geometry and then adding them together, the sought-after seismic reflections are reinforced whereas unwanted signals or noise tend to cancel out because the shots, detector sites, and travel paths are all different. What multiplicity of coverage to use is, of course, governed by the severity of the noise problem.

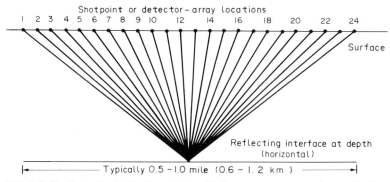

FIG. 5.2. Vertical plan indicating relationship between a common deep reflection point and locations of seismic detectors and shot-points when shooting a "CDP" line.

[15] This assumes proper corrections have been made to place all shot and detector locations on the same horizontal datum.

The major problem in implementing the CDP method was in making all the necessary corrections prior to stacking the CDP traces. In particular, the correction for path geometry, called the "normal moveout" correction, increases with increasing shotpoint-to-geophone distance and decreases continuously with increasing reflection times. Hence, it could not be accomplished without development of very elaborate and highly unique analog magnetic-tape playback equipment. As a result, the Petty firm was not able to reduce the CDP method to practice until 1958. In the meantime, Shell Oil Company and Phillips Petroleum Company had quietly and independently developed the same general approach to seismic recording and processing and soon became early licensees of the Petty patent. Phillips' playback center for CDP processing completed in 1957 under the leadership of Harold L. Mendenhall was crude but ingenious. Its processor employed 12 magnetic tape storage drums (right-hand side of the console) and a field-tape playback drum (left side of the storage drums), all mounted on a common drive shaft. In operation, first individual traces from 14 different shots were read successively and distributed among the storage drums according to shotpoint-to-geophone distance. Then sets of CDP traces were picked off the storage drums and combined after applying a different constant time correction to each drum. These corrections were chosen to yield the proper normal-moveout correction for each trace for a specific travel time. Then by repeating the process to produce an accurate correction at various other travel times, strongly reinforced, interpretable reflections could be obtained at any record times where reflections from key geological horizons normally occurred, only to be masked by extraneous seismic energy.

The complexity of the CDP technique and its requisite processing equipment is a clear indication of how far the geophysical industry would go in order to obtain useful subsurface information in poor shooting areas, as well as why the convenience and versatility of digital processing would be such a boon to the industry just a few years later.[16] In fact, as soon as military restrictions were removed from private use of digital electronic computers at the end of World War II, both seismologists and meteorologists had been anxious to experiment with them. One of the first programs to do so was a

---

16. Up until about 1970, the two most important developments in reflection seismology are considered to have been magnetic recording (analog and digital) and Mayne's introduction of CDP processing. In the latter case, as experience built up, it was evident that almost all "poor" seismic areas yielded improved data once the CDP method was employed. As a result, the technique essentially made all previous seismic data obsolete for petroleum-exploration purposes. Since that time, much more sophisticated digital processing techniques have also made earlier digitally-processed CDP results obsolete.

small research project initiated in 1950 within MIT's Mathematics Department by Enders A. Robinson, a graduate student, and Professor George P. Wadsworth to study the possibility of using digital computers and time series analysis for enhancing the quality of reflection seismograph records.[17] So encouraging were the results of these preliminary efforts that MIT organized in 1953 the "Geophysical *Analysis Group*" (GAG) within its Department of Geology and Geophysics. Fiscal and technical support was provided by 20 companies, including most of the major oil companies plus United Geophysical Company and GSI.[18] GAG research was aggressively pursued by MIT graduate students between 1952 and 1957, first with Robinson in the lead role (1952–1954) and then with Stephen M. Simpson, Jr. (1954–1957).[19] Eleven reports totalling some 1,100 pages were issued during the project, although the work was not publicly summarized until the June, 1967 issue of *Geophysics* (Silverman, 1967; Shrock, 1982).

Of all the work done, the most important proved to be the basis of Robinson's doctoral dissertation (Robinson, 1967). Although highly mathematical, his work found a very practical application in the important problem of resolving overlapping reflections. This problem stems basically from the fact that the downgoing acoustical wavelet from an impulse-type energy source such as dynamite is typically over 100 m long. This wavelet is reflected with varying amplitudes, called "reflection coefficients", from each of hundreds of subsurface reflections horizons which are frequently as close as a few meters apart.[20] Hence, a single seismic trace recorded at the

[17.] Early mathematical work on analyzing a time series (of which a seismic trace is a good example) was done by G. Udney Yule, a British statistician; by A. Kolmogorov and A. Khintchine, two Russian statisticians; by Harold Cramer and his student, Herman Wold, both Swedish statisticians, and by MIT's Norbert Weiner, who published a classic paper in 1930 entitled "Generalized Harmonic Analysis".

[18.] Dr. Daniel Silverman of Stanolind Research and Development Company chaired the advisory group, while Magnolia Petroleum Company and Atlantic Refining Company provided seismic records on photographic paper from the field for hand digitization and processing on MIT's one-of-a-kind "Whirlwind" vacuum-tube computer. The donors as a group provided $134,254 over the project's life.

[19.] After graduation, GAG participants went in varied directions. Enders Robinson first went with Gulf Research and Development and subsequently became a private consultant. Simpson also went into consulting. Mark K. Smith, who earned his Ph.D. in 1954, joined GSI and became Director of Research in 1955, Vice President in 1963, and President in 1966. Freeman Gilbert, who received his Ph.D. two years after Smith, eventually joined Walter Munk's group at Scripps, while Sven Treitel (Ph.D. in 1958) went first with Chevron and then with the Pan American research group in Tulsa, Oklahoma. Milo Backus also received his Ph.D. degree from MIT in 1956 but did not, as commonly believed, work as a member of the GAG team.

[20.] The reflection coefficient of an interface, which can be plus or minus, is governed by the acoustic impedance (velocity times density) of each of the two rock media bordering

surface as a function of two-way travel time to these interfaces consists of a continuous series of highly overlapping reflections. What Robinson did was to develop a processing technique aimed at resolving the travel times and amplitudes of individual reflections.[21] This method produces a much sharper picture of the subsurface than would otherwise obtain, and it (or equivalent methods) has become one of the most important modern processing steps.

In order to check the results of seismic processing techniques in the vicinity of a well, one needs a continuous velocity log (and density log too, if available) in order to compute the reflection coefficients. Continuous velocity logs have been quite easy to obtain because the appropriate down-hole tool was developed in the late 1940s and early 1950s by three different oil companies — Humble, Shell, and Magnolia (now Mobil) (see Kean and Tullos, 1948; Vogel, 1952; and Summers and Broding, 1952). After being licensed to Seismograph Service Corp., the Magnolia version was the first to be offered commercially — in 1954. All well-logging companies now offer essentially the same type of log, which produces a continuous recording of travel time per foot of hole (velocity) versus depth.[22]

---

the interface. Density is inversely related to velocity, and velocity is related to rock type, porosity, and fluid content. The reflection coefficients for all interfaces may be represented as a series of positive or negative spikes of proper amplitude, placed on a time scale at the appropriate two-way travel times. This time series is called the "reflection coefficient function". Then the seismic trace is formed by replacing each spike by a reflection wavelet with the same sign and amplitude as the spike. This transformation of the reflection coefficient function into a seismic trace is known as "convolution". It is one of a general class of filtering operations which also includes many ordinary types of electronic filtering.

21. Robinson's method, which he named "predictive decomposition", strives to undo the wavelet-filtering process to recover the reflection coefficient function. Any technique which starts with the output from a convolution filtering process and tries to determine one of the two input functions is more commonly known as "deconvolution". Several such approaches, all mathematically equivalent, are now in use. The two greatest problems in deconvolving seismic traces are: (1) that the wavelet must be accurately known before the reflection coefficient function can be determined precisely, and (2) that deconvolution greatly amplifies any frequency components missing in the wavelet, principally high-frequency noise (see Rice, 1962). The main drawback in Robinson's method is that it assumes that the wavelet has the nice mathematical property of being "minimum-phase". This is usually not true, particularly because of the distortions introduced by recording systems. Hence, many data-processing companies now perform initial "wavelet processing" to correct some distortions prior to deconvolution. Furthermore, because of noise problems, only partial deconvolution is effected, leaving some closely-spaced reflection horizons unresolved.

22. Another very useful type of well-velocity survey is made by lowering one or more geophones down a well and systematically firing surface shots at specified geophone depths, generally at about 150-m intervals. Such a survey permits one to check and correct the average velocity values obtained by the continuous down-hole logging tool. Between 1944 and 1960, *Geophysics* regularly included indexes of wells shot for velocity. Today, with continuous velocity logs run in almost every well, well-velocity surveys are rather infrequently conducted even though this limits the usefulness of the continuous velocity log because the well-velocity survey is still needed to correct it.

This log can then be integrated to obtain *total* travel time or *average* velocity versus depth.[23] The detailed velocity versus total travel time along with an assumed downgoing wavelet can be used to compute so-called "synthetic seismograms". Such synthetic seismograms are usually quite similar to actual seismic traces near the well in question if the assumed wavelet matches the actual wavelet closely enough, and if the field traces are relatively free of noise. If so, then the synthetic traces plotted in terms of both time and depth permit positive identification of the time of occurrence and depth of key geological markers on the field seismic traces.

Non-dynamite Energy Sources for Seismic Exploration

Because both sports and commercial fishermen complained loudly that dynamite explosions killed too many fish (a claim not borne out in the field), the industry began to use slower detonating nitro-carbonitrate in marine shooting during the early 1950s. Later in the decade, the increased number of explosions required per mile, whether on land or at sea, for multifold CDP shooting led to the creation of other, cheaper energy sources. At sea, such sources had to be capable of being towed and fired at frequent, regular intervals in a series of high-energy, uniform "explosions". In addition, the system had to minimize the effect of the explosion's air bubble moving up to the surface and collapsing, thereby creating a secondary source of energy which complicates interpretation. Finally, the optimal marine energy source needed to be capable of being assembled into an array, frequently of different signal strengths, and actuated at exactly the same time in order to reinforce the downgoing signal and to aid in canceling noise energy propagating laterally across the towed hydrophone array.[24] By the 1970s, the most common marine energy source was an "air gun", essentially a steel container charged with very high-pressure air from the towing vessel and then, on command, released rapidly to form a pressure

---

23. Travel time per foot provides a sufficiently detailed measure of velocity for computing a reflection coefficient function, assuming density remains constant. Or if a density log is available, the product of velocity and density gives a more accurate calculation of reflection coefficients. The reflection coefficient function is convolved with the assumed wavelet to give a synthetic seismogram for comparison with a field trace before deconvolution. A deconvolved field trace can also be compared to the reflection coefficient function as a check on the effectiveness of the deconvolution.

24. In the early days of marine shooting, a typical hydrophone array consisted of five to 20 channels of multiple hydrophones housed in a semi-buoyant oil-filled streamer cable from several hundred to 1000 m long. Today 96 to 500 channels are used, and a typical streamer length is 3200 m.

pulse in the water. Since then, other marine sources have been developed to generate seismic energy by means of high-voltage "electrical sparkers" and superheated steam bubble generators.

To provide an alternative land energy source to dynamite, John M. Crawford, William E. N. Doty, and Milford R. Lee of Continental Oil Co. developed the unique Vibroseis® system during the later 1950s (see Crawford, Doty, and Lee, 1960). In this case, a massive ground plate carried by a heavy truck or trailer is hydraulically vibrated for many seconds according to a given signal pattern consisting of a sweep through a broad range of seismic frequencies much as in a "chirp". The long incoming trains of seismic signals are then processed in a rather complicated fashion to collapse the long signals and define the travel times to subsurface horizons. The method has some inherent advantage in detecting signals in the presence of noise, is a cheap way of providing multifold coverage in CDP shooting, and can be used in urban areas and other localities where the use of dynamite is prohibited.[25] Another surface energy source was developed by geophysicists with the Sinclair Refining Co. (subsequently merged into the Atlantic Richfield Corporation in 1969) under the research direction of James F. Johnson. This impulse-type source uses gas exploding in an enclosed chamber above a ground plate and was called Dinoseis®.*

Burton McCollum, a founder of the pioneering Geological Engineering Co., achieved further recognition in the 1950s with the development of a workable weight-dropping scheme employing a three-ton weight. In order to achieve an adequate signal-to-noise ratio, records from 20 to 200 weight drops were combined into a single record by means of McCollum's elaborate analog magnetic-tape recording and compositing equipment (Sterne, 1954). In fact, McCollum was the first to obtain usable reflection records in many parts of the difficult surface limestone area of West Texas' Llano Estacado even without the use of CDP methods.

A Seismic Search for St. Peter's Tomb

After St. Peter had been crucified head downwards by the Emperor Nero in A.D. 64, the body of Christ's first disciple (and the first Pope) was taken to the nearest burial ground on a hill (now called

---

® Registered trademark of Continental Oil Co.

25. Within the U.S. in the early 1980s, dynamite remained the preferred land energy source because of better downgoing energy penetration and the information such shots give about the thickness and velocity of the weathered, low-velocity surface layer. However, the use of Vibroseis® ran a close second.

*® Registered trademark of Atlantic Richfield Oil Corporation.

Vatican Hill) which rose out of the flood-plain just north of the Tiber River. In A.D. 323, Emperor Constantine built a major basilica over this grave, but in A.D. 1504 that basilica was torn down to make way for the greater Vatican basilica that stands there today. While building both cathedrals, the builders took care to fill in, rather than destroy, the tombs and to leave the crematory urns, sarcophagi, and altars intact. Quite by accident, in 1939 a workman broke through a wall and found evidence of these early tombs. Pope Pius XII asked for additional excavations, hoping to find the actual tomb of St. Peter. But the work went slowly, and after World War II the Jesuit seismologists, Professors J. Joseph Lynch of Fordham University and Daniel Linehan of Boston College, were asked to conduct mini-geophysical surveys, both electrical and seismic, in the basilica as an aid to the supervising archeologist, Monsignor Ludovicus Kaas (Linehan, 1956).

Employing special equipment provided by Century Geophysical Company of Tulsa, Oklahoma, the two Jesuits first made a reconnaissance survey outside the cathedral by using very small charges of dynamite buried in alleyways to a depth of slightly more than a meter. Their first instrument spread, when shot and reversed, indicated that the average depth to natural clay was about 2.7 m, an accuracy which held to within about 0.3 m when excavations were made later. Afer moving under the basilica, their field technique switched to using blows of a sledge hammer against an iron plate as the energy source and arrays of small geophones spaced at 0.3 to 3.0 m between detectors. In this case, the arrays stretched along the "itinera" (the passages among the tombs) that had been dug to a depth of as much as nine m below the present level of the cathedral's main floor. These passages typically ran east—west, and the surveys indicated earth fill and man-made structures even further below. Upon working up the mini-records, Lynch and Linehan did not find St. Peter's tomb but did find such features as a cistern and even a step cut into the original hillside clay that had since been covered over. The greatest challenge came in exploring the unexcavated portion below the massive Michelangelo Basilica started back in 1546. To do this, the seismologists laid out an array over the marble floor using a three-m spacing between geophones and dropped a 13.5-kg weight onto an iron plate from a height of about 1.2 m. Based on these data, they were able to advise the Vatican authorities where buried rooms and walls still exist within this unstudied section of the cathedral.[26] However, the acquisition of these seismic record-

---

[26.] Subsequent excavations did find a simple tomb and a skeleton that might — or might not — be St. Peter's.

ings proved to be possible only because Lynch and Linehan learned to shout out the all-important phrase, "Stand still!", in so many different languages that the ever-present crowd of tourists and worshipers would actually quiet down at "shot-time".

Project Magnet — A Global Airborne Survey of the Earth's Magnetic Field

Although the first "modern" magnetic declination charts of the earth were drawn by Sabine back in 1845, man's knowledge of the magnetic field at sea dwindled as wooden ships disappeared from the face of the sea. To counteract this deficiency, Dr. L. A. Bauer had the Carnegie Institution of Washington construct a non-magnetic ship, *Carnegie* (crew members even ate off aluminum plates) and sail her throughout the world's oceans starting in 1909. However, the *Carnegie* blew up in Apia Harbor, Samoa during November, 1929, leaving the geophysical community again with the inability to map changes in the earth's magnetic field over ocean areas.[27] By 1951, the earth's magnetic charts were so flawed that the U.S. Naval Hydrographic Office (Hydro) undertook Project Magnet to fill this gap by using airborne magnetometry. Flight operations of the specially equipped long-range P2V, Pineapple Special, began in 1953 under the technical direction of B. C. Byrnes.[28] Since that time, the project has gone through a progression of improved aircraft and magnetic sensing devices and has more than achieved its goal — routine global measurement of the earth's magnetic field.[29]

Should there be a Nuclear Test Ban?

After the Bikini atomic bomb tests of mid-1946, the United

27. If this cruise had been successfully completed, the Carnegie Institution would have chartered the *Carnegie* to Scripps for $60,000 a year for the purpose of conducting detailed oceanographic observations throughout the eastern Pacific Ocean during 1931—1932.

28. Five of Hydro's six original airborne geomagneticians moved into high-ranking managerial slots within the successor Naval Oceanography Command. One (W. H. Geddes) eventually became Director, Oceanographic Department, of the Naval Oceanographic Office, while three others (B. C. Byrnes, R. W. Seaton, and G. R. Lorentzen) became Division Chiefs in the same Office. V. J. Covello remained in the Washington, D.C. area and became a high-ranking staff assistant in the Office of the Oceanographer of the Navy.

29. After flying a series of P2-V, NC-54 and NC-121 aircraft, the project acquired a special RP3-D Orion aircraft with a range of 9,000 km and equipped with a "Geophysical Airborne Survey System" fabricated by the Applied Physics Laboratory, Johns Hopkins University. Magnitude of the earth's magnetic field is measured by a metastable helium magnetometer placed in the non-magnetic tail boom of the aircraft, while vector quantities of the magnetic field are measured by a fluxgate vector magnetometer corrected for aircraft attitude by input from three inertial navigation systems.

States continued a widely publicized program of testing nuclear devices. The Soviet Union soon countered with its own testing program, the first fission device being detonated in August, 1949, followed by their first fusion device (H-bomb) during August, 1953.[30] Then, in December, when President Eisenhower gave his "Atoms for Peace" speech before the United Nations proposing an international atomic energy pool supervised by that body, the Soviet Union picked up this concept in just four days. It agreed to discuss Eisenhower's offer if the Western Powers would also agree to discuss a total ban on use of nuclear weapons. At this time, there still was no clearly defined public consensus whether the use of nuclear weapons should be totally prohibited or whether only nuclear weapon tests should be outlawed.[31] However, public attention soon focused on the test ban issue after the United States conducted a spectacular and quickly notorious series of thermonuclear explosions in the Pacific Testing Area during March–April, 1954.

The first of these detonations — *Bravo,* a 15-megaton surface shot on Bikini Atoll — turned out to be twice as powerful as expected. In addition, an unforeseen wind shift caused nuclear fall-out, much of it contained in pulverized coral dust, to drift over several neighboring atolls — Rongerik, Alinginae, Utirik, and Rongelap –- on which were both Polynesian natives and American weathermen. Their evacuation was then delayed because the test director did not immediately credit off-scale monitor readings on these atolls as indicative of anything more than instrumental malfunctions caused by the tropical environment.[32] But the incident which caused the most international commotion occurred when a Japanese fishing vessel, the *Fortunate Dragon,* moved into the danger area without being detected at shot time. As a result, the ship did not live up to her name for she received a distinct coating of "white snow". Although the fishermen had seen the premature dawn caused by the explosion 120 km away and had also heard the ensuing "bang",

---

[30.] The first British fission device explosion took place in October, 1952, and their first fusion detonation during May, 1957.

[31.] The complex give-and-take between 1953 and the mid-1970s concerning the banning of nuclear tests is well described in these two excellent books: *Nuclear Ambush — The Test Ban Trap* by Earl H. Voss (1963) and *Nuclear Explosions and Earthquakes — The Parted Veil* by Bruce A. Bolt (1976). The latter provides the more even-handed treatment of this emotional subject. It also lists all publicly announced underground nuclear detonations by the Soviet Union and the United States, plus all announced nuclear detonations by France, Great Britain, and the People's Republic of China.

[32.] The 264 people involved received doses ranging from 14 Roentgens (R) at Utirik to 175 R at Rongelap. None appeared to have suffered bad long-term effects. However, if the wind had shifted a bit more to the south, all would have received deadly doses of over 1,000 R.

they did not correlate these events with the "snow" which appeared three hours later. Consequently, they did not immediately wash down the ship to reduce exposure. When the trawler came into Yaizu, Japan, two weeks later, 23 of the crew were sick, and one eventually died.[33] Their case became a *cause célèbre* with the Japanese public and government, for it became known early on that the fish aboard the vessel were contaminated by radioactivity.

The international hue-and-cry about this event became tremendous. In late April, 1954, the Asian prime ministers, while conferring in Colombo, Sri Lanka, adopted a resolution urging suspension of H-bomb tests pending negotiation by the United Nations of a nuclear weapons test ban. However, U.S. weapon designers were just as convinced that the Soviets' H-bomb might be quicker and cheaper to build than the U.S. version. Led by mercurial Professor Edward Teller of the University of California and "father of the H-bomb", they were not about to let themselves be frozen into a position of technical inferiority by diplomatic maneuverings of a group of communistic and socialistic nations. The U.S. weapon designers argued at all levels that continued weapons testing was necessary not only to achieve weapons improvement but also to assure reliability of aging weapons. Moreover, they argued that should there be a nuclear test ban, there must simultaneously be an iron-clad method for policing the ban (see Table 5.1).

Between 1954 and 1956, there was limited public concern regarding the efficacy of national nuclear test ban monitoring networks. Thus, when the United States announced that there had been seven nuclear tests during Operation Redwing in the Pacific Ocean area between 5 May and 22 July 1956, Japanese scientists rather perversely reported that they had noted ten tests (Voss, 1962). In rebuttal of Governor Adlai Stevenson's campaign speeches supporting an immediate nuclear test ban, President Eisenhower noted in October, 1956 that only tests of large H-bombs could be detected with certainty. That seemed to be the essence of the matter until mid-1957 when Secretary of State John Foster Dulles raised the detection issue on both radio and television as follows:

> It is technically possible to devise a monitoring system which would detect significant nuclear tests and make evasion a highly risky business. But the possibility of concealment is such that inspection teams will have to be numerous and located near to possible test areas. The problem is not so simple as many have believed.

In this case, he was replying to a statement of Soviet Ambassador

---

[33.] The fishermen received doses of about 200 R. All but Aikichi Kuboyama, who died not long after, returned to work. Whether Kuboyama's death was primarily caused by radiation or by natural causes will never be known.

TABLE 5.1.
Capabilities for and Limitations of Detection of Nuclear Explosions

Method	Desired sensitivity of equipment	Frequency or period range	Extreme range of detection for 1 kt under good conditions	Location accuracy	Detection ability
Acoustic microphones (air)	0.1 dyne/cm²	0.5 to 40 sec	2,000–3,000 km downwind, 1,300 km crosswind, and 500 km upwind	100 km radius for 3 stations	Air burst below 30 km. Shallow underground or underwater burst
Acoustic hydrophones (water)		Various bands in range 1 hz to several khz	Over 10,000 km	High	Underwater burst not shielded by islands or shallow seas
Radioactive debris collection	Sample containing 10^8 fissions $(4 \times 10^{-14}$ g)		2,000–3,000 km station interval	Poor	Air burst below 10 km. Shallow underground or underwater burst
Electromagnetic (radio) signal	2 millivolt/meter 10 millivolt/meter	10–15 khz for direction finding 500 cycles to 200 kc for wave form analysis	Over 6,000 km	±2° in azimuth (≈30 km per 1000 km)	Air burst only. Tests shielded by shallow depth of burial
Seismic	10^{-7} centimeter displacement 10^{-5} centimeter displacement	0.5 to 2 sec for first arrival 10 to 50 sec for surface waves	1,000 km first zone, 2,000–3,500 km 2nd zone (Shadow zone between 10° to 20° of Latitude in areas of new mountain chains)	100 to 200 square km	Ranges refer to completely contained underground explosions. Decrease rapidly for aboveground shots

Source: Haskell (1959).

Zorin at the ongoing London nuclear disarmament talks proposing that the Soviet Bloc would agree to convening a "conference of experts" to work out test-ban controls if the Western Powers would suspend the next series of nuclear weapon tests. A month later, Governor Harold Stassen replied with a proposal on behalf of France, the United Kingdom, Canada, and the United States for a 12-month nuclear test ban, provided "inspection posts with scientific instruments (were) located within the territories of the Soviet Union, the United Kingdom, the United States, in the area of the Pacific Ocean and at such other places as may be necessary with the consent of the Governments concerned". But by that time the Russians had returned to supporting an unmonitored test ban. Thus deadlocked, the London disarmament discussions recessed after the 157th meeting on 6 September 1957 *sine die.*

By 1956, the American public was becoming increasingly critical and apprehensive of federal policy which permitted testing nuclear weapons in the atmosphere, no matter whether it be fission devices in Nevada or fusion devices in the Pacific.[34] The photogenic Nobel biochemist, Dr. Linus Pauling, was even predicting on television that there would be 200,000 mentally or physically defective children in the next twenty generations if nuclear testing continued, and that the life expectancy of a million people might be reduced by five to ten years. To respond to this issue, Dr. Teller collaborated with an UCLA geophysics professor, Dr. David T. Griggs, in preparing the report, *Deep Underground Test Shots,* which argued that going underground would eliminate the need to wait for favorable weather, avoid airline traffic interruptions and, most important of all, skirt the nuclear fall-out problem (Teller and Griggs, 1956).[35] In this paper, the authors predicted that a one-kiloton nuclear detonation at a depth of 300 m in the dry alluvium of the Yucca Flat test site 65 km north of Las Vegas, Nevada, would not vent any radioactive gas to the atmosphere; moreover, there would be little chance of triggering an earthquake in such material for it was "peculiarly suitable to attenuation of seismic impulses". They further claimed that if the nuclear test were conducted in solid rock, the outgoing seismic wave train would not be as damaging as that from an equally energetic

---

[34.] Even C. Bates' two small daughters ate some radioactive hailstones in Washington, D.C. on 26 May 1953 just 29 hours after the first atomic cannon shell was fired at the Nevada Test Site (Jaffe, Wittman, and Bates, 1954).

[35.] Griggs started his war service at the MIT Radiation Laboratory in 1941. By 1945, he was Chief, Science Advisory Group, Far East Air Forces. Though a civilian, he flew on many combat missions to improve weapon-delivery techniques; in so doing, he won the Purple Heart Medal for being wounded in battle (1944) and the President's Medal for Merit (1946). He later served as Chief Scientist for the Air Force during 1951–1952.

earthquake because most earthquake damage came from shear and surface waves which were either minuscule or absent in the case of a buried explosion. Finally, they noted that the tremendous heart of such an explosion would create a molten rock liner that would trap released radioactive material within recrystallized silicates which, when solidified, were highly resistant to leaching by percolating ground water.

The first announced underground nuclear test came on 19 September 1957. This was Rainier, a shot of 1.7-kiloton yield fired in a tunnel dug into a mesa at the Nevada Test Site (NTS) in such a way that the shot chamber was 270 m below the surface and 240 m back in from the feature's face. The surrounding country rock was a water-soaked rhyolitic tuff selected because its porosity and mineralogic composition tended to absorb shock waves and radioactive contaminants readily. Test objectives included (1) determination of the depth for burial of a nuclear device of specified yield in order to provide full containment of released radioactivity, (2) observation of resultant ground motion, and (3) ascertaining whether ground water would be contaminated to any notable degree. The international seismological community was alerted of the upcoming test by the U.S. *C*oast and *G*eodetic *S*urvey (C&GS), which also later supplied the actual shot time, 1659:69.5Z. The shot proved to have an equivalent Richter body-wave magnitude of 4.6 (see Fig. 5.3).[36] Forty-six seismic stations between 166 and 3,680 km distant detected the event, with the most distant detection being at College, Alaska, where the incoming wave train had a signal-to-noise (S/N) ratio of 2:1.

The initial U.S. *A*tomic *E*nergy *C*ommission (AEC) press release did not note that the distant College, Alaska, seismographic observatory had detected Rainier. Test ban advocates accordingly accused the AEC of intentionally suppressing that particular detection in order to make the detection problem appear more difficult than it really was. In rebuttal, Rear Admiral Lewis Strauss, USNR, as the AEC's Chairman, charged that some of the foreign stations claiming to detect Rainier had actually detected a moderate-sized earthquake which occurred in the remote Tonga Islands of the Southwest

---

[36.] Although widely used in the media, the Richter scale is strictly a measure of ground motion as recorded on seismographs, with each increase of one number indicating an increase of 30 times as much energy. In terms of damage, an earthquake of 3.5 could cause slight local damage, one of 4.0 moderate damage, that of 5.0 considerable damage, and 6.0 severe damage depending upon the local geology and the depth of the earthquake. Earthquakes of 7.0 are considered major events capable of widespread, heavy damage, 8.0 are great quakes causing severe damage and of notable historic interest, and 9.0 are unknown.

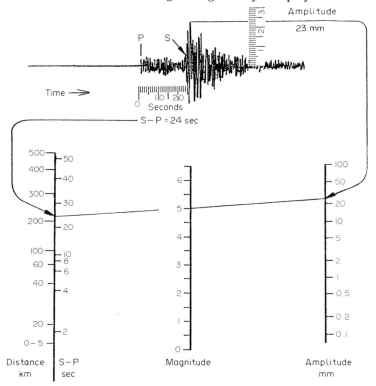

FIG. 5.3. Diagram for determining the Richter body-wave magnitude of an earthquake. To do so, lay a line from (1) the maximum amplitude recorded by a standard seismometer to (2) the distance of that seismometer from the epicenter of the earthquake, or use the difference in times of arrival of the compressional (P) and shear (S) waves. The point at which this line crosses the center scale specifies the event's magnitude. In this instance, the magnitude (m_b) is 5.0.

Pacific some 30 minutes later. However, the test-ban advocates were pleased to learn that at distances of 300 km or less, six seismic stations reported that the first seismic motion observed was an outward (or upward) compressional signal. This, of course, supported those who argued that distinguishing between the linear fracture of an earthquake, with its "push—pull" relationships in different observing quadrants, and an underground explosion's symmetrical "outward shove" in all directions, should be a relatively simple matter. Unfortunately, this logic does not always hold up in practice. The "first motion" is a rather small "pip" compared to the next half cycle of the sinusoidal shot signal, making it quite easy to lose the "first arrival" in background noise and to read the "second arrival" with its opposite sense of direction as "first motion". Just the same, all of Rainier's key test objectives were met. Gaseous radioactivity, and that minor, reached only half-way to the surface, while the chance of poisoning ground water proved negligible because the radioactive solids were contained in a highly insoluble fused silica. Surface motion also proved to be so small that it was barely noticeable four km from ground zero.

By 1958, the tug-of-war between the supporters for and against continued nuclear testing became even more spirited. To shed light on the issue, Senator Hubert Humphrey (who would be elected Vice President six years later) held three months of hearings before his Disarmament Subcommittee on the nuclear weapons issue. Half-way through the hearings on 28 February 1958, scientists of varying disciplines were asked to express their personal beliefs about the controversial subject. Professor Harrison Brown, a geochemist at CalTech and eventually Foreign Secretary of the U.S. National Academy of Sciences, suggested placement of test-detection stations using a spacing of about 1,600 km (ten stations in the Soviet Union and ten in North America) and further noted:

> ... there would be a minimum cutoff, depending upon the nature of the land, below which in energy you would not be able to detect the test. Tests in that area should be quite permissible ... there is little point in having a ban on something where you have no control over the ban.

The physicist, Hens Bethe, soon presented arguments which indicated that he, too, knew a lot about seismology:[37]

> ... Earthquakes are never reasonable. They produce strange patterns of compression and rarefactions around them. It is not a clean-cut quadrant where these are all positive and those are all negative. You will find in your positive quadrant where you will have three or four out of perhaps fifteen that will be negative, and you start to worry why they were negative. ... Looking at that close-in Rainier data, you would strongly suspect that it was an explosion since, let's say, nine out of ten of stations were compressional. But that one station would still worry you somewhat, and you would not know but what if you had a better distribution of stations around Rainier you might have picked up a few negatives. You would strongly suspect it enough so that you would report to your headquarters, your inspection headquarters, something suspicious had occurred, and you would certainly want to investigate that spot for further confirmation.

Nevertheless, even if there were over a thousand earthquakes a year in Russia and China generating seismic signals larger than that of Rainier, he argued that once a *bona-fide* on-site inspection system

---

37. Bethe, who had headed the Theoretical Physics Division at the Los Alamos Scientific (Weapons) Laboratory during World War II, received the Nobel Prize in physics during 1967.

were in place, the odds of discovery would be with the inspectors, given the complexity and surprises typically associated with nuclear testing. In contrast, Professor Edward Teller argued that this was not necessarily so for scientists had been working for just a few months on improving methods for hiding nuclear explosions.[38]

To expand upon the oral testimony, Chairman Humphrey queried by mail 31 seismologists chosen almost equally from academia and from government circles on the seismic aspects of an underground test-ban-control system.[39] The replies varied widely in content and indicated no consensus on the number of necessary seismic monitoring stations nor the amount of manpower needed to operate these stations and analyze the ensuing flood of seismograms. Haskell, for example, noted there should be 26 to 73 stations in the Soviet Union and China, depending on whether adequate event identification could be obtained within a radius of 900 km or of 1,760 km. Other replies indicated that as many as 200 stations would be needed in the Soviet Union alone. Manpower estimates were equally variant. One respondee suggested that the network could get by with five men per station, plus a central analysis force of 230 persons, while another seismologist argued that it would be difficult to man the necessary stations properly because the world supply of seismologists was not large enough. System cost estimates also ranged from five million to 100 million dollars.[40]

In the meantime, President Eisenhower proposed to the new Soviet premier, Nikita Kruschev, on 8 April 1958 that the long-postponed "conference of experts" be held to study the technical requirements needed to monitor adequately a nuclear test ban. Although Kruschev had held the top post for less than two weeks, his reply, back within a month, stated:

[38.] The leading proponents for a research program to conceal underground nuclear explosions were a pair of highly competitive and convincing brothers, Richard and Albert Latter. While working for the Air Force's "think tank", the RAND Corporation, "Dick" became noted for his work regarding the effects of nuclear explosions. "Al", who had been writing ghost stories for radio, also decided to study physics and came up with the idea of decoupling a nuclear explosion by firing it in a cavern hundreds of meters across. By 1979, the brothers were operating a highly successful "think-tank" of their own — R&D Associates of Los Angeles, California.

[39.] Responses were submitted, in part, by Dr. Carl Romney, now with AFTAC, Leslie Bailey with the C&GS, Dr. Norman Haskell of AFCRL, Dr. Thomas C. Poulter of the Stanford Research Institute, and the CalTech seismologists Drs. Beno Gutenberg, Charles F. Richter, and Frank Press.

[40.] In a more elaborate systems analysis performed during 1960 by Dr. Raymond A. Peterson, Vice President of United Geophysical Corporation, the cost of installing an appropriate detection network in the Soviet Union came to 1×10^9 dollars and involved three to five years of work. A global network was estimated to cost five times the value just quoted (Peterson, 1960).

The Soviet Government agrees to having both sides designate experts who could immediately begin a study of methods for detecting possible violations of an agreement on the cessation of nuclear tests with a view to having this work completed at the earliest possible date, to be determined in advance.

A week later on 15 May, Eisenhower proposed Geneva, Switzerland, as the conference site and work to start three weeks after arrangements had been agreed on. With such a high level degree of interest, it was possible to convene the conference on 1 July 1958 with Soviet Bloc experts from the Soviet Union, Czechoslovakia, Romania, and Poland and experts from Great Britain, Canada, France, and the United States for the Western Bloc.

Although the Western nations hoped to keep the "Conference of Experts" strictly technical, leaving diplomats to work out acceptable treaty language later, the Soviet experts were expected to press for a final report as political and binding as possible. Thus, selection of the Western negotiators was particularly important. The British "Experts" were Sir John Cockcroft, Director of the Atomic Energy Research Establishment and a Nobel Prize winner (physics – 1951), and Sir William Penney, Director of the Atomic Weapons Research Establishment.[41] The Canadian "Expert" was Dr. O. M. Solandt, Director of the National Research Council, while the French "Expert" was Professor Yves Rocard, a physicist with the École Normale Supérieure (the CalTech of France) who ran the French atomic-test-detection service on the side.[42] Selecting the three "Experts" from the United States was not easy. Chairman Strauss of the AEC wanted Professor Teller to be one, but this caused so much controversy that President Eisenhower accepted instead Dr. Ernest O. Lawrence, founder of the Lawrence Radiation Laboratory at the University of California (Berkeley).[43] With Teller out, his adversary, Dr. Hans Bethe, had to be relegated to an advisory post, and his "seat" assigned to Dr. Robert F. Bacher, a nuclear physicist who had headed the "Experimental Physics Group" at Los Alamos before moving on to direct the Bridge Laboratory at CalTech. The third "seat" and that of neutral chairman went to 47-year-old Dr. James Fisk, vice presi-

41. In the early part of World War II, Penney wrote a perceptive Ministry of Supply research paper dealing with diffraction of waves moving past the tip of a breakwater. Bates subsequently used the paper in estimating how much ocean swell would impinge on Utah Beach in the lee of the Cherbourg Peninsula. Later on, Penney joined the Los Alamos group and was on board an aircraft monitoring the nuclear explosion at Hiroshima. He also served as "Coordinator of Blast Measurements" at Operation Crossroads in 1946.

42. A small, droll man reminiscent of the comedian, Charles Chaplin, Rocard had been a distinguished member of the French Underground. For this war service, he received the "Légion d'Honneur" Medal from General De Gaulle – whom he personally disliked.

43. Two weeks into the conference, Lawrence developed ulcerative colitis and died back in Berkeley on 22 August 1958, age 57.

dent of the Bell Telephone Laboratories and an expert in both communication systems and the design of nuclear reactors.[44] None of the Western Experts were geophysicists, so they were backed up by an assortment of "spear-carriers", many of them seismologists.[45] Within the group was an equally large number of nuclear weapon specialists. These included Dr. Harold Brown, Deputy Director of the *Lawrence Radiation Laboratory* housed at Livermore, California (LRL), Doyle Northrop, Technical Director of AFTAC, the omnipresent Dr. Richard Latter of the RAND organization, and Dr. Wolfgang Panofsky, Director of Stanford University's High Energy Physics Laboratory.[46,47]

The Soviet Bloc delegation was equally high-powered. Its chairman was Dr. E. K. Federov, Director of the Institute of Applied Geophysics, Soviet Academy of Sciences. Other members included Dr. Nikolai N. Semenov, a chemist who won the Nobel Prize (chemistry) in 1956, I. P. Pasechnik, a nuclear explosion seismologist, and A. Zatopek, a noted Czech seismologist. Contending with the Soviet team was onerous and nervewracking. In commenting on the conference to a subcommittee of the Joint Committee on Atomic Energy, U.S. Congress, on 19 April 1960, the 32-year-old Dr. Harold Brown observed:

> During the Conference of Experts it became apparent, and was no surprise to anyone, the Soviets were in favor of as small a number of stations and as minimal accessibility by the control organization to the Soviet Union as they could get the Western delegation to agree to. Their action in this direction took the form of exaggeration of the capabilities of each of the methods for detection. By producing experimental evidence or more accurate theories than the Soviets, the Western delegation was able to get its evaluation of the individual capabilities agreed to by the Soviets, although the language was not always as clear as could have been hoped had the Soviets been operating on a purely technical basis.

Following 30 formal sessions ending on 21 August 1958, the Conference of Experts issued a report which concluded that it was

---

[44.] The Bell Telephone System also operated the major atomic weapons laboratory known as the Sandia Corporation at Albuquerque, New Mexico.

[45.] Seismic experts included Carl Romney and Norman Haskell of the U.S. Air Force, Frank Press of CalTech, Perry Byerly of the University of California (Berkeley), Jack Oliver of the Lamont Geological Observatory, Edward Bullard from Cambridge University, and Patrick Willmore of the University of Edinburgh.

[46.] Washington political tradition in the post-World War II period called for leading defense officials to be from the West Coast defense complex. Thus, Dr. Brown served as Director of Defense Research and Engineering (1962—1965), Secretary of the Air Force (1965—1969), and Secretary of Defense (1977—1981). He also was CalTech's President during 1969—1977.

[47.] Among the scientific professions, Dr. Panofsky was jocularly termed the "Bright Panofsky". As an undergraduate student at Princeton University, he received only "A" grades; his brother, Hans, while studying meteorology at Pennsylvania State University, also received straight "A" grades except for one "B" grade, thereby becoming known to his many friends as the "Dumb Panofsky"!

". . . technically feasible to establish with the capabilities and limitations indicated . . . a workable and effective control system to detect violations of an agreement on the worldwide suspension of nuclear weapons tests" (see Table 5.2).[48]

Within two months of the report's completion, two underground nuclear explosions took place at the NTS — Logan (five kt) and Blanca (19 kt). Both appeared to have generated far smaller seismic

TABLE 5.2

Major Components of Proposed Geneva Test Ban Monitoring System

Component	Remarks
Detection stations	
Continental	100 to 110 stations equipped with acoustic, seismic, and radio-signal detectors, plus radioactive debris-catchment gear. Spacing would range from 1,000 km in seismic areas to 1,700 km in non-seismic areas.
Islands or coastal points	Continental site equipment plus hydro-acoustic detectors on 20 large islands and 40 small islands. Spacing would be about 1,000 km in seismic areas and between 2,000 and 3,500 km in non-seismic areas.
Ships	Ten ships equipped to collect radioactive debris and detect hydroacoustic signals.
Aircraft	Regular oceanic flights to check levels of atmospheric radioactivity, plus special flights if indicated.
Central Control Organization	Internationally operated with capability to: (1) develop, test and accept technical equipment, plus specify site criteria; (2) conduct continuous observation program; (3) provide communications within network; (4) transport system personnel; (5) process and analyze rapidly data generated by the monitoring network; (6) carry out timely on-site inspection of unidentified events; (7) staff the control system with specialists; (8) assist in improving the system's scientific standards.

48. For the complete text of this report, see U.S. Department of State (1961) or Voss (1962).

signal levels than had been anticipated from the results of the Rainier shot. A special study panel immediately convened under AFTAC's Carl Romney, and on 5 January 1959, the *President's Science Advisory Committee* (PSAC) reported:

> ... about twice as many natural earthquakes equivalent to an underground explosion of a given yield as had been estimated by the Geneva Conference of Experts.... The total number of unidentified seismic events with energy equivalents larger than 5 kilotons might be increased 10 times over the number estimated by the experts.[49,50]

These new conclusions were then given to the reassembled Geneva Conference.[51] The Soviet reaction was immediate and negative, for their spokesman, Ambassador Tsarapkin, charged that these new findings were being introduced to allow an increased number of inspection parties:

> ... Automatic action will result in the frequent despatch of inspection groups when there is no need for them, and inspection then, to put it bluntly, becomes intelligence work. In other words, inspection group after inspection group will be roaming all over a country. You visualize sending out scores and even hundreds of inspection groups every year....

On 22 January, the Soviets further complicated the situation by claiming that ". . . these Powers (the United Kingdom and the USA) do not want an agreement to end atomic and hydrogen tests and are looking for an excuse to wreck the current Geneva talks".

By this time, President Eisenhower's mood was one of irritation and frustration with both the American seismological community and the Soviet leadership because he had been maneuvered into an informal nuclear test ban without having adequate monitoring systems in place. He had accomplished this by announcing on 22 August 1958, the day after the Conference of Experts had ended, that he welcomed their report and would initiate a year-long ban on American nuclear testing on 31 October 1958 if other nations would do the same and participate in treaty negotiations starting at

---

[49.] Besides Romney, the study panel included: Billy Brooks (Geotechnical Corporation), Perry Byerly (University of California), Frank Press (CalTech), Jack Oliver (Lamont Geological Observatory), James T. Wilson (University of Michigan), Hans Bethe (Cornell University), David Griggs (UCLA), Dean S. Carder (U.S. Coast and Geodetic Survey), Kenneth Street (Lawrence Radiation Laboratory), and Carson Mark (Los Alamos Scientific Laboratory).

[50.] Both the Soviet and Western Blocs used Gutenberg and Richter's *Seismicity of the Earth* (1954) to provide the statistical basis for arriving at the annual number of earthquakes of a given magnitude in various geographic locations.

[51.] These conclusions derived from having to downgrade the Richter magnitude of Rainier from 4.6 to 4.1 because Logan had to be assigned a magnitude of 4.4 and Blanca one of 4.6. Equally bad news came relative to the "first motion" identification criterion, for Logan's first motion was recorded as "down" or "dilational" at several seismograph stations, indicating a natural, rather than a man-made, origin.

Geneva on the same date. The British soon made a similar commitment, and the Soviets eventually agreed. Then started a race to get requisite nuclear tests off before the cessation date. On the day of Eisenhower's announcement, the British fired a kiloton-range device from a balloon tethered above Christmas Island, followed by two megaton devices and one kiloton device during September, 1958 (the Grapple series). The AEC moved up its Hardtack-II shot series by six months and executed 37 tests between 12 September and 30 October 1958, including the troublesome tests, Logan and Blanca. The Russians, somewhat irked, then fired two "catch-up" devices during the week that treaty negotiations started on schedule in Geneva. But after these shots, a self-imposed test ban went into effect that the three powers honored until September, 1961.[52]

The Berkner Panel on Seismic Improvement

Once it was publicly apparent that the scientific basis for a permanent nuclear test ban treaty was inadequate, the President's Science Advisor, Dr. Thomas Killian, established a "Panel on Seismic Improvement" to: (1) reanalyze the Hardtack-II seismic data, (2) determine what was needed to upgrade the proposed Geneva system for detecting and accurately identifying seismic events, natural and man-made, and (3) recommend a program in basic and applied research that would insure the optimal contribution by the seismological community to the detection problem. As was so typical of numerous other *ad hoc* review groups for nuclear test-ban controls, the new panel was a "mixed bag", including seismologists familiar with the problem, representatives of the nuclear weapons laboratories who stood the most to lose if the seismologists became too competent, and outside scientists of note. The chairman, Rear Admiral Lloyd Berkner, USNR, had splendid credentials, both scientifically and militarily, and had just finished directing the U.S. participation in International Geophysical Year.[53]

---

52. The only nuclear detonations during this test ban were by the French, who fired their first device on 13 February 1960 from a 106-m-high tower near Reggan in the Sahara desert, followed by three more small atmospheric tests during the remainder of 1960 and in early 1961.

53. An electrical engineer, during World War II Berkner had served as Head, Radar Section, Bureau of Naval Aeronautics and then as Director, Electronic Material Branch. Immediately after the war, he had been the Executive Secretary of the newly formed Defense Research and Development Board. Other panel members included five participants in the Geneva test-ban talks: Drs. Hans Bethe, John Tukey (a top expert in statistics), David Griggs, Frank Press, and Carl Romney. Additional seismological expertise was provided by Hugo Benioff of CalTech, Maurice Ewing of Lamont, and Jack Hamilton of the Geotechnical Corporation. Weapons specialists were Dr. Kenneth Street of Livermore and Dr.

In addition to its own resources, the Berkner Panel had available to it the work of an unadvertised Air Force seismic research panel which had existed between 1948 and 1960.[54,55] During this 12-year period, the Air Force research program had quietly spent about $8,000,000 on nuclear-detection research at universities, government laboratories, and private industrial laboratories (as an example, see Teledyne Geotech, Inc.). The Berkner Panel also found that the National Science Foundation was spending about $0.3 million per year in pure seismological research. Thus, despite the urgency of the test-ban-monitoring problem, the entire U.S. government was spending less than a million dollars per year to improve seismic technology, or less than one-tenth of the industrial research expenditure for improving exploration geophysics.[56] The Panel was also of the opinion that the Soviets ". . . in recent years had emphasized seismological research to such an extent that (they) enjoy a position superior in many respects to our own".

In view of all this, the Panel concluded:

> The strategic requirements of detection, together with the need to maintain a competitive position in one of the most significant fields in the earth sciences, make it a matter of urgency to institute a high level of support of seismological research.

Within American science, the euphemism, "institute a high level of support", actually meant, "Please send large amounts of money". As a starter, they suggested $22.8 million for the first year (see Fig. 5.4). Nearly half would be for a program of nuclear and high-explosive detonations which would be carefully monitored, one-fifth for system development, and the remainder for basic and applied research. The Panel also suggested that the entire program be centrally funded and directed, and that the high-priority systems development effort be assigned to one central laboratory familiar with designing large systems.

Julius P. Molnar, President of the Sandia Corporation. Outside specialists were Dr. Walter Munk of Scripps and Dr. John Gerrard of Texas Instruments, Inc.

54. The Geneva Conference of Experts had recommended an array of ten vertical-component seismographs which peaked at a response of one hertz for each seismic detection station; the basic Benioff device peaked at three hertz.

55. As of early 1960, the Air Force panel was chaired by Dr. Roland Beers, a founder of the Geotechnical Corporation. Members included such familiar names as Dr. Perry Byerly, Dr. Norman Haskell, Father Macelwane, Dr. Maurice Ewing, Drs. Benioff, Gutenberg, and Richter of CalTech and Dr. Louis Slichter of UCLA. Federal members included Drs. Frank Neumann and Dean Carder of the Coast and Geodetic Survey and Dr. John Adkins of the Office of Naval Research.

56. *Fortune* Magazine in April, 1960 estimated that the U.S. Atomic Energy Commission had spent $19 billion ($10^9$) during the first 14 years of its existence to develop and stockpile nuclear devices. In addition, the *Defence Atomic Support Agency* (DASA) often spent $10 million or more for instrumenting just one nuclear test.

Two-year program of Seismological Research
and Systems Development

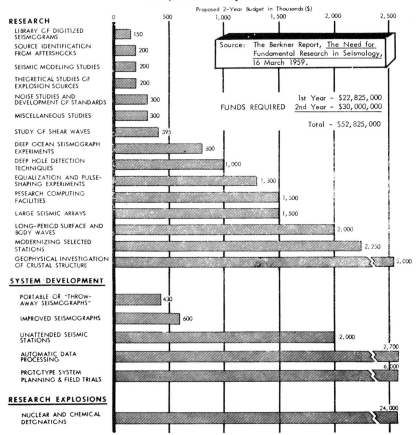

FIG. 5.4. Graphical presentation of research and development program proposed
by the Berkner Panel on Seismic Improvement.

In order to implement the recommendations of the Berkner Panel quickly, Dr. Killian called in the Deputy Secretary of Defense, Dr. Donald Quarles, and the new AEC Chairman, Mr John McCone, on 23 April 1959. At this meeting, it was decided that the Department of Defense (DOD) would pay for implementing the Berkner Report if the AEC would fund and conduct necessary nuclear and chemical explosions at the NTS. DOD initially assigned its responsibility internally to AFTAC. However, it soon became apparent that AFTAC, as a member of the intelligence community, was not a good choice for running a major national — and even international — research program. As a result, the Director of Defense Research and Engineering, Dr. Herbert York, established his own *ad hoc* Advisory Panel on Seismology under Dr. Frank Press.[57,58] The Panel met on 24 August 1959, and on 2 September, just a week later, the Secretary of Defense pulled back the AFTAC delegation and gave it to Dr. York's operating arm, the *A*dvanced *R*esearch *P*rojects *A*gency (ARPA). The only catch in this decision was that ARPA had been primarily a missile agency and had no geophysicists on board to direct the much needed program in seismological research.

Meteorological Fallout from Escalating Nuclear Test Programs

The spectacular air and water detonations of fission nuclear devices produced 760 kilotons of chemical yield equivalent during the first seven years that mankind tested such instruments of destruction. These tests also created 220,000 curies of radioactive strontium-90 (Sr^{90}) and caesium-137 (Cs^{137}), the two elements of greatest biomedical concern because of their long half-lives and close chemical similarity to normal body components (calcium and potassium), which made for relatively high rates of uptake by man and other animals (see Table 5.3).

With the advent of hydrogen fusion (H-bomb) testing in 1952, the equivalent chemical yield of nuclear explosions jumped to 90,700 kt during the next seven years, or by a factor of 119 times over that of the first seven years. The total amount of Sr^{90} and Cs^{137} released into the biosphere also jumped by a factor of 116 for the same period. Just how dangerous these — and the many other radioactive fallout products — were to man became the subject of great argument during the 1950s.[59] This debate continues even into the early 1980s, particularly as to whether there is a threshold

---

[57.] In 1952, Dr. York became the initial Director of LRL, followed by Dr. Edward Teller in 1958 when York moved to Washington.

[58.] Press's panel had many familiar names: Dr. Norman Haskell, Dr. Walter Munk, Dr. Jack Oliver (of Lamont Geological Observatory), Dr. James T. Wilson (of the University of Michigan), Dr. Hugo Benioff, and Dr. J. Gerrard of Texas Instruments. Only one member was from the geophysical industry — Dr. F. Gilman Blake, Supervisor of Geophysical Research, California Research Corporation, a subsidiary of the Standard Oil Company of California.

[59.] Every 20 kilotons of explosive energy in a fission detonation produces approximately a kilogram of a variety of radioactive materials composed of more than 100 isotopes with half-lives ranging from a fraction of a second to many years. Cs^{137}, for example, has a half-life of about 140 days and is responsible for most of the radiation load from fall-out which affects gonads, although it also distributes itself more or less evenly throughout all the body tissue. Sr^{90}, with a half-life of about 25 years, concentrates chiefly in the bones, thereby affecting the bone-marrow, particularly of children whose new bone is being formed continuously.

TABLE 5.3

*Nuclear Weapon Fission Yields and Production Rate of Key Radioactive
End-Products (Strontium-90 and Cesium-137) Between 1945 and 1958
(Inclusive)*

| Year | Fission yield (kilotons) | | Rate of production (megacuries)* (all nations) | |
	USA & United Kingdom**	USSR***	Strontium-90	Cesium-137
1945	60	0		
1946	40	0	0.08	0.14
1947	0	0		
1948–51	600	60		
1952–54	37,000	500	3.75	6.75
1955–56	9,200	4,000	1.32	2.38
1957–58	19,000	21,000	4.00	7.20
TOTAL	65,900	25,560	9.15	16.47

Source: Joint Committee on Atomic Energy, U.S. Congress, Hearings of 5–8 May 1959 (pages 23 and 1074).

*Megacurie may be defined as one million curies, a curie being that quantity of a radioactive nuclide disintegrating at the rate of 3.70×10^{10} atoms per second. In terms of a real distribution, it may be noted that every two megatons of fission energy creates about one millicurie of strontium-90 per square mile of the earth's surface (if it were distributed in such a fashion — which it is not).

**The initial nine British detonations in the kiloton or lower yield range took place at three different Australian test sites — Monte Bello Islands, Woomera, and Marlinga. These occurred between 3 October 1952 and 22 October 1956. Testing of the new thermonuclear devices had to be done in the Christmas Island area south of the Equator, a practice which started on 15 May 1957 with the Grapple series.

***Most of the Soviet testing was done at their Semipalatinsk site in East Kazakhstan, but the megaton shots were normally conducted at the Arctic Test Site in Novaya Zemlya.

of minimal radioactivity below which there is no noticeable effect on man as an individual, on the total population, and on the gene pool determining the character of future generations. Between 5 and 8 May 1959, the Special Subcommittee on Radiation of the U.S. Congress's Joint Committee on Atomic Energy held a comprehensive set of hearings on this subject, paying particular attention to a report on "The Effects of Atomic Radiation" issued the year before by the United Nations General Assembly. The high-quality UN report concluded that the estimated mean dosage to a person's bone marrow, calculated on a world-wide average and over a 70-year lifetime, would be that shown in Table 5.4.

Starting in 1952, this massive injection of radioactive material into the upper atmosphere (and in a few cases directly into the ocean) provided geophysical specialists, particularly those in meteorology, oceanography, geochemistry, and hydrology, with unparalleled

TABLE 5.4
*Estimated Mean Dose of Radioactivity to a Person's
Bone Marrow over a 70-year Lifetime*

Source of radioactivity	Dose (in Roentgen equivalents)
Natural sources (from cosmic rays and soil of residence)*	7
Man-made sources (such as medical X-rays and luminous watch dials)	7
Occupational exposure**	0.1–0.2
Environmental contamination (primarily nuclear weapon tests prior to 1958)***	0.1

*Natural external radiation from outside the earth increases from about 0.11 R per year at sea level to about 0.23 R per year at an elevation of about 1.8 km, i.e. to more than the global fallout level in 1958. Surface materials, because of their natural content of uranium, thorium, and potassium, contribute about 0.06 R per year through local gamma radiation.

**As of 1965 the U.S. record-holder for exposure to radioactivity without adverse affects may well have been Dr. Alvin C. Graves, Science Advisor, NTS Operations Group, who had received a dose between 300 and 500 R equivalent during a Los Alamos Laboratory accident in the 1944–1945 period. He died, age 56, of a heart attack on 29 July 1965 while vacationing at his cabin in Colorado. Among geophysicists, the oceanographer, Willard Bascom, probably holds the radiation record. During the U.S. nuclear testing program at the Pacific Test Site in 1952–1954, he repeatedly dove to plant and retrieve instruments in radioactive water, as well as monitored certain events by helicopter. This field work gave a whole body exposure of at least 50 R. In addition, he also "bought and paid for 36,000 R for cancer treatments over a period of three years. . . . Thus, my lifetime estimate of whole body radiation is of the order of 80 to 100 Roentgens. . . . My health, by the way, (at age 65) is excellent" (Bascom, 1981).

***With the U.S. Armed Forces experiencing approximately 1,400 fatalities per year from accidental causes, U.S. weapon testers were of the opinion that the world could probably tolerate ten megatons of atmospheric nuclear tests split between the nuclear powers.

tracers which traversed the entire globe. The most famous of these scientists was Dr. Willard F. Libby, who eventually received the Nobel Prize in chemistry (1960) for demonstrating that naturally occurring radioactive carbon-14 (C^{14}) could be used to date carbonaceous material back some 40 millennia. In 1954, Libby moved to Washington, D.C. to serve jointly as an AEC commissioner and a staff member of the Geophysical Laboratory, Carnegie Institution of Washington. Just a year before, Libby prodded the AEC into measuring the global extent and possible hazards of atomic fallout by means of Project Sunshine, a label which caused much criticism in later years. As originally conceived, the project was a secret study for determining the amount and type of fallout from a nuclear war. Two major contract laboratories were involved, Libby's former laboratory at the University of Chicago and Columbia University's five-year-old Lamont Geological Observatory, which although oceanographic in nature, had been measuring the rate at which strontium was deposited in human bones.[60]

As soon as the Pacific Bravo H-bomb test series raised a major international furor during the spring of 1954, Project Sunshine was declassified and quickly expanded to include hundreds of laboratories, hospitals, and sample collection points throughout the world. Even so, on 29 March 1959, the *New York Times,* under the by-line of meticulous John W. Finney, bannered:

FALL OUT SURVEY HELD MAKESHIFT — AEC SPURS URGENT REVIEW
OF SPLIT AUTHORITY AND "CONFUSION" IN RESEARCH

What was happening was that the AEC was finding it impossible to select a project director satisfactory to both Dr. Libby and to Dr. Charles Dunham, the AEC's Director of the Division of Biology and Medicine.[61] Despite the confusion, Libby was actively advising the world community just what they could expect relative to global fallout. For example, in a speech to the Swiss Academy of Sciences on 27 March 1958, Libby pointed out that the troposphere was the key medium for disseminating fallout from kiloton explosions, but that the stratosphere was both a major storage and a distribution facility for fallout from megaton detonations. For example, particulates in the troposphere would have an average residence time of only 30 days, although they could be washed out earlier if there were numerous water droplets in that part of the atmosphere. On the other hand, particulates in the stratosphere might stay up to ten years before returning to earth. In a somewhat "Strangelovian" tone, Libby then noted (Libby, 1958):[62]

60. As noted earlier, the Lamont Geological Observatory was always searching for new sources of funding. In this case, the facility's lead geochemist, Dr. J. Lawrence Kulp, discovered that the AEC would provide both a new laboratory building and extensive funding for subsidizing young geoscientists in the study and use of radioactive tracers, both natural and man-made (see vignette by M. Ewing). In 1964, however, Kulp left to become President of Isotopes, Inc. (a Teledyne company) and be active in the analysis of nuclear fission debris acquired by high-flying aircraft, including U-2s, of the U.S. Air Force.

61. During 1959, Libby moved to the West Coast and became Director, Institute of Geophysics and Planetary Physics, at the University of California (Los Angeles).

62. Many other fallout models were also postulated during this period. Two of the most attractive were developed by the U.S. Weather Bureau under L. Machta and by the Air Force Cambridge Research Center under E. A. Martell. Martell's model specifically called

The more recent data, particularly on bomb carbon-14, when taken together with the earlier data on bomb fission products and tritium, give us some confidence in our present understanding of the fallout mechanism. All of these observations and considerations afford unprecedented opportunities for the study of meteorology and geophysics, particularly in an international cooperative effort such as the International Geophysical Year.

The International Geophysical Year

Back in 1882–1883, there had been an "International Polar Year". This was then followed by another "Polar Year" on its 50th Anniversary. But then Dr. Lloyd Berkner had the audacity to propose in April, 1950 at a cocktail party hosted by Dr. James van Allen that there should be a third "Polar Year" just 25 years after the last one. The party's guest of honor, Dr. Sydney Chapman, the dean of British geomagneticians, quickly agreed for such a "Year" would permit probing the ionosphere with rocket-launched sensors during a period of maximum sun-spot activity.[63] Eventually Chapman and Berkner obtained the support of the International Council of Scientific Unions for this event. However, by mid-1952, it was obvious that the academies of science in the large nations were not enthusiastic about the concept. Rather than being discouraged, these two men then expanded their concept into a geophysical study of the entire earth. To do this, they formed a new group in 1953 labeled the *Comité Special de l'Année Géophysique Internationale* (CSAGI) and sent invitations to all of the world's nations to participate in an "*International Geophysical Year* (IGY)" that would last from 1 July 1957 to 31 December 1958.[64]

The concept of a common study of the earth for the benefit of all mankind during an "18-month year" quickly caught on in many places. As Professor Lawrence M. Gould has noted, "There was an almost magical quality about the IGY in the way it lowered international barriers and opened closed doors" (Gould, 1981). In the end, 67 nations joined the effort which rapidly became the largest, most complex, and costliest of any international scientific effort

---

for radioactive contamination from the Soviet H-bomb tests to remain in northern latitudes and to settle out in less than one year (Martell, 1959).

[63.] Chapman, like many other geophysicists, was noted for expending tremendous amounts of energy. In 1939, for example, he rode a bicycle between scientific meetings in Chicago, Illinois, and Washington, D.C. after noting that the two gatherings were spaced about a week apart. At the time, he was 51 years of age.

[64.] CSAGI's president, of course, was Dr. Chapman, and its vice-president, Dr. Berkner. The young physicist, Dr. Hugh Odishaw, performed much of the requisite managerial and report editing functions while posted as Executive Director, U.S. National Committee for the IGY.

ever undertaken (see Table 5.5). The Chapman—Berkner ground rules covering the types of earth science to be included were entirely catholic. Working groups soon existed for thirteen specialties: meteorology, oceanography, seismology, gravity, geomagnetism, aurora and airglow, ionosphere, solar activity, cosmic rays, nuclear radiation, glaciology, longitude and latitude, and last, but not least, because they came to dominate this international effort, rockets and satellites.

Some 40,000 scientists, volunteer satellite observers, and support technicians participated, manning 4,000 principal IGY observation points and an equal number of sub-stations.[65] For example, on Antarctica, the last major land-mass not owned by any nation, 12 countries established 48 new IGY stations.[66] The total cost of

TABLE 5.5
Countries Participating in the International Geophysical Year, 1957—1958

Argentina	France	Pakistan
Australia	German Democratic Republic	Panama
Austria	German Federal Republic	Peru
Belgium	Ghana	Philippines
Bolivia	Greece	Poland
Brazil	Guatemala	Portugal
Bulgaria	Hungary	Rhodesia and Nyasaland
Burma	Iceland	Romania
Canada	India	Spain
Ceylon	Indonesia	Sweden
Chile	Iran	Switzerland
Colombia	Israel	Thailand
Cuba	Italy	Tunisia
Czechoslovakia	Japan	U.S.S.R.
Denmark	Korea (Democratic People's	Union of South Africa
Dominican Republic	Republic)	United Kingdom
East Africa	Malaya	U.S.A.
Ecuador	Mexico	Uruguay
Egypt	Mongolia (People's Republic)	Venezuela
Ethiopia	Morocco	Vietnam (Democratic Republic)
Finland	Netherlands	Vietnam (Republic)
Formosa (National Republic	New Zealand	Yugoslavia
of China)*	Norway	

*Because of the continued IGY participation by Formosa, the IGY Committee of the People's Republic of China formed in September, 1955 withdrew in June, 1957.

65. The IGY spawned numerous publications. Among the more interesting were Sydney Chapman's *IGY — Year of Discovery* (1959), J. Tuzo Wilson's *The Year of the New Moons* (1961), and Walter Sullivan's *Assault on the Unknown — The International Geophysical Year* (1961). Technical results were summarized in *Annals of the International Geophysical Year* initiated by Pergamon Press in 1959 and in the two-volume *Research in Geophysics* issued by MIT Press to record the papers presented at UCLA's special conference on the IGY held in August, 1963.

66. Inasmuch as the Antarctic ice cap contains 90 percent of all the fresh water on earth,

Berkner's proposal, once it was fleshed out, has been estimated at 2×10^9. Within the United States, the National Science Foundation (NSF) allocated \$43.8 million for "IGY science", while the Department of Defense spent an equal amount, primarily for field logistics and operational support of stations in the Arctic and Antarctic. Most of the IGY funding was "extra money", with most all of it going to academic institutions in accordance with the NSF's standard policy.[67] For example, the \$2,000,000 IGY allocation to the academic oceanographic community raised its level of research funding by nearly 30 percent.

The biggest impact of the IGY on mankind was not new scientific discovery but rather the start of the "space race". Once the tiny "beep" of SPUTNIK-I, the initial Soviet IGY satellite, was heard circling the world on 4 October 1957, President Eisenhower had the delicate task of reassuring his nervous nation that a Soviet "basketball" in orbit was nothing to be alarmed about. But when the Navy's rocket failed to get the IGY VANGUARD satellite into orbit and the Russians launched an even larger IGY satellite with the dog, Laika, on board, news commentators did not treat the situation lightly. To dampen this concern, Eisenhower appointed Dr. James R. Killian, the President of MIT, as his Special Assistant for Science and Technology on 7 November 1957. However, it eventually took some 40×10^9 before America could counterbalance the lost lead in IGY satellite technology by finally landing man on the moon (20 July 1969) and by flying the first space shuttle, Columbia, capable of repeatedly orbiting the earth and then making a wheels-down landing (April, 1981).

Laying the Foundation for Earth-oriented Satellites

By the mid-1950s, meteorologists were indicating that reliable

---

hundreds of IGY scientists turned their attention to mapping this unique feature and determining its effect on the earth's weather. Rather than being the creator of major storms in the Southern Hemisphere, the ice cap proved to be the world's largest heat sink. The IGY also led to The Antarctic Treaty, signed on 1 December 1959, which recognized: ". . . that it is in the interest of all mankind that Antarctica shall continue forever to be used exclusively for peaceful purposes and shall not become the scene or object of international discord." Because it is international policy to keep this remote continent in the "deep-freeze" from the social and economic point-of-view, this book does not further pursue the large amount of geophysical work that is being accomplished there year in and year out.

67. The Antarctic research program subsequently developed into a continuing effort by the participating nations. For example, as of 1980, the U.S. National Science Foundation's program, which costs about \$63 million per year, permits nearly 2,000 people to work there during the Antarctic summer, although the number drops to less than 100 during the winter months.

three- to five-day weather forecasts could only be made if weather data were available from the entire Northern Hemisphere, and that comparable forecasts of greater duration required continuous global coverage. Such coverage, it appeared, could most readily be obtained by placing one or more meteorological observatories in outer space to observe and report back on the earth's atmospheric envelope in its totality. As far back as March, 1947, the U.S. Army and the Naval Research Laboratory (NRL), by employing a team effort at the White Sands Proving Ground, New Mexico, obtained primitive cloud photographs from altitudes up to 160 km by using German V-2 rockets and then the new Navy Viking rocket. However, these early pictures showed little more than some streaky cloud patterns and the earth's curvature.[68] In October, 1954, NRL, using two 16-mm motion-picture cameras aboard an Aerobee rocket, finally did acquire sufficient photographic coverage from a retrieved nose cone to assemble a mosaic of spiraling cloud bands associated with a tropical storm centered near Del Rio, Texas.

In the meantime, Drs. Stanley Greenfield and William Kellogg of the RAND Corporation had been secretly pointing out to the Air Force as early as August, 1951 the advantage of performing weather reconnaissance by satellite.[69] Then in 1953, while on duty with the U.S. Navy's research office in London, the 29-year-old Dr. S. Fred Singer presented a paper to the British Interplanetary Society which called for launching MOUSE (Minimal Orbital Unmanned Satellite of the Earth). Singer claimed such a device could determine upper atmospheric densities by changes in orbital parameters and also give some indication of gross weather patterns, atmospheric radiation budget, and ozone distribution.[70] Upon returning Stateside, Singer had sold, by the summer of 1955, Colonel William Davis, USAF, and Dr. Morton Alperin of the Air Force Office of Scientific Research (AFOSR) on the idea of a four-stage booster rocket, HARVIE (High Altitude Research Vehicle), that could lift a sensor package to the altitude of 6,000 km *vice* the conventional 100 km by launching from a large balloon floating at an altitude of 25 km. In mid-1956, it was evident that the American IGY politicians had frozen both the U.S. Army and Air Force missileers out of the VANGUARD

---

68. During January, 1949 Major (ultimately Brigadier General) Delmar Crowson, USAF, published a comprehensive paper, "Cloud Observations from Rockets" in the *Bulletin, American Meteorological Society*.

69. Greenfield became the first Assistant Administrator for Research of the U.S. Environmental Protection Agency in 1970, while Kellogg became Assistant Director, National Center for Atmospheric Research, in 1966.

70. Singer subsequently served as the first Chief of the U.S. Weather Bureau's Weather Satellite Center between 1962 and 1964.

satellite program by giving the booster responsibility to the Navy. In rebuttal, Brigadier General H. F. Gregory, USAF, Commander of AFOSR, began selling the modest HARVIE concept to his superiors on the basis that its payload might even reach the back side of the moon at a cost of only $719,000. To be sure, such a payload would be just a thin-walled geiger counter plus telemetry. Just the same, six unsuccessful launches were made from Eniwetok Atoll in September–October, 1957. Shortly after the second failure, the Soviets placed SPUTNIK-1 into orbit, and, as the AFOSR historian notes (Komons, 1966): "As a scientific probe, Project Farside was a failure. But as a political act, it was a disaster. The Air Force was made to look like a modern Don Quixote."[71]

After several other unsuccessful launches involving meteorological sensors, America's first successful meteorological package went into orbit aboard Explorer VII on 13 October 1959. Known as the "Suomi heat-budget package", its telemetered data indicated that heat loss at equatorial latitudes averaged about 25 per cent more than at polar latitudes. It also discovered non-periodic variations almost as large as the latitudinal variations across the United States; because the non-period fluctuations had durations of a few days, Suomi inferred that they might well be associated with major weather patterns. Nevertheless, such broad radiational patterns were of little use to the operational weather forecaster.[72]

Fortunately for the practical side of the house, the U.S. Air Force and the U.S. Army were found to be building duplicative reconnaissance satellites as of early 1958. The Air Force was allowed to proceed with its plans, but the Army was told that its payload must be converted into a "*T*elevision and *I*nfra-*R*ed *O*bservational *S*atellite" (TIROS) for meteorological purposes. Project direction was assigned to the DOD's *A*dvanced *R*esearch *P*rojects *A*gency (ARPA), and the payload was to be assembled by the Astro-Electronics Division, Radio Corporation of America (RCA) at Hightstown, New Jersey, not far from the Army's Signal Corps Research and Development Laboratory at Fort Monmouth.[73] Because ARPA did not have a meteorologist on its staff, it created a "Committee on Meteorology"

---

71. Because he was never shown any of Farside's telemetered data, Singer does not know even today whether the experiment found any of the trapped radiation so typical of the Van Allen radiation belt discovered by Explorer I in February, 1958.

72. Suomi was awarded in 1977 the National Medal of Science, the highest federal award of its type. The following year he became the initial director of the University of Wisconsin's Space Science and Engineering Center.

73. Fort Monmouth had assembled an excellent meteorological satellite team involving Dr. W. G. Stroud (team leader), Drs. W. Nordberg, Rudy Stampfl, R. A. Hanel, J. Licht, and Mr. Herbert Butler, with the latter serving as the federal contract representative to RCA.

to advise on the innumerable decisions involved in designing and launching such a sophisticated "bird".[74] At one time, it even appeared that the payload would weigh but ten kg for it would have to be launched by an Army Jupiter-C missile. However, skilful maneuvering by Warner broke loose an Air Force Thor-Able II booster configuration and thereby provided a payload of about 127 kg. With such a capability in hand, system planning and construction went steadily ahead towards a prospective launch date in early 1960.

Military Oceanography Thrives During "Peacetime"

In the early 1950s, one no longer had to own a yacht to be considered an "oceanographer", and the field was growing whether "old-timers" liked it or not. At the Naval Hydrographic Office, "professional oceanographers" increased in number from nine in 1947 to 47 in 1950 and then to 143 in 1952, even though the entire country had only 80 students in graduate oceanographic studies at a given time. However, not over five of these post-graduate oceanographers ever joined "Hydro" for the facility was still primarily a "chart factory" far from salt water. To fill the employee gap, Hydro aggressively recruited meteorologists, chemists, geologists, and marine biologists, preferably those with military experience, and provided a "school of hard knocks" that eventually developed many of Washington's top-ranking oceanographic bureaucrats over the next two decades.[75]

Much of Hydro's initial growth relative to the field of military oceanography was associated with turning out scores of data sheets, special studies, and literature surveys for intelligence purposes. When combat once again flared in Korea during 1950, Hydro's Division of Oceanography developed a comprehensive team of specialists capable of working around the clock to assemble within a five-day period a 50-page document incorporating all that was known "Stateside"

---

74. Committee co-chairmen were Roger Warner, ARPA, and William W. Kellogg, RAND; members were: Edward Cortwright, National Advisory Committee on Aeronautics (the forerunner to the National Aeronautics and Space Administration); Dr. Sigmund Fritz, U.S. Weather Bureau; Dr. William Widger, Jr., AFCRL; Dr. Ernst Stuhlinger, Army Ballistic Missile Agency; Dr. William Stroud, Army Signal Research and Development Laboratory, and Dr. Charles C. Bates, Office of the Chief of Naval Operations.

75. Such persons included John Lyman (Bureau of Commercial Fisheries), Robert Abel (Office of Sea Grant Programs), Murray Schefer (Bureau of Aeronautics), John Ropek (Bureau of Ordnance), Thomas Austin (National Environmental Data Service), Gilbert Jaffee (National Oceanographic Instrumentation Center), Wayne Magnitsky (Deputy Chief of Naval Operations — Development), Charles Bates (U.S. Coast Guard), and Boyd Olson (U.S. Naval Oceanographic Office).

about the hydrographic, tidal, astronomic, climatic and oceanographic conditions of a potential assault area.[76] However, field activities gradually began to receive support as well. Once the "*A*coustic-*M*eteorological-*O*ceanographic *S*urvey" (AMOS) cruises were under way in the North Atlantic Ocean under the technical direction of T. S. Austin, it became possible to start an "Inshore Survey Program" for sea approaches to U.S. harbors as part of a large mine defense effort.[77] In addition, Dr. Richard Fleming, as the Division's Director, believed there was nothing wrong with the third ocean bordering the United States — the Arctic Ocean — except that it was solid on top. Although no funds were specifically assigned to polar research, by 1948 efforts were being made to improve ice-observing techniques by studying aerial photographs of the ice pack obtained during Air Weather Service Ptarmigan flights to and from the North Pole.

The accelerated build-up of the Canadian—United States polar weather station program during the summer of 1950 also made it possible for Hydro's Murray Schefer to fly ice-reconnaissance missions with the *R*oyal *C*anadian *A*ir *F*orce (RCAF). However, Schefer's role was secondary, and Colonel Charles Hubbard, USAF (Reserve), soon pulled rank on Schefer as to which of the two would accompany a Canadian air-supply mission from Thule, Greenland, to the station at Alert. It proved to be an excellent flight to miss, for a supply parachute tangled in the tail assembly and the plane crashed, killing all on board. Nevertheless, Schefer acquired an able understudy, Henry Kaminski, for the 1951 ice season, and the two were soon deeply involved in America's largest polar effort yet — Operation Bluejay.[78] The project's goal was to build a complete Strategic Air Command base at Thule, Greenland, during just one summer. Unfortunately, Schefer and Kaminski were not assigned ice-reconnaissance aircraft of their own and had to make do by flying aboard cargo planes high above the cloud cover. As a result, the task force with thousands of engineering troops and highly paid civilian technicians on board

---

[76.] Both Marine Corps and Naval headquarters had disbanded their beach intelligence groups once World War II ended. In the case of the Navy, not much was lost for the captain in charge once told Bates, "I have the best possible group for this work — all my officers are from Harvard, the best school in the country; in addition, they are all English majors and therefore know how to write!"

[77.] Because Hydro's Chief Engineer, Guillermo Medina, was opposed to "in-house" oceanographic research, Hydro's first oceanographic laboratory was labeled a "Chemical Weighing Room".

[78.] Kaminiski was killed north of Thule when a Navy P4Y2 ice reconnaissance aircraft crashed on 16 April 1954. Once a year, the U.S. Naval Oceanographic Office makes an award in his name for the best in-house work of science accomplished during the prior year.

lost its way in the ice maze to the south of Thule and squandered weeks of valuable time. To keep this from happening again, the cost-conscious *Military Sea Transport Service* (MSTS) asked Hydro to expand its ice-observation program by using specially assigned Canadian and U.S. Navy patrol aircraft during the upcoming 1952 Arctic supply season.[79] Moreover, MSTS asked for short- and long-range ice forecasts as well, and the first experimental long-range ice forecast was accordingly issued during March, 1952.

Sufficient external pressure now existed for establishing an "Experimental Oceanographic Central" at Hydro, and by mid-1953, the Central, in conjunction with the Naval Weather Central at Kodiak, Alaska, was issuing ice forecasts of one, five, and 30 days duration, plus a 150-day seasonal outlook (Bates, 1958). Although Hydro ice forecasts utilized Russian and Austrian formulae for calculating heat loss from a water column, the Central's oceanographers introduced many improvements keyed to the availability of electronic computers and extensive environmental data from the field (Lee and Simpson, 1954).[80,81] End products included rate of freeze-up in coastal harbors, the position of the offshore ice boundary, and thickness of seasonal pack ice to a few cm. The total operation soon proved so successful that the Office of the Chief of Naval Operations formalized the effort in January, 1954:

> Aerial ice reconnaissance has been vital to the success of these (Arctic) operations in that it has enabled the task force commanders to utilize all possible time available for movement of ships during the very short shipping season. The aircraft, ice observers, and the ice forecasts required for support of these operations have been provided on a temporary basis from year to year. Experience has proved the desirability of providing for these services on a routine and continuing basis in order that firm planning can be completed by responsible commanders well in advance. . . . To ensure the availability of ice observers when needed, the present civilian ice observers, who have been serving on a volunteer basis, must be replaced by naval personnel.[82]

[79.] Eight hundred seventy-five hours of ice-reconnaissance flights were flown in 1952 by Lancasters of Canada's 404 Maritime Squadron and by P4Y2's of U.S. Naval Patrol Squadron 23.

[80.] Mr. Jack J. Schule headed the Oceanographic Prediction Branch. Co-workers included Major A. R. Gordon, Jr., Air Weather Service liaison to Hydro, John F. Ropek, Howard French, Edward Corton, George Hansen, Ray McGough, Walter Wittman, Richard James, Owen Lee, and James Winchester.

[81.] Several Canadians, and particularly Moira Dunbar of the Arctic Section, Defence Research Board, and Wing Commander K. R. Greenaway, RCAF, were also active at this time compiling a comprehensive pattern of ice distribution in the Arctic waters of Canada (Dunbar, 1954).

[82.] Additional Hydro civilian ice observers who participated without any kind of flight or danger pay included: 1951 — Robert Parker; 1952 — Frank Gaetano, William Petrie, and Robert Parker, and 1953 — William Butscher, Edward Corton, Edward Goldstein, James Harrington, Charles Kammer, Owen Lee, Ray McGough, A. Russell Mooney, Edward Nadeau, Bernard Swanson, James Tapager, and Walter Wittman.

International recognition came quickly. By late 1953, the British Government asked that Hydro's ice forecasts be supplied to the large British trawling fleet operating west of Central Greenland in the Disko Bay region. During the summer of 1954, tall, bouncy Commodore Robertson, RCN, was able to circumnavigate North America during just one season aboard the icebreaker, HCMS *Labrador,* and soon reported that his ship's forward progress was speeded up by as much as 50 percent when using ice reconnaissance and forecasting services in the Canadian Arctic. As a consequence, Dr. Andrew C. Thomson, Chief, Meteorological Service of Canada, assigned five of his staff to an ice-forecasting course offered by Hydro in early 1955.[83] Domestic pressure also continued to build for additional ice-information services, for Canada and the United States decided in late 1954 to build a *Distant Early Warning* (DEW) Line of some 50 radar stations across the top of the North American Continent during the next two summer shipping seasons.[84]

An Armada of 126 ships (52 Canadian and U.S. Navy fleet vessels and 74 MSTS cargo vessels) was assigned to polar supply operations for the summer of 1955. An elaborate ice-information program was also activated which resulted in the expenditure of 4,436 hours of aerial ice observation and preparation of 969 ice forecasts, over half of which were issued in the field. This massive ice observation and forecasting effort proved most timely, for the permanent polar pack ice stubbornly refused to move back from the north-eastern coast of Alaska during mid-August. With 42 ships working as far as 2,700 km east of Point Barrow, the situation looked so grave relative to extrication before permanent freeze-up that VADM F. C. Denebrink and Dr. Bates briefed the U.S. Armed Forces Policy Council concerning the situation. Fortunately, on 5 September 1955, the ice forecasters finally spotted the long-awaited shift in storm tracks and so advised VADM Denebrink and his task-force commander, RADM "Bull" Towner, USN.[85] By 16 September, all 42 vessels had delivered their

---

[83.] A Canadian student, William E. Markham, eventually became Head, Ice Branch, a position he still held in 1981. Another student, Captain Wilhelm Pedersen, of the Danish shipping firm, J. Lauritzen, proceeded to East Greenland to assist in shipping ore out of the mine at Mestersvig. On 1 January 1960, the firm's *Helga Dan* opened the first regularly scheduled winter shipping service between Europe and Quebec City, Canada.

[84.] A leading proponent for building the DEW Line was RADM L. Berkner. He had directed a study project, East River, in 1951 and learned that the United States had only ten minutes of guaranteed warning prior to a Soviet bomber attack.

[85.] Denebrink subsequently wrote Bates: "If I live to be a million, I will never forget the treacherous days of September, 1955 when we were trying to find one nice little storm to blow the damned ice pack away from the Alaskan coast!" What worried everyone was that the operation was utilizing most of the *Landing Ship* (*Tanks*) (LST's) of the Amphibious Force, Pacific at a time when they might be needed to countermand a move by Communist China against Taiwan.

cargo and were west of Point Barrow with only four ships having suffered severe ice damage.[86]

VADM Denebrink and his deputy for the North Atlantic Area, RADM "Dutch" Will, USN, now thought so highly of oceanographic forecasting that they approved a joint MSTS-Hydro "Wave Forecast Ship Routing Project" based on a scheme in Richard James's doctoral thesis, "Application of Wave Forecasts to Marine Nagivation" (James, 1957).[87] In the experiment, MSTS notified Dr. James whenever two of its identical cargo ships were about to start on a trans-oceanic crossing using the same departure date. One of the two masters was asked to follow James' predicted optimal sailing track, while the other master proceeded on his own best judgment in time-honored fashion. James' results were startling.[88] VADM Will, USN, who had taken over the MSTS command by then advised the Chief of Naval Operations on 22 June 1957:

> During this test twenty-two MSTS ships were routed in the Atlantic Ocean with indications of a saving of 44 hours maximum. Ten MSTS ships were routed in the Pacific Ocean with a maximum voyage saving of 64 hours. These preliminary results have . . . demonstrated a marked saving in steaming distance, time, weather damage and passenger comfort. . . .
>
> The past success of this method of ship routing warrants continued operation. It is requested that a permanent ship routing system be established utilizing the methods developed by this test, and that it be made available to all MSTS and fleet shipping. . . .[89]

This was subsequently done and proved to be an extremely cost-effective program. For example, between October, 1956 and April, 1959, 919 vessels were assigned optimal routes with out-of-pocket costs to Hydro of $44,600, yet MSTS estimated its ships saved $1.38 million in operating and fuel costs for an effectiveness to cost ratio of 30:1.

---

86. When an oil company consortium mounted a massive marine supply effort to Prudhoe Bay in 1975, they had to contend with an equally severe ice season. President Gerald Ford finally intervened and authorized Coast Guard ice-breaker support. Even so, a dozen tugs and barges had to winter over.

87. James built on age-old concepts such as those of Benjamin Franklin and Matthew F. Maury, updated by the subsequent introduction of "pressure pattern navigation" for aircraft (Hall, 1945).

88. James eventually won three incentive awards totaling $10,000 from the U.S. Navy for his innovation, the highest payment ever made to a staff member during Hydro's 150 years of existence. A B-17 pilot during World War II, James had acquired extensive industrial weather forecasting experience before getting his Ph.D. Yet his starting salary at Hydro was but $6,235 a year when hired in May, 1955.

89. The first MSTS shipload of Hungarian refugees brought to the United States after the unsuccessful revolt of late 1956 was not weather routed. Handling a load of seasick passengers incapable of communicating because of language barriers soon caused MSTS to phone James: "Weather route all refugee ships — and we don't care where you send them as long as they don't get seasick!"

Geophysical Research Penetrates the Polar Basin

Dr. A. P. Crary, one of Professor Maurice Ewing's earliest graduate assistants, was the first American to make extensive geophysical measurements seaward of the Arctic Ocean's continental shelf. A member of the Geophysical Directorate at the Air Force Cambridge Research Laboratories, Crary became a guest of Colonel Bernt Balchen's 10th Search and Rescue Squadron when it established a small base on the pack ice 100 km north of Barter Island, Alaska, during February, 1951. Although the ice broke up within a month, Crary persuaded squadron pilots to land him at six different points within the Polar Basin where he measured ocean depth, gravity, and water temperature and current profiles.[90] At the same time, Dr. Gordon Lill and others back at the Office of Naval Research finally had the Naval air arm mounting Operation Skijump out of Point Barrow, Alaska. This effort utilized three aircraft — an R4D (a Dakota in British parlance) equipped with skis, oceanographic winch, gravity meter, and a seismic water-depth indicator, and two P2V escort aircraft, also outfitted with skis and auxiliary fuel pumps capable of refilling the R4D's 550-gallon supplementary fuel tank when alongside on the ice pack.

During 1951, the P2V's of Operation Skijump practiced landing on an artificial ice strip built by repeated pumping of seawater onto a revetted area of natural ice. Beginning in March, 1952, the Skijump team, with oceanographers John Holmes and Val Worthington of Woods Hole and William Kielhorn of ONR on board, started making landings and oceanographic observations as far as 500 km offshore. But when they attempted to occupy oceanographic stations over half-way to the North Pole, everything came apart. The R4D cracked up, one of the P2V escorts similarly cracked up while landing to assist, and things ended up in a bad way (see vignette by Holmes).

Even as the Navy faded from the scene, the Air Force's polar observation program was accelerating. As far back as July, 1946, the 46th Strategic Reconnaissance Squadron (Photographic) based at Ladd Air Force Base, Fairbanks, Alaska, had been regularly flying over the Arctic Ocean. On 14 August 1946, a radar observer was surprised to see an enormous mass of ice, vastly thicker than pack ice, showing up on his radar scope less than 500 km north of Point Barrow. Labeled "Target X", the feature was promptly classified

---

[90.] Crary was noted for being a prodigious field worker. For example, after a hard morning sledging along the polar shore ice, he stopped his group for lunch. He then suggested, "Take five deep breaths!" They did. Next came the order, "Let's get going!" "But how about lunch?" And then came Crary's punch-line, "You just had it!"

"Top Secret". Although tabular icebergs were common in Antarctic waters, they were unknown in the Polar Basin. Those in the know continued to puzzle over its origin, for it moved steadily along and when last sighted on 6 October 1949, was over 2,400 km from the initial position. Once energetic Colonel Joseph Fletcher, USAF, a combination seismologist—meteorologist—pilot, took command of the 375th Weather Reconnaissance Squadron at Eilesen Air Force Base, Alaska, in May, 1950, he started a serious effort to find more "ice-islands". On 20 July 1950, a second was spotted and labeled "T-2" (with Target-X becoming T-1). Nine days later, T-3 was found by radar and visually verified on 24 August 1950.[91] But no matter how hard Fletcher and his squadron reconnaissance officer, Major Lawrence S. Koenig, USAF, searched, they could find no more. While convalescing from an operation for appendicitis so inept that he became conscious in the middle of the surgery, Fletcher read the book, *Nearest the Pole,* and found RADM Robert Perry, USN, noting that the ice shelf of northern Ellesmere Island had rolling terrain that he called "ice waves". A special overflight during March, 1951 verified Perry's story. A check flight on 1 August 1951 then found T-1, which had been "lost" for two years, residing 60 km north of Ward Hunt Island. With this much evidence at hand, Fletcher and Koenig concluded that the ice islands broke off long ago from the Ellesmere Ice Shelf and had since been drifting in a giant closed gyre between the North Pole and the North American Continent at the rate of about 1.7 km/day (see Fig. 5.5) (Fletcher and Koenig, 1951).[92]

Because the "ice islands" were larger than many strategic islands in the Pacific Ocean, Fletcher was anxious to occupy the "high ground" in the best military tradition. At first, no one would listen to his suggestion that the Air Force should establish a base on T-3, which had drifted to within 150 km of the North Pole. After being shown photographs of the Papanin drift station set up at the North Pole in 1937 by the Russians using four-engine aircraft, the Commander, Alaskan Air Command — Major General W. D. Olds, USAF — finally agreed.[93] On 19 March 1952, with the surface air temperature at $-45°C$ and three rescue aircraft overhead, a ski-equipped C-47

[91.] Radar dimensions of the islands were: T-1 — 24 × 29 km, T-2 — 27 × 29 km, and T-3 — 7 × 14 km. All had long, parallel corrugations and an associated drainage system.

[92.] Upon being apprised of Fletcher's work, the noted Canadian polar navigator, Wing Commander Keith L. Greenaway, RCAF, observed that he had originally spotted and photographed T-3 as "land ice" on 27 April 1947 as it lay near Isachsen Island, three years before the Americans found it deep in the Polar Basin.

[93.] Papanian's party was extricated by a Soviet icebreaker nine months later while drifting to the south off eastern Greenland.

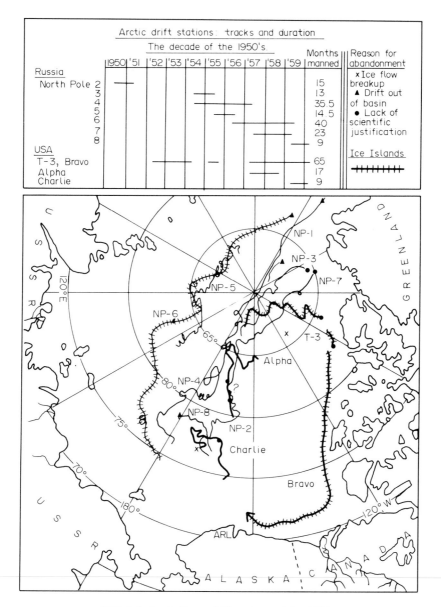

FIG. 5.5. Arctic Ocean drift stations of the 1950s.

aircraft of the 10th Search and Rescue Squadron set down on T-3 after a very bumpy landing.[94] Fletcher stayed behind with Captain Brinegar, USAF, and Dr. Kaare Rodahl, a Norwegian-born polar expert, and made camp on what became known as "Fletcher's Ice Island". Within two weeks, "Bert" Crary was on board to serve as "Chief Scientist", and Captain Green, USAF, as "Communicator". Before everything was shipshape, one of Skijump's P2V's landed during the first week of April and was unable to depart because of engine trouble. As a result, the Air Force permanent party was outnumbered eleven to five even before two weeks had gone by! However, island activities gradually smoothed out, meteorological reports were regularly transmitted every six hours, and the island found to be 48 m thick and ablating at a rate of 0.3 m per year.

Arctic drift stations became quite common during the 1950s, the Soviet Union establishing seven and the United States, with the help of Canada, three. During the International Geophysical Year, Ice Island T-3 was relabeled Ice Station Bravo to coincide with two new Canadian–U.S. ice stations, Alpha and Charlie. Although the ice islands proved to be excellent oceanographic platforms because of stability and weight-carrying capability, they also had some disadvantages. They drifted slowly but certainly away from their aerial supply bases. During the ice melt season of June to September, it also was impossible to land aircraft on them, although a blimp and an icebreaker did visit T-3.[95]

Geophysical exploration of the Polar Basin by submarine had been a long time scientific goal even though Sir Hubert Wilkins and Sir Harald Sverdrup almost killed themselves trying to do this back in 1931. In 1946, Dr. Waldo Lyon, an underwater physicist at the Naval Electronics Laboratory, began pushing the Navy hard relative to taking submarines under the polar ice pack.[96] When the world's first nuclear submarine, the USS *Nautilus,* came on the line in early 1954, the Navy would not let her locate and surface in polynyas (open areas in the ice pack) until the autumn of 1957. This field

94. General Olds served as co-pilot in the landing. Upon catching up to Fletcher in the snow, his first remarks are said to have been: "I don't see how any man can live on this thing. We're going to take off now — if we can!"

95. When a University of Wisconsin geophysicist, Jerome Hirschman, fell ill on *A*rctic *R*esearch *L*aboratory *I*ce *I*sland (ARLIS) II in July, 1963, the best that could be done was to have a medical doctor fly overhead and discuss his symptoms. He suddenly died nineteen days later while eating a can of peaches. Similarly, when a cook went berserk on T-3 in the late 1960s and killed the camp commander, the murderer's evacuation had to wait for freeze-up.

96. Dr. Lyon received the "President's Award for Distinguished Federal Civilian Service" from John F. Kennedy in 1962 for ". . . persevering in believing that transarctic submarine navigation could become a reality . . .".

trial went rather poorly, and the gyrocompasses also acted up because of the high latitude. However, in 1958, both the USS *Nautilus* and the USS *Skate* were outfitted with inertial navigational devices from the defunct Navaho missile and directed to proceed to the North Pole. Sailing from the West Coast with Dr. Lyon on board, the *Nautilus* (Commander William Anderson in charge) squirmed past the down-hanging ice ridges of the shallow Chuckchi Sea, quickly crossed the North Pole, and reached the open North Atlantic Ocean on 3 August 1958.

Skate sailed from Groton, Connecticut, on 30 July 1958, but found it impossible to surface at the North Pole on 12 August because of heavy ice cover.[97] Upon turning to her second task — a visit to Ice Station Alpha 500 km away she learned that Alpha did not know its current position because of cloud cover. However, after acquisition of a line of bearing by use of a radio direction finder, *Skate* finally homed on the sound of an outboard motor driving a skiff about a polyna next to Alpha's camp. Upon surfacing to periscope depth, her captain, Commander James Calvert, USN, found the U.S. flag flying just a few meters away.[98] After an overnight stay involving the exchange of polar bear steaks for ice cream, *Skate* submerged in order to reach the North Pole and to survey, under Art Malloy's direction, the Lomonosov Submarine Ridge as she proceeded south to open water. By 20 August, *Skate* had transited 3,800 km, acquired 652,000 soundings, and discovered enough underwater peaks, ridges, and canyons to name one after each member of the 97-man crew. But the *Skate* still had one other important task to do — return to the Arctic Ocean during the dead of winter and punch her way upwards to the surface by brute force at the North Pole. This she did on 17 March 1959. While there, her captain conducted the funeral service for Sir Hubert Wilkins, who had died shortly after visiting the *Skate* the previous October. Once the ensigns of Australia, the United Kingdom, and the United States had been raised, Commander Calvert spoke:

> "Unto Almighty God we commend
> The soul of our brother departed,
> and we commit his ashes to the deep."

---

97. Ranking scientists on board were Eugene La Fond, a Naval Electronics Laboratory oceanographer, and Walter Wittmann and Arthur Malloy of Hydro, the former being an ice specialist and the latter a hydrographer.

98. Alpha's resident party numbered 29 — 16 scientists and 13 U.S. Air Force support personnel. Chief Scientist was the University of Washington's likeable Dr. J. Norbert Untersteiner, who would eventually conceive, promote, and direct the complex *Arctic Ice Dynamics Joint Experiment* (AIDJEX) during 1975 and 1976.

A rifle squad then fired the last salute to this valiant explorer who had tried and failed to take an earlier *Nautilus* to the North Pole. This done, his ashes were finally committed to the polar void as both he and his wife had wished.

The Oceanographers also Prefer Large-scale Projects

The early 1950s saw the last of the cruises typical of the 19th-century voyages of discovery that covered all facets of natural history and lasted a year or more. After World War II ended, the noted geographer, Professor Hans Pettersson, was first off the mark and mounted a full-scale global expedition aboard the Swedish training ship, *Albatross,* during 1947 and 1948. Using the newly invented Kullenberg coring device capable of obtaining cores up to 20 m in length from clay deposits in the deep sea bed, Pettersson probed the sea floor in a totally unique fashion. He also was careful to measure the thickness of the soft sediment by exploding small chemical charges in the water and recording the ensuing reflection from the base of the sediment layer. As a result, the *Albatross's* collection of cores and sedimentary thickness measurements was the most extensive and best ever taken and provided the basis for some of the first definitive geo-chronology regarding the recent geological history of the deep sea floor during the recent Ice Age. The *Albatross* also used echo sounders capable of obtaining continuous depth profiles in waters up to seven km deep. The Danes, too, mounted a post-war, world-wide deep sea expedition aboard the *Galathea* with the famed biologist, Anton Brunn, as Chief Scientist. As might be expected, the *Galathea* specialized in collecting animals from the sea, although deep-sea dredging was also carried out in the classical 1872 manner of HMS *Challenger.* In the meantime, the British continued regular voyages of the RRS *Discovery* to the Southern Ocean for whale-behavior studies, which they supplemented by the regular taking of depth soundings and oceanographic stations.

This period also saw the resolution of the age-old question of where was the deepest spot in the ocean — and how deep was it? Between May, 1950 and October, 1952, HMS *Challenger,* a British Admiralty survey vessel under the command of Captain Stephen Ritchie, RN, and with Dr. Gaskell on board as Chief Scientist, steamed 120,000 km through all of the world's major oceans.[99] In addition to surveying coastal waters off Bermuda and Jamaica, *Challenger*

---

[99] After serving during 1965 as an Aide to the Queen, Ritchie was promoted to Rear Admiral and named Director, Hydrographic Department a year later.

conducted one-ship seismic refraction surveys on a global basis using techniques and equipment by Maurice Hill and John Swallow at Cambridge University (Gaskell, 1954).[100] Secondary observational programs included streaming the bathythermograph whenever possible and making traditional scientific studies of sea birds and plankton.

However, Captain Ritchie's personal interest lay in sounding the deeper parts of the ocean. Thus, during a bathymetric survey of the Mariana Trench 100 km east of Guam while in search of a suitable seismic refraction measurement site, those in the ship's charting room were thrilled when the echosounder began recording travel times of 14.5 seconds which corresponded to a depth in excess of 10,725 m, the previous world's record for oceanic depth. This new value was immediately verified by the old fashioned method of lowering a chunk of iron on piano wire borrowed from the taut-wire survey gear used to measure horizontal distances. Then, following a visit to New Zealand, *Challenger* returned and surveyed the area for some kilometers around to reconfirm its original finding and published the results in the very first issue of *Deep Sea Research* (Gaskell, Swallow, and Ritchie, 1953). Russian survey ships subsequently visited the same region and claimed depths greater than the ultimate 10,807 m depth measured by *Challenger*. However, they derived their values by using a velocity value for sound in sea water higher than did the *Challenger* scientists, and the British authors still believe that the value published in 1953 remains valid.

By the time that the IGY appeared, European oceanography had recovered from the difficult war years. As a result, arrangements were made by the International Council for Exploration of the Sea in Copenhagen, Denmark, to mount, in the summer of 1958, the first major synoptic survey of the complex boundary zone between the warm, highly saline extension of the Gulf Stream and the cold, low salinity water of polar origin lying between Greenland and northern Europe. Because the mixing of these two water masses was known to impact on both European weather and fisheries, 22 research vessels from eight nations participated in the effort (see Table 5.6). Although the total effort was not truly simultaneous, it did acquire vertical temperature and salinity profiles at more than 3,000 stations and delineated the same complex Gulf Stream pattern

---

[100.] The technique used radio sonobuoys to record seismic events generated by 23 kg explosive charges dropped by *Challenger* up to 32 km from the detector. Because of the relative short distances between charge and detector, the seismic wave trains observed did not normally penetrate to the 8.0 km/sec velocity rock lying below the Mohorovičić discontinuity. However, this deficiency was counterbalanced by the wide scope of the measurement program utilizing identical apparatus throughout.

TABLE 5.6

Research Ships Participating in the IGY's Atlantic Polar Front Survey
(1958)

Country	Ship	Country	Ship
Canada	*Investigator* *Sackville*	Iceland	*Aegir*
Denmark	*Dana*	Norway	*Helland Hansen* *Johan Hjort* *G.O. Sars*
France	*Aventure* *Calypso* *Eveillé* *Emporté* *Le Verrier*	United Kingdom	*Discovery II* *Explorer* *Sarsia* *Vidal*
Federal Republic of Germany	*Anton Dohrn* *Gauss*	USSR	*Ekvator* *M. Lomonosov* *Poljarnik* *Sebastopol*

of meanders and isolated warm and cold core eddies that had been found off eastern North America by Operation Cabot nine years before (Roll, 1979).

The 1950s also saw the introduction of a powerful new method for measuring deep ocean currents directly, a particularly desirable thing to do for most strong surface currents died out rather rapidly at depths of 200 m or so. Dr. John C. Swallow, the inventor of the technique, had, in fact, been on the global cruise of HMS *Challenger*. By 1954, he had earned his doctoral degree from Cambridge University and joined the NIO staff. Soon he was at sea with the "Swallow float", a long, neutrally buoyant cylinder designed to float at any depth and made trackable by being equipped with an ultra-sonic "beeper" that could be monitored by hydrophones aboard the mother ship at the surface.[101] Because of the simplicity and ruggedness of design, the unique float was used extensively for tracking intermediate depth currents, as well as determining the rate at which bottom water moved towards the tropics from the polar regions, an important bit of information for those who dump radioactive waste on the deep sea floor (Swallow, 1957).

[101.] During the 1970s, oceanographers at the Woods Hole Oceanographic Institution, particularly Drs. William Schmitz, Jr. and James F. Price, expanded on the Swallow technique by building neutrally buoyant, pinger-equipped floats which intermittently emitted "chirps" that were then monitored by hydrophones placed in the SOFAR acoustic channel hundreds of km distant. If the "chirp" time is known, only two acoustic reception points are needed, as off Bermuda and Nova Scotia, to track a field of five floats (Richardson *et al.*, 1981).

During the 1950s, the American oceanographic research effort was dominated by the policies and funding of ONR's Geophysics Branch.[102] The Branch dispersed about $85 million of federal research funds during the decade, most of it directed towards expanding nine existing oceanographic laboratories and starting two new oceanographic departments (University of Washington in 1951 and Oregon State University in 1959).[103] The laboratories' seagoing fleet was also expanded quickly and cheaply by arranging for the institutions to obtain World War II surplus vessels either on bailment or outright donation. As time went on, the oceanographic laboratories acquired additional funding from other sources, including the Navy's Bureau of Ships, the Atomic Energy Commission, the Bureau of Commercial Fisheries, and the National Science Foundation, which had been formed in the ONR pattern during 1950.

At long last, American oceanographers could go to sea on a large scale. For example, during the decade, Scripps sent out fourteen oceanographic expeditions, each assigned an intriguing and often exotic name.[104] On just one cruise — the Mid-pac Expedition of mid-1950 under the direction of Roger Revelle and utilizing the *Horizon* of Scripps and the PCE(R)-587 of the U.S. Naval Electronics Laboratory — a bevy of major discoveries were made. Using Professor Bullard's probe, Revelle and a graduate student, Arthur Maxwell, found that outward heat flow is the same in ocean basins as on continents. The cruise also located a new submarine mountain range (labeled the "Mid-Pacific Mountains"), learned that bottom sediments of the Pacific Basin were typically but a few hundred meters thick, and traced, under the direction of an apprentice marine geologist, Dr. H. William Menard, a vast fracture zone which lay seaward of Cape Mendocino which, as the cruise went on, proved to be the longest fault trace on earth.[105]

102. The "Golden Rule" operates in Washington, D.C. as follows: "He who hath the gold setteth the rules!"

103. Although the University of Washington had been one of the original recipients of Rockefeller Foundation oceanographic funds in 1931, the school's collegial approach fell apart during World War II, and the "Oceanographic Building" was shunted off to other uses. Later on, Drs. Richard Fleming and Clifford Barnes returned from Hydro and started a conventional "Department of Oceanography" in 1951. Finally, in 1981, the University began finalizing arrangements to consolidate its varied oceanographic and fishery groups, plus the Applied Physics Laboratory, the Office of Sea Grant Programs and the Institute for Marine Studies, into a "College of Ocean and Fishery Science".

104. The cruises were: Northern Holiday, Shellback, Capricorn, Transpac, Midpac, Norpac, Chubasco, Eastropic, Chinook, Acapulco Trench, Mukluk, Downwind, Dolphin, Doldrums, and Naga.

105. Revelle became Acting Director of Scripps in 1950 and the permanent Director a year later. In 1964, he left for Harvard University to become the Saltonstall Professor for Population Studies, but the University of California was thoughtful enough to name one of

Although ONR's Geophysics Branch had only a half-dozen scientists, they proved to be the most innovative group of oceanographers around. For example, in 1958, they contracted with General Dynamics, Inc. to design a radio telemetry system capable of transmitting between permanently anchored oceanographic buoys on the high seas and a shore station. This effort quickly led to "monster buoys" now used to replace manned lightships as well as to monitor environmental conditions offshore. At Dr. Robert Dietz's suggestion, they arranged for the Navy to purchase the bathyscaph, *Trieste,* from its Swiss developers, Professor Auguste Piccard and his son, Jacques, for $267,000. After a new pressure cabin was added, young Piccard and Lt. Donald Walsh, USN, dove in it to the deepest part of the Mariana Trench on 23 January 1960.[106] By making the world's deepest possible dive early in the project, the "ocean spectacular" phase was quickly out of the way.[107] Consequently most of the effort went into developing sensing instruments, lights, and cameras needed to make this prototype "*Deep Sea Research Vehicle*" (DSRV) a reliable working tool for oceanographers.

The third major technical innovation by ONR's geophysicists was that of deep-sea drilling. Because so many research proposals did not fit into any proper scientific category, Gordon Lill and Carl Alexis whimsically formed the "*American Miscellaneous Soci*ety"

---

the first subordinate campuses at the University of California (San Diego) after him. Between 1959 and 1965, Maxwell was Chief of ONR's Geophysics Branch and then, in 1971, Provost of the Woods Hole Oceanographic Institution. In addition, he was the American Geophysical Union's President during 1976—1978. Menard did equally well, serving as Director of the prestigious U.S. Geological Survey from 1977 to 1981.

106. Being Swiss-built, the *Trieste*'s initial diving trials had been in the fresh water of Lake Geneva. As a result, when the *Trieste* finally reached the deepest part of her dive, the depth gauge read considerably more than the 10,807-m depth measured by HMS *Challenger*. Once the reading was corrected for the greater density of sea water over that of fresh water, the data agreed and provided a check on the value of sound velocity in sea water at great depths.

107. After Lt. Walsh advised Admiral Arleigh ("31 Knot") Burke, USN, the Chief of Naval Operations, in January, 1959 about what the ONR team proposed do, Burke exploded:

"Look! I'm taking a lot of heat for the Navy 'space spectaculars' down at Cape Canaveral that are not so spectacular. They get all those cameras lined up, and all they shoot is a big splash from another MARINER. Now you squirrels come in here with the grade of Lieutenant and tell me we ought to dive to the deepest part of the ocean. I haven't heard of your program before today. Now I've heard more than I want to know! How many people are in your program? How many military officers?"

Walsh replied, "Just two officers — Lieutenant Shoemaker and myself". So Burke asked, "How many of you dive in it at one time?" Walsh noted, "Just one. The other stays topside to supervise." Burke then gave this sage bit of advice, "Young man, I want to tell you something! If one of you doesn't make it back to the surface, the guy who stayed down is the lucky one for I'm going to have the other guy's 'balls'!" (Walsh, 1980).

(AMSOC) in 1952.[108] While the oceanographic "in group" was having a wine breakfast during April, 1957 on Walter Munk's patio, RADM Harry Hess, USNR (still a Princeton professor), and Munk reaffirmed that all research proposals up for consideration were unimaginative. Why not do something truly revolutionary? Munk then proposed drilling a hole somewhere in the ocean to the earth's mantle just below the Mohorovičić ("Moho") Discontinuity in order to determine what the mantle consisted of.[109] Hess was intrigued and soon showed up at ONR with a proposal to do so. ONR informally transmitted the proposal over to the National Science Foundation, which turned it down as being "crazy". But Hess was not to be denied, for soon the National Research Council (NRC), the operating arm of the National Academy of Sciences, created a formally constituted "AMSOC Committee on the Moho".[110]

This time around, the NSF saw fit to transmit $15,000 to the NRC for a planning study directed by brilliant, individualistic Willard F. Bascom.[111] By 1959 Bascom and his sponsors agreed that the way to go was the construction of a vast floating drilling platform dynamically positioned by placing servo-controlled propellers at each corner. As a consequence, in March, 1961 the first offshore hole was drilled in this manner 40 km west of Scripps in 943 m of water using the barge, *Cuss-1,* held in position by four 200-horsepower

---

108. Because oceanographers were not eligible to win the Nobel Prize, AMSOC's principal property became a mangy stuffed seagull given regularly to some member of the clan as the "Albatross Award of the Year". Oceanographers forced to carry the scruffy bird home by 1960 included Roger Revelle, Walter Munk, Sir Edward Bullard, John Knauss, Victor Vacquier, and John Swallow.

109. Among the various opinions held as to what the mantle, with its compressional velocity of 8 km/sec, might consist of were stony meteoritic material, peridotite, and eclogite, the latter being but a phase change from basalt because of high temperature and pressure at depth. Based on seismic refraction work at sea, Maurice Ewing was claiming by 1949 that the "Moho" interface lay about 13 km below the sea floor while the comparable depth under continental land masses was 45 km. A year later, Gaskell found some evidence that the "Moho" occurred but 11 km below the sea surface at a point 430 km north of Bermuda during the global cruise of HMS *Challenger*. The peridotite school of thought also argued that fragments of this rock found in the diamond-bearing kimberlite volcanic plugs had actually been transported from the upper mantle to near the surface, thereby negating the need to drill at all.

110. The Committee's chairman was Gordon Lill, with additional membership by Roger Revelle (Scripps), Harry Ladd, Joshua Tracey and William Rubey (U.S. Geological Survey), Harry Hess (Princeton), Arthur Maxwell (ONR) and, in self-defense, Maurice Ewing (Lamont).

111. Trained as a mining engineer at the Colorado School of Mines, Bascom served during 1954–1956 as Technical Director, Committee on Civil Defense, National Research Council where he worked with such luminaries as Nelson Rockefeller and Henry Kissinger. The thought at this time was that the nation should create a large number of underground facilities in case there should be an exchange of deadly nuclear-tipped rockets.

outboard motors operated from a central joystick.[112] On 2 April 1961, *Cuss-1* pulled a long core of basalt from what marine seismologists termed the "second layer", a strong reflecting horizon 200 m below the sea floor in a water depth of 3,537 m off Baja California. President John F. Kennedy immediately telegraphed a commendation to the National Academy of Sciences, noting that this action was ". . . a remarkable achievement and a historic landmark in our scientific and engineering progress" (Bascom, 1961).[113]

The fourth major action by ONR's geophysicists during the late 1950s was strictly political. In 1956, Lill and his fellow Potomac River oceanographers created an "Informal Coordinating Committee on Oceanography". After a couple of meetings, it was decided to fund a "Topic-X" type committee on oceanography at the National Academy of Sciences. Although the Academy's third committee on this subject was not as distinguished as the comparable Rossby—Berkner "Panel on Meteorology", the oceanographers did complete by 1959 a 12-chapter document, *Oceanography: 1960 to 1970,* with all the proper justification for an expanded research program, including military urgency, intellectual challenge, and inadequacy of facilities.[114] The price tag for overcoming these deficiencies was spelled out in detail, including the need for 70 new research and survey ships at a probable cost of $213 million.[115] ONR also succeeded in having Admiral Burke endorse its own report, *"Ten* Years in *Oc*eanography" (TENOC), which called for additional naval ship and laboratory construction, plus incrementally adding $2 million more each year to ONR's oceanographic research funding level of $7.6 million (as of 1959). Looking back two decades later, Lill stated at the Second International Congress on the History of Oceanography held in Edinburgh, Scotland (Lill, 1971):

[112.] Operated by Global Marine Exploration Company, *Cuss-1* was named for its original sponsors — the Continental, Union, Shell and Superior Oil Companies — who used it to drill while anchored in place. At that time, offshore drilling was limited to water depths of 100 m or less, while on-land drilling had penetrated to a depth of just seven km.

[113.] After the completion of the *Cuss-1* drilling exercise, Bascom eventually entered private industry and formed two companies of his own — Ocean Science and Engineering, Inc. and Seafinders, Inc.

[114.] Dr. Harrison Brown, a geochemist from CalTech, was named chairman. Members were: Maurice Ewing (Lamont), Columbus Iselin (Woods Hole), Roger Revelle (Scripps), Athelstan Spilhaus (University of Minnesota), Fritz Koczy (University of Miami), Gordon Riley (Yale University), Milner Schaefer (Inter-American Tropical Tuna Commission), Colin Pittendrigh (Princeton University), and Sumner Pike, a former Vice President, Standard Oil Company (New Jersey) and also former Commissioner, U.S. Atomic Energy Commission. Mr. Richard Vetter served as the extremely able Executive Secretary of the group (nicknamed "NASCO") for the next quarter of a century.

[115.] The Committee also proposed spending an additional $438.9 million over the ten-year period for new shore facilities, equipment, education and general operational expenses.

It was, in fact, our intention to get the Congress interested in the problems of the oceans, and we succeeded. There are some who still contend that we made a mistake of throwing oceanography into the political arena, but one wonders where we would be today if we hadn't done so.

Lill's attitude was typical of that held by most of the world's leading oceanographers. Consequently, of all the geophysical sciences, oceanography was the first to be featured in an "International Congress" held in the massive, beautiful General Assembly Hall of the United Nations in New York City between 31 August and 7 September 1959.[116] This congress, which drew 800 delegates from 38 countries, had five themes — History of the Oceans, Populations of the Sea, Nature of the Deep Sea, Boundaries of the Sea, and Cycles of Organic and Inorganic Substances in the Sea. Ironically, some of the low-key salesmanship generated by this conference came back to haunt the academic oceanographers. Once the diplomats and international lawyers became convinced that there were "untold treasures of the sea", they began talking of the "Global Commons" and eventually started the long-running "Law of the Sea" conferences.[117]

[116.] Dr. Roger Revelle served as Congress President, and Dr. Mary Sears edited the "Preprint Volume".

[117.] The Third Law of the Sea Conference convened in 1973 and still had not agreed on an updated maritime treaty as of 1981. However, there was agreement that oceanographic research ships of one nation could not explore the continental shelf of another without obtaining diplomatic permission and honoring certain other restrictions.

CHAPTER 6

Science in Government and Government in Science—the 1960s

We are rapidly moving into an age in which research is the intellectual equivalent of sex. Philosophy, literature, history, poetry are all *passé* — it is research that is admired, promoted, and discussed. In point of fact, as with sex, in certain circles it is more talked about than performed since to be really satisfactory its practice requires not only interest and pleasure but hard work. There are other analogies such as the fact that the probability of something productive resulting is closer to 1% than to 100%.

Anecdote by Dr. Lloyd Berkner at Honors Night, 40th Anniversary Meeting of the American Meteorological Society on 20 January 1960

On a scale never attempted until this decade, scientists and governments in many countries are joining hands across national boundaries to serve the entire human community. Their example should be instructive for all of us as we pursue lasting peace and order for our world.

President of the United States Richard Nixon, 13 March 1969

The Exuberant 1960s

By the decade of the 1960s, the public was certain that science and technology could provide the answers to most of the world's problems. To be sure, between 1900 and 1964, the average life expectancy of man had risen from 49 years to 70 years. In addition, he enjoyed a broad spectrum of new wonders — transistor radios, television, jet aircraft, artificial fibers, plastics, unusually high crop yields, and cheap electricity produced from nuclear power and low-priced oil from overseas. Consequently, scientific leaders were proud of their accomplishments and the presence of strong public support. As Dr. Donald Hornig, Science Advisor to President Nixon, observed (Hornig, 1965):

> . . . Our federal expenditures (for research and development) have increased some two hundred times since the beginning of World War II. Put differently, the size of the effort has doubled every seven years, measured in dollars, or every twelve years, measured in numbers of people engaged.[1]

1. What Hornig did not say was that something had to change; otherwise the federal

160

Dr. Andreas Rechnitzer and Jacques Piccard await pick-up from *Trieste I* after setting world-record dive of 5,670 meters during November, 1959. (U.S. Navy photograph.)

Captain Don Walsh, USN, who commanded *Trieste I* during her dive to the deepest part of the ocean in December, 1959. (Courtesy of Don Walsh.)

The giant Scotland Sea Buoy off Sandy Hook, New Jersey, in late 1967; it not only replaces a manned lightship but also transmits meteorological and oceanographic data. (U.S. Coast Guard photograph.)

Aerial view of Ice Island T-3 showing summer-
time drainage pattern filled with snow. Upper
third of picture is typical Arctic ice pack.
(Courtesy of Alaskan Air Command.)

Colonel Joseph Fletcher and M. Brinegar outside
their recently established camp on Ice Island
T-3. (Courtesy Military Airlift Command.)

Dr. A. P. Crary operating a gravity meter
on Ice Island T-3, April, 1951. (Courtesy
Military Airlift Command.)

Waldo Lyons monitoring overhead ice
canopy while transiting below the Polar
Pack. (U.S. Navy photograph.)

Commander J. R. Calvert, USN, reads funeral service for Sir Hubert Wilkins at the North Pole on 17 March 1959.

U.S. Coast Guard icebreaker *Storis* attempting to free a bargeload of seismic equipment caught in the Polar Pack 25 kilometers southwest of Point Barrow, Alaska.

Winterized surveyor and shothole drilling rig. (Courtesy of Western Geophysical Company.)

Seismological observatory at La Paz, Bolivia, one of the participants in the Worldwide Standardized Network. (Courtesy USGS.)

Louis Pakiser and Ronald Willden inspecting a seismic record beside a USGS crustal studies recording van. (Courtesy L. Pakiser.)

Dr. William C. Kellogg, Co-chairman of ARPA's Technical Steering Group for development of the TIROS meteorological satellite. (Courtesy National Center of Atmospheric Research.)

Dr. S. Fred Singer, initial Director of the U.S. Weather Bureau's Satellite Meteorology Section. (Courtesy S. F. Singer.)

Set of instruments utilized in the Worldwide Standardized Seismograph Network. (Courtesy of Teledyne-Geotech.)

Professor Yves Rocard at his field station, Chateau Renardière, in the Loire River Valley.

Downhole seismometer used in the USGS Seismic Research Observatory Program. (Courtesy USGS.)

Professor Bruce Bolt, Director, Seismological Station, University of California (Berkeley). (Courtesy B. Bolt.)

Seismic Advisory Panel of AFOSR, plus guests, about to overfly Alaska's seismic belt during April, 1964. Clockwise from lower left: L/C R. Smith, W. J. Best and L. A. Wood (all AFOSR); Eduard Berg (U. of Alaska); C. Bates (ARPA); L/C J. Brennan (AFOSR): John Steinhardt, Carnegie Institution; Roy Hansen (NSF): Jack Oliver (Lamont): USGS Representative;James T. Wilson (U. of Michigan); D. Carder (US C&GS); William Stauder (St. Louis U.), and Sidney Kaufman (Shell Development). (Courtesy Alaskan Air Command.)

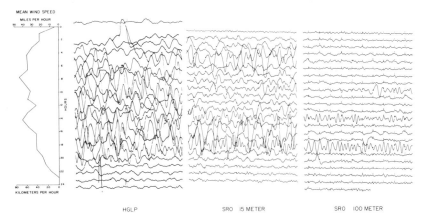

Comparison of seismic records from high-gain, long-period surface seismometer (on left) and from Geotech broad-band downhole seismometers operating at 15 and 100 meter depths. (Courtesy USGS.)

Mark III version of Texas Instruments' deep-sea seismometer *circa* 1964. (Courtesy of Texas Instruments.)

"Relax, Bing! Sooner or later, one of them is bound to come up with something!" (Sketch by Priscilla Robinson based on an idea expressed by Major Robert Harris, USAF.)

Howard Simons, Managing Editor of the *Washington Post*. (Courtesy *The Washington Post*.)

Dedication Day (6 April 1963) at the Berkner seismic array facility, Tonto National Forest, Payson, Arizona.

Dr. Robert Frosch experimenting with a "forked stick" at Flers, Normandy, during November, 1963.

For scientists and engineers, the 1960s were truly a "Golden Age". If you conceived something new, its development was likely to be funded.

Bad News and Good News for the Geophysical Industry

In sharp contrast to Dr. Hornig's optimistic report cited above, the economic and technical problems facing the petroleum industry in the late 1950s continued right into the 1960s. In fact, the Society of Exploration Geophysicists (SEG) lost membership (under three percent) between 1960 and 1963. Within the United States, the number of seismic land crews fell from about 380 in 1960 to a low of 190 in 1970. Oil from Venezuela, North Africa, and the Middle East, was so cheap that it could be landed for approximately $2.88 a barrel in New York City. Fortunately, some companies believed in continuing the search for new oil and gas. Consequently, rank wildcat wells drilled on geophysical evidence turned up two startling finds during the decade — one on land at Prudhoe Bay, Alaska, in sight of the Arctic Ocean and the other in the stormy North Sea.

The Prudhoe Bay story goes back to 1944 when seismic reflection crews of the United Geophysical Company began exploring the desolate tundra of the U.S. Navy's Petroleum Reserve Number Four to the landward of Point Barrow, the northernmost point in contiguous North America. Getting seismic equipment to the region was a continuing battle with the harsh environment, and the subsequent field work was the same. During the first five years, just enough oil was found at Umiat near the forbidding Brooks Range and natural gas near Barrow Village to keep interest high in further exploration of the deep Tertiary basin which lay between the mountain range and the Arctic Ocean 300 km to the north.[2] Seismic surveys for industry began in 1962 first in the foothills and then in the coastal region. By 1963, a Western Geophysical Company crew working for the Sinclair Oil and Gas Company far to the east of the Naval Petroleum Reserve mapped some westerly and southwesterly dips in the Prudhoe Bay area where regional geology indicated dips should be to the east and north.[3] Sinclair's Chief Geophysicist,

---

expenditures for research and development would exceed the nation's gross national product by the mid-1980s.

[2.] The South Barrow gas accumulation proved to be tricky to drill for it had formed in a geological structure created in part by a meteorite's impact (approximately the Cretaceous period) onto the crest of a great domal structure, the Barrow Arch (Lyons, 1981).

[3.] Western was using single-fold recording with a split spread of seismometers centered on a single down-hole explosive source. Despite the simplicity of the technique, good data were obtained because of few problems with weathering in the surface layer (actually the

Paul L. Lyons, quickly worked up the seismic picture (Lyons, 1981).[4] His initial map drawn in 1963 indicated the tremendous thinning of the Kingak shale (Jurassic period) over a massive dome and correctly indicated the structure's crest. Subsequent shooting confirmed this interpretation and added a major fault that created part of the easterly dip.

When the state during July, 1965 put up the Prudhoe Bay area for lease during its second sale of North Slope oil and gas rights, Lyons and Sinclair's Exploration Manager, L. L. Ware, could not convince their New York headquarters to bid as high as $100 an acre for this remote caribou pasture. After all, such a structure would have to produce at least 300 million barrels of oil to be commercial![5] But their exploration partner, British Petroleum Ltd. (BP), was willing to gamble and bid this amount. Even so, the Atlantic Refining Company, Richfield Oil Corporation, and Humble Oil and Refining Company outbid BP for choice leases over the structure's crest, leaving BP with extensive down-flank acreage. During the next four years, nine wells were drilled in the general area, including one by BP towards the south end of the structure which found only oil and gas shows. Another was a Sinclair well on the crest of the large Colville "High" 67 km west of Prudhoe Bay that had to be abandoned during 1966 because it would have made only 200 to 500 barrels of oil per day. Finally, in 1967, the now combined firm of Atlantic-Richfield (ARCO) faced a major decision along with its partner, Humble. Do we leave this expensive area or drill one more hole?[6] Because the costs of departure and of drilling one more hole were about the same, ARCO went ahead to drill their Prudhoe Bay State Number 1 well which came in from the Triassic Sadlerochit formation making 2,025 barrels of oil per day, plus large amounts of gas, during January, 1968. By June, a confirmatory well 10 km to the southeast along the structure's crest indicated a like amount of oil and gas, and the word was soon out that a giant oil and gas field was in the making.[7] As drilling expanded, BP found the structure to be a

---

soil was permanently frozen to a depth of as much as 300 m) and the presence of highly reflective interfaces at depth (Specht *et al.*, 1981).

4. Lyons served as President of the SEG during 1954—1955. When Atlantic Richfield took over Sinclair in 1969, Lyons remained in Tulsa and became an independent oil operator with his son in the firm of Lyons and Lyons, Inc.

5. Sinclair's advertising symbol was the dinosaur; the firm disappeared along with its symbol when most of Sinclair's resources were merged into ARCO during 1969.

6. As of that time, ARCO and Humble were reputed to have spent about five million dollars exploring for oil within Alaska's North Slope region. Sinclair's exploration costs in the same area were recovered many times over by sales of seismic records within the firm's exploration library.

7. Prudhoe Bay is the largest oil field in North America, containing over 10 billion

vast structural-stratigraphic trap with the BP acreage containing a majority of the oil, while ARCO and Humble had mainly purchased the field's massive gas cap (Jamison *et al.*, 1980).

During the same time frame as the Alaskan oil excitement, the international oil industry was also drilling for the purpose of determining whether any of the numerous geophysical anomalies mapped below the North Sea might have commercial value. The existence of small oil fields in Holland and the British Midlands, together with many boreholes drilled to prove coal reserves near the coastline in the southern part of the North Sea, had earlier suggested that a thick sequence of sedimentary rocks existed in the region. However, the indications also were that this sequence would contain only uneconomic small-scale petroleum deposits. Then, in 1959, Shell and Standard Oil (New Jersey) (now Exxon) drilled a geophysical anomaly near Groningen, Holland, and discovered the largest gas field ever found in Europe, containing 39×10^{12} cubic feet of natural gas. This discovery changed the North Sea prospects overnight. By this time, the 1958 Geneva Convention on the Outer Continental Shelf provided general guidelines for dividing up this area between the seven nations bordering this body of water — the United Kingdom, Norway, Denmark, Federal Republic of Germany, the Netherlands, Belgium, and France. Finally, in 1964, Great Britain began to award offshore drilling licenses, followed by Norway a year later.

In 1965, BP brought in the first offshore gas discovery — the West Sole Field — from a fixed platform. By using the semi-submersible platform, Ocean Viking, the Phillips—Norway Group was able to drill in deeper water and, by 1968, found the barely commercial Cod gas condensate field in Norwegian waters.[8] By this time, 20,000 km of seismic reflection lines criss-crossed the North Sea and indicated several promising regions.[9] During 1969, the Phillips—Norway Group, with some time left on its Ocean Viking charter, decided to drill an attractive but complex geophysical anomaly very near the center of

---

(10×10^9) barrels of producible petroleum, as well as some 26 trillion (10×10^{12}) cubic feet of natural gas. These figures do not include additional oil reserves in the Pennsylvanian—Mississippian Lisburne group below the Sadlerochit or in the Early Cretaceous Kuparuk River formation to the immediate west of the Prudhoe Bay field where some 1.2 billion barrels of recoverable oil makes the Kuparuk field also one of the 10 largest oil discoveries in the United States.

8. The Group consisted of: Phillips (37%), Norske Fina (30%), Norsk AGIP (13%), Norsk Hydro Produksjon (7%), Elf Norge (5%), Total Marine Norge (4%), and Aquitaine Norge (2.7%). Percentages of less than one percent each were held by Eurafrep Norge, Coparex Norge, and Cofranord.

9. Much of this seismic coverage was duplicative, being shot north—south and east—west along lines specified as to exact "minutes" of latitude and longitude.

the North Sea. Drilling proved difficult, and after encountering both an oil show and a well "kick" in the Miocene carbonate at a depth of 1.68 km, it was decided to start over again a km to the south-southeast (Bark and Thomas, 1980). This time the drill penetrated all the way to the 200-m-thick Danian limestone (a rock comparable to that in the "White Cliffs of Dover") at a depth of 3.0 km and found it saturated with oil and gas and capable of producing at the rate of 10,500 barrels a day. Despite the highly rigorous physical environment, such a flow rate was highly profitable and led the way to development of the Ekofisk field with ultimate recoverable reserves in excess of 3×10^9 barrels of oil.[10] Since then, as many as 201 wells have been drilled yearly in the North Sea, for the region between Scotland and Norway has been found to contain some excellent oil and gas-bearing structures. By the time a decade had passed, the North Sea offshore play was employing, directly or indirectly, 100,000 persons, of which 12,500 worked offshore. Today the British side of the North Sea alone produces as much oil (1.8 million barrels daily) as did Kuwait before its recent production cut-back, and Great Britain actually achieved net oil sufficiency in oil supply during 1981, a situation likely to last for the rest of the decade. In addition, the southern sector of the North Sea proved to be gas-prone and now plays a major role in providing Great Britain with natural gas.

The Geophysical Industry Shifts to Digital Recording and Processing

As with most technological revolutions, full-scale acceptance of the capabilities and merits of digital recording and processing by the oil exploration industry did not come easily.[11] No one climbed to the mountain top and announced that the millennia had arrived in exploration geophysics. To be sure, as early as 1960, Carl Savit of Western Geophysical had published in *Geophysics* seismic traces which showed reflection amplitudes to be greater if oil or gas was present within a given geological horizon (Savit, 1960).[12] Yet, even as MIT's Geophysical Analysis Group closed up shop during 1957, stored-

---

10. Field development costs are tremendous, however. For example, the new production platform, Statfjord "B", to be installed in a field 580 km north of Ekofisk, will cost at least 1.8×10^9 when finally operating in January, 1983.

11. Many technical firms consider there is more money to be made by being "second" in a new field, letting the pace-setter make the mistakes up front and then, hopefully, lose out in the stretch.

12. It was more than a decade later before an oil company official could say, when reviewing the results of a lease sale, "We entered bids on nine gas fields and thirteen oil fields", rather than bids on 22 geophysical prospects (Taylor, 1974).

program, general-purpose digital computers had become commonly available. Although very slow and small in capacity by today's standards, these machines permitted simpler types of seismic analyses to be performed on a practical basis (see Fig. 6.1). Examples of such analyses included the computation of synthetic seismograms from continuous velocity logs, calculation of the noise attenuation characteristic to be derived from various types of geophone arrays, and mathematical filtering using convolution and deconvolution concepts (Rice, 1962).

Nevertheless, the geophysical industry preferred to improve what it understood best — analog magnetic tape recording and play-back systems. One service company, GSI Inc., however, did take the lead in developing seismic digital recording and processing for regular

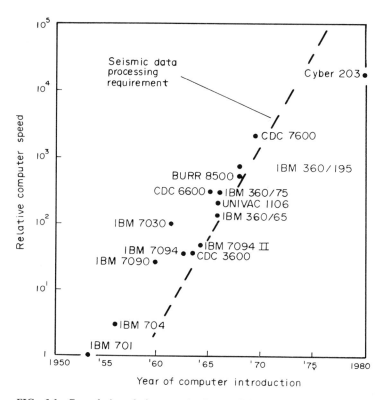

FIG. 6.1. Growth in relative speed of top-of-the-line computers versus demand for enhanced seismic data processing rates. For reference purposes, note the CDC-7600 performs about 10 million operations per second. (Diagram based on information supplied by George P. Cressman and Carl H. Savit.)

commercial use — and that due in large part to Dr. Kenneth E. Burg's enthusiasm for the approach and GSI's aggressive company policy that stressed revolutionary rather than evolutionary technical developments. GSI's first vacuum-tube digital field system was field tested in 1958–1959.[13] At the same time, a compatible "DARC" computer was developed and programmed to handle various correction and filtering processes for comparison with analog results. This led to the design of GSI's "9000" digital field recording system plus solicitation of financial and consulting support from Mobil Oil Co., and Texaco in exchange for a two-year proprietary information and field experience lead on the rest of the industry. Because of Dr. Elmer Eisner's enthusiasm for the approach, Texaco received the first set of digital field recording equipment in 1962.

Meanwhile, Texas Instruments had been working on the new, all-transistorized *T*exas *I*nstruments *A*utomatic *C*omputer (TIAC) specifically designed to process seismic data from GSI's digital recording equipment. After initial TIAC tests, production of multiple systems began in late 1962. The first transistorized "TI-10,000" digital field recording system was then completed for company use during 1963. When this new technology was announced to the trade press a year later, many GSI crews already had the "TI-10,000 System" and five GSI data-processing centers the TIAC computer-processor. Such a transition represented a very large investment and complete commitment by GSI to digital technology. After the announcement, GSI leased digitally-equipped crews to various companies for experimental tests of this new approach in areas chosen by the client. Differences of opinion sometimes arose as to how best to use this new technology; despite this and the fact that the improvements achieved were usually not very dramatic, the results still typically left the client convinced that a new revolution in seismic data recording and processing was indeed under way.[14] GSI was not content with just doing digital work on land and soon initiated digital recording in the North Sea (1964), offshore California (1965), the Gulf Coast and other coastal areas

---

13. Much of this field testing was done by three young geophysicists who had earned doctorate degrees at MIT — Mark K. Smith, Milo Backus, and Lawrence Strickland.

14. Instead of developing its own special computer system, Western Geophysical, under the astute direction of Carl Savit, chose to go with standard digital recording and processing systems. Their first digital recording was on board ship in November, 1964 using equipment built for them by Redcor using standard, seven-track, gapped tape with a density of 556 bits per inch. This tape could then be read directly by an IBM computer. By May 1965, Western was turning out digitally-processed CDP-stacked data that was superior to CDP analog results. Also, because their software was written in "floating-point" format from the beginning, they were easily able to produce seismic sections with relative amplitudes preserved to look for "bright spots". Although immediate attempts were made to sell the concept, Western's first request for such processing did not come until 1972.

(see Fig. 6.2). To complement the digital approach, GSI also began making extensive use of the Bolt air gun, which readily introduced controlled pulses of seismic energy into the water column. Finally, GSI also introduced an integrated navigational system, Geonav®, which would position the survey vessel to within 100 m of the true position no matter where the vessel was on the high seas and how bad the reception was of radio-positioning signals.[15, 16]

As late as 1965, there remained a sharp difference in opinion whether the large cost of more than $100,000 for a set of digital field

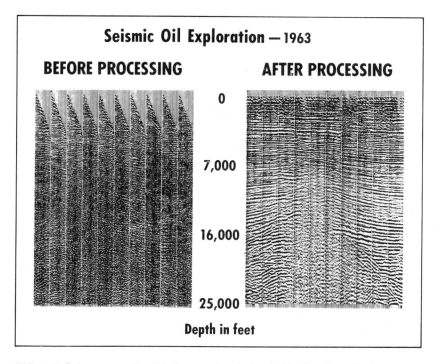

FIG. 6.2. Enhancement by digital processing during 1963 of analog seismic traces to demonstrate the presence of a salt dome. (Courtesy of Texas Instruments.)

[15.] The Geonav® (registered trademark of Texas Instruments) used a combination of inertial navigation, Loran-C, Omega, doppler sonar, and Transit satellite inputs.

[16.] By 1966, there were 35 different types of position-fixing techniques offered to coastal survey ships, each with advantages and disadvantages. As a result, the field was extremely competitive and continued so into the 1980s. One of the more highly publicized feuds in the area closed on 14 April 1960 when the Seismograph Service Corporation transmitted a cheque of $80,000 to Hastings Instrument Company, Portsmouth, Virginia, in payment of federal court costs and attorney fees spent in patent litigation concerning the interference of SSC's Lorac navigational system with the Hastings Raydist patent rights.

recording equipment was technically justifiable even if there was now a much greater dynamic recording range (at least 84 db versus about 50 db) (Dobrin and Ward, 1962; Levin *et al.*, 1966).[17] Despite the industry's economic slump, the transition did occur, but in four gradual stages: (1) analog magnetic tape recording and minimal analog data processing, (2) analog recording and analog preparation of fully-correct record sections, (3) analog recording and digital processing, and (4) digital recording and digital processing. The technology shifted so dramatically that when a firm moved from one stage to another (or leap-frogged a stage), the outmoded equipment had to be completely discarded. Suffice it to say that the industry constructed some very ingenious analog magnetic tape playback equipment to fully correct traces for static and dynamic (normal moveout) corrections — for playing out time-corrected record sections in several forms, including the variable density type of presentation first used by Rieber — and for automatically digitizing analog magnetic recordings for input to digital computers. There was even one service company that would digitize old paper records. The digitizing was greatly complicated by the fact that there were different analog tape widths and formats, and, of course, different input formats were required for different digital computers. The format problem still exists today in spite of considerable efforts made by SEG committees to effect some degree of uniformity in digital recording and transcription formats.

By the time the second generation of general-purpose digital computers became available in the mid-1960s, most oil companies had digital field recordings to work with and some started to do a considerable amount of processing — for research purposes and/or to try to improve upon contractor processing results.[18] Core memories of up to 32,000 words or more were available along with multiplication times in the microsecond range, so that processing rapidly became more sophisticated. The cost per record was reported to still be higher

[17.] Invention of the binary gain ranging amplifier early in 1961 by H. C. Hibbard of Exxon Production Research Company finally provided actual use of the dynamic range potential inherent in digital recording. The basic patent, applied for in March, 1960 and issued during 1967, was licensed to five manufacturers who incorporated the approach in some 450 early digital field recording units. Subsequently, the industry shifted over to using the "Instantaneous Floating Point" method developed by Texaco at about the same time.

[18.] The first IBM seismic array processor (IBM 2938) came on the line during 1966. After receiving a list of algorithms from Western Geophysical in 1965, John Koonce and Byron Gariepy at IBM conceived the device, worked out the basic design, and pushed the IBM organization into carrying through the development. As a result, Western had a substantial advantage in processing efficiency until the use of array processors became widespread at the end of the decade.

than for analog processing (Levin *et al.*, 1966). Fortunately, costs could be gradually reduced through the development of the "Fast Fourier Transform" technique for frequency analysis of seismic data, by a faster digital method of filtering, and by special purpose high-speed array processors (Silverman, 1967).

Processing of data from the North Sea, with its complex and highly variable geology, stretched to the limit all the capabilities of state-of-the-art digital processing techniques. In a paper in *Geophysics*, Rockwell indicated that the new art of digital geophysical data processing, although but five years old commercially, was being employed to solve these and other perplexing problems by: (1) reducing high-amplitude reverberations ("multiples") within the water layer caused by the hard ocean bottom (this problem involved a combination of CDP stacking and deconvolution); (2) determining the velocities from CDP data for accurate stacking, depth conversion and migration; and (3) migration of dipping reflection horizons using automated ray-path techniques (Rockwell, 1967).[19] The velocity-analysis methods required the utmost ingenuity because of structure and the sandwiching of chalk and salt beds between sand—shale sequences. The digital computer was also effectively employed in making numerous adjustments to velocity functions used in correcting CDP traces for normal moveout prior to stacking, in making statistical studies of velocity variations, in computing interval velocities from average stacking velocities, and for ray tracing. The latter two calculations were frequently required to analyze data from extremely long spreads (7.5 km or more) in order to obtain more accurate velocities for deeper zones. In concluding his paper, Rockwell accurately predicted the coming trends in future digital processing, ranging from automated velocity interpretation of CDP data to machine picking, plotting, and migration of seismic reflections. In other words, the era of the most significant technical advances yet achieved in processing and interpreting seismic data was about to begin.

The Search for Appropriate Underground Nuclear Test Ban Language

Although the three major nuclear powers — the Soviet Union, the United Kingdom, and the United States — renewed Geneva Test Ban Conference negotiations on 31 October 1958, little progress was being made to develop agreeable treaty language relative to the

---

[19.] Migration is the process of moving observed dipping reflection horizons to their correct positions in three-dimensional space.

technical monitoring of surreptitious underground nuclear tests. Accordingly, Technical Working Group Two (TWG-2, also known as the Seismic Panel) convened at Geneva between 25 November and 19 December 1959 to do the following: (1) review the Group's past studies and update them, including an analysis of the Latter decoupling theory and the troublesome results obtained while monitoring the Hardtack underground nuclear shot series at the Nevada Test Site, and (2) develop, if possible, objective seismic instrument readings that could serve as the basis for initiating on-site inspections of suspicious seismic events.

This time around, the Russians were more difficult to deal with than the year before. They agreed to the premise that the original Geneva seismic control system should be upgraded, such as adding additional seismometers at each control post, but they sharply disagreed with the American premise that sorting out seismic signatures of secret underground nuclear explosions from those of natural earthquakes had become more difficult. The Soviet experts now argued that the annual number of continental earthquakes equivalent to the yield of any specific underground nuclear explosion was less by a factor of two or three than the 1958 Geneva estimates. In contrast, an American working paper submitted back on 5 January 1959 held that the same Geneva estimates should be upped by a factor of at least two. Views of the British delegation didn't help, for its revised figures suggested a spread of values ranging from the low to the high sides of the Geneva estimates (see Table 6.1). Other disagreements included: (1) evaluation of first motions generated in the Hardtack shot series; (2) validity of the Latter decoupling concept;

TABLE 6.1

*Postulated Numbers of Continental Earthquakes Equivalent to or Larger than Underground Explosions of a Given Yield (Assuming Rainier-type Coupling)**

Yield (kilotons)	Comparable U.S. shot	Geneva estimates (1958)	USA estimates (1/5/59)	United Kingdom estimates (11/59)	Soviet estimates (11/59)
1.7	Rainier	5,000	13,000	3,750–15,000	1,500
5	Logan	1,900	2,900	1,250–5,000	750
19**	Blanca	750	800	500–2,000	400

*To estimate the total number of global earthquakes, including those at sea, multiply by two.

**The Hiroshima explosion of August, 1945, often classed as a "nominal nuclear explosion", had a yield of about 20 kilotons. The U.S. working paper assigned a magnitude of 4.8 ± 0.4 to Blanca, but the Soviet annex of 18 December 1959 assigned a magnitude of 4.7 ± 0.1 to Rainier, thereby creating an order of magnitude difference in interpretations by the two sets of seismologists.

(3) evidence regarding the existence of long-period surface waves caused by Hardtack events; (4) adequacy of seismographs (particularly Benioff's two-second period vertical instrument) used by the Americans to observe distant Hardtack wave-trains; and (5) the very principles by which objective criteria could be developed to define the eligibility of suspicious seismic events for on-site inspection.

In response to new seismic information provided by the American team, the Soviet experts brusquely wrote in an annex to their final report:

> ... On the basis of careful analysis of all the material presented, the Soviet experts categorically repudiate the way in which primary (seismic) data have been handled ... by some members of the U.S. delegation in the Working Group. Having uncovered many errors as mentioned above, and even some misrepresentation, in U.S. statements and documents, the Soviet experts note that they all tend in a single direction — towards reducing the estimates of the control system's effectiveness. The Soviet experts therefore cannot regard these shortcomings as resulting from carelessness or coincidence. ...
>
> It is clear from this that the next task is not to engage in endless discussion about the potentialities of the control system but to conclude an early agreement on the cessation of tests, establish a control system, and improve it on the basis of practical experience.

Nevertheless, by 20 January 1960, the 15-month-old "Conference on the Discontinuance of Nuclear Weapons Tests" had agreed on 17 articles and one annex of the proposed treaty but not agreed on five articles and two annexes that had also been drafted. In the latter were two key sections: "Procedures of the Nuclear Weapons Test Commission" and the definition of the "Detection and Identification System".

Meanwhile, certain members of the United States Congress were anxious to see their nation move more quickly to resolve these technical issues in order that the diplomatic corps could write a totally acceptable treaty text. To this end, Senator Hubert Humphrey, in his role as Chairman, Subcommittee on Disarmament, held a hearing, "Technical Problems and the Geneva Test Ban Negotiations", on 4 February 1960.[20] In the course of the testimony, he observed:

> ... All I hear about are the problems involved. What about the solutions? ...
>
> But where is the enthusiasm for overcoming the difficulties? Can you list for me 25 research projects which we are undertaking at the moment? Can you list five or two or one or three that we are really excited about as to the possibility of overcoming some of these difficulties?
>
> If we cannot, then why are we talking about continuing the (Geneva) negotiations? It seems to me that if we are going to be experts in the big hole theory, which is the big bad wolf now, and if we are going to be experts on the fact that the Russians can

---

20. Another much less active member of the Subcommittee was John F. Kennedy, President of the United States between 1961 and 1963. Humphrey also became Vice President under Lyndon B. Johnson between 1965 and 1969.

cheat, which is the second big bad wolf, there ought to be some good fairy coming along very soon. There ought to be some good news around here, or somebody ought to be looking for it. . . . Do we really have somebody who is alive, alert, and after it . . . or are we just taking this in an easy stride?

Dr. Wolfgang Panofsky, Deputy Chief of the American team of experts, testified in reply:

During the negotiations we felt that as responsible representatives of the U.S. government we had to do a scrupulously honest job in terms of the scientific facts as they were available. . . . I might say from the point of view of scientific honesty, I think we have completely and entirely a clean record.

On the other hand, I agree with you that there does not exist at present an efficient enough mechanism within the Government for carrying out the necessary scientific work to back up the constructive aspects of these problems. . . . Our response regarding such further research is, to me, at least, discouragingly slow. In many instances I believe that the statements which were made by the Berkner Panel, that research will definitely restore the capability of the system back to the 1958 experts' estimate, is probably only a statement which can never be proved unless you try to do the research. I would like to emphasize if you do research, it is dangerous to assume that you will necessarily get results. So I believe that we have taken the correct position in our negotiations of not taking for granted that such research will give (positive) results.

On the other hand, I personally — and I am speaking only for myself here — share your concern about the lack of rapid and enthusiastic progress in this field. I believe that this is a situation which is, to some extent, due to the fact that this whole tie-in reaction between science and foreign relations is a very new problem and, therefore, its solution seems to be taking a great deal of time, and this may very well be unavoidable.

Project Vela Uniform — A Potential Solution to the Treaty Language Problem

In view of Panofsky's reference to the 11-month-old Berkner Panel Report, Humphrey was quick to ask the *Advanced Research Projects Agency* (ARPA) where Project Vela stood at the moment relative to generating the much needed technical solutions.[21] Unfortunately for Humphrey's adrenalin level, ARPA's Director, Brigadier General Austin (Cy) Betts, USA, had to advise the Senator that Vela had only a few research projects under way at the moment. To be sure, $825,000 had been transferred to the *Air Force Technical Applications Center* (AFTAC) on 2 October 1959 to get work under way in the Vela Uniform area, followed by an additional $7.0 million on 7 January 1960. Even so, juggling money did not answer Senator Humphrey's basic question, "Who will provide aggressive research leadership for

21. Project Vela was the centralized research effort directed at improving the nation's techniques to detect and identify nuclear explosions at very high altitudes by space satellites (Hotel) and by surface-based detectors (Sierra), as well as nuclear detonations underground and underwater (Uniform). The Greeks first used the term, Vela, to designate a stellar constellation. Its American pronunciation was never standardized; some called it "Vayla" as in "say", and others termed it "Veela" as in "see".

the program?" AFTAC's Chief Seismologist, Dr. Carl Romney, was already involved in two full-time tasks — first, directing AFTAC's own seismological detection service and second, providing technical support to the treaty negotiators.

To fill the gap, ARPA's Assistant Director for Test Detection, Carlton Beyer, invited Dr. Bates to move across the Pentagon Building from the Navy and become Chief, Vela Uniform Branch, at ARPA. Bates had barely settled in when Prime Minister Macmillan and President Eisenhower issued a communiqué on 29 March 1960 which read in part:

(a) There are great technical problems included in setting up a control system which would be effective in detecting underground nuclear tests below a certain size, i.e. less than seismic magnitude 4.75.

(b) An agreed program of coordinated research undertaken by the three countries (Soviet Union, the United Kingdom, and the U.S.A.) will lead in time to a solution of this problem.

(c) The United States and the United Kingdom would be ready to institute a voluntary moratorium of agreed duration on tests below magnitude 4.75 as soon as the treaty was signed and arrangements have been made for a coordinated research program for the purpose of progressively improving control methods for events below seismic magnitude 4.75.

(d) The United States and the United Kingdom invite the USSR to join at once in making arrangements for a coordinated research program and in putting it into operation.

In early May, President Eisenhower formally announced that Project Vela was under way, and that its underground components would soon be spending funds at the rate of $65 million per year. Unfortunately for those who dreamed that this effort might be a tri-national research program, the Soviet scientists said "Nyet" on 11 May 1960 at the reconvened TWG-2 at Geneva, claiming that the control system's upgrading agreed on last December was quite adequate as a basis for signing a nuclear test ban treaty.

The time was now appropriate to mount a national — even an international — effort to move "low-frequency" seismology out of what certain politicians believed to be the "scientific Stone Age".[22] ARPA's overall annual budget of about $200 million was actually quite flexible.[23] Leadership was also high quality but subject to high turn

---

[22.] An excellent and readable account of the entire international effort in this field during the 1960s may be found in Professor Bruce A. Bolt's *Nuclear Explosions and Earthquakes: The Parted Veil* (Bolt, 1976).

[23.] Because ARPA's staff numbered only 70, it did not operate an in-house laboratory. Instead, it drew on the best technical expertise available, whether in government, academia, or industry. ARPA preferred to work through "ARPA Orders" issued to interested governmental entities which then did the actual contracting and detailed technical monitoring of the work effort. ARPA on-going projects were ballistic missile defense, suppression of guerilla warfare, enhancement of materials technology, and development of interactive computer networks.

over. Thus, within the first five years of Vela, there were four ARPA Directors — Brigadier General Betts, and Drs. Jack Ruina, Charles Herzfeld, and Robert Sproull, and four Nuclear Test Detection Office Directors — Carlton Beyer and Drs. George F. Bing, Jack Ruina (who simultaneously served as ARPA's Director), and Robert Frosch.

In the best Washington tradition, ARPA quickly established its own *ad hoc* Advisory Group on the Detection of Nuclear Detonations with no less than Dr. Richard Latter, the "big-hole advocate", as Group Chairman. Sub-Group Chairmen were much more neutral, for Dr. W. Panofsky took the lead role for high-altitude test detection, while Dr. Frank Press took on the challenge of seismic detection.[24] Press's group was particularly useful, for it helped "fine-tune" the Vela Uniform master plan (see Fig. 6.3) and made certain that ARPA, rather than AFTAC, would have the final say on how the money would be spent (see Table 6.2). Nevertheless, the advisors outnumbered

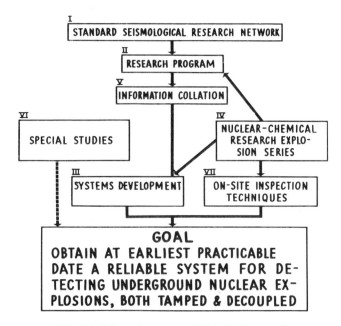

FIG. 6.3. Schematic pattern of Vela Uniform tasks.

24. Other members included such regular seismic advisors as Dr. Hugo Benioff (CalTech) and Dr. Jack Oliver (Lamont Geological Observatory) as well as two from the geophysical industry, Dr. John Crawford, Chief of Geophysical Research, Continental Oil Company (co-inventor of Vibroseis®), and Dr. F. G. Blake, who held the same post in the field research laboratory of the Standard Oil Company of California. (*Note*: Vibroseis is a registered trademark of the Continental Oil Company.)

TABLE 6.2
Funding Patterns of ARPA's Vela Uniform Program (in millions of dollars)
The Build-up Period, FY 60—63

Type of effort	FY 60	FY 61	FY 62	FY 63	Total
World-wide seismic network	0.8	2.6	1.5	1.5	6.4
Basic and applied research	3.6	9.7	9.4	10.4	33.1
Systems development	0.9	8.2	5.0	10.8	24.9
Explosions and their monitoring*	2.3	10.3	18.4	9.7	40.7
Information collation	0	0	0.3	0.2	0.5
Special studies	0.3	0.3	0.05	—	0.7
On-site inspection techniques	0.1	2.0	1.3	1.0	4.4
Total $:	8.0	33.1	36.0	33.6	110.7

The Era of Maturity, FY 64—71

	FY 64	FY 65	FY 66	FY 67	FY 68	FY 69	FY 70	FY 71	Total
Detection	1.1	2.5	1.5	0.9	1.0	0.6	0.4	0	8.0
Location	1.4	1.9	2.3	1.1	1.2	1.3	0.7	0.2	10.1
Identification	15.2	34.2	17.6	22.6	13.1	14.3	17.0	9.6	143.6
On-site inspection techniques**	0.4	2.6	5.3	3.5	2.2	0.1	0.03	0.02	14.1
Evasion Research***	0	1.9	8.8	0.3	5.2	6.7	4.1	5.7	32.7
Total $:	18.1	43.1	35.5	28.4	22.7	23.0	22.2	15.5	208.5

Grand Total for entire program, FY 60—71: $319.2 millions

Source of information: Joint Committee on Atomic Energy's Hearings of 27—28 October 1971 by its Subcommittee on Research, Development, and Radiation.

*In FY 61, the U.S. Atomic Energy Commission also received a $33 million appropriation to prepare and conduct major explosions for Vela Uniform.

**FY 65, the "On-site inspection techniques" task was granted sub-program status and labeled Vela Cloud Gap.

***"Evasion Research" became a specific task within Vela Uniform in FY 65, and in FY 67 was transferred to the Prime Argus Project but is included here for continuity's sake. Funding primarily went for the following special nuclear detonations designed to test seismic evasion hypotheses: Sterling of 0.4 kilotons on 3 December 1966 in the Salmon cavity in Mississippi, Scroll of low yield on 23 April 1968 (probably tamped with carbon), and Diamond Dust and Diamond Mine, both of low yields, and shot on 12 May 1970 and 1 July 1971, respectively, to test a cavity decoupling concept. The last three were detonated at the Nevada Test Site.

the implementors for ARPA's Vela Uniform Branch had the services of only six professionals during the 1960—1963 period.[25]

The Vela effort was formally launched at a wide-ranging, classified conference held in the Pentagon on 4—5 October 1960 to spell out the issues and solutions sought to the entire American technical community. Nearly 300 persons were present, including not only representatives of major geophysical contractors but also of major oil companies such as Standard Oil (New Jersey), Continental Oil, Gulf

25. Besides Bates, these were: Theodore H. George, Donald Clements, Rudy Black,

Oil, and Texaco. In addition, ARPA quickly established the *Vela Seismic Information and Analysis Center* (VESIAC) at the University of Michigan's Institute of Science and Technology.[26] Although the cost of operating VESIAC was less than 0.5 percent of Vela Uniform's budget through 1964, the Center was an invaluable tool in keeping 60 contractors, 12 ARPA agents, and seven governmental departments (Defense, State, Air Force, Navy, Commerce, Interior, and the Atomic Energy Commission) posted on the total effort and its achievements (see Table 6.3).

Creation of a Modern Global Seismological Network

During the 60 years following Professor John Milne's initiation of a worldwide network of seismological observatories in 1897, a hodge-podge of over 700 seismic stations had grown up of widely varying quality and instrument types. As a result, if one wanted to study a major seismic event, his initial effort was a frustrating exercise in long-distance correspondence, followed by an equally frustrating time of trying to get the uncalibrated, highly varied instrument responses reduced to a common base.[27] To overcome this difficulty, ARPA arranged for the *National Academy of Sciences* (NAS) to

---

Dr. Joseph Berg (a former professor of geophysics from the University of Utah on contract via the Institute of Defense Analyses), and Major Robert Harris, USAF, a former AFTAC seismologist who served as the group's informal poet laureate. Here is his "A Seismologist's Lament":

> "Mine eyes have seen the wonders of the works of Livermore,
> They've conjured up a program to last forevermore.
> They are digging bigger caverns than a million moles can bore.
> Decoupling marches on!
>
> They have 20 million miners working three shifts 'round the clock.
> They're using all the shovels that the nation has in stock.
> And they've drafted all the railroads just to haul away the rock!
> Decoupling marches on!
>
> They're heaping up the rubble into mountains that tend to go so high,
> That the people in the valleys cannot even see the sky,
> And they do this all in a manner clandestine that'll fool the shrewdest spy.
> Decoupling marches on!
>
> They'll be able to test some big ones ranging up to several hundred meg,
> But the blast wave in that cavern won't even crush an egg.
> And all of us who want the word now have to kneel and beg.
> Decoupling marches on!"

26. VESIAC was the creation of three people — the Institute's Director, Professor James Wilson, one of Professor Perry Byerly's first students, Dr. John Denoyer (a seismologist who would replace Bates as the Vela Uniform manager in 1965), and hard-driving Thomas Caless, who kept his staff in a continual uproar.

27. Record feeds varied widely — some seismic traces went from top to bottom, others vice versa, some read from right to left, others left to right.

TABLE 6.3
Key Program Managers within the Vela Uniform Effort

Task	Agency	Key contract monitors
Worldwide Standardized Seismograph Network	U.S. Coast and Geodetic Survey	Leonard Murphy and R. J. Brazee
Basic and Applied Research	USAF Cambridge Research Laboratories	Norman Haskell and Captain Robert Gray, USAF
	USAF Office of Scientific Research	Major William Best, USAF
	USAF European Research Office	Lt. Colonels Albert K. Stebbins and Charles Allard, USAF
	USAF Rome Air Development Center	First Lt. John Entzminger, USAF
	U.S. Geological Survey	Louis C. Pakiser and George Keller
	Office of Naval Research	James Winchester
	Naval Electronics Laboratory	Charles Johnson
	Naval Radiological Defense Laboratory	Kenneth Sinclair
	U.S. Naval Bureau of Ships	Vincent Saitta and Lt. (j.g.) A. H. Sobel, USN
Systems Development	Air Force Technical Applications Center	Carl Romney, Major Robert Meeks, USAF, Captain Clent Houston, USAF
	Lincoln Laboratories of MIT	ARPA Staff
Research Explosions	Defense Atomic Support Agency	John Lewis and Brigadier General Leo Kiley, USAF
	Atomic Energy Commission	James Reeves
On-site Inspections	Air Force Technical Applications Center	Harry Woo

establish a special panel under Professor James Wilson that would specify a set of "standard seismic instruments" and propose appropriate observatories throughout the world to receive this equipment if they would agree to share the records obtained (Romney and Beyer, 1960).[28] Joining the network was not a simple matter, for the recipient station had to provide, at its own expense, a lightproof seismic vault, electric power, and routine operational and maintenance services.

The Wilson Panel did an excellent piece of work and recommended that the cooperative program stay within the state of the art and use

[28.] Other Panel members included the familiar personages of Professors Perry Byerly, Jack Oliver, and Frank Press. However, there was one new name — that of a personable, productive young seismologist from St. Louis University — Dr. Carl Kisslinger.

analog light traces on photographic paper. By March, 1961, the *U.S. Coast and Geodetic Survey* (USC&GS) had let a contract with the Geotechnical Corporation to build 125 sets of equipment at the cost of $15,394 each.[29] Shortly thereafter, the Survey let a contract with Itek, Inc., to construct an unique record-copying camera capable of converting 240,000 original seismographic records a year to easily readable 70-mm film chips.[30] Interest in joining this new worldwide standardized seismograph network was extremely high outside the Soviet Bloc (see Fig. 6.4), and installation got under way during July, 1961.[31]

The "care and feeding" of this cooperative network was handled by the C&GS's Albuquerque Seismological Laboratory.[32] Although original planning called for 125 stations, only 120 were installed when the network was considered completed in 1967. Equipment installation, which cost $1.8 million, was a challenging task handled by two men working a month at each station, followed up by a "trouble-shooting" inspection team some time later.[33] Many of the stations were in remote and often primitive spots, making travel to and from such points an adventure in itself. For example, among the travel conveyances used were military and cargo aircraft, seaplanes, coastal freighters, railroad hand-cars, rickshas, and dog sleds. And even when the installers arrived at an observatory, the equipment never seemed ready for installation, as when one team arrived at the South Pole and found several instruments frozen inside a large block of ice. In addition, many locations had special siting problems. In the case of Trieste, Italy, for example, the seismometers were placed 130 meters below the surface in a grotto subject to tourist traffic. As a result, mountain climbers were used to lay the connecting cables behind stalactites in such a fashion that they would not harm the picturesqueness of a cavern 90 m high.

[29.] As finally manufactured, the short-period instrument system (0.2- to 10-second period) could operate at magnifications between 3,125 and 400,000, while the long-period system (10- to 30-second period) was capable of magnifications between 375 and 6,000.

[30.] By the mid-1970s, the copy center provided 2.5 million chips a year to the community at a cost of $0.55 a chip or $40.00 for a complete month of records from a given station.

[31.] Canada upgraded its seismic network on its own but maintained close cooperation with the C&GS's global effort.

[32.] The Laboratory has fortunately had only three directors (H. Judd Wirz, 1961–1965; Jon Peterson, 1965–1978, and William Green, 1979–1981) in 19 years of existence, thereby stabilizing the dealings with foreign observatories.

[33.] The second member of these teams was supplied on contract by Texas Instruments. Installers possessing large amounts of tact, perseverance, and technical skill included O. J. Britton, L. Parks, H. N. Meeks, and T. J. Scherbel. Highly respected trouble shooters and inspectors were Jon Peterson, William Green, and Harry Whitcomb.

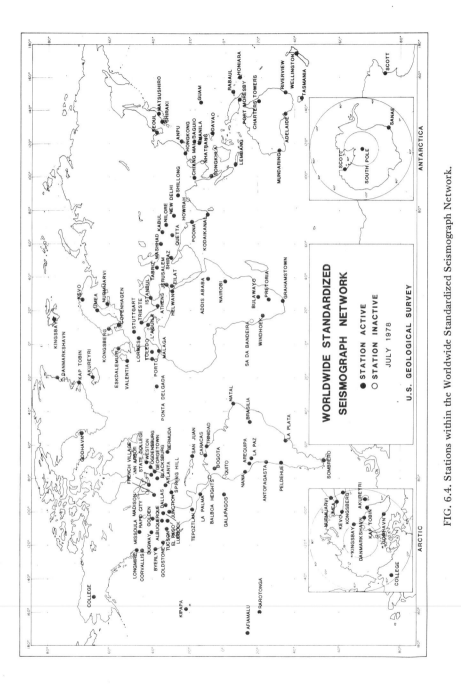

FIG. 6.4. Stations within the Worldwide Standardized Seismograph Network.

Even today, most academic seismologists agree that the advent of the standardized global network was the major "happening" in their lifetime. Father William Stauder, SJ, explained this feeling at the formal dedication of the network in Washington, D.C. on 18 April 1963:[34]

> Just a month ago I had occasion to request a group of records for ten earthquakes from selected stations of the World-Wide Standard Station Network. About ten days later I received a package, a shipment which included the film chips for 300 of the records I had requested. The records fitted easily in the palm of one hand. As I unpacked these records and extended them to an assistant, I realized that a new era indeed had begun in observational seismology.

After the late 1960s, the standardized network began to lose co-operative stations until, in 1980, only 105 observatories regularly transmitted photographic records to the central data center for copying.[35] Nevertheless, the network was so valuable that, by the mid-1970s, it was decided to upgrade the system by creating 13 "*S*eismic *R*esearch *O*bservatories" (SRO's), typically sited at a network station and equipped with the latest in borehole seismometers and digital data recording units (Peterson *et al.*, 1976).

Vela Uniform's Research Program

The basic problem facing the Vela Uniform program was one of communications theory and practice. For example, an underground explosion typically converted about two percent of its total energy into an impulsive, sinusoidal, symmetrical seismic wave train of less than 0.001-second duration. But the complexities of the paths surrounding the point of detonation were such that one no longer saw a single "pip" on a seismic trace once the detector was offsite. In fact, half an earth away the incoming wave train (see Fig. 1.1) would be a series of complex, often overlapping, combinations of direct, refracted, reflected and diffracted pulses with wave periods up to 100 seconds and a duration of several minutes. Moreover, each of the different types of incoming waves could be further scattered, distorted,

---

[34.] As of 1980, Father Stauder had become Dean of the Graduate School at St. Louis University but still found time to continue seismological research on the side.

[35.] The network has had an erratic administrative life. In 1966, ARPA, because of its location within the Department of Defense, had to cease funding the net. Fortunately, the National Science Foundation, through some skilful arrangements by Dr. Roy Hanson, funded the network for a few critical years. In 1974, the Albuquerque laboratory and its network were transferred to the Branch of Global Seismology, U.S. Geological Survey, while the Seismic Data Center remained in Boulder, Colorado, as part of the Environmental Data Information Service, National Oceanic and Atmospheric Administration. Inflation has also severely impacted the network. With each station using six sheets of photographic paper per day, the cost of this paper has jumped from $0.11 a sheet in 1962 to $1.14 each in 1980.

dispersed, selectively attenuated according to wave period, and even mode converted. In addition, such incoming signal energy was probably mixed with a generous portion of unrelated background noise. There was also considerable question as to just how homogeneous the major paths were that took signal energy from the point of detonation to the detector site. For example, did a slight velocity inversion exist everywhere at a depth of 80 to 150 km, thereby causing a "signal shadow" at distances of 600 to 1,600 km from the source (see Fig. 6.5)? In fact, how many major layers were there in the deep crust itself?[36] In this instance, some believed that the United States was likely to be two layered at depth, while three comparable layers existed below Alberta and Iceland. Certain Soviet seismologists also believed that there were five crustal layers in central Russia, with the possibility of three more layers in the supposedly homogeneous mantle underneath.

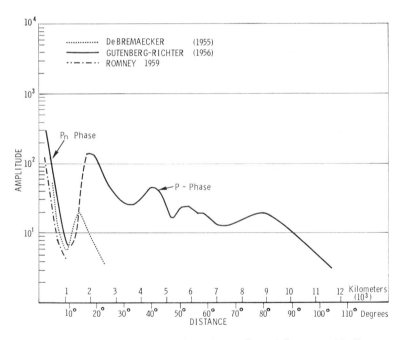

FIG. 6.5. Amplitude attenuation of maximum *Pn* and *P* waves with distance from the western United States. (Courtesy of the Seismological Society of America.)

[36] As of 1959, only 20 sets of reliable deep-crustal measurements had been made within the United States.

To place the science on a more reliable footing, the Vela Uniform staff mounted a carefully considered multi-path attack on the basic problems cited above. By June, 1961, 66 contractual efforts were under way or in the final stages of negotiation (see Table 6.4). A major hope in using so many diverse contractors was that closely-held expertise in the new types of signal processing (including velocity and directional filtering, cross-correlation and running summation, real-time spectral analyses, optical correlation, and audio interpretation of time-compressed seismic signals) might provide a breakthrough

TABLE 6.4

Vela Uniform Research Contractors as of June, 1961

Seismic Detection

Source mechanisms	Propagation phenomena	Signal analysis and display
University of California Berkeley	Columbia University (Lamont)	University of Michigan
Pennsylvania State	CalTech	University of California (Scripps)
St. Louis University	Weizmann Institute (Israel)	CalTech
Columbia University	Stanford Research Institute	Columbia University
United Electro-Dynamics, Inc.	University of Oklahoma	MIT
Sandia Corp.	California Research Corp.	Texas Instruments, Inc.
Stanford Research Institute	Pennsylvania State University	Bell Telephone Laboratories
U.S. Bureau of Mines	Southern Methodist University	Armour Research Foundation
American Machinery and Foundry Co.	National Engineering Science	U.S. Coast and Geodetic Survey
Engineering-Physics Corp.	*Signal detection*	
	University of Michigan	
Propagation path	Columbia University	
U.S. Geological Survey	California Institute of Technology	
St. Louis University	University of California (Scripps)	
UCLA	Geotechnical Corp.	
CalTech	Jersey Production Research Company	
Uppsala University (Sweden)	Texas Instruments, Inc.	
University of Witwatersrand (S. Africa)	Dresser Industries (SIE)	
University of Alberta (Canada)	Electro-Mechanics Co.	
University of Saskatchewan (Canada)	Rensselaer Polytechnic Institute	
	Century Geophysical Corp.	
	Radio Corporation of America	
	Electromagnetic detection	
	U.S. Geological Survey	
	Edgerton, Germeshausen & Grier, Inc.	
	Space Electronics Corp.	
	Underwater detection	
	U.S. Navy Electronics Laboratory	
	U.S. Naval Radiological Defense Laboratory	
	Columbia University	

in distinguishing earthquake from underground-explosion signals.[37] University researchers, of course, were involved where their personal interests and those of Vela Uniform happened to intersect. Three of the academic groups (University of California, Weston Observatory of Boston College, and St. Louis University) even experimented with the design, construction, and operation of low cost monitoring networks in which unmanned, remote seismometers continuously transmitted by phone line to a central recording and analysis center. For example, Professor Bruce Bolt started updating the eight-station Berkeley network in 1963 and engineered a processing facility so sophisticated that the station analyst could take the incoming data, run it through a computer, and come out with the locations of both local and teleseismic events, singly and in groups, from the same set of incoming signals. Similarly, Father Linehan, SJ, at Weston started with four remote stations in 1964 and built up the network until, in 1980, it covered all New England with 40 seismological installations.

For nearly a decade, Romney had thought that it might be possible to obtain a greatly improved signal-to-noise ratio by placing seismometers below the noisy surface layer.[38] According to theory, most of the persistent surface noise was from several modes of surface waves trapped in the upper strata of the earth's crust. By 1962, seven firms, including the research subsidiaries of Standard Oil Company (New Jersey) and the Shell Oil Company, were hard at work on the problem. Then, at a VESIAC bore-hole seismometry conference during March, 1964, several bits of good news were reported.[39] A short-period seismometer at the bottom of a 2.7-km well near Pinedale, Wyoming, had actually operated at a gain of three million, giving a signal-to-noise improvement of ten over that obtainable at the surface. But the most exciting results were reported by Geotech's Richard M. Shapee, for he had been able to magnify long-period wave trains in the

[37.] The quality of industrial participation is indicated by two of the contractor's principal investigators later becoming President of the Society of Exploration Geophysicists, namely, J. Dan Skelton of Jersey Production Research (1974—1975) and Milo Backus of Texas Instruments and the University of Texas (1979—1980).

[38.] The signal level Romney wished to detect was of the order of one millimicron with a one-second period. For example, typical underground nuclear explosions at the NTS gave these signal levels when measured in millimicrons at a distance of 1,000 km; Antler (2.4 kt) — 6.5 and Logan (5 kt) — 2.8, when fired in volcanic tuff; Hardhat (5 kt) — 32 when fired in granite, and Fisher (13.5 kt) — 2.7 when fired in dry alluvium. On the other hand, background noise levels on hard rock in quiet, well-sheltered sites typically ran five to ten millimicrons and could run 30 millimicrons in less favored locations.

[39.] Speakers for the major contractors included: Century Geophysical Company — R. A. Broding; University of Michigan — John DeNoyer; Geotechnical Corporation — E. J. Douze and R. M. Shappee; United Electrodynamics — R. L. Sax; Mandrell Industries — G. C. Phillips and L. M. Mott-Smith; and Shell Development Company — J. H. Rosenbaum and A. J. Seriff.

20-second period realm by 50,000 times, or 15 times that previously available. By building on that result — and on some radically new ideas of broad-band seismometer design by Dr. Paul Pomeroy and others at the Lamont Geological Observatory — Geotech placed on sale a decade later an orthogonally oriented, force-balanced accelerometer type detector capable of measuring three components of earth motion at a gain of 100,000 while implanted within an 18-cm borehole 0.1 km deep (Pomeroy *et al.*, 1969; Peterson and Orsini, 1976). But the most important contribution by this device was that it was essentially broad-band over the frequency range from zero to one hertz, meaning that just one sensor detected short-period, long-period, very long-period, and earth-tide motion simultaneously. As a result, the device was quite capable of replacing all six seismometers in a "Standardized Station", plus a gravimeter and two tiltmeters, while still responding to ground motions that varied in amplitude by 100,000 or more.

In the Berkner Report, Professor Jack Oliver had indicated that "one of the most significant experiments open to the field of seismology is the determination of the seismic motion of the floor of the ocean". By May, 1962, Oliver's associates at Lamont had actually operated an "*Ocean Bottom Seismometer*" (OBS) from the drifting ice station, Arlis I, when it was but 750 km from the North Pole and found the background noise level on the sea floor to be but one to five millimicrons. Moreover, the device had recorded compressional waves from an Alaskan earthquake of magnitude 4.75 some 2,500 km distant with a signal to noise ratio of 100:1 in the latter part of the wave train. The second academic contractor, Scripps, was much slower getting started, but *Texas Instruments* (TI) put together an excellent engineering team and actually turned out three generations of OBSs between 1961 and 1964.[40, 41] Looking very much like a package to be used in outer space, TI's Model-3 could operate for 30 days at a time in depths of as great as 7.3 km and then be "popped-up" to the surface by either a set timer or by acoustic command.[42]

[40.] The Scripps effort by Walter Munk and Frank Snodgrass eventually produced a long-period, accelerometer-type device responding primarily to free oscillations of the earth. By the mid-1970s, fourteen of these were deployed to various spots on the globe, including two in the Soviet Union, as part of Project IDA (*I*nternational *D*eployment of *A*ccelerometers) funded, in part, by Ida and Cecil Green.

[41.] Key TI engineers included Paul Davis, Buford Baker, J. R. Baird, G. Dale Ezell, J. T. Thomson, and William A. Schneider. One offshoot was the creation of TI's RA-2 parametric amplifier product line.

[42.] Housed in an aluminium sphere one m in diameter, the device consisted of a three-component seismometer capable of measuring movement of less than one millimicron in the one to ten hertz pass-band, a pressure transducer, a 40-day digital clock with 0.1 second accuracy, sub-audio reactance amplifiers, a digital magnetic recording unit of 72-db range,

With such a device at hand, it would have been quite possible to maintain a seismic monitoring network on the high seas just off the highly seismic Kamchatka Peninsula of the Soviet Union. Monitoring close in would have definitely been called for because some of the earthquakes generated in that region give outward first pulses in all directions, just as did explosions, when monitored at teleseismic distances. Although this specialized use never came about, the OBS has since become an important working tool for the marine seismologists interested in close-up monitoring of seismic activity created by sea floor spreading.

To overcome the American lack of knowledge regarding the nature of the lower crust and upper mantle of the earth, the Berkner Panel recommended drawing on the skills, equipment and manpower of the geophysical exploration industry and probing the earth seismically with geophone layouts up to 400 km in length. Such a task appeared to be a logical one for the U.S. Geological Survey. However, the Survey did not have a "Crustal Studies Branch". On the other hand, Dr. Thomas Nolan, the Survey's Director, advised that he'd be glad to create one if Vela Uniform would pay the bill. Nolan's choice for Branch Chief, Mr. Louis C. Pakiser, Jr., proved to be a superb manager as well as an excellent geophysicist. As a result, much of the Survey's present-day eminence in global seismology, crustal studies and earthquake prediction stems from "Lou's" ability to select and motivate junior scientists.[43, 44]

By late 1961, Pakiser's group was outfitted with ten special recording trucks built by the SIE Division of Dresser Electronics and capable of recording variations in signal voltages of as much as 1,000 without automatic gain control in the frequency range of one to 200 hertz. Thus, his field units could readily monitor a wide range of seismic events, including small earthquakes, earthquake aftershocks, underground nuclear explosions and the scores of five-ton explosions that

---

an anchor release mechanism triggered by sonar command or at a preset time, and a 23-kg internal battery power supply. Should the need have arisen, this type "OBS" device could have been modified into a "black-box" of the type advocated by arms-control advocates in the late 1950s and early 1960s for use as "gap-fillers" in nuclear-test-control networks.

43. A graduate of the Colorado School of Mines, Pakiser had worked as a computer for the Carter Oil Company between 1942 and 1949 (except for a break for service in the U.S. Army). Married but childless, he became a major advocate for the training and employment of women and ethnic minorities in the field of geophysics.

44. Among Pakiser's most effective collaborators were D. B. Hoover, R. E. Warrick, W. H. Jackson, J. C. Roller, G. Eaton, Alan Ryall, S. W. Stewart, D. P. Hill, J. H. Healy, and G. B. Mangan.

were detonated under contract by United Electrodynamics for the crustal study program.

As a Distinguished SEG Lecturer during 1964, Pakiser pointed out that the Survey's findings to date repeatedly verified the hypothesis made during 1958 by Dr. Merle Tuve of the Carnegie Institution of Washington wherein Tuve stated:

> . . . the idea of a single, world-wide value for wave velocity for the outermost portion of the mantle is probably an erroneous simplification . . . there appear to be non-uniformities and regional geographic differences in the mantle of the earth, just as there are in the crustal rocks.

Lou's main points in his intriguing lecture included: (1) the mean velocities of the deep crust and the upper mantle could vary by as much as ten percent; (2) within the United States, the earth's crust was thickest in the Great Plains (over 50 km) and thinnest in the Great Valley of California (less than 20 km); (3) mountains have "roots" in some localities and not in others; and (4) the nature of the "Moho" discontinuity remained a matter for speculation.

Getting a comparable crustal study under way in Europe was not easy. Seismologists within the Federal Republic of Germany had not yet recovered from the disastrous war years, and French classical seismologists were not to be involved in an American project. Fortunately, Professor Yves Rocard, Director of the Laboratory of Physics at the prestigious École Normale Supérieure (France's version of CalTech), had the stature and desire to accept ARPA support.[45] The ARPA funding permitted Rocard to outfit five mobile seismic vans with which to conduct field investigations into the nature of seismic propagation within France. Such equipment, it might be added, supplemented the test detection seismic network that Rocard operated for the French Government. This capability consisted of three very simple tripartite arrays in eastern, southern, and northwestern France (the Lormes, Lorgue, and Flers areas) and a field headquarters at the beautiful Chateau de la Renardière not far from the River Loire. Eventually no trip to France by a ranking American seismologist was complete unless it included a trip to one of these facilities where, after a fine lunch, he would be initiated into the ancient art of water

---

[45.] During World War II, after having observed a new German radio navigational beam station scheduled to control night-fighters, Rocard escaped to England with the information. By 1944, he was serving as Director of Scientific Research, Free French Navy. After the war, the British Government gave him the signal honor of designating him a "Commander of the British Empire". Subsequently, his only son, Michel, became the National Secretary of the French Socialist Party in 1967 and was named Minister of Planning when the Party won the national elections during May, 1981.

witching.[46,47] By mid-1962, Rocard, too, was advising that the amplitude of seismic signals throughout France was strongly dependent on geological features (Allègre, Mechlér, and Rocard, 1962). In other words, here was more evidence that the earth was definitely not as homogeneous as had been assumed during the Geneva talks back in 1958.

Are Bigger and Better Seismic Arrays the Route to Go?

Although Vela Uniform's research program received a very high degree of acceptance and cooperation within the scientific and technical community, getting the parallel "Systems Development" effort under way was far from easy. Washington politics required that AFTAC be ARPA's sole agent in this area. However, AFTAC had a very high "*N*ot *I*nvented *H*ere" (NIH) factor, for it operated its own seismic detection stations in Wyoming, Oklahoma, and Alaska and wished to maintain its local monopoly in nuclear-test-detection expertise.[48] When asked by ARPA during September, 1960 to build an "Experimental Seismic Control Network" of five Geneva type seismic detection stations four km across that surrounded the *Nevada Test Site* (NTS), a "Berkner type" seismic detection array of as many as 100 elements and nine km across, and a "Seismic Data Analysis Center", AFTAC moved at a slow, deliberate pace with the results shown in Table 6.5 (also see Fig. 6.6). Another problem with these arrays was in their inherent technical design, for their horizontal dimensions were but a fraction of a wavelength of the incoming seismic signal even though communications theory held that a good antenna should be at least two times the length of the incoming signal.[49]

46. Because Rocard did not speak English, one of his protégés typically did. These up-and-coming young Frenchmen included Pierre Mechlér, Claude Allègre, and Jean Lebel. A descendant of a noted French engineering family, Lebel married one of the Schlumberger grand-daughters. While a youth, he was a member of the Maquis (Resistance Forces) in southern France. Assigned as a messenger, he was equipped with a motorcycle and a "Tommy-gun". As he expressed it, "What more could a boy of 17 ask for?"

47. Rocard's explanation of water witching is explained in his book, *Le Signal du Sourcier* (Rocard, 1962). His theory maintains that tightly holding the forked stick places pressure on nerve trains within one's arm in such a way that rapidly walking across a localized magnetic field of force induces an electrical current. This current neutralizes one's own nerve signals, the tendons go flaccid, and the wand drops no matter how hard the experimenter wishes it did not.

48. Working level policy at AFTAC was dominated by its Technical Director, Doyle Northrop, who also served as "Technical Consultant" to the Joint Congressional Committee on Atomic Energy.

49. In the "near zone" of detection, a one hertz seismic signal would have a wavelength of one to ten km, while a similar signal in the "far zone" beyond 1,800 km would have a wavelength of ten to thirty km.

TABLE 6.5
Seismic System Development Efforts Funded by ARPA

System effort	Geographic location	Contractor for construction	Completion date	Ultimate fate
Seismic detection stations				
Tonto Forest (Berkner type)	Payson, Arizona	United Electro-dynamics	4/63	Phased out 3/75
Wichita Mountains (Geneva type)	Fort Sill, Oklahoma	Geotechnical Corporation	10/60	
Blue Mountains (Geneva type)	Baker, Oregon	Texas Instruments	9/62	Phased out 2/76
Uinta Basin (Geneva type)	Vernal, Utah	Texas Instruments	12/62	Phased out 9/73
Cumberland Plateau (Berkner type)	McMinnville, Tennessee	Texas Instruments	2/63	Phased out 3/77
Large Aperture Seismic Array (LASA)	Eastern Montana	United Electro-dynamics	9/65	Conditioned data telemetered to Alexandria, Virginia and Lexington, Massachusetts until phase-out in mid-1979
Alaskan Long Period Array (ALPA)	North of Fairbanks, Alaska		8/71	
Norwegian Seismic Array (NORSAR)	South Central Norway	Royal Norwegian Council for Scientific and Industrial Development	3/71	Raw data telemetered to Alexandria, Virginia
Data Analysis and Technique Development Center	Alexandria, Virginia	United Electro-dynamics	1962	Now called Seismic Data Analysis Center and operated by Teledyne Geotech*
Research and Analysis Center	Lexington, Massachusetts	Lincoln Laboratory, Massachusetts Institute of Technology	1964	Now called Applied Seismology Group

*In 1965, Teledyne, Inc. bought United Electro-dynamics from the United Geophysical Corporation; in 1966, Teledyne purchased (via an exchange of stock) the Geotechnical Corporation and then merged these two seismic entities into one called "Teledyne Geotech".

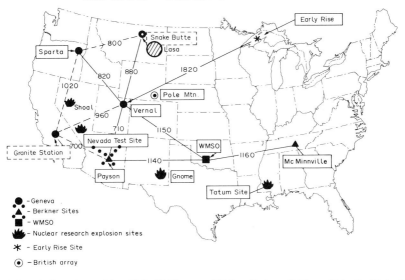

Legend:
- ● – Geneva
- ▲ – Berkner Sites
- ■ – WMSO
- 🌿 – Nuclear research explosion sites
- ✳ – Early Rise Site
- ◉ – British array

FIG. 6.6. Locations of key Vela Uniform and Tabor Pluto activities within the United States (distances in kilometers).

Almost as soon as construction contracts were let, the political infighting grew hotter as to just how good a technical effort the Vela Uniform program should be. Although Senator John F. Kennedy, a liberal, became President on 20 January 1961, his Director of Defense Research and Engineering proved to be brash, brilliant Dr. Harold Brown, who had become the Director of the *Livermore Radiation Laboratory* (LRL) in place of the ebullient Professor Edward Teller. Brown soon had an LRL associate, Dr. George Bing, a Teller protégé, on board to run ARPA's Nuclear Test Detection Office.[50] Bing spent most of the summer of 1961 in the company of Doyle Northrop watching the frustrated nuclear test ban talks drone on at Geneva and by late August was refusing to spend any of the $11 million that had just become available for continuing Vela Uniform's research program. Then, when the Soviets resumed nuclear testing on 1 September 1961, Bing directed Bates to have AFTAC place an indefinite "hold" on building the seismic detection stations now under contract. Dr. Berkner immediately dropped in at the Pentagon and reviewed the situation with Drs. Brown and Bing. Nevertheless, Bing persevered and canceled two Geneva type control posts (Granite, California, and

50. For the broad picture, the reader is referred to the article, "How Livermore Lab Survived the Test Ban" by George Harris (*Fortune* Magazine, April, 1962). Two decades later, Bing illustrated Teller's book, *Energy from Heaven and Earth,* with humorous sketches (Teller, 1979).

Snake Butte, Montana), downgraded the "Berkner" far-zone detection station in Tennessee to a Geneva-type station, and squashed the concept of having a "Vela Research and Evaluation Center" established along the lines of an unsolicited research proposal submitted by a consortium of California universities led by Dr. Frank Press.

During Bing's arbitrary and opinionated tour at ARPA, many of the top newspapers in the Western World, including the *New York Times,* the *Washington Post,* and the London *Sunday Times* followed the underground test ban treaty in great detail. The *Post*'s Howard Simons, with a background in military intelligence, was particularly perceptive in his coverage of this issue.[51] In addition, John Finney of the *New York Times* and Earl Voss of the *Washington Star* could also be depended upon to write straightforward stories. Meanwhile, Nikita Kruschev, Harold Macmillan, and John F. Kennedy were rapidly becoming amateur seismologists. Fortunately for ARPA, Bing returned to the greener pastures of LRL in January, 1963, and ARPA's new Director, Dr. Jack Ruina, decided to "double-hat" himself as Acting Director of the Nuclear Test Detection Office until someone better could be found. This someone turned out to be Dr. Robert Frosch, who would prove to be one of the most brilliant and diplomatic scientists on the Washington scene for the next 17 years.[52] In the meantime, the foreign ministers of the Soviet Union, Great Britain, and the United States went ahead on 5 August 1963 and signed in Moscow a "Treaty Banning Nuclear Weapons Tests in the Atmosphere, in Outer Space, and Under Water". In view of the continuing hassling about on site inspections, underground testing was specifically omitted from the treaty except for prohibition of any resultant radioactive debris moving across national borders (see Table 6.6). Even so, eventually over 100 nations did ratify the treaty.

After being on board five months and having visited the British counterpart to the Vela Uniform program, Dr. Frosch began to move aggressively during mid-February, 1964. Calling a special "seismic systems study" together at the Pentagon, he asked the group to answer two questions:[53]

[51.] In 1962 and again in 1964, Simons won the Westinghouse Award of the American Association for the Advancement of Science for outstanding writing in the field of science. By 1973, he became Managing Editor of the *Post* and played important roles in the "Watergate cover-up operation" exposé in 1974—1975 and the bogus Pulitzer Prize "Drug Addict Jimmy" story in 1981.

[52.] A Columbia University graduate who had specialized in theoretical physics, Frosch became Director of the University's Hudson Laboratories at age 28 during 1956. His major effort there was the conception and building of a massive underwater array, Artemis, off Bermuda to determine whether long-range sonar could detect and track submarines.

[53.] The participants were: C. Romney (AFTAC), Herbert Scoville (CIA), Henry Myers

TABLE 6.6

Typical Press Play During Vela Uniform's Critical Years (1961—1963)

Date and Year	Source	Reporter	Gist of reporting
8/31/61	*New York Times*	White House Statement	"Soviet Resuming Atomic Tests; Kennedy Sees Peril to World."
11/22/61	*Washington Post*	Howard Simons	"Prospect of New Test Ban Talks Puts Focus on Detection Again."
3/10/62	*Washington Post*	Howard Simons	Prime Minister Harold Macmillan reports that remarkable advances in scientific instruments will make test detection easier.
3/25/62	London *Sunday Times*	Tom Margerison	"Britain's New H-blast Detector Is Ready."
6/2/62	*New York Times*	Walter Sullivan	"Underground Test Detection Splits Scientists — Opinion Divided on Ease of Spotting an Explosion by Earth's Tremors."
7/23/62	White House Press Conference	John F. Kennedy, President	"As you know, there has been additional information gathered as a result of our underground tests — the ability to detect an underground test, its range, and to distinguish between an underground test and an earthquake ... the national government considerations of this information should be concluded by the end of this week."
12/19/62	The Kremlin (Private Letter to J. F. Kennedy)	Nikita Krushchev, Premier	"... the Soviet Union has made lately an important step toward the West and agreed to installing automatic seismic stations . . . we stated our agreement that three such stations be installed on the territory of the Soviet Union in zones most frequently subjected to earthquakes."
12/28/62	The White House (Private reply to N. Krushchev)	John F. Kennedy, President	"... your implication that on-site inspections should be limited to seismic areas also gives us some difficulty . . . an unidentified seismic event coming in an area in which there are not usually earthquakes would be a highly suspicious event."
3/7/63	White House Press Conference	John F. Kennedy, President	"I want to say that we have made substantial progress as a result of a good deal of work by the United States Government in recent years in improving our detection capabilities."
3/20/63	*Washington Evening Post*	Earl H. Voss	"U.S. Reduces Its Terms for Atomic Test Treaty"; U.S. advised the Kremlin that seven on-site inspections a year satisfactory instead of eight to ten. Russians still offer two or three inspections a year on take-it-or-leave-it basis.
8/5/63	The Kremlin	—	Foreign ministers of United Kingdom, Soviet Union and U.S. sign limited test ban treaty (outlawing all but underground nuclear tests).
8/15/63	*Washington Post*	Howard Simons	"Teller Has New Reasons for Opposing A-Test Ban."

(1) How could one design a seismic detection system five times more sensitive than any in existence and thereby identify 90 percent, rather than 50 percent, of the Soviet·earthquakes of magnitude 4.0 and above?

(2) Could such a capability be provided by a steerable seismic array several hundred km across and interlinking its output with those of several other comparable steerable arrays thousands of km away and in totally different directions from the Soviet Union?

Before a minute was up, Dr. Romney rebutted with AFTAC's standard party line relative to the Vela Uniform Program: "It won't work. It would cost too much. And even if it worked, we wouldn't use it!" But Frosch knew what he wanted. When the Air Force began a heel-dragging operation with regards to constructing the "*L*arge *A*perture *S*eismological *A*rray" (LASA), Frosch quietly observed that the Office of Naval Research would be happy to take on the task if the Air Force didn't want it. From then on, the Air Force was totally receptive for they knew that Frosch truly did understand how to get things done in the "five-sided puzzle palace".[54] Moreover, if AFTAC couldn't figure out how to do it, Frosch had his own trouble-shooter, Harry Sonneman, who typically generated suitable technical solutions within a day or so.[55]

As it worked out, LASA was built on the open, windy plains of eastern Montana. The array actually consisted of 21 subarrays (25 detectors each) planted in a circular area 200 km across, or in an area approximating the size of Belgium. The first two subarrays were a Texas Instrument project using specially modified refraction geophones (Hall-Sears 10-1/ARPA) implanted within an 18-cm diameter casing, the center detector being at a depth of 150 m and the other 24 detectors at a depth of nearly 60 m. This type of burial proved to reduce the surface noise level by 25 db, and *U*nited *E*lectro*d*ynamics (UED) was immediately placed under contract to build and hook-up

---

(U.S. Arms Control Agency), Robert Frosch and Charles Bates (ARPA), H. I. S. Thirlaway and Eric Carpenter (British Tabor Pluto Project), Paul Green and Edward Kelly (Lincoln Laboratory of MIT), Albert Rubenstein (Institute of Defense Analysis), John DeNoyer (University of Michigan), and Victor Anderson (Marine Physical Laboratory, University of California).

[54] Frosch became Deputy Director of ARPA in 1965, Assistant Secretary of the Navy for Research and Development (1966–1972), and Administrator of the National Space and Aeronautics Administration (1977–1981).

[55] By this time, Texas Instruments' "multiple array processor" had been installed for testing at the McMinnville, Tennessee, site. Conceived in large part by Dr. Milo Backus, this processor could manipulate seismic signals from 19 detectors in 10 different ways using processing delays of as much as 12 minutes.

the remaining 19 subarrays before 1 June 1965. As a result, scores of field personnel from UED and Mountain Bell Telephone suffered through the worst Montana winter of the past 43 years to complete the array on schedule. To do this, 900 km of telephonic circuitry cable were buried to avoid lightning strike and livestock problems. Signals from these cables then passed through 800 km of new telephone lines and a specially built microwave relay circuit capable of passing on the 150,000 bits of seismic data generated each second for processing at the master data reception center at Billings, Montana, or for further transmission to the Vela Seismic Analysis Center in Alexandria, Virginia, and the Lincoln Laboratory at Bedford, Massachusetts.[56, 57]

After the expenditure of $10 million, LASA was formally dedicated on 12 October 1965 with most of the enthusiasts for large seismic arrays present. In one of the keynote speeches, Dr. Eric Carpenter of the British Tabor Pluto program suggested that perhaps the time was now ripe for Dr. Frosch to design the ultimate "world array" using seismic data generated in many different parts of the world and fed by satellite into a single multiple array processor. In fact, he and his associate, Dr. H. I. S. Thirlaway, had digitally processed in a crude sort of way — but still in array format — the signal from the underground Bilby explosion (235 kilotons at the Nevada Test Site) as recorded in Norway, Germany, Bolivia, Hawaii, Canada, and Tennessee and found good signal coherence along such deep paths. As a result, they were very much of the opinion that the main problems remaining were ones of careful beam steering and surface noise eradication (Carpenter, 1965). Although simultaneously working so many detectors and subarrays proved to be a "beast" in terms of complexity and computer time, LASA did prove to be a very helpful tool in advancing the state of the art in seismic detection. For example, by July, 1966, E. J. Kelly of Lincoln Laboratory reported that with a beamformer program, the threshold of detection for 75 percent of observed events $30°$ to $90°$ (3,300 to 10,000 km) away was moved

---

56. Details on the construction and initial achievable results may be found in the first six papers included in the December, 1965 *Proceedings*, Institute of Electrical and Electronic Engineers. Top governmental troubleshooters for this massive project were Major Robert Meek, USAF, of AFTAC and Mr. Harry Sonneman of ARPA. Design and field engineers turning in notable performances were C. B. Forbes, R. Obenchain, and R. J. Swain of UED and R. V. Wood, Jr., R. G. Enticknap, C. S. Lin, and R. M. Martinson of Lincoln Laboratory.

57. UED Vela Seismic Analysis Center personnel deeply involved in this project included Drs. W. O. Dean, R. L. Sax, E. A. Flinn, and C. B. Archambeau plus Messrs. J. C. Bradford and P. W. Broome. Overall direction of the Center at that time (and still so in 1981) was by Dr. Robert Van Nostrand, who edited *Geophysics* during 1964 and 1965 and was named an "Honorary Member" of the SEG in 1979. At that time, the Center's UED staff numbered 48, supplemented by 16 AFTAC personnel directed by Captain Clent Houston, USAF.

down from magnitude 4.2 for a single detector to 3.7 when using the entire LASA array (Kelly, 1966). But let us now turn to what the British had been doing, for they had been working with steerable seismic arrays as far back as 1961.

Tabor Pluto — The British Equivalent to Vela Uniform

Once the British *A*tomic *W*eapons *R*esearch *E*stablishment (AWRE) at Aldermaston, near Reading, ceased testing in late 1958 (as did the Soviet Union and the United States), it appeared appropriate to organize a technical program specifically supporting negotiations at the continuing test ban talks in Geneva. AWRE accordingly retreaded some of its smarter scientists into the field of nuclear-test detection under the direction of the laboratory's field test director, a wiry Welshman by the name of Ieuan Maddock.[58] The title of his project was, appropriately, Tabor Pluto.[59] In contrast to the American approach dominated by geophysicists, the initial Maddock team did not even have a nationally recognized seismologist on board. Instead, the team was composed of specialists in electronics, radio, optics, communication theory, and radiation measurement. As a consequence, the problem to them was definitely one of communications, with special emphasis on the nature of the signal form and what happened to it along the transmission path. Hence, one of the group's first efforts led to creation of a wide-aperture, tunable, steerable seismic array long before the Americans quit squabbling among themselves on how the "Berkner array" should be built.[60]

Locating a relatively quiet site for seismological studies in the heavily populated islands of Great Britain was not easy. However, a small experimental array was laid out on Salisbury Plain not far from Stonehenge for at night this was the quietest area of England. Then, in March, 1961 under the direction of Frank Whiteway, the system was tested on a very foggy night by having a Royal Navy frigate cruise up the English Channel dropping 136-kg depth charges along the way. By this time, Maddock had brought aboard as over-all project director Dr. Hal I. S. Thirlaway, the Cambridge-trained geophysicist who had just completed installing a seismological network in Pakistan for a

---

58. Eventually raised to the rank of knighthood, Maddock became Chief Scientist, Ministry of Trade and Industry, in the early 1970s and then moved on to the post of Executive Secretary, British Association for the Advancement of Science.

59. In Greek, "Tabor" means "drum", while "Pluto" was lord of the underworld.

60. As in many other technical developments, the British effort was a tightly managed in-house effort using few but highly capable scientists, engineers, and support technicians. By being flexible, economic, and highly productive, Tabor Pluto kept the more comprehensive but ponderous American program very much on its toes.

United Nations technical aid program. Slim and persuasive, Thirlaway soon concocted the term, "Forensic Seismology", to indicate the application of seismology for aiding the law in "fixing the perpetration of a crime" (the detection of treaty violations) and for "rescuing an innocent person from a falsely imputed crime" (the avoidance of calling a natural earthquake an underground nuclear explosion).

Although late on the scene, Thirlaway totally agreed with the AWRE team's concept of employing an array system capable of inserting signal delays before summing the individual seismic inputs. He soon arranged for a second AWRE team directed by Dr. Eric Carpenter to join the Whiteway group as Carpenter and his associates had already experimentally verified the Latter decoupling theory first publicized by Dr. Hans Bethe at the Geneva negotiations two years earlier.[61] By firing a series of chemical explosions and specially shaped charges in a mine near Callington, Cornwall, they had actually been able to reduce the signal level by factors of up to 100 when the detonations occurred in spherical excavations within the granite.[62] In fact, they found that signal amplitude could be reduced by a factor of four simply by adding a thermal absorber such as a water shield around the high explosive.

In March, 1961 Thirlaway also arranged for the combined teams to move three km up the road to a converted country house, Blacknest, where, in an open environment, they were free to invite anyone who might help with the seismic discrimination problem.[63] Once in the new location, the Tabor Pluto group launched the construction of two improved steerable arrays of seismometers to acquire "first zone" data for work directed towards improving the resolution of "first motion" and of "second arrivals" that might allow better estimates of source depth. Once the Soviets broke the test ban in early September, 1961, the British raced against the clock and had their first array in place on Pole Mountain, Wyoming, between Laramie and Cheyenne in time to monitor the unique Gnome nuclear detonation on 10 December 1961. Intentionally located 1,000 km from both the Gnome event and the Nevada Test Site, the installation lay at the distance where first motion amplitudes of seismic signal in the Hard-

---

61. This theory predicted that if an explosion of given yield was detonated in a hole of such radius and depth that it would deform elastically, the amplitude of the radiated seismic signal would be reduced by a maximum factor of 300 (Latter *et al.*, 1959).

62. The U.S. Atomic Energy Commission had the Coast and Geodetic Survey and the National Geophysical Company conduct a comparable series of experiments, termed Cowboy, in the Carey salt mine near Winfield, Louisiana, during late 1959 and early 1960. The results obtained were comparable to those acquired by the British.

63. VESIAC, for example, held an international conference during June, 1963 at Blacknest on "The Role of Seismic Arrays in the Detection of Underground Tests".

tack shot series had fallen to a minimum. The second seismic array, also roughly L-shaped and with ten short-period, vertical seismometers spaced at intervals of about one km along each arm, was installed in early 1962 at Eskdalemuir in the Scottish Borders.[64] Array apertures for both installations were thus roughly equal to the maximum wavelength of initial seismic signals recorded within the first zone of detection.

Though improvements in signal/noise and signal identification were clearly demonstrated with these arrays, the difficulties of first zone verification were not solved; indeed, the improved resolution and detection capacities just revealed more unidentified earthquakes. Further, once the Soviets had arbitrarily broken the moratorium on nuclear tests, the original concept of 180 or so first zone monitoring stations became less and less realistic in a political sense. In consequence, the British began advocating the possibility of verifying compliance with a controlled test ban by use of a detection network sited entirely outside the Soviet Union. They advanced two principal arguments: (1) because of the nature of the earth's layering, the amplitude of the first seismic arrival at 5,000 km is roughly equal to that at 500 km, and (2) the shape of the seismic signal when recorded at distances of 3,000 to 9,000 km appeared less distorted than when recorded closer in. These arguments were sustained by one of the first Soviet underground tests near Semipalatinsk, East Kazakhstan, on 2 February 1962. Thus, when top-ranking British scientists led by Sir Solly Zuckerman (Science Advisor to the Prime Minister), Sir William Penney, and Ieuan Maddock met a month later in Washington, D.C. with their American counterparts, Drs. Jerome Weisner (Science Advisor to the President), Franklyn Long, and Frank Press, the British indicated their intention to drop all "first zone verification" research and to concentrate on "third zone" (teleseismic) research.

Records obtained of the French underground explosion on 2 May 1962 in the Sahara Desert convinced the Tabor Pluto group even further of the correctness of their decision. The six-element array at Pole Mountain was therefore enlarged to an aperture of 18 km or roughly the wavelength of the first arrival from seismic events 3,000 to 9,000 km away. In the meantime, negotiations with the Canadian Government had been successful, and work was going forward to create a 25-km aperture array near Yellowknife in the Northwest Territories.[65] Despite many installation difficulties, including the

---

64. Recording at both sites was by a specially developed 24-channel analog magnetic tape deck.

65. Canada's Department of Energy, Mines, and Resources was particularly helpful to both the Tabor Pluto and Vela Uniform Programs during the 1960s and the 1970s. The

proclivity of many animals to chew on the telemetry cables, the array came on line in December. Thus, by late 1962, the British were well along in detecting and analyzing seismic signals recorded 3,000 to 9,800 km from the source for, as Thirlaway observed, "This broad zone is, as it were, a window through which seismic waves pass without distortion".

The initial results using steerable arrays were extremely exciting. To highlight the difference between the relatively sharp, short-term wave train of a buried explosion and the longer, more ragged wave train from a typical earthquake, the Tabor Pluto group developed a "correlogram" technique which graphically portrayed the relative rate of arrival of distant seismic energy recorded within the area covered by the array.[66] As the data built up, they found that 90 percent of the recorded earthquakes had a rather complicated appearance, often with pulses of energy following one after another for as much as 40 seconds or having a pronounced tail of low level but still coherent energy following the large initial pulse. In contrast, underground explosion correlograms exhibited a single hump of energy (see Fig. 6.7). Later they found that the rate of energy arrival from earthquakes exhibited a definite dependence on azimuth and distance to the event, a particularly important consideration for they were soon observing that some earthquakes of shallow origin gave seismic signatures at certain stations that were as simple as those of distant underground explosions.

To study these phenomena further, the Tabor Pluto group, via its scientific contacts within the Commonwealth, was able in 1965 to arrange for cooperative installations of British-type seismic arrays at Gauribidanur in southern India 80 km north of Bangalore and at Warramunga near Tennant Creek in the remote Northern Territory of Australia. In the former instance, the array is operated by the Bhabha Atomic Research Centre at Trombay, and in the latter case by the Australian National University at Canberra. As far as South America was concerned, the burly, jovial British seismologist, Dr. Patrick L. Willmore, was able to persuade the Brazilian Government to install a similar array near Brasilia in cooperation with the United Kingdom's Institute of Geological Sciences.

Department's Earth Physics Branch (formerly the Dominion Observatory under Dr. John Hodgson) created a Seismic Applications Section in 1964 and became very active in the area of seismological verification of a controlled test ban. In the late 1960s, the Branch also took over the operational responsibility for the Yellowknife array (Manchee and Cooper, 1968).

66. A correlogram is created by multiplying the separately summed and phased signals from the two array lines in real time and then smoothing this result by integrating over a time interval of 1.5 to 2.0 seconds.

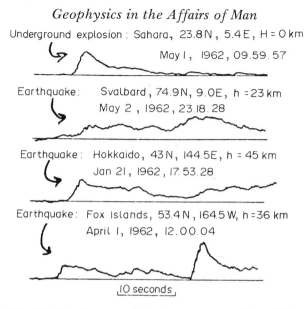

FIG. 6.7. Correlograms of seismic energy recorded at Tabor Pluto array, Pole Mountain, Wyoming. (From McGraw-Hill *Yearbook of Science and Technology.* Copyright 1965 by McGraw-Hill and used with their permission.)

Five of the British-style arrays (Australia, Brazil, Canada, India, and Scotland) remain operational as of 1981, with the first two being the only steerable seismic arrays in the Southern Hemisphere.[67] Inasmuch as the first four of these unique observatories are located on low noise Pre-Cambrian Shield rocks, this well-distributed, skeleton network forms an integrated recording system that provides high signal-to-noise seismograms — in which signal-generated (reverberation) noise has been cleaned up by virtue of delay and summation processing. In addition, with the help of modern computers, the system is capable of yielding wide-band recordings in the spectrum of 10 hertz to 0.1 hertz. Such recordings have become of increasing importance for Alan Douglas at Blacknest, in cooperation with John Hudson of Cambridge University, pioneered in formulating a scheme for reliably producing synthetic seismograms in this band. This source modelling technique, as improved on by Robert Pearce of Blacknest, is now capable of deciding, within appropriate confidence limits, whether a seismic event in the far region was generated by an explosion or by an earthquake that looks like an explosion. Before this stage was achieved, the best way of recognizing an underground

67. The Pole Mountain installation was closed out during September, 1963.

explosion was by comparing the relative amounts of high- and low-frequency energy within the broad-band recording — the so-called m_b/M_s (body wave versus surface wave magnitude) criterion developed to a considerable degree by Tabor Pluto's Peter Marshall.

Research Explosions and Their Impact

By late 1958, about the only thing that pro- and anti-nuclear-test-ban proponents could agree on was that there should be a series of carefully monitored chemical and nuclear underground explosions to determine the true seismological state-of-the-art.[68] Within ARPA, all of 1960 and part of 1961 was spent with the assistance of the Latter Committee in debating and planning the minimal number of explosions necessary for firing under differing yield, depth of burial, surrounding media, and coupling situations. Three interlinked shot series proved necessary — Shade in Nevada using underground nuclear explosions, Dribble involving partially and totally decoupled nuclear explosions, within a massive salt bed, and Groundhog, which would be a series of 100-ton or smaller chemical explosions in granite, sandstone, volcanic tuff, and limestone. President Kennedy then invited the Soviet Union to participate in the research effort but quickly abandoned the idea when the Russians indicated they would like to do so but needed to inspect each nuclear device before firing to make certain that they were not contributing to a surreptitious nuclear weapons test program.

Meanwhile, AFTAC arranged for Geotech to equip and man 40 "*L*ong *R*ange *S*eismic *M*easurement" (LRSM) vans, half to be placed in semi-permanent locations and the other half continually repositioned to suit the particular monitoring requirements of each research explosion.[69] By late 1964, some of these units were deployed as far away as the Aleutian Islands, Canada, Bolivia, Germany and Norway.[70] Inside the 200-km monitoring zone, the measurement program was far more complex. For example, Vela's initial five-kiloton nuclear explosion, Orchid, was to involve 40 different projects. Despite this elaborate planning, Orchid was never fired, for when the Soviets restarted the nuclear testing race on 1 September 1961, the two

---

[68.] During September, 1957, the quiet-spoken Australian seismologist, Dr. Keith Bullen, co-author of the revered Jeffreys—Bullen travel time curves, urged in his presidential address to the International Association of Seismology and Physics of the Interior of the Earth that one or more nuclear devices should be fired for seismological purposes.

[69.] The three-man LRSM units were outfitted with seismometers comparable to those used in the Worldwide Standardized Seismograph Network; however, signal recording was upgraded to magnetic tape decks and Geotech-developed "Helicorders" and "Develocorders".

[70.] Canada allowed the LRSM vans to roam within its borders but Mexico never did.

weapons laboratories of the University of California — Livermore and Los Alamos — raced to see which would get off the first new American test.[71] Livermore won, firing the 2.6-kiloton device, Antler, within Rainier Mesa exactly two weeks later and nosing Los Alamos out by a day.

However, the underground nuclear explosion that, seismically speaking, was heard around the world came on 10 December 1961. This was the 3.1-kiloton Gnome event, the first in LRL's highly touted Project Plowshare series designed to demonstrate that nuclear explosions could actually be beneficial to mankind.[72] Gnome, emplaced in bedded salt within the potash mining district 40 km southeast of Carlsbad, New Mexico, was designed to determine whether such an explosion generated enough steam for electrical power, fractured sufficient rock to be used in freeing up tight oil and gas sands, and manufactured useful radio-isotopes. At the press conference the night before, Dr. Edward Teller, the "Father of the H-Bomb", advised the audience of 350 persons from ten countries that they would likely see the "miracle of the decade", although he hastened to add: ". . . if everything goes right, we will see nothing. However, I do think there might be a little cloud of dust if the ground jumps."

Unfortunately for Teller, when the shot went off the next morning, the observers saw plenty for a growing plume of radioactive steam quickly rose from the mine shaft.[73] Seismically, however, the explosion did prove to be a miracle. Through the good auspices of the SEG's Committee on Cooperation with Governmental Agencies, 30 volunteer geophysical exploration crews had taken to the field to monitor the event even though it occurred on Sunday.[74] Their data,

---

71. The Russian test series was an interesting one for it included a 55,000-kiloton explosion on 30 October 1961 (the Hiroshima explosion was about 3,000 times smaller). As Kruschev explained to the press, the test series was needed "to deter capitalists from threatening the Soviet Union". This was, of course, the month that the Soviets began building the infamous "Berlin Wall". (*Note*: Explosions in the million-ton chemical equivalent yield category are generally stated in "megatons", with 1,000 kilotons being equal to one megaton.)

72. The term, Plowshare, came from the biblical Book of Isaiah, ". . . and they shall beat their swords into plowshares". Livermore was particularly anxious to experiment with building a sea-level Panama Canal and harbors in Alaska and Australia by the use of nuclear excavating techniques.

73. On this same morning, the famed philosopher, Sir Bertrand Russell, was urging 50,000 fellow citizens to conduct a massive "Ban-the-Bomb" invasion of assorted United States Air Force bases in Great Britain. Nevertheless, this particular invasion never did occur because of bad weather and massive security precautions.

74. The Committee's initial membership consisted of: L. D. Ervin, Chairman (Tenneco), and Members: L. Howell (Humble Oil), S. Kaufman (Shell Development), W. E. N. Doty (Continental Oil), F. G. Blake (California Research Corporation), L. Mott-Smith (General

supplemented by that from the LRSM vans and conventional seismological observatories, provided arrival times at 157 stations (Romney *et al.*, 1962). When Romney and others calculated the supposed depth and location of the Gnome shot using conventional travel timetables, they were surprised to find that the shot apparently occurred 30 km high in the atmosphere and 40 km to the northeast of the actual shotpoint.[75] Just as surprising was a tremendous difference in signal strength depending upon which direction the detector unit lay (see Fig. 6.8).[76] However, savvy Professor Eugene Herrin at Southern Methodist University was quick to point out that there really was no problem if one would drop the assumption that seismic velocities were constant in the earth's upper mantle and be willing to admit that such velocities could be as low as 7.5 km/sec in the Nevada—Utah region and as high as 8.5 km/sec in southeastern Oklahoma. In fact, by using his own HYLO computer program, the epicentral fix for Gnome came to just 1.5 km west of ground zero (Herrin and Taggart, 1962).

Classical seismological assumptions were shattered even further during the summer of 1963 when the U.S. Coast Guard cutter

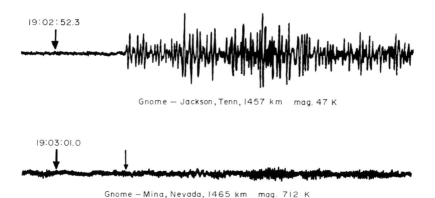

19:02:52.3

Gnome — Jackson, Tenn, 1457 km   mag. 47 K

19:03:01.0

Gnome — Mina, Nevada, 1465 km   mag. 712 K

FIG. 6.8. Example of directional effect on seismic signal level for the Gnome event (Tennessee to the northeast and Nevada to the northwest of ground zero).

Geophysical), B. Rummerfield (Century Geophysical), C. Savit (Western Geophysical), R. Peterson (United Geophysical), J. Hawkins (Seismograph Service Corporation), F. Romberg (Texas Instruments), H. Mendenhall (Phillips Petroleum), T. J. O'Donnell (Gulf Research and Development), D. Silverman (Pan American Petroleum), and Ira Cram, Jr. (Pure Oil).

75. The Geneva test ban negotiations assumed that practically all near-surface seismic events of approximate magnitude 4.0 and up could be correctly assigned to an area about 200 square kilometers in size.

76. Gnome's Richter body-wave magnitude ranged from 3.1 in a northwesterly direction to 6.5 in a northeasterly direction. Within the United States, initial signal arrivals were up to seven seconds slow in Nevada and up to four seconds fast in the Great Lakes area.

Woodrush began firing a series of chemical explosions, mostly one ton in size, along the long axis of Lake Superior as part of the International Upper Mantle Project then getting under way.[77] The resulting seismic signals not only showed up locally but also at all the Geneva-type seismic detection stations and many of the LRSM vans.[78] Such a unique propagation phenomenon was too good to pass up, and the *Woodrush* repeated the exercise in 1964 and again in 1966.[79] Upon working up the 1963–1964 shot series, the Vela Seismic Analysis Center reported that a ten-ton Lake Superior chemical explosion should be assigned a body-wave magnitude of 3.7 if measured in the western United States and one of 6.3 if measured in the eastern United States (Mansfield and Evernden, 1966).[80] Moreover, the central part of the North American continent did not have a seismic shadow zone at distances of 600 to 1,600 km as did the mountainous western United States (Iyer *et al.*, 1969).

Finally, on 26 October 1963, the Vela Program was able to get its first nuclear research explosion fired outside the Nevada Test Site. This 12-kiloton event, Shoal, was designed to compare equivalent man-made and natural seismic events on identical instruments and over comparable signal transmission paths. The site picked was the Fairview Park area east of Fallon, Nevada, and within 40 km of the epicenter of an earthquake of magnitude 4.4 which had occurred within 10 km of the surface on 20 July 1962.[81] The comparison worked well, for Shoal proved to have a magnitude of 4.9 and a calculated epicenter within one km of the actual shot point (Van Nostrand and Helterbran, 1964). The data supported several of the

---

77. Instigated by Dr. John Steinhart of the Carnegie Institution of Washington and Dr. Robert Myer of the University of Wisconsin, the experiment involved thirteen research institutions studying the nature of the basement rock in the Pre-Cambrian Shield of North America. Although it had been assumed that the region's make-up would be relatively uniform and simple, the actual stratigraphy proved to be quite complex (Steinhart, 1974).

78. Stations receiving clear signals on many occasions included Houlton, Maine (1,320 km), Beckley, West Virginia (1,165 km), Wichita Mountains Seismological Observatory (1,487 km) and the Tonto Forest Seismological Observatory (2,304 km). An underwater shot of 0.12 ton actually produced a measurable signal at Glendive, Montana (1,128 km).

79. The results of this trio of field operations are compiled in the 43 paper monograph, *The Earth Beneath the Continents*, published by the American Geophysical Union in 1966 to honor the 65th birthday of the crustal pioneer, Dr. Merle Tuve.

80. As of 1966, the U.S. Geological Survey was still not certain why the western United States was such an absorber of seismic energy. The phenomenon could be caused by a mineralogic difference in the deep crust, by a lesser increase in velocity with increased depth, or by a combination of the two. Nevertheless, the experiment showed that ten tons shot in Lake Superior could give a P-wave signal equivalent to that generated by a ten-kiloton detonation at the NTS in volcanic tuff out to distances of 2,000 km in many instances.

81. Carl Romney and his faculty adviser, Professor Jack Evernden of the University of California (Berkeley), knew the region well as Romney's doctoral dissertation, "The Dixie Valley–Fairview Peaks Earthquakes of December 16, 1954", studied this very area.

postulated identification criteria, including (1) the earthquake's initial P-wave shape was more complicated than that of the explosion, (2) the ratio of compressional to shear wave amplitude was 0.35 for the earthquake and 1.7 for Shoal, and (3) the earthquake generated much larger surface waves, with the "Love waves" having amplitudes six times greater than those from Shoal.

Now that the Tabor Pluto and Vela Uniform research programs showed signs of progress, the anti-test ban forces were anxious that the Latter "big hole" decoupling theory be tested using nuclear explosions. After screening the 200-plus salt domes between the Mexican border and southern Alabama, ARPA decided to use the Tatum salt dome 30 km southwest of Hattiesburg, Mississippi, a city of 35,000, for the Dribble shot series.[82] Although the AEC had already fired nearly 200 nuclear explosions within the domestic United States, only three — Trinity, Gnome and Shoal — had been outside the remote Nevada Test Site. Yet once the problem had been explained to the appropriate state officials, Governor Ross Barnett readily went on statewide television to advise his citizens of what was planned. Field work started in 1962, and it soon became evident that the shot series should be reduced from six to two events. The first of these, Salmon, would be a fully coupled five-kiloton shot at a depth of 830 m to create a spherical void 36 m across, followed by Sterling, a 0.4-kiloton event to be shot in a fully decoupled format within the Salmon cavity.

Although hundreds of workers were employed at the Tatum site, the earliest Salmon could be detonated proved to be late September, 1964, right in the midst of the hurricane season. Hurricanes Hilda and Isbell did appear, and it took nine countdowns during the month after 22 September before the firing circuits could be closed. The initial and key press conference held on 27 September went off extremely well, for the chief briefer was the avuncular James E. Reeves, Director of the AEC's Albuquerque Operations Office. After pointing out that the site had been intensively studied for the past two years relative to wind trajectories, ground water conditions, and background levels of radioactivity, Reeves concluded:

> ... we have never before fired a nuclear device underground so near to where people live. ... We don't expect to hurt anyone. ... Property damage we can pay for if we need to; human injury is not readily resolved. The fact is that we are asking everyone who lives within about a mile and a half of the shot point to leave before shot time, and others out to four and a half miles to stand outside; we are blocking roads to prevent unauthorized persons entering the hazard area, and we have installed bracing and reinforcement in about 70 structures near the shot point.

82. Ground-zero lay within eight km of the state's second largest oil field and the near-by hamlet of Baxterville.

So we're about as prepared, on about as sophisticated a basis, as I think we can be, in protecting the safety and the interests of the public . . . the calculations have been made with great care, we have the experience of more than 100 underground nuclear detonations behind us, and I think we're ready.

Reeves' prediction proved correct. At ground-zero, some wag left a Confederate flag and a sign, "The South will rise again!" This it did by about ten cm amidst a cloud of dust. Observers five km away heard a "whoomph" and felt a shock comparable to jumping off a street curb. However, the surface ground roll did not die down as quickly as expected, and later homeowners as far away as Hattiesburg asked to be reimbursed for cracked plaster.[83]

As predicted, the Salmon detonation left a nearly spherical cavity 33 m across. The decoupled shot, Sterling, was fired in this cavern on 3 December 1966 and indicated a decoupling factor of 70 ± 20 (Springer *et al.*, 1968). Despite this second event, the cavity stood up well and was subsequently used for two methane—oxygen explosions of about 100 tons equivalent yield as part of the Miracle Play shot series during 1969 and 1970. In other words, it was evident that several explosive tests could be conducted in a single salt-dome cavity if the cavity were not severely overdriven (Rodean, 1971).

The anti-test ban forces also worried about whether British and American seismologists could really classify correctly the frequent earthquakes in the Kamchatka Peninsula of the Soviet Union. ARPA therefore scheduled the 80-kiloton nuclear event, Longshot, at a depth of 700 meters in the basalt of remote Amchitka Island 600 km east-southeast of Kamchatka. When Longshot was fired on 29 October 1965, the seismic results again were surprising. Travel times were off by as much as five seconds, making epicentral determinations inaccurate. On the other hand, the detonation was very clearly a buried explosion, for its bodywave magnitude was 6.0, as expected, but its surface wave magnitude was a lowly 4.0, a definitely atypical situation for shallow earthquakes.[84] Tabor Pluto's intercontinental seismic array network was also in operation at the time of the shot, and the British analysts were tremendously impressed by the simplicity and symmetry of the energy radiated by Longshot as compared to the asymmetrical wave patterns generated by earthquakes occurring in the same region.

---

83. The total cost of the Salmon event was approximately $15.48 million. Except for one false alarm during 1979, there never has been any problem with radioactive contamination at the site.

84. Once Professor Evernden replaced Dr. John DeNoyer as Director of ARPA's Underground Nuclear Test Detection Branch in 1969, he established an *Alaskan Long Period Array* (ALPA) to test further the proposition that shallow earthquakes generated more surface wave motion than did comparable underground nuclear explosions.

Ever since the negotiations started concerning a controlled underground-nuclear-test ban, mention had been made of the need for at least some on-site inspections of suspicious seismic events. To make certain that adequate inspection techniques and associated handbooks existed, the Vela Uniform effort supported a large number of projects to determine what methods worked — or did not work. Some twenty techniques proved to have potential in this regard. For example, did aerial photography show a localized wilting of vegetation or an unusual depression? Was the seismic aftershock pattern typical of a natural event? Did borehole logging show any unusual thermal, acoustic or neutron levels? All were field tested, and the results compiled by *S*tanford *R*esearch *I*nstitute (SRI) in an easy-to-use format should the need ever arise for deploying such teams.[85]

International Seismology with a Neutralist Flavor

The *U*nited *N*ations *E*ducational and *S*cientific *C*onsultative *O*rganization (UNESCO), operating as if the underground-nuclear-test ban issue did not exist, called the "First Intergovernmental Meeting of Experts on Seismology and Earthquake Engineering" at UNESCO House in Paris during April, 1964.[86] Ten technical issues were chosen for discussion by representatives of the 33 participating countries. These issues included: (1) world-wide standardization of seismometers and observatory practice, (2) fostering collection and diffusion of seismic data, (3) measurement of close-in ground motion from strong earthquakes, (4) provision of on-site inspection teams for areas that had just experienced destructive earthquakes, and (5) enhancing education and training of seismologists. Although such topics pertained directly to the goals of Vela Uniform, the cultural side of the U.S. State Department acted as if that program did not exist and appointed academics to all the slots on its delegation except one for its own representative, Dr. J. Wallace Joyce.[87] As a result, the three federal agencies most intensively involved in international seismology — the Coast and Geodetic Survey, the National Science Foundation, and ARPA — were only permitted observer status. Compounding the

---

[85.] Leading SRI specialists in this field of endeavor were: Richard M. Foose, Sylvan Rubin, Warren Westphal, Robert B. Vaile, Jr., Harvey Dixon, and Lawrence Swift.

[86.] UNESCO's seismological budget for 1963–1964 was a minuscule $45,000, increasing to $117,000 in the 1965–1966 biennium. Approximately one-third of UNESCO's total budget was supplied by the United States.

[87.] The U.S. delegation was chaired by the engineering seismologist, Professor D. E. Hudson of CalTech. Other members were Professors Clarence Allan (CalTech), Ray W. Clough (University of California at Berkeley), Jack Oliver (Lamont Geological Observatory), and Daniel J. Linehan (Weston Observatory), plus Dr. Joyce of the State Department.

problem was Joyce's tardy arrival at UNESCO and the U.S. Delegation's inability to submit formal resolutions in advance.

In contrast, the Soviet delegation was headed by Professor V. V. Beloussov, past President of the International Union of Geodesy and Geophysics, who promptly submitted formal resolutions regarding all technical issues on the agenda. Other countries were also strongly represented, including Dr. H. I. S. Thirlaway for Great Britain, Dr. John Hodgson for Canada, and Baron C. H. G. von Platen for Sweden.[88, 89] By the time the conference closed, all resolutions of greatest import to Vela Uniform had been passed, largely because of the good efforts of Dr. Hodgson and the Soviet delegation. In fact, it was quite evident that the Soviets were in a "sweetness and light" phase and taking every opportunity to export ideas, handbooks, and manpower to the "Third World".

After the UNESCO conference was over, Sweden and Canada began to play increasingly important roles in international forensic seismology. Under Dr. Ulf Ericsson, the Swedish Research Institute for National Defense built its own seismic research array with long-period seismometers near Hagfors (about 150 km due east of Oslo, Norway) in May, 1969. Even before, the government-financed "Stockholm International Peace Research Institute" called week-long conferences in April, 1968 and again that June in Tallberg, Sweden, to thresh out objective methods for identifying the true cause of a seismic signal. The conference's end-product, *Seismic Methods for Monitoring Underground Explosions,* was ably edited by the British geophysicist, David Davies, and signed by knowledgable scientists from both sides of the Iron Curtain.[90, 91] One of the most important conclusions reached

88. As a youth, Dr. Hodgson assisted his father, Dr. Ernest Hodgson, with seismological field work each summer. In 1939, the senior Hodgson noted that energy released by a large rockburst in a deep mine 500 km away at Kirkland Lake, Ontario, was appearing on seismograms in Ottawa. Young John followed through and by using a selection of rockbursts as the energy source, developed the first deep crustal profile made in Canada. Upon his father's retirement in 1951, John became Chief, Seismological Division, and then, after becoming Director-General, Observatories Branch, Department of Mines and Technical Surveys, took on a UNESCO assignment during 1974 in Asia.

89. The Baron, by then Ambassador to France, had been Sweden's "Number Two" delegate to the eighteen-nation disarmament conference in Geneva, Switzerland, during 1962 and 1963. Other highly able Swedish delegates to the UNESCO conference were Professor Markus Båth, Director, Seismological Observatory, University of Uppsala, who was usually the first to advise the world press that the Soviets had fired another nuclear device, and two other veterans of the Geneva talks, Ulf Ericsson and J. Pravitz.

90. Report signatories were: I. P. Pasechnik (Soviet Union); L. Constantinescu (Romania); V. Karnik (Czechoslovakia); U. Ericcson (Sweden); S. Miyamura (Japan); T. G. Varghese (India); Pierre Mechlèr (France); Kenneth Whitham (Canada); H. I. S. Thirlaway (United Kingdom), and Frank Press and Eugene Herrin (United States).

91. Swedish interest in test-ban verification issues continued during the 1970s. One useful end-product was the 440-page book, *Monitoring Underground Nuclear Explosions,*

was that data from the Worldwide Standardized Seismograph Network clearly showed a significant difference in the generation of body and surface wave energy by explosions and earthquakes. Besides Evernden of ARPA, Peter D. Marshall of the Blacknest group, Peter W. Bashman and Kenneth Whitham of the Seismological Service of Canada, and I. P. Pasechnik of the Soviet Union became deeply involved in establishing the reliability of this criterion (see Fig. 6.9).[92]

Thus, the decade of the 1960s closed with the geophysical profession in reasonably close agreement as to what their science could and could not do relative to controlling a formal underground-nuclear-test ban. To be sure, various individuals still had their own interpretation of the situation. For example, the caustic, humorous editor of

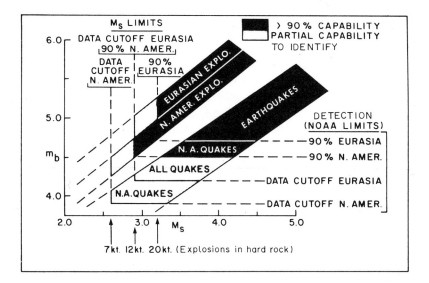

FIG. 6.9. Diagram of capability to detect and identify underground explosions according to difference in body-wave magnitude (m_b as determined with 1-second waves) and surface-wave magnitude (M_s as determined with 20-second waves.) (Reproduced with permission of Canadian Department of Energy, Mines and Resources.)

---

by Ola Dahlman and Hans Israelson of the National Defence Research Institute, Stockholm (Elsevier Scientific Publishing Company, 1977).

92. After observing some anomalies in this spectral ratio (m_b/M_s) criterion, Marshall devised a method, while on loan to the Livermore laboratory, that predicted the attenuation factor in differing geophysical provinces, thereby providing an improved method for normalizing the seismic magnitude—nuclear yield relationship. Being able to make such an adjustment is extremely important for the success of any test-ban treaty which introduces nuclear-yield thresholds below which testing is authorized.

Geophysics, Dr. F. A. Van Melle, commented in the first of two special Vela Uniform issues (Van Melle, 1964):

> Vela Uniform has given a powerful impetus to seismic research. . . . It has thus protected the Western World from hastily agreeing to an absolute test-ban treaty, and from heavily investing to the tune of thousands of technical people and billions of dollars in a Geneva-type, worldwide detection network.

And Professor Carl Kisslinger, as President, Section of Seismology, American Geophysical Union, observed (Kisslinger, 1971):

> The experience of seismologists in carrying out their part of Vela Uniform offers lessons of general applicability. We have seen how mission-oriented research, properly planned and administered, can lead to the upgrading of a whole branch of science. Seismology moved from an active and productive, but modestly supported science, in the '50s, to a vigorous, well-funded endeavor in the '60s, adequately equipped with manpower and tools to carry out its responsibility to the acquisition of knowledge and to meeting the needs of the nation that fall within its scope. We have further seen the immense value of interdisciplinary cooperation, an approach now widely recognized and employed.

Industry, too, found Vela Uniform worthwhile. Because this massive project came at a time when geophysical exploration was at a very low ebb, participation in the effort permitted keeping some of the corporate groups at nearly full manning levels. In addition, Vela Uniform's vigorous search for better methods of enhancing signal-to-noise ratios proved useful in the commercial world as the digital revolution gradually affected one and all.

The Substitution of Plate Tectonics for Continental Drift

While one group of earth scientists was deeply involved with nuclear-test-ban issues, another large group was engaged in an equally exciting debate — were the continents drifting apart right before their very eyes?[93] As far back as 1620, the then Baron Francis Bacon had proposed that North and South America were once joined to Africa and Europe. Then, in 1756, the German theologian, Theodor Ilenthal, concluded there had been a tearing apart of the earth's surface after "The Flood". The defense of this claim lay in biblical statements and the reasonable fit of continents when shoved back together.[94] As time went on, others expanded slightly on the idea, and in 1915, Dr. Alfred Wegener published *The Origin of Continents and Oceans*

[93.] For a well-written, detailed account by a first-rate observer of the clash of ideas, read Walter Sullivan's *Continents in Motion — the New Earth Debate* (McGraw-Hill, 1974).

[94.] The only geologist to "pound rocks" on the moon, Dr. (now Senator) Harrison H. Schmitt, radioed on his way out, "I didn't grow up with the idea of drifting continents and sea-floor spreadings, but . . . when you look at the way the pieces of the northeastern portion of the African continent seem to fit together, separated by a narrow gulf, you could almost make a believer out of anybody."

which elaborately drew on many fields within the earth sciences to support the theory of continental drift.[95] For every believer in Wegener's hypothesis, there was a score of vociferous critics.[96] Sir Harold Jeffreys, in his book, *The Earth: Its Origin, History and Physical Constitution,* dismissed the theory by stating (Jeffreys, 1924): ". . . quantitatively insufficient and qualitatively inapplicable. It is an explanation which explains nothing which we wish to explain." Nevertheless, during the next 33 years, astute observers wrote of certain earth features that required better explanations than the authors could give at the time (see Table 6.7). Even as late as July, 1960 the major school of thought, represented by Professor V. V. Beloussov, President of the Upper Mantle Commission of the International Union of Geodesy and Geophysics, held that more field data needed to be taken. In fact, at the Union's XIIIth General Assembly in Helsinki, Finland, Beloussov proposed launching a major new project on "The Upper Mantle and its Influence on Development of the Earth's Crust" that could be a profitable follow-on to the now defunct International Geophysical Year.

But even as Beloussov was calling for continued tree-gazing, Professor Harry Hess was submitting an "Essay in Geopoetry" to the Office of Naval Research which viewed the entire forest. His thinking ran as follows:

> The mid-ocean ridges could represent the traces of the rising limbs of convection cells, while the circum-Pacific belt of deformation and volcanism represents descending limbs. The Mid-Atlantic Ridge is median because the continental areas on each side of it have moved away from it at the same rate — 1 cm/yr. This is not exactly the same as continental drift. The continents do not plow through oceanic crust impelled by unknown forces; rather they ride passively on mantle material as it comes to the surface at the crest of the ridge and then moves laterally away from it. On this basis the crest of the ridge should have only recent sediments on it, and recent and Tertiary sediments on its flanks; the whole Atlantic Ocean and possibly all of the oceans should have little sediment older than Mesozoic.

He then smoothed up his report and published a 19-page paper which further noted that the ocean basins had to be temporary features while the continents were permanent, although the latter might be torn apart and rewelded, and their margins considerably deformed

---

[95.] Wegener was Director of Research, Deutsche Seewarte (German Marine Weather Service) before World War I and assisted his father-in-law, Dr. W. Köppen, in writing *Die Klimate der Vorzeit* (Climate Through the Ages). Wegener died in late 1930 while returning by ski from a critically needed supply expedition to "Eismitte", a meteorological research station on the Greenland Ice Cap.

[96.] Geological arguments for and against the theory were expounded at a special symposium on the subject held by the American Association of Petroleum Geologists in New York City on 15 November 1926 (AAPG, 1928).

TABLE 6.7

Isolated Observations during the 1920s—1950s Leading to the Modern-day Theory of Plate Tectonics

Time period	Observer and affiliation	Observation of note
1925—1927	German Meteor Expedition	Use of the new echo-sounder indicated that the North and South Atlantic Oceans were split by an axial Mid-Atlantic Ridge
1935—1936	J. D. H. Wiseman and R. B. S. Sewell aboard British Research vessel *John Murray*	Carlsberg and Murray Ridges in Indian Ocean link up with Red Sea Cleft and perhaps with East African Rift Valley; these submarine ridges are also split by axial rift valleys
1949	John W. Graham, Carnegie Institution of Washington and Johns Hopkins University	Remanent magnetism in 200-million-year-old rocks of West Virginia indicate position of North Magnetic Pole different at time of deposition than that of today
1949	Hugo Benioff, CalTech	Based on location and depth of earthquakes, oceanic crust appeared to be thrusted under island arcs to an angle of $33°$ from the vertical to a depth of 300 km, followed by a steeper dip of $60°$ to a depth of as much as 700 km
1950	P. M. S. Blackett assisted by J. A. Clegg, P. H. S. Stubbs, and Mary Almond, University of Manchester*	Remanent magnetism in English rocks points to different magnetic pole than that of Graham's; Spain had apparently rotated away from France, opening up Bay of Biscay and forming the Pyrenees Mountains by crushing along the hingeline
1952	R. Revelle, A. Maxwell, and E. Bullard, Scripps	Thermal probes into sea floor indicated that only 20 percent of heat being emitted could be ascribed to a normal cooling earth, leaving excess to be generated by localized radioactivity or upward convectional flow from mantle
1953	J. P. Rothe, Bureau Central International de Séismologie, Strasbourg	Based on epicenters of past 30 years, an almost continuous seismic zone over 30,000 km in length runs close to the median line of the Indian and Atlantic Oceans
1955	S. Keith Runcorn, University of Manchester	Traced positions of North Magnetic Pole as far back as the Pre-Cambrian Period and found European-derived polar positions lay always to east of polar positions indicated by American rocks
1955—1958	A. D. Raff and R. G. Mason, Scripps	Magnetometer tow behind C&GS *Pioneer* off western United States finds pronounced north—south magnetic intensity lineaments which, along the Mendocino—Pioneer Fracture Zone, prove to be offset to the northwest by 1,100 km
1957	Bruce Heezen and Marie Tharp**	Continuous submarine ridge system 65,000 km long and often split by an axial rift wends through all of the world's major

Time period	Observer and affiliation	Observation of note
		oceans and is associated with Rothe's seismic belt. Cause of the rift is uncertain but might result from an expanding earth caused by gradual decay of the earth's gravity field
1958	H. William Menard	Crests of "Mid-Ocean Ridges" closely correspond to median lines of many oceans, this cannot be accidental

*Blackett was awarded the Nobel Prize in physics during 1948 for having demonstrated the presence of the positive electron back in 1932.

**Heezen died with "his boots on" from a heart attack suffered while doing bathymetric work aboard the U.S. Navy nuclear research submarine *NR-1* south of Iceland in 1979 at age 55.

(Hess, 1962).[97] With such a beautiful hypothesis now in hand that would have done Professor Sverdrup proud, the race was on to see who could prove or disprove it.[98, 99] How the proof was developed during the past two decades is one of the great success stories in geophysics. Because the story has been so frequently told and elaborated on in a flood of publications, we choose to provide only a brief synopsis here for those not familiar with the sequence of events.

One of the earliest bits of support for Hess's concept came from

[97.] Hess picked up the term "sea-floor spreading" from his good friend, Robert Dietz (Dietz, 1961). Hess's and Dietz's concepts were initially opposed by many American scientists, including that inveterate sea explorer, Maurice Ewing. However, the idea caught on quickly in Europe, and by 1966, Hess was awarded the Italian Feltrinelli Prize that carried an honorarium of $32,000. As a passing comment, it may be noted that earth scientists felt slighted because none of them could ever earn a Nobel Prize. Fortunately, during the late 1950s, a Scandinavian ship owner created the Vetlesen Prize of $50,000, an award won since by such geophysicists as Maurice Ewing, Sir Harold Jeffreys, Sir Edward Bullard, J. Tuzo Wilson, and S. Keith Runcorn, the latter sharing it with Allan V. Cox and Richard D. Doell.

[98.] Although he lived only nine more years, Hess got to see the proof of his concept. A chain smoker who chose Princeton over Harvard University because the latter had "No Smoking" signs, Hess died of a heart attack, age 63, while chairing a Space Science Board meeting during August, 1969 at Woods Hole, Massachusetts. His death was similar to that two years before of RADM Berkner's, who died at age 62 while attending a committee meeting at the National Academy of Sciences. Since 1978, Hess's memory has been commemorated by the USNS *Hess* (AGS-38), the largest and fastest of the survey ships assigned to the U.S. Naval Oceanographic Office. By September, 1981, she had completed 385,000 survey miles. When operating, her basic crew consists of 14 officers and 40 seamen, plus Oceanographic Unit Three (three officers and 25 enlisted technicians) and a Navoceano civilian complement of eight geophysical and bathymetric specialists.

[99.] Even in 1963, Belousov was claiming he could not perceive of ". . . horizontal crustal movements against the mantle; if the crust moves, it must move with the upper mantle. . . . The supporters of horizontal displacement must tell us how all this is possible" (Beloussov, 1963). It certainly was possible, for in October, 1970, the President of the International Council of Scientific Unions was writing various national academies of science that they should support the new "Geodynamics Project" designed to study the basic mechanisms of earth deformation.

Canada's "benign cyclone of science", Dr. James Tuzo Wilson, who had taken his doctoral degree at Princeton just four years after Hess had earned his back in 1932. After cogitating on the "geological conveyor belt" concept, Wilson observed in 1963 that it was quite likely that the Hawaiian Islands had formed one or two at a time over a fixed hot spot in the earth's mantle and then moved away to the northwest to become older and lower with time (Wilson, 1963). However, he was unable to provide a numerical indication of just how fast this motion could be, and he also had to admit that the Hawaiian "hot spot" was nearly half an ocean away from Hess's "mid-ocean ridge".

Quantitative data on the rate of movement came about from a totally unsynchronized British—American—Canadian effort. While on board the survey ship HMS *Owen*, in the Indian Ocean during 1962, D. H. (Drum) Matthews and Fred Vine of Cambridge University noted that the unexplained geomagnetic stripes of Raff and Mason also occurred near the Mid-Ocean Ridge in the Indian Ocean, with half of the stripes reversely magnetized (Vine and Matthews, 1963). But why? Fortunately, the state of the art now permitted assigning radioactive dates to these magnetic reversals by determining the potassium-40/argon-40 ratio in samples taken from adjacent rocks of the same geological formation but of differing polarity. By mid-1964, enough careful work had been done by Allan Cox and his associates at U.S. Geological Survey's Menlo Park, California, for them to state when the earth's magnetic polarity had reversed itself several times during the past several million years (Cox, Doell, and Dalyrumple, 1964).[100] Now that some dating bench-marks, much like tree rings, were available, the scene shifted to Cambridge University's "G and G Department" at Madingley Rise where several of the "crustal drifters" came together in early 1965. This group included Harry Hess, J. Tuzo Wilson, Drum Matthews, Fred Vine and, of course, Sir Edward Bullard. By playing computer games, Sir Edward and some co-workers came up with a "best-fit" of what the original "supercontinent" of 200 million years ago would have looked like if the land-masses bordering the Atlantic Ocean were rotated back together from a point at $44°$ North Latitude, $30.6°$ West Longitude (Bullard, Everett, and Smith, 1965). Wilson and Vine also played computer games with the width

---

100. Two Canadians, Lawrence W. Morley and A. Larochelle, came close to earning scientific precedence, but the editor of *Nature* rejected their paper as inconceivable and it took some time to find another publisher (Morley and Larochelle, 1964). Be as it may, by 1979 Cox became Dean, School of Earth Sciences, at Stanford University (which had not started teaching geophysics until 1947), and Morley, a year later, was Scientific Counsellor, Office of the Canadian High Commissioner, in London.

of geomagnetic stripes paralleling the Juan de Fuca Submarine Ridge off the Puget Sound region. By using the Cox type of timetable for dating reversals in the magnetic field and assuming a symmetrical spreading rate away from the Ridge's axis of two cm/yr, their computer-derived magnetic profile almost exactly duplicated the measured profile of geomagnetic stripes in width and polarity (Vine and Wilson, 1965). A Lamont Geological Observatory team including James Heirtzler, Walter Pitman, Xavier le Pichon, and Manik Talwani then verified this hypothesis by demonstrating that good matches could be obtained between the axially symmetrical geomagnetic stripes mapped by Project Magnet during 1963 over the Reykjanes Submarine Ridge south of Iceland and computer-derived geomagnetic stripes based on the assumption that the stripes were spreading apart at one cm/yr (Heirtzler *et al.*, 1966).[101]

Once the rate of motion away from the rising limbs of convection cells was firmly established, embellishments came rapidly, converting the limited concept of "sea-floor spreading" into the more comprehensive and highly fashionable one of "plate tectonics". As early as 1965, J. Tuzo Wilson had recognized that the earth was divided into "plates" which could either separate via the spreading process or collide to form mountains or deep-sea trenches. In addition, such plates, when moving away from the mid-ocean ridge, could move at different speeds, thereby setting up offsets in the mid-ocean ridge and a line of "transform faulting" where they rubbed against each other (Wilson, 1965). By 1967, using seismograms from the Worldwide Standardized Seismograph Network and 25 stations of the upgraded Canadian seismological network, Lynn Sykes at Lamont verified the sense of plate motion predicted by Wilson and in the next year wrote with Jack Oliver and Bryan Isacks the classic paper, "Seismology and the New Global Tectonics" (Isacks *et al.*, 1968).

Now it turned into another race to see who could delineate and name all the plates of rigid lithosphere (including not only the crust but the upper mantle) that were moving about upon a less viscous asthenosphere deeper in the mantle. There appeared to be eight very large plates, plus several smaller fragments, moving principally at right angles to and away from the axis of the mid-ocean ridge

---

[101.] Heirtzler directed the Hudson Laboratories from 1967 until their abolition in 1969 because of student hostility to academic research on behalf of the military. University management was particularly "wishy-washy" during this period. Heirtzler's predecessor, Dr. Alan Berman, who moved on to be Scientific Director of the U.S. Naval Research Laboratory, was even formally accused of "anti-Semitism", even though with his black, bushy beard he could have been taken for a rabbi. Upon hearing of Berman's problem, his predecessor at Hudson Laboratories wondered, "Al, why didn't you drop your pants and show them your circumcision?"

(see Figs. 6.10 and 6.11). Closer scrutiny indicated that the plates could move relative to each other in the following ways: (1) parallel, causing great fault zones such as the San Andreas fault system in western California which eventually will cause San Diego to be seaward of San Francisco, (2) apart, as in the oceanic axial spreading centers, (3) converging, requiring one plate to dive under the other along a Benioff fault zone in a process of subduction and causing volcanoes to form when the submerged crustal rocks melted sufficiently at depth to form magma which then returned to the surface, and (4) converging and creating great zones of crumpling (nappe folds), as in the case of the Swiss Alps.

But J. Tuzo Wilson, in his jovial way, was not about to let the situation crystallize, for he further reiterated that "ocean basins went in and out like concertinas", forming a new line of mountain ranges each time a basin came back together.[102] As a result, continental fragments could be left far behind, and it might just be that part of Florida came from Africa, while parts of Nevada, British Columbia, and Alaska might be left-overs from Asia (Wilson, 1968). This concept, in fact, did set off a search for ancient "suture lines" inside continental land masses and for exotic geological fragments far from their point of origin (Ben-Avraham, 1981; Ben-Avraham *et al.*, 1981; Moores, 1981). Most geologists, however, did not trust *avant-garde* thinking by the geophysicists and waited for good old-fashioned paleontological evidence that could only be provided by deep-sea drilling.

Project Mohole Loses Out to the Deep Sea Drilling Project

After the drilling barge *Cuss-1* successfully drilled to the upper basaltic layer of the deep sea floor by April, 1961, the next question was which of two routes should be taken for drilling to the Moho-rovičić (Moho) discontinuity. The conservative route, urged by Willard Bascom, was to use an intermediate size drillship of an existing type and make technical progress in finite increments. However, the National Science Foundation had taken over direct project management from the somewhat specious AMSOC Committee of the National Academy of Sciences and decided to "go for broke". On 27 July 1961, the NSF held a public briefing to set forth its "Request for Proposals" which invited contractors to bid on the largest drilling ship ever built. The successful contractor would then not only design

---

102. Even today, Wilson remains as active as ever. While a Princeton student back in the 1930s, he had earned extra money by working up seismological computations for Maurice Ewing over at Lehigh University. Thus, it was fitting that the SEG awarded him the Ewing Medal at the society's annual meeting in late 1981.

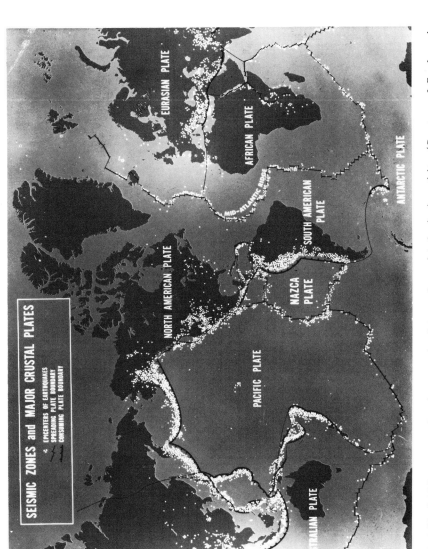
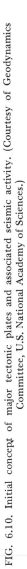

FIG. 6.10. Initial concept of major tectonic plates and associated seismic activity. (Courtesy of Geodynamics Committee, U.S. National Academy of Sciences.)

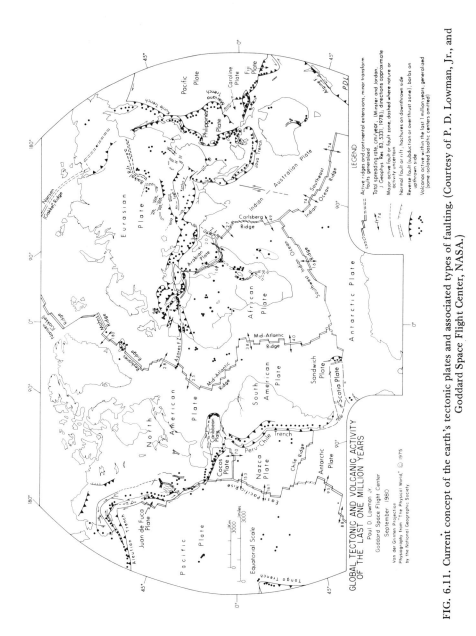

FIG. 6.11. Current concept of the earth's tectonic plates and associated types of faulting. (Courtesy of P. D. Lowman, Jr., and Goddard Space Flight Center, NASA.)

and build the ship but also locate suitable drilling sites and drill a series of holes, one of which would penetrate the earth's upper mantle. Such a contractor, the NSF specified, must have "necessary experience, organization, technical qualifications, skills and facilities or the ability to obtain them", plus "interest and enthusiasm in undertaking the Mohole project". A further requirement was that proposals must be in by 11 September, 1961, an extremely short time frame for spelling out how one would meet the greatest challenge yet within the field of ocean engineering. In spite of all this, eleven entities involving many of America's greatest corporations and oceanographic institutions submitted proposals (see Table 6.8).[103]

Contractor selection which involved a three-step process rapidly became politically complex and messy.[104] By the time it was over, Dr. Allan Waterman, the Foundation's Director, chose Brown and Root, Inc., the leading offshore engineering firm, to be the Mohole contractor even though it had previously avoided doing major technical pioneering. Waterman stoutly defended his actions just the same. The major oil companies, he argued, despite their great technical expertise, were still more interested in oil exploration than in science. Both Willard Bascom and Gordon Lill dissociated themselves from the project which they had started, with Lill's role being assumed by Dr. Hollis D. Hedberg, Vice-president (Exploration) of the Gulf Oil Corporation and an adjunct professor in geology at Princeton University.[105] However, Hedberg leaned toward the original hopes of Professors Thomas Jaggar and Maurice Ewing that called for drilling a large number of holes to moderate depths rather than immediately drilling one fantastically deep "Mohole".[106] Hedberg began selling

---

[103.] Even America's largest industrial corporation, General Motors, wanted in. As late as the 1950s, General Motors had eschewed doing development work for the Federal Government. However, a change in company policy caused the establishment of a "Defense Research Laboratory" at Santa Barbara, California, in 1961. Dr. David Potter, an oceanographer, was brought in to be its Deputy Director. During 1967 Dr. Potter replaced Dr. Frosch as Assistant Secretary of the Navy (Research and Development) and moved up to Under Secretary of the Navy in 1969. As of 1981, Dr. Potter is now Vice President (Governmental Affairs) for General Motors.

[104.] Details of this intriguing episode in "political science" can be found as Daniel S. Greenberg's contribution, "Mohole: Geopolitical Fiasco", to the book entitled *Language of the Earth* (Rhodes and Stone, editors, 1981) and Herbert Solow's article, "How NSF Got Lost in the Mohole", in the May, 1963 issue of *Fortune* Magazine.

[105.] Even when in his mid-70s, Hedberg was a major critic of the U.S. State Department's handling of "Law of the Sea" negotiations for delineating offshore boundaries between the sea bottom owned by the United States and foreign nations. Hedberg's argument was that such boundaries should be based on geological and geophysical considerations, rather than on a magic "200-mile" limit or "200-meter" isobath.

[106.] In 1943, Professor T. A. Jaggar, founder of the Volcano Observatory on Hawaii's Mount Kilaeua, wrote Professor "Dicky" Field at Princeton to suggest than an ideal postwar project would be drilling a thousand holes in the world's oceans to obtain core samples to a depth of approximately 0.3 km.

TABLE 6.8

National Science Foundation Contractor Selection Process for Final Stage of the Mohole Project

Date	Action involved
7/27/61	Issuance of "Request for Proposals" and associated public briefing (not attended by a Brown and Root representative)
9/11/61	Proposals on hand from: Socony Mobil, Standard Oil of California, Texas Instruments, and General Motors Global Marine Exploration, Aerojet General, and Shell Oil Company* Zapata Offshore, General Dynamics (Electric Boat Division), Dresser Industries (Continental Oil Company added later)** General Electric, Kerr–McGee Oil, Youngstown Sheet & Tube, Petty Geophysical Engineering Company, W. S. Nelson & Company, and Columbia University (Lamont Geological Observatory) Brown and Root of Houston, Texas Westinghouse Air Brake Corporation (Melpar Division), Bethlehem Steel Corporation, Geonautics, Inc., and assorted drilling companies (four to be exact) Litton Systems, Inc. Systems Development Company and the Offshore Company (a drilling firm) National Engineering Science Company University of California (Scripps Institution of Oceanography) Batelle Memorial Institute supplemented by Scripps, Texas Instruments, Ocean Drilling and Exploration Company (ODECO), and Cornell Aeronautical Laboratory
10/20/61	NSF contract review by six staffers reports Socony Mobil's proposal in class by itself (score of 936), followed by Global Marine (902 points). Brown and Root ranked fifth (score of 801 points)
12/61	NSF senior official review board rates Global Marine – 968, Socony Mobil – 964, and Brown and Root – 899
2/28/62	NSF announces Brown and Root wins the competition

*Dr. Thomas F. Barrow, when President of the Humble Oil and Refining Company, advised Bates that he considered Global Marine the best firm in existence for experimenting with new deep sea drilling techniques

**Zapata's President at this time was George H. W. Bush, who became Vice President of the United States on 20 January 1981.

this intermediate approach around Washington, D.C. and by November, 1963 had to be replaced as chairman of the AMSOC Advisory Committee to NSF.

Two months later, the NSF announced a new "Mr. Mohole" with full authority to move the project quickly ahead. Once again, the "major domo" was Dr. Gordon Lill, late of the Lockheed Aircraft Corporation. But Lill found a problem. When he and his AMSOC associates had blithely proposed the project back in 1959, they thought the effort could be done for less than $10 million. On returning to Washington, Lill now found his current year budget to be $8 million and escalating to $25 million on 1 July 1964 for a probable total overall project cost of about $68 million. Outside of some field surveys, not too much happened during the rest of 1964, but in early 1965, the announcement was finally made that the "Moho" drill site lay to the south of the Hawaiian Islands. However, a year later, the project's "sugar-daddy" on Capitol Hill (Congressman Albert Thomas of Houston, Texas) died of cancer. Congressman Joe Evins of Tennessee, as the new chairman of the authorizing committee, then moved to squelch the effort, for total project costs had begun to look as if they could go to as high as $125 million. After some further political shenanigans, the NSF finally issued the kill order on 25 August 1966 and turned this strange project into Project "Nohole" once and for all.[107]

Once control of the Mohole project was lost by the academic community, several scientific entrepreneurs asked the NSF to fund an informal university "*Long Core*" (LOCO) consortium which would charter a commercial oil drilling vessel. This didn't pan out, so a "*Consortium for Oceanic Research and Exploration*" (CORE) was launched in 1963 involving Drs. Roger Revelle and Maurice Ewing. This didn't work either. Finally, in early 1964, a consortium of the "*Joint Oceanographic Institutions for Deep Earth Sampling*" (JOIDES) appeared on the Scripps campus and incorporated Woods Hole, the Lamont Geological Observatory, and the University of Miami as well as Scripps. Their request for funding by the NSF was accepted, for it involved only a modest amount needed to pay Global Marine's dynamically positioned drill-ship, *Caldrill*, for boring a few holes into the puzzling Sigsbee Plateau off eastern Florida while the vessel was on her way to the Grand Banks of Newfoundland under contract to an oil company.

Both the contractual and scientific aspects of the Sigsbee Plateau drilling effort worked out well, and in January, 1967 the NSF awarded JOIDES an 18-month drilling program in both Atlantic and Pacific Oceans that would cost $12.6 million. Ten months later, Global Marine, Inc. again appeared on the JOIDES payroll with the proviso that the job should be done with the new drill-ship, *Glomar Challenger*, equipped with over seven km of 12.6-cm diameter drill pipe.[108] This

---

107. By then, Brown and Root was a major contractor in building military facilities for the escalating war in South Vietnam. However, the firm's "acres of engineers" approach left a legacy of advanced ideas for others to exploit in such areas as stable deep-sea drilling platforms, dynamic positioning at sea, drill-hole re-entry on the deep sea floor, and turbocorers.

108. To save money and technical complexity, the *Glomar Challenger* was purposely without a "blow-out preventer" or "riser pipe"; thus, if the drilling operation hit oil or gas accumulations, there was no way to keep a blowout from occurring. To prevent such a situation, as soon as the drill-ship found indications of gas at drill sites in the deep Gulf of Mexico (the Sigsbee Knolls), in the Barents Sea, and off Antarctica, she immediately withdrew her drill pipe and left the area.

time around, academia, industry and government worked smoothly together and were able to achieve many technical "firsts" that set the pace for consequent deep-water drilling by the international oil companies (see Table 6.9).[109, 110]

TABLE 6.9
Achievements of the First Eighty-one Cruises for the Deep Sea Drilling Project (8/68—10/81)

Type of achievement	Remarks
Holes drilled	922 holes at 555 sites
Cores recovered	16,686 cores
Total distance drilled below sea floor	210,634 meters
Interval cored	143,061 meters
Cores recovered and in repositories	77,266 meters
Deepest hole penetration	1,741 meters in water depth of 3,900 meters in Atlantic Ocean (Site 398 on Leg 47)
Maximum penetration of basaltic layer	623 meters in Pacific Ocean (Site 448A on Leg 59)
Deepest water drilled in	Depth of 7,044 meters in Mariana Trench near Guam (Site 461A on Leg 60)
Longest drill string ever suspended	Length of 7,060 meters (Site 461A on Leg 60)
Re-entry into holes	First operational re-entry on 12/25/70 in 3,062 meters of water within Venezuelan Basin (Site 146 on Leg 25)
Distance traveled	604,257 km
Utilization of vessel (average hours/year)	Operating: 4,500; Cruising: 2,700; Port-time: 1,100; Mechanical breakdown: 400; Bad weather: 100

After a month's shakedown during August—September, 1968 in the Gulf of Mexico (Leg 1) and drilling a few holes for Lamont across the Mid-Atlantic Ridge of the North Atlantic Ocean (Leg 2), the *Glomar Challenger* was ready to move to the South Atlantic Ocean and prove out Hess's "Essay in Geopoetry". To do this, nine drill sites were selected across the Mid-Atlantic Ridge along an east—west line 750 km south of the latitude of Rio de Janeiro and assigned dates according to the Heirtzler—Pitman numbering scheme for geo-

109. With JOIDES based at Scripps, the contractual principal investigator has been Scripps' Director, Dr. William Nierenberg, a nuclear scientist and the second director of the Hudson Laboratories (1953—1954). Dr. Melvin Peterson, a sedimentary geologist, became the top-notch project manager and co-principal investigator. Scientific results are summarized in a recent publication by the Society of Economic Paleontologists and Mineralogists (Warme *et al.*, 1981).

110. The 1958 Geneva Convention on use of the continental shelf authorized contiguous coastal nations to exploit resources below the sea bed out to a depth of 200 m or as much further as was "economically feasible". As a result, by 1981, oil companies had drilled holes with blow-out preventers installed in waters as deep as 1,478 m off eastern Canada.

magnetic stripes (Dickson, Pitman, and Heirtzler, 1968). Working in water depths of 2.1 to 4.7 km, the skilled drill crews penetrated all of the sedimentary column, which ranged up to 180 m in thickness at each location, and brought to the surface a wealth of paleontological, lithological, sedimentary, and geophysical material. Published results were not long in coming, for on 29 May 1970, the lead article in the journal *Science* announced: "Cores from the Deep Sea Floor in the South Atlantic Strongly Support the Hypothesis of Sea Floor Spreading."[111] At a distance of 191 km from the ridge axis, the cores had magnetic and paleontological dates of nine and eleven million years, respectively, while at the drill site 1,079 km farther away from the ridge axis, the same types of dates worked out to be 70 and 67 million years, respectively.[112] Moreover, fully 95 percent of the sea floor of the South Atlantic Ocean that was studied lay within ten km of the predicted position. Maxwell and his co-workers then confidently stated (Maxwell *et al.*, 1970):

> It is concluded that the sea floor of the South Atlantic has been spreading at an essentially constant rate for the past 67 million years. Further, the spreading half-rate of 2 centimeters per year (determined from the drilling) is in agreement with and has provided a critical test for both sea-floor spreading and magnetic stratigraphy.

Thus, once the exotic Mohole drilling project had been shunted aside, it had proved possible to check out the sea-floor-spreading concept with just a year's drilling effort costing about $15 million.[113]

A Decade of Ocean Politics and Ocean Policy[114]

Man has long had a continuing love affair with the sea since the

---

[111.] The paper's authors were Arthur E. Maxwell (by now Provost at Woods Hole), R. P. von Herzen, also of the Woods Hole staff, K. J. Hsü, Swiss Federal Institute of Technology, J. E. Andrews, University of Hawaii, E. E. Milo and R. E. Boyce of Scripps, and T. Saito of Lamont—Doherty Geological Observatory ("Doherty" was added to the title in 1969 in recognition of a major financial gift).

[112.] The oldest sedimentary rocks yet recovered by *Glomar Challenger* were obtained in early 1981 about 500 km east of Fort Lauderdale, Florida, in a water depth of five km. They proved to be of mid-Jurassic age (145 to 155 million years ago) and were laid down just as North America began to move away from Africa. Subsequently, they were covered by an additional 1.6 km of sedimentary rock. Co-Chief Scientists for this particular drilling effort were Dr. Felix Gradstein of the Geological Survey of Canada and Dr. R. E. Sheridan of the University of Delaware.

[113.] Starting in 1973, foreign scientific organizations were able to buy into the continuing project at the cost of $1 million a year, and the Soviet Union, Japan, France, Great Britain, and the Federal Republic of Germany soon did so. Data from the project have had such a major impact on man's understanding of climatic change, oceanic circulation, geological events, and the evolution of living organisms that by June, 1978, 98,227 samples had been supplied upon request to 231 organizations, 48 percent of which were foreign.

[114.] For a highly readable and authoritative account of this decade, read Dr. Edward Wenk's *The Politics of the Ocean* (Wenk, 1972).

time of the Phoenicians. In Great Britain, admirals have long held greater public esteem than captains of industry and military generals. In contrast, many presidents of the United States, including George Washington, Abraham Lincoln, Harry Truman, and Dwight Eisenhower, held Army commissions. In 1960, this trend was broken with the election of the naval reservist, John F. Kennedy, followed without a break by election of fellow naval reservists Lyndon B. Johnson, Richard Nixon, Gerald R. Ford, and Jimmy Carter. By the time of Kennedy's inauguration, America's merchant marine and fishing fleet had seriously deteriorated while its offshore oil exploration and production technology was the world's best. In addition, America had become demographically more like Great Britain, for by then, fully 45 percent of the United States' urban population lived near the coast (assuming the Great Lakes to be the "fourth sea coast").[115]

As a result, Congressman Hastings Keith, whose district included the Woods Hole Oceanographic Institution, found it timely to note with alarm in the *Congressional Record* for 9 March 1959:

> The United States is losing to the Soviet Union the biggest and most important sea battle in mankind's history — the contest to unlock the ocean's secrets for use in peace and war. . . .[116]
> Now that the United States has drawn alongside the Soviets in the race for outer space, it is essential that we concentrate as well on developing to the fullest our capacity for probing the oceans.[117]

To get the ball rolling, Senator Magnuson introduced and then held hearings on Bill 2692 (The Marine Sciences and Research Act of 1959) during April, 1960. Formal resolutions ". . . that we (must) encourage and support Federal and private agencies concerned with oceanography" from such diverse groups as the American Legion, the Veterans of Foreign Wars, and the District of Columbia Federation of Women's Clubs were then entered into the hearing's record.

Despite the orchestrated hue and cry, the 22 Federal agencies involved in marine matters preferred to be left alone. Thus, when the U.S. Senate held hearings on Bill 2482 during January, 1960 to lift restrictions limiting the Coast and Geodetic Survey to charting only the coastline and the continental shelf of the United States and its Territories, the U.S. Navy formally protested, urging that the restric-

---

115. As of 1960, the United States had 15 million registered pleasure craft.

116. Other equally alarmed politicians included Senators Hubert Humphrey (Minnesota), Warren Magnuson (Washington), and Robert Bartlett (Alaska), plus Congressmen George Miller (California), Alton Lennon (North Carolina), and John Dingell (Michigan).

117. In his book, *The Age of Discontinuity*, published in 1969, the noted business management expert, Peter Drucker, notes that the Egyptians nearly 4,000 years ago produced two great achievements — the erection of the Pyramids and the invention of the plow. Relative to publicity campaigns of the time, he believed that Americans might find the "Space Program" to have been its "Pyramids" and ocean exploration its "plow".

tion be kept in place to avoid duplication of the Department of Defense's high seas surveying and charting efforts. After admitting that its existing ocean charts were comparable in detail and accuracy to land maps published two centuries before, the Navy lost its case, and the Coast and Geodetic Survey, plus, for good measure, the Coast Guard, were authorized to conduct oceanographic research wherever opportunity permitted.

The U.S. Congress had less luck in persuading the federal establishment to use the capabilities of the geophysical survey industry. For example, during oversight hearings, the Honorable George P. Miller, Chairman of the Subcommittee on Oceanography and a former civil engineer, observed on 20 June 1961:

> One could imagine from some of the presentations we have heard that oceanography is confined to Government operations and non-profit institutions. Much to my surprise, and I should have known, I found that the continental shelves of a large part of the world have been surveyed in detail. . . . These surveys have been conducted by the geophysical industry with a profit motivation. . . .
> If we are going to accomplish a national objective in ocean surveys and the development of proper instrumentation for a synoptic look at our aquatic environment, we cannot fail to use the industry in our over-all program.

But the quartet representing the Society of Exploration Geophysicists was not so sanguine about this possibility.[118] After noting that academic capabilities for conducting marine geophysical surveys lagged industry's by at least a decade, Dr. F. Gilman Blake bluntly alleged:

> Now, I am afraid that in certain fields in the Government, people tend to stand too much in awe of the National Academy, to feel that "the king can do no wrong"; not to question anything they propose. Now this is not true of all Government agencies. . . . For example, in the fields of electronics, aircraft, missiles and so on, the Government makes very extensive use of industrial R & D facilities. But not in the earth sciences. Not in oceanography. Not in other branches of the earth sciences.
> I do not know why this tight little clique exists in this field, but it seems to. I think it highly undesirable. People tend to review their own proposals. . . . I would suggest that some of the universities' ethics in this area are not so strict as that of industry, not the industry I am in, anyway.

Carl Savit's follow-up testimony further emphasized the logic and cost-effectiveness of using the offshore survey industry for federal ocean surveys:

> Because we cannot afford to waste men on routine research work (not that we could get them into routine work if we tried), our instruments have to be capable of operation, adjustment, and maintenance by technicians and other nonprofessionals. We cannot and would not send the inventor to sea with an item of equipment.

118. The quartet consisted of Dr. F. G. Blake (Supervisor, Geophysical Research, Standard Oil Company of California), Mr. Carl Savit (Director, Systems Research, Western Geophysical Company), Dr. Fred Romberg (Texas Instruments), and Dr. Lewis Mott-Smith (Director of Research, General Geophysical Company).

But such frank talk did not win friends or influence people within the "oceanographic establishment", so nothing came of this initiative except for one "Moho" site survey awarded to the Western Geophysical Company.

Even the staid U.S. Coast Guard had great trouble fitting the geophysical industry into its pattern of operation. As late as May, 1965, when queried by John M. Drewry, Chief Counsel of the Subcommittee on Oceanography, whether the industry's survey craft could be included in the waiving of certain manning requirements for "oceanographic research vessels operating in the public interest", Commander William Benkert, USCG, Assistant Chief of the Merchant Vessel Inspection Division, adamantly advised (Benkert, 1965):[119]

> We do not feel that this type of operation would be in the public interest, sir. . . . We definitely do not feel, for example, that a commercial oil drilling-type operation or seismographic type of operation, which is being operated by a private corporation for their own utilization, their own determination of locating oil — we do not consider these as being oceanographic research vessels engaged in a public interest operation, sir.

As a result, the geophysical industry was forced to use craft of 299 or less gross tons to avoid the hundreds of pages of Coast Guard regulations associated with operating vessels of 300 tons or more.

As soon as Senator Kennedy became President in January, 1961, he and his sailing partner and science advisor, Dr. Jerome F. Weisner of MIT, began to make certain that oceanography received a greater share of the federal research and development "pie". As a starter, the President sent Executive Communication Number 734 of 29 March 1961 to the Speaker, House of Representatives asking that oceanographic funding for 1962 be nearly doubled over that appropriated in 1961.[120] In addition, he proposed that construction begin on ten new oceanographic vessels during the coming fiscal year.[121] The President's enthusiasm for the sea was backed up by that of the Interagency Committee on Oceanography led by Dr. James H. Wakelin, the Navy's first Assistant Secretary for Research and Development.[122] After some hard staff work by the up-and-coming "go-fers",

---

[119.] After reaching the rank of Rear Admiral, Benkert resigned his commission during 1979 and became President, American Institute of Shipping.

[120.] The federal budget for oceanographic research and surveys was $54.97 million in Fiscal Year (FY) 1960, $62.07 million in FY 1961, and $104.79 million in FY 1962. In FY 1960, "oceanography" received only 0.5 percent of federal research funds, but the percentage grew to 2.5 percent by FY 1976.

[121.] As of 1961, the American oceanographic fleet consisted of 27 research and 17 survey vessels.

[122.] The lion's share would be spent by Navy (36%), National Science Foundation (22%), Bureau of Commercial Fisheries (15%), and Coast and Geodetic Survey (12%). The remaining 13 percent would be split largely among the Public Health Service, Atomic Energy Commission, Bureau of Mines, Coast Guard, Bureau of Sports Fisheries, Office of Education, Smithsonian Institution, the Army, Maritime Administration, and Weather Bureau. Primary

Drs. Edward Wenk and Robert Abel, the Committee produced a report, *Oceanography — the Ten Years Ahead*, during June, 1963 which said the government needed to spend $2.32 billion on the oceans during the coming decade.[123]

Although the Committee's master plan called for Department of Commerce agencies to spend only a sixth of this total allocation, some persons thought differently. The aerospace industry wanted a "Wet NASA" as well as a "Dry NASA", and RADM Lloyd Berkner, USNR, had long advocated combining the old-line Weather Bureau and the Coast and Geodetic Survey to effect better end-products. After articulate, aggressive Dr. J. Herbert Hollomon became Assistant Secretary for Science and Technology at "Commerce" in 1962, he was anxious to place the Berkner concept into effect, as was Dr. Robert White, who took over at the Weather Bureau a year later. The actual consolidation took time to arrange, but finally during mid-July, 1965 President Lyndon B. Johnson promulgated his "Reorganization Plan Two" combining the two agencies and a few related U.S. Bureau of Standards units into the "*E*nvironmental *S*cience *S*ervices *A*dministration" (ESSA).[124] A year later on 13 July 1966, President Johnson followed up his vote of confidence and appeared at the commissioning of ESSA's *Oceanographer.* While at the podium, he announced that he would release a new presidential study, *Effective Use of the Sea,* once he got back to the White House.[125] In addition, he noted that federal support of marine science and technology had grown to $320 million per year and that the country's research fleet currently totaled 115 vessels.

Even as the federal ocean agencies were arranging their own expansion, the academically oriented oceanographic laboratories instigated a new international program of their own to replace the International Geophysical Year. In this case, Roger Revelle took the initiative in encouraging the international oceanographic community to launch an "*I*nternational *I*ndian *O*cean *E*xpedition" (IIOE). Although the IIOE was said to be "organized chaos" by Wenk and a "Woodstock-

---

goals were: improving national defense (36%), strengthening basic science (23%), charting and understanding marine resources (27%), and protection of lives and property (2%).

123. Dr. Wakelin started his federal career as a young naval reserve officer in ordnance with the Coordinator for Research and Development in the Office of the Secretary of the Navy during World War II.

124. The commissioned officer corps which had run the Survey for over a century was gently shunted aside by promoting its director to Vice Admiral and leaving him out of critical policy decisions.

125. Chaired by Dr. Gordon J. F. MacDonald, a Munk protégé, the study panel recommended that ESSA, the U.S. Geological Survey, the Bureau of Commercial Fisheries, and oceanographic activities of the Coast Guard and the Bureau of Mines should be consolidated into one "super" ocean agency.

type happening" by Chapman (see Chapman vignette on the IIOE), the NSF saw fit to provide nearly $20 million to the effort between 1962 and 1966. Many other countries, including the Soviet Union, were nearly as generous (see Fig. 6.12). After all, who could pass up a scientific opportunity to help feed the starving millions in Asia?

By then, each of America's major oceanographic laboratories had developed their own style of operation.[126] For example, when asked to explain the difference between Scripps and Lamont-Doherty, the perceptive marine seismologist, Dr. George C. Shor, explained it this way (Shor, 1976):[127]

> Scripps is a summer hotel in which the director did his very best to provide good conditions and pleasant surroundings for people to do their thing — but he had no more control over what they did than the manager of a summer hotel has over his guests. . . . Lamont is the world's greatest widget factory. It makes absolutely magnificent widgets and makes them exceedingly well. And everybody there works for Maurice Ewing, making his widgets. (See vignette by Ewing.)

No sooner had the IIOE closed down than a new method appeared for providing federal support to oceanographic academicians. During September, 1963, while giving the keynote address to the annual meeting of the American Fisheries Society, Dr. Athelstan Spilhaus asked: "Why shouldn't there be 'Sea Grant Colleges' funded in the same manner as are the highly respected 'Land Grant Colleges' started by President Abraham Lincoln back in 1862?" The concept picked up the enthusiastic support of Senator Claiborne Pell of Rhode Island ("The Ocean State") and was subsequently enacted into law on 15 October 1966. Under the aegis of the National Science Foundation, the Sea Grant Program, according to Spilhaus, was to be conducted in such a way that it placed "oceanographers in hip-boots" with the working fishermen, rather than just expanding an introvertive community of academicians. In many states, this hope was fulfilled because of the skilful efforts of articulate Dr. Robert Abel, who shifted over to manage the effort from the staff of the National Council on Marine

---

126. Walter Munk once threatened to leave Scripps for the more scholarly atmosphere of Harvard University. To lure him into staying, Revelle suggested that Munk form a branch of the "Institute of Geophysics and Planetary Physics" at La Jolla with the university paying half of the building's cost of $486,000. While raising his half, "Walter" discovered "Munk's Law of Giving". This law states that as prospective donors wish to pay only half of the remaining cost, an infinite series is created in which even the most modest donor can be a patron of science. In Munk's case, he first raised $125,000 from the Air Force Office of Scientific Research, then obtained $65,000 from the National Science Foundation, followed by $33,000 from the Fleischman Foundation. Fortunately, the Research Corporation eliminated his misery by donating $20,000 to close out the series.

127. After majoring in seismology at CalTech, Dr. Shor served as a party chief for Seismic Explorations, Inc. in 1948—1950 before joining the Marine Physics Laboratory at Scripps following three more years at CalTech while earning a doctor's degree.

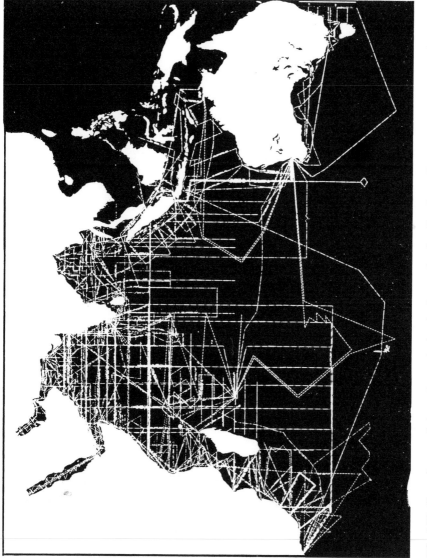

FIG. 6.12. Plot of some of the cruise tracks for the International Indian Ocean Expedition, 1959–1965. (Extracted from International Oceanographic Commission Technical Series Number 2, 1966; copyrighted by and used with permission of UNESCO.)

Resources and Engineering within the Executive Office of the President.[128]

The Marine Council, on whose staff Abel had been serving, was another of the U.S. Congress's initiatives taken during 1966 relative to marine affairs, as was the establishment of the "blue-ribbon" Commission on Marine Science, Engineering and Resources chaired by Dr. Julius Stratton, Chairman of the Ford Foundation and Past President of MIT.[129] Although the Commission worked diligently, its four-volume report, *Our Nation and the Sea: A Plan for National Action*, was issued just before the Johnson Administration went out of office in early 1969. About all that Vice President Hubert Humphrey, in his role as Chairman, Marine Resources Council, could do was to transmit the report to the incoming Republicans and particularly to his successor, Spiro Agnew, with the following advice:

> This is an imaginative study made by a distinguished Commission, deserving study by all interests, in and out of the Federal Government. I have a strong conviction on the promise of the oceans, on the deep seabed as a legacy for all mankind, and on the importance of America's stake therein. The rapid growth of technology makes these problems acute. The next Administration should give Commission recommendations immediate attention, especially as to Federal organization and Coastal Zone management.

In just four sentences, Humphrey had touched on three issues that became increasingly debatable during the 1970s — ownership of the deep sea floor, formation of a "super ocean agency", and federal management of the coastal zone.

Internationally, the political battle for control of the "boundless riches of the sea" was also accelerating. In July, 1960 UNESCO created an *I*ntergovernmental *O*ceanographic *C*ommission (IOC) to provide a formal scientific arena for discussing oceanographic issues of international import. During 1967, the situation began to become truly complicated as the General Assembly of the United Nations established a 42-nation "Standing Committee on the Seabed" to ensure, as both President Lyndon B. Johnson and Ambassador Aldo Pardo of Malta maintained, ". . . that the deep sea and the ocean

---

[128.] Starting with a token appropriation of $1 million in Fiscal Year 1967, by 1980 the Sea Grant annual budget had grown to $38.7 million and supported some 700 cooperative projects at 45 institutions. By then, 16 educational institutions had been raised to the rank of "Sea Grant College" and acquired a modest degree of permanence in their annual funding allocations. The program director was Dr. Ned Ostenso, a former student of Professor George Woollard at the University of Wisconsin and one of the most personable, persuasive geophysicists based in Washington, D.C.

[129.] Vice-chairman was Dr. Richard Geyer, who held a geology—geophysics degree from Princeton University. During the immediate post-war period, he had been the "in-house" oceanographer for Humble Oil before moving to Texas Instruments and later to Texas A&M University, where he became the Chairman, Department of Oceanography (1963—1976). He also served as Editor of the journal *Geophysics* between 1949 and 1951.

bottoms are, and remain, the legacy of all human beings". Furthermore, within the year, 15 UN entities claimed a bit of the action, creating as much confusion as action when it came to using the ocean's resources in an optimal sort of way (see Table 6.10).

TABLE 6.10

United Nations Bodies with Marine Involvement (as of 1968)

General Assembly	Economic and Social Council	Specialized agencies
Committee on the Seabed	UN Development Program	UNESCO (IOC)
Resource and Transport Division	UN Children's Fund	Food and Agriculture Organization
International Atomic Energy Agency	Economic Commission for Asia and the Far East	World Meteorological Organization
Conference on Trade and Development		Intergovernmental Maritime Consultative Organization
International Law Commission		International Bank for Reconstruction and Development
		International Telecommunication Union
		World Health Organization

By now, geopoliticians who never went to sea claimed that the deep ocean held untold riches, but marine geologists took a more cautious view, pointing out that "the ocean's bottom is no better known than the moon's behind". While everyone else emphasized vertical probing of the sea floor using conventional echo-sounders, bottom cameras and small, manned submersibles, the British National Institute of Oceanography chose to be different. During 1964, NIO's Director, Dr. George Deacon, gave the go-ahead to another Cambridge University graduate, Dr. A. S. (Tony) Laughton, to build the "grandfather" of all side-looking sonars. Nicknamed GLORIA (*G*eological *L*ong *R*ange *I*nclined *A*SDIC), by 1969 the Mark-I model was demonstrating the ability to acquire useful acoustic pictures out to a range of nearly 22 km when the sensor was towed at a depth of approximately 50 m in water 5 km deep.[130] Despite skilful engineering by R. J. S. M. Rusby and his associates, the device was difficult to use at sea for it weighed six tons, measured 1.75×10 m, took four hours or more to stream in nearly calm sea conditions, and could only be retrieved in the lee of an island or headland. Consequently, once GLORIA was streamed, she stayed in the water until near the cruise's end if at all possible. To overcome these problems, M. L. Somers came up with a

---

130. One of GLORIA's earliest successes was the spotting of the active fault valley lying to the east of the Azores which separates the African and Eurasian tectonic plates. Also of passing interest is the observation that GLORIA's mentor, Dr. Laughton, moved up to the Directorship of what was now called the National Institute of Oceanographic Sciences during 1978.

Mark-II device which has now cut the sensor's weight and diameter by a factor of three while upping the tow-speed to ten knots.[131] As a result, by 1980 the two "monster fish" had been out on 26 cruises (all but two in conjunction with RRS *Discovery II*) and insonified more than one percent of the ocean's floor (Laughton, 1981).

As soon as Laughton's team began to deploy GLORIA, they learned that much of the continental shelf, or at least that around the United Kingdom, was acoustically dull and provided little new information. However, when working on the shelf west of Scotland, schools of herring up to five km in length were detected from ten km away and could be easily tracked by referencing their movement to rocky outcrops. As soon as the researchers moved to the continental slope, however, they obtained a rich harvest of sonograms portraying canyons, erosional channels, fault faces, slumps and diapirs (Belderson *et al.*, 1972). The device proved equally useful in the deep sea, for it readily mapped sedimentary bedforms, submarine channels, submarine deltas, compressional tectonics within thick sediment beds lying outside island arcs, volcanoes, seamounts, and the tectonic fabric associated with mid-ocean ridges and transform faults. GLORIA also generated considerable information of commercial interest, including the rapid mapping of surface expressions of potential oil- and gas-bearing geological structures, delineating slump areas capable of breaking underwater pipelines, cables, and platform supports, and clarifying the nature of the sea floor where radioactive waste is either being or likely to be dumped. As a result, a GLORIA Mark-III is now under design which will utilize the latest in digital recording and image enhancement techniques to improve the quality of each picture and to remove the lineation formed when joining several pictures into a mosaic.

During the 1960s, the U.S. Navy also had the need to learn much more about the sea floor. On 10 April 1963, the nuclear-powered submarine, USS *Thresher*, was lost during a training dive in 2500 m of water 400 km east of Boston, Massachusetts.[132] Because the rescue ship, USS *Skylark*, was present during the dive, the sunk position was known to within an 18-km square.[133] The task of locating and photo-

---

[131.] GLORIA Mark-II uses two rows of 30 acoustic transducers on each side of a 7.75-m-long fish. Its operating frequency lies between 6.2 and 6.8 khz, and the actual acoustic pulse is a swept frequency one of 100 hz, making the actual length of the outgoing signal approximately 5.3 m. Angular beam widths are $2.5°$ in the horizontal and $30°$ in the vertical at a fixed inclination of $20°$ below the horizontal. Peak acoustic power introduced into the water per side is 10.5 kilowatts. Areal sweep rate is about 1,100 square km per hour.

[132.] The cause of the sinking may have been loss of control after unexpectedly penetrating a large solitary internal wave (soliton). Knowledge of the exact cause, however, went down with those in the ship.

[133.] A spacing of 0.25 km between search tracks was utilized, with positioning a combination of Decca and the newly available LORAN-C of the U.S. Coast Guard.

graphing the remains was assigned to Task Group 89.7 with technical aid by a special *Thresher* Advisory Group (Spiess and Maxwell, 1964).[134] A large number of sensors were deployed to spot the wreckage. These included precision depth sounders, bottom-towed detectors (magnetic, electrical potential, and nuclear radiation), underwater photography and television, dredging and viewing through the portholes of the submersible, *Trieste I*. By late May, a debris streak about one km wide and four km long which looked much like an automobile junk yard was spotted. Then on 12 June, Dr. J. Lamar Worzel, Lamont's Assistant Director, while using a deep-towed magnetometer of his own design, found a magnetic anomaly of nearly 100 gammas, suggesting that the hull would be found largely at one location. By mid-August, the *Trieste I*, under the command of Lt. Commander Don Keach, USN, was able to make eye-ball contact on additional wreckage during two out of five dives. The following summer, *Trieste II*, commanded by Lt. Commander J. B. Mooney, Jr., was able to find the *Thresher*'s hull sections and eventually Secretary of the Navy Paul Nitze announced: ". . . the area of search is now better known to oceanographers than any other area of similar depth in the world."[135]

The new-found Navy expertise in locating objects on the ocean floor was soon needed again in the Mediterranean Sea off Palomares, Spain. Here, an unarmed "hydrogen bomb" with a deployed parachute was observed falling into the sea by a local fisherman following a refueling accident by two U.S. Air Force aircraft on 17 January 1966.[136] As it turned out, the device had fallen into a deeply fissured, steeply dipping part of the continental slope that made acoustic search infeasible. Before the weapon was finally extricated on 7 April 1966 from a water depth of 860 m, Task Force 65 had used 25 Navy ships, four research or commercial vessels, and four submersibles (*Alvin* from Woods Hole, the *Aluminaut* of Reynolds Metals, Inc., a Perry *Cubmarine*, and a "*C*ontrolled *U*nderwater *R*ecovery *V*ehicle" (CURV) from the Naval Ordnance Test Station, China Lake,

---

134. Captain Frank Andrews, USN, who later chaired the Department of Ocean Engineering at Catholic University, Washington, D.C., commanded the task group which included a bevy of Navy-owned or financed research vessels. These included the *Conrad* of Lamont, *Atlantis II* of Woods Hole, the *Gibbs, Allegheny* and *Mission Capistrano* of Hudson Laboratories, the *Mizar* of the Naval Research Laboratory, and the *Gillis* of the Naval Oceanographic Office. Dr. Arthur Maxwell of ONR served as Chairman, Technical Advisory Group.

135. Mooney later coordinated the use of deep submersibles in the search for the lost hydrogen bomb off Palomares, Spain. During July, 1981, by which time he had reached the rank of Rear Admiral, Mooney took command of all the Navy's geophysical activities in his new role as "Oceanographer of the Navy".

136. Books describing the situation in detail are *The Bombs of Palomares* by Ted Szulc (1967) and *One of Our H-Bombs Is Missing* by Flora Lewis (1967).

California). Surface positioning was by Decca HI-FIX, but submerged navigation was largely hit or miss. After being spotted by *Alvin*, the CURV made the final lift to reduce hazards to life created by the possibility of the parachute shrouds entangling themselves in propellers of the manned submersibles (Chief of Naval Operations Technical Advisory Group, 1967).

Before the decade was out, the U.S. Navy was also involved in two other searches for radioactive debris. One incident was naval in origin, being caused by the failure of the nuclear attack submarine, USS *Scorpion*, to arrive in the United States after sailing from Naples, Italy, during 21 May 1968. Eventually she, too, was located and photographed on the deep sea floor. The other incident, which occurred four months earlier, was a U.S. Air Force *Broken Arrow* incident in which a B-52 bomber with four fusion weapons on board crashed while on fire 12 km west of Thule, Greenland. Here the 84,000-kg aircraft with 100,000 kg of fuel on board impacted bay ice about 0.8 m thick. The ice shattered locally but quickly refroze after the burn which covered an area of about 150 by 750 m. Fortunately, most of the aircraft and bomb fragments, ranging in size from that of a small coin to that of a cigarette pack, proved to lay in the blackened snow area (Sundstrom, 1970).

Despite 24 hours of darkness each day, the U.S. Air Force quickly mounted Operation Crested Ice which eventually involved 700 people from 70 Danish and American agencies before the effort came to a successful conclusion, namely, locating all the radioactive debris and shipping it back to the United States for permanent storage. Ten days after the accident it was possible to start the all-important ice-coring survey of the crash site.[137] This effort eventually showed that almost all the plutonium oxide dust was firmly attached to metallic debris and snow crystals above the ice layer, thereby making it mechanically possible to remove over 99 percent of all the freed plutonium and bring the region back to a normal background radiation level. However, to make certain that none of the bomb debris had reached the sea floor, a U.S. Naval Oceanographic Office oceanographic team worked the area with underwater cameras during late March.[138] After the ice went out, the Danish fisheries research cutter *Aglantha* also checked the area, as did General Dynamics' submersible *Star III* which was flown to Thule and eventually set a "Farthest

---

[137.] The ice-coring specialists were Dr. B. Fristrup, a Danish glaciologist, and Dr. Gunter B. Frankenstein of the U.S. Army Cold Regions Research and Engineering Laboratory.

[138.] Working in air temperatures that dipped down to $-32°C$, the team of Drs. Lloyd Breslau and James Welsh and Messrs. Leo Fisher and F. M. Daugherty finished the survey within a two-week period.

North" diving record for small submarines of $76°30'$ North Latitude. In all instances, no crash debris was found.

In view of the national hullabaloo concerning the need to do more about the ocean, in 1962 the U.S. Congress changed the name of the Navy's Hydrographic Office to the "Oceanographic Office" (NAVOCEANO). Later in the decade, the Office grew to the largest marine survey organization in the Western World, employing a staff of some 2,800 and providing technical control for 15 surface vessels, one gravity-measuring submarine (USS *Archerfish*), and four survey aircraft (two for Project Magnet, one for year-round ice reconnaissance and one for airborne oceanography). New ships, such as the USNS *Bent* with a $3.6 million integrated survey system developed by Texas Instruments, also came on the line. In addition, NAVOCEANO took the unprecedented step during March, 1965 of contracting out a number of major marine geophysical surveys to Alpine Geophysical Associates and to Texas Instruments.[139, 140] At this time, NAVOCEANO was deeply involved in conducting surveys for the Fleet in South Vietnamese waters, maintained a branch office in Saigon, and provided research oceanographers, such as Dr. Lloyd Breslau, to work on riverine warfare problems.[141]

In sharp contrast to the problems of Vietnam were those in the Arctic. The discovery in 1968 of the Prudhoe Bay oil field in sight of the polar sea stimulated Humble Oil and Refining to investigate the "marine alternative" for moving the oil to the east coast of the United States, even though its partners, ARCO and BP, were far from enthusiastic. To do this, Humble chartered the largest ship ever built in the United States — the 115,000-deadweight-ton tanker, *Manhattan* — and refitted her with a special icebreaking bow and other unique

[139.] Alpine's award was a $5.8 million turn-key contract with a potential profit of $522,650, while TI's was for $5.6 million with a potential profit of $515,000. The goal of the project was to provide acoustical information (transmission path and bottom loss, ambient noise, and reverberation levels) needed to plan the operational deployment of long-range sonar systems.

[140.] As far back as 1959, TI had offered the U.S. Navy a free demonstration in the use of a seismic streamer to detect a submerged NATO submarine. Because of patent considerations, the field test was never held. By 1977, the Navy accepted the concept and, in 1981, Secretary of Defense Harold Brown stated: ". . . the towed array sonar was the most important surface-ship ASW development in a generation". As a result, the Navy plans to spend $355 million on the technique, about half of which is for tow ships and half for sensor packages. By late 1981, Hydroscience, Inc., a sister firm of the Seismic Engineering Company, was under contract to supply eight SURTASS arrays at the cost of $58 million.

[141.] Breslau was an ideal field geophysicist. The holder of four degrees (electrical engineering, geology, geophysics, and oceanography) from MIT, he was among the very first (1962) to use an electronic computer for processing geophysical data in real-time at sea — in this instance, computing the power spectra of bottom bounce signals. During 1972, he became Assistant Technical Director of the U.S. Coast Guard Research and Development Center.

features.[142] She sailed from the Delaware River on 24 August 1969 for Prudhoe Bay via the fabled Northwest Passage (see Fig. 6.13). On 11 September, she tried a direct run into the Arctic Ocean via McClure Strait but came to a dead halt. However, her problem was soon diagnosed from studying side-looking radar photographs dropped by parachute from a U.S. Coast Guard C-130 aircraft overhead.[143] Unhampered by prevailing fog and low cloud, this unique radar showed the ship to be stuck within an old ice floe of polar origin, with even worse conditions ahead. Discretion being better than valor, Captain Roger Steward turned the *Manhattan* south, ran down the sheltered Prince of Wales Strait, and was ready to take on board the token barrel of Prudhoe Bay oil five days later. After the *Manhattan* returned to New York City via the Northwest Passage, Humble stated tersely that the marine route was technically, but not economically, feasible. This was undoubtedly true in 1969. However, the ice-breaking-tanker concept remains very much alive, and one may yet see such unusual vessels going to and from the Canadian Arctic during the 1990s if the economic need for additional supplies of oil and natural gas continues to build.

Earth-oriented Satellites at Last

Once payload weight and orbital parameters had been established during mid-1959 for the TIROS meteorological satellite, Sidney Sternberg and his resilient group at the RCA Astro-Electronics Division quickly constructed three "birds" — one for "dry runs" and two for potential flight models.[144] Each satellite carried two vidicon television cameras with lenses such that one field of view measured approximately 1,200 km on a side and the other approximately 100 km on the side. Each of the cameras then fed a two-channel tape recorder, one channel holding 32 pictures taken at 30-second intervals and the other holding data indicating the satellite's rotational posi-

[142.] The overall project director was a sensible, cooperative chemical engineer, Stanley Haas, who drew heavily on the ice-breaking expertise resident within the Coast Guards of Canada and the United States. The bow shape, for example, was that developed by Lt. Roderick White, USCG, as a doctoral research project at MIT. The now Captain White is presently Dean of Faculty, Coast Guard Academy.

[143.] The photographs were generated by an AN/DPD-2 side-looking radar that had been specially modified for the project by Dr. Lloyd Breslau, Lt. Commander James A. McIntosh, USCG, and Mr. Dennis Farmer (Johnson and Farmer, 1971). A Canadian C-54 ice-reconnaissance aircraft equipped with a laser-beam profiler and operated by the Ministry of Transport also provided vital information on ice conditions along the route.

[144.] For details, readers should read *Meteorological Satellites* by William Widger, Jr. (1966) and NASA's *Final Report on the TIROS-I System* (Technical Report R-131 of 1962).

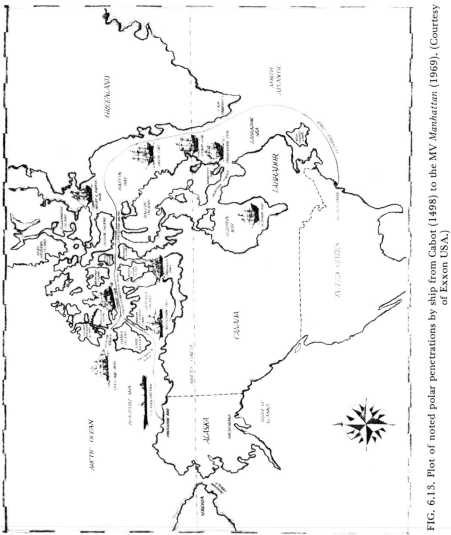

FIG. 6.13. Plot of noted polar penetrations by ship from Cabot (1498) to the MV *Manhattan* (1969). (Courtesy of Exxon USA.)

tion.[145, 146] Additional details of TIROS's working parameters are given in Table 6.11.

TABLE 6.11
Significant Parameters of the TIROS-I Weather Satellite

Parameter	Type of camera	
	Wide angle	Narrow angle
Field of view	104 deg	12.7 deg
Lens aperture	f/1.5	f/1.8
Shutter speed	1.5 msec	1.5 msec
Lines per frame	500	500
Frames per second	0.5	0.5
Video bandwidth	62.5 kc	62.5 kc
Power consumption (average)	9 w	9 w
Resolution	2.4 km	300 m
Peak spectral sensitivity (in visible spectrum)	0.7 to 0.9	0.7 to 0.9
Orbit	99-minute orbit ranging between $50°$ North Latitude and $50°$ South Latitude at altitude of 750 km	
Infra-red radiometers	Omitted on TIROS-I but flown on TIROS-II which was landed 23 November 1960	

Two weeks before TIROS-I was launched, President Eisenhower shifted the responsibility for peaceful satellite programs from ARPA to the newly labeled *N*ational *A*eronautics and *S*pace *A*dministration (NASA). As a result, Dr. Morris Tepper of NASA Headquarters, rather than Roger Warner and William Kellogg, proudly briefed the Washington, D.C. press corps on 1 April 1960 about the highly successful TIROS-I launch that morning down at Cape Canaveral, Florida. Subsequently the satellite made 1,302 useful earth orbits and transmitted nearly 23,000 pictures before failing technically on 29 June 1961, nearly 15, rather than three, months after launch.[147] Although 60 percent of these pictures had usable meteorological information, orienting and positioning the picture elements geographically proved to be a major problem.[148] Nevertheless, man could now, from outer

145. Because the state-of-the-art was too primitive to keep the satellite permanently oriented towards the earth, TIROS was spin-stabilized and looked at the earth only half of the time.

146. Acceptable optical parameters for photographing the cloud field were developed largely by Dr. Arnold Glaser of Harvard University's Blue Hill Meteorological Observatory, with refinements at RCA by Mr. Jules Lehman.

147. The first operational nephanalysis of TIROS-I derived cloud patterns was by Major James B. Jones, USAF, and Commander John Mirabito, USN, of the two military weather services. Their analysis used data from the Mediterranean Sea region barely four hours old, although the satellite by that time had been in orbit for 60 hours.

148. Even before TIROS-I was launched, Dr. William Stroud and his associates, who later moved as a team to NASA's Goddard Space Flight Center, were designing the NIMBUS "bird" capable of keeping its cameras continuously aimed at earth.

space, watch the evolution of large vortices within the earth's atmosphere, chart the position of snow and ice fields and flooded river valleys, and trace the globe-encircling frontal patterns that Jakob Bjerknes and his cohorts had first plotted forty years before. The world press promptly predicted a new era in weather forecasting, while Dr. Reichelderfer adroitly adjusted the Weather Bureau's budget in order to have about a fifth of his financial resources going into satellite meteorology.

Although NASA's Office of Space Science and Applications favored the fields of telecommunications, planetary science, and meteorology, it was willing to experiment with other areas. For example, during 1965, NASA lofted "Polar Orbiting Geophysical Observatory Two", to map the earth's magnetic field at some distance from the earth and repeated the experiment during 1967 and 1969. Moreover, as early as 1964, NASA saw fit to sponsor a week-long conference at Woods Hole under the direction of air-minded Dr. Gifford Ewing on potential uses of spacecraft for oceanographic purposes. By late 1965, NASA and the U.S. Navy had signed an agreement making NAVOCEANO NASA's agent in a government-wide effort to identify oceanographic experiments involving manned or unmanned space platforms.[149] On 2 January 1966, bright, aggressive Arthur G. Alexiou opened NAVOCEANO's "Spacecraft Oceanography Branch" and formally asked agencies for suggestions on how NASA's money could best be spent in this area.[150] Flight experiments were particularly sought for: (1) sea ice behavior and mapping, (2) wave and current fields, (3) coastal and shallow-water features, (4) air—sea interaction processes, and (5) mapping of conditions affecting schooling of fish and migration of marine mammals. Promising sensing techniques for such work included active radar (scatterometry, imagery, and altimetry) and four passive systems (photography, infra-red radiometry, micro-wave radiometry, and infra-red imagery) (see Fig. 6.14). To

---

[149]. The NASA planner in charge of this effort was Dr. Peter C. Badgley, formerly a pilot officer with the Royal Canadian Air Force (1943—1945), a Princeton University doctoral graduate (1952), and author of a scholarly geological book, *Structural and Tectonic Principles*. At the time, it appeared that NASA would have an extra 20 Saturn boosters not needed for the moon-landing project.

[150]. Advisory Committee principals were: National Science Foundation, Dr. William E. Benson; Bureau of Commercial Fisheries, Dr. John Lyman; Atomic Energy Commission, Dr. John Wolfe; Federal Water Pollution Office, Mr. John E. McClean; Geological Survey, Dr. Joshua Tracey; U.S. Coast Guard, Captain Leroy A. Cheney; Office of Naval Research, Commander Milton Gussow; National Academy of Sciences, Dr. Gifford C. Ewing; Naval Research Laboratory, Dr. Wayne Hall; ESSA, Mr. Raymond Nelson; U.S. Army Cold Regions Research and Engineering Laboratory, Mr. Ambrose Poulin; Bureau of Naval Weapons, Murray Schefer; Smithsonian Institution, Dr. Sidney R. Galler; U.S. Army Coastal Engineering Research Center, Mr. Thorndike Saville; and NAVOCEANO, Dr. Charles Bates, who also served as Committee Chairman.

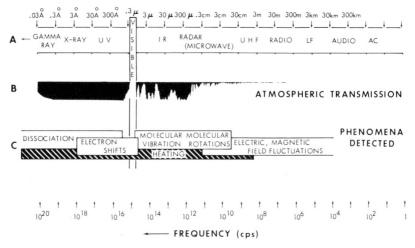

FIG. 6.14. Characteristics of the electromagnetic spectrum which can be of significance in remote sensing of the earth. (Courtesy of NASA.)

support this experimentation, NASA offered qualified investigators flight time on specially configured earth-resource survey aircraft. Before a year had passed, all of NASA's initial $900,000 had passed to Scripps, New York University, University of Miami, Texas A&M University, Philco-Ford, Barringer Research, University of Kansas, Autometrics, Inc., U.S. Bureau of Commercial Fisheries, and the U.S. Army's Cold Regions Research and Engineering Laboratory.[151]

By April, 1967, enough new data were at hand for the American Society of Oceanography to hold a symposium on the oceans from space (Badgley, Miloy, and Childs, 1968). In addition, the National Council on Marine Resources and Engineering Development issued a beautifully illustrated booklet, *United States Activities in Spacecraft Oceanography*, the following October, largely through the hard work of the balloonist, Kurt Stehling. By 1969, however, Commerce's ESSA had prevailed upon NASA to allocate spacecraft oceanography funds to ESSA rather than to the Navy. As a consequence, interest in

---

[151.] While at Texas A&M University in 1966, Lt. Commander Dan Walsh, USN, the former skipper of the *Trieste-I*, needed high-level color photographs for his doctoral research on "Experimental Use of Airborne Sensors in the Measurement of Mississippi River Outflow". NASA's earth-resource aircraft were not yet available, so the ever-inventive and persuasive Walsh talked several astronauts into flying him upside down at high altitude over the Mississippi Delta in order that he could frame the entire feature in just one photograph.

the area remained at a low key until 1977, when plans to fly the specially dedicated oceanographic satellite, SEASAT, finally jelled.[152] Although SEASAT suffered a catastrophic failure after being aloft only 60 days during the summer of 1978, it did generate unique synthetic-aperture radar data from 800 km above the earth that not only has merit for wave and sea ice studies but also for geologic mapping (Sabins *et al.*, 1980).

At the same time that Dr. Badgley was fostering the spacecraft oceanography initiative, Dr. William Pecora, Director of the U.S. Geological Survey, and Badgley were also actively pursuing the possibility of mounting a major "*E*arth *R*esources *O*bservation *S*ystem" (EROS) project to apply remote sensing techniques to resource inventory, monitoring and management practices. As in the case of spacecraft oceanography, ESSA under Dr. Robert White made a play to acquire this mission; after all the pieces were sorted out, however, the U.S. Department of Interior was allowed to establish the EROS Data Center under Geological Survey management during 1972 in Sioux Falls, South Dakota, the same year that LANDSAT-I was successfully launched.

In looking back at the 1960s, it is evident that it was a decade of great geophysical accomplishments, for now earth scientists could readily roam to all parts of the earth and probe deeply into the ocean and monitor the earth from outer space. But like the cloud that starts no larger than a man's hand, a new force called "environmentalism" was beginning to emerge. How this proposed change in man's ethics and sets of values impacted on the profession of geophysics becomes a major theme in the next chapter as we address the 1970s and proceed into the 1980s.

---

152. NASA's revived interest in oceanography may have come in part when its new Administrator as of 21 June 1977 was Dr. Robert Frosch, who had just finished a tour as Associate Director of the Woods Hole Oceanographic Institution. One of the few academic oceanographers who "kept the faith" during the lean years of spacecraft oceanography was Professor Willard J. Pierson of the Institute of Marine and Atmospheric Sciences, City University of New York. A pioneer in the interpretation and implications of ocean wave spectra, he served as an invaluable consultant to NASA on how satellites could best map the ocean. As a result, he was the sole non-NASA employee to receive NASA's Exceptional Scientific Achievement Medal during that agency's annual award ceremony on 21 October 1980.

CHAPTER 7

Geophysics Interacts with the Environmentalists and OPEC—the 1970s and the Early 1980s

Upon this gifted age, in its dark hour
Rains from the sky a meteoric shower of facts . . .
They lie unquestioned, uncombined.
Wisdom enough to leach us of our ill
Is daily spun, but there exists no loom
To weave it into fabric undefiled.

Huntsman, What Quarry? Edna St. Vincent Millay (1939)[1]

A Period of Controversy

Until now, the tone of this book has been positive, for the practice of geophysics, whether applied or basic, was both exciting and productive during the first half of the 20th century. This approach fitted nicely with the expansionist, consumption-oriented, problem-solving, optimistic outlook which dominated this period. But, starting quietly in the 1960s, divergent approaches began to be taken by scientists and humanists when attempting to analyze and solve the same set of human problems. As a result, a sharp cultural rift set in throughout the Western World. Two dramatically different sets of ethics competed for the mind of youth. One choice that could be followed — and the traditional one — was the "Judaeo-Christian ethic" which gave man stewardship over the earth, the concept on which our modern industrial society is based. The alternate was the "Humanist ethic" which urged that there should now be a "Post-Industrial Society" where "Smaller is Better", the quality of life more important than economic productivity; society, rather than the individual, is the prime source of critical social adjustment prob-

1. From *Collected Poems,* Harper & Row. Copyright 1939, 1967 by Edna St. Vincent Millay and Norma Millay Ellis. All rights reserved. Reproduced through the courtesy of Norma Millay Ellis.

Aerial view of a single LASA subarray which utilizes 25 buried seismometers, four along each black line and one in the center. Twenty-one such subarrays were then linked to a recording center in Billings, Montana, in order to form a "LASA". (Courtesy of Lincoln Laboratory.)

One of 40 Long-Range Seismic Monitoring Units operated for the Vela Uniform Program. (Courtesy Teledyne-Geotech.)

Drills readying instrumentation holes at the Dribble nuclear explosion test site, Tatum Salt Dome, southwest Mississippi.

Certificate of participation in the Dribble project.

Dr. H. I. S. Thirlaway (on left) and Professor I. P. Passechnik (center) representing the United Kingdom and Soviet Union respectively at Geneva meeting of UN Disarmament Committee's *ad hoc* Group of Scientific Experts. (Courtesy H. I. S. Thirlaway.)

Members of the British seismic decoupling experimental team during 1959. Left to right: Dr. Eric Carpenter, (name not known), and Peter Marshall. (Courtesy H. I. S. Thirlaway.)

Blacknest near Brimpton, Berkshire, home of the British Tabor Pluto effort. (Courtesy H. I. S. Thirlaway.)

Photograph spoofing announcement of new British seismic array method *circa* 1962. (Copyrighted by and used with permission of *Daily Mirror,* London.)

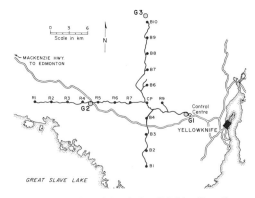

Layout as of 1974 of the British—Canadian seismic array near Yellowknife, Northwest Territories. (Courtesy Energy, Mines and Resources Canada.)

Control board for the seismic array at Yellowknife. (Courtesy Energy, Mines and Resources Canada.)

Geophone emplantment and associated radio telemetry link used in Yellowknife seismic array. Outmoded overhead wire linkage shows in background. (Courtesy Energy, Mines and Resources Canada.)

Dr. John H. Hodgson, successor to his father as Director, Seismology Branch, Dominion Observatory. (Courtesy J. H. Hodgson.)

Collapse craters caused by underground nuclear explosions at the Nevada Test Site. Near-by roads are about 20 meters wide. (Courtesy Las Vegas Office, U.S. Department of Energy.)

Captain Edward Snyder, USN, debating a point with RADM Harry Hess, USNR, during 1966. Shortly after, Snyder commanded the battleship, USS *New Jersey*. (U.S. Navy photograph.)

Drs. Arthur Maxwell and Thomas Gaskell discuss the merits of drilling to the "Moho" discontinuity. (Courtesy of A. Maxwell.)

Dr. Bruce Heezen, leading American bathymetrist of the 1950s and 1960s. (Courtesy of Lamont-Doherty Geological Observatory.)

Dr. J. Tuzo Wilson, Canada's leading geophysicist. (Courtesy of J. T. Wilson.)

Dr. James Heirtzler, former Director of the Hudson Laboratories, Columbia University. (Courtesy J. Heirtzler.)

Dr. F. Gilman Blake. (Courtesy of F. G. Blake.)

Willard Bascom, one of America's most innovative ocean engineers. (Courtesy W. Bascom.)

The Honorable James Wakelin, the first Assistant Secretary of the Navy (Research and Development). (Courtesy of J. Wakelin.)

Dr. George Shor inspecting a seismic record obtained in the tropical Pacific Ocean. (Courtesy of SIO.)

President and Mrs. Lyndon Johnson visit ESSA's *Oceanographer* on 13 July 1966. Commerce Secretary John Connor on extreme left flanked by Dr. J. Herbert Hollomon. (NOAA photograph.)

Vice President Hubert H. Humphrey *en route* to Europe during 1968 accompanied by Dr. Edward Wenk, Jr. (Courtesy of E. Wenk, Jr.)

Last meeting during December, 1968 of the U.S. Commission on Marine Science, Engineering and Resources. Clockwise starting in lower left corner: David A. Adams, Charles F. Baird, Leon Jaworski, Taylor A. Pryor, Richard A. Geyer (Vice Chairman), Julius A. Stratton (Chairman), Samuel A. Lawrence (Executive Director), George A. Sullivan, Jacob Blaustein, John A. Knauss, Lewis M. Alexander (Deputy Director), Robert H. White, George E. Reedy and Carl A. Auerbach. Staff members in the rear from the left are Stuart Ross, William Beller, Timothy Coleman, Holmes S. Moore, Sheila Mulvihill, and Harold C. Goodwin. (U.S. Coast Guard photograph.)

lems; culture is permissive rather than authoritarian; it is smarter to be a debtor than a saver, and there is little individual accountability or responsibility for the prosperity and health of oneself, his family, or society as a whole.[2]

As a consequence, much of the wisdom and technology that had been accumulated by mankind for improving the world's standard of living was rudely shunted aside in the early 1970s by a group, modest in size, calling themselves "environmentalists".[3] So alarming were the claims of this group that, by exploiting concepts expressed in the best-seller, *Silent Spring* (Carson, 1962), their thinking soon dominated the media and political circles. Their sales pitch was keyed to the assumption that modern industrial society was contaminating the globe with dangerous, even lethal, chemicals in such quantity that the chain of life was threatened in a manner that was, for the most part, irreversible.[4] Moreover, these substances were already everywhere — in mother's milk, in one's food and drinking water, and in the air one breathed — and continuing to accumulate in a most insidious fashion. In fact, the group's extremists argued that there was an urgent need to restructure industrial production, to redistribute global wealth, and to make drastic changes in life-style and value systems if various types of disasters were to be avoided.[5]

[2.] *Time* Magazine, the leading news journal in the United States, reports that during the 1970s, their best newsstand sales came as the result of "cover stories" on sex, rock-and-roll music, and the drug culture.

[3.] In 1980 the sixteen largest American environmental organizations had a membership of 5.2 million, over four million of whom belonged to the National Wildlife Federation. Membership of the more activist groups included: Sierra Club, 183,000; Cousteau Society, 110,000; Wilderness Society, 70,000; Environmental Defense Fund, 45,000; National Resources Defense Council, 42,000; Defenders of Wildlife, 42,000; Friends of the Earth, 22,000; Environmental Action, 20,000; and New Directions, 15,000.

[4.] Among the pollutants given extensive public attention besides pesticides were mercury, lead, sulfur dioxide, carbon monoxide and dioxide, nitrous oxide, phosphates, nitrates, nitrites, asbestos, radioactive materials, ozone, polychlorinated biphenyls, phthalate plasticizers, petroleum hydrocarbons (particularly the polycyclic aromatic hydrocarbons), and fluorcarbons. Strangely enough, heat released in the effluent from electrical power plants also came to be classified as a "pollutant" rather than a low-level energy supplement useful for purposes other than power generation.

[5.] What the old-timers thought about this new concept of ecology is described in a round-robin letter written on 4 May 1970 by Dr. W. M. Chapman to the oceanographic community during the nation's first "Earth Week":

"Roger Revelle and I just spent a most frustrating and discouraging week in discussion of world-wide ecological problems at the Center for the Study of Democratic Institutions in Santa Barbara. (We were) in company of several of what are now called ecologists, the more prominent members of which are becoming book and article salesmen rather than scientists. The new theology of ecology demands (a) that we have zero economic growth, and (b) that we have zero population growth. The rest of the new truths fall inexorably from these two revealed truths. There is thus need for stopping the manufacture of automobiles, the ceasing of extraction of fuel from the earth, and doing away with the Judaeo-Christian philosophy which has set us

To be sure, there were still leading geophysicists like Athelstan Spilhaus who argued for establishing "eco-librium", a situation within which there is a rational balance between economic and ecological considerations.[6] "Spilly's" thinking was echoed by the Under Secretary of the Interior, Dr. William T. Pecora, who pointed out in the Congressional Record for 21 March 1972:[7]

> A conservation ethic requires a better understanding of the natural base line before rigorous actions are taken out of apprehension and ignorance. . . . As the most intelligent species on earth, man can certainly provide for himself and yet prudently protect the total ecosystem from unnecessary and unacceptable degradation.

In view of all this, most of the world's political leaders agreed to the United Nations holding a massive "Conference on the Human Environment" in Stockholm, Sweden during June, 1972 that generated 109 recommendations, including establishing an "Earth Watch". Another resulted in forming a full-fledged "United Nations Environmental Program" (UNEP) based in Nairobi, Kenya (after all, African leaders had been promised the next major United Nations agency even though pollution was the least of their worries).[8] In the United States, the environmental movement joined forces with anti-nuclear and anti-business groups and paraded under the self-anointed label of "public interest" organizations.[9] With such a broad coalition announcing that man was "an endangered species on an endangered planet", the U.S. Congress began passing a steady flow of new

---

on this terrible track. An interesting thing about this new theology is that it has accepted the apocalypse from the more rabid fringe of the ecologists. (In this regard), there appears to be some contesting between Paul Ehrlich, Barry Commoner, and such lesser lights as Ken Watts who is also striving mightily. If we sinners do not stop what we are doing at once, there will be a horrible catastrophe in which we will all justly perish. We must, according to this theology, be frightened out of our wits, we must grovel in the blinding glare of the new revelation, and we must not be diverted by facts which are tools of the devil."

[6] Spilhaus first floated this concept in his presidential address to the American Association for Advancement of Science in December, 1971.

[7] A noted geochemist, Dr. Pecora served as Director of the U.S. Geological Survey from 1965 to 1971. He subsequently died in office during 19 July 1972.

[8] UNEP's first director was the accountant, Maurice F. Strong, who eventually returned to set up the Canadian state-owned oil company, Petro-Canada. An initial Assistant Executive Director was Dr. Robert Frosch (1973—1975), who had just completed a six-year stint as Assistant Secretary of the Navy (Research and Development).

[9] Contributing to this "eco-mania" were not only such photogenic events as the stranding of the *Torrey Canyon* off Cornwall (1967) and the Santa Barbara, California, offshore oil blow-out (1969), but also improvements in analytic chemistry that permitted detection of compounds with concentrations as low as one part per thousand million and even, in some cases, to one part in a million million. In many cases, the agitators and the agitated had no sense of ratios and simplified their argument further by claiming that there was no such thing as a threshold level below which a "contaminant" did not affect the ecology or one's health to some degree.

legislation in this area (see Table 7.1). Similar legislative actions took place at lower echelons of government as well.

With such legislation in hand, the "no-growth" advocates could challenge any major United States action, whether federal or industrial, involving the earth's environment via the court system and its multiple appeal levels almost *ad infinitum.* Thus, once the seismic prospect at Prudhoe Bay, Alaska, proved to be, in 1968, the largest oil field yet discovered in North America, the lease owners immediately applied for right-of-way permits. Three long years later, the U.S. Department of Interior, after hearing 12,000 pages of public testimony, released its favorable "*E*nvironmental *I*mpact *S*tatement" (EIS). However, it still took special congressional legislation during 1974 before court injunctions were lifted against pipeline construction. When it was all over in 1977, the paper work had taken six years and the pipe laying but three!

TABLE 7.1
Key Environmental Legislation Enacted into U.S. Law
(1969–1979)

Year	Short title of Legislative Act
1969	National Environmental Policy Act (NEPA)
1970	Clean Air Act
	Occupational Health and Safety Act
1972	Federal Water Pollution Protection Act (Amendments)
	Federal Insecticide, Fungicide, and Rodenticide Act
	Marine Mammal Protection Act
	Ocean Dumping Act
	Coastal Zone Management Act
	Noise Control Act
1973	Endangered Species Act
1974	Safe Drinking Water Act
	Energy Supply and Coordination Act
1976	Resource Conservation and Recovery Act
	Toxic Substances Control Act
	Federal Land Policy and Management Act
	National Weather Modification Policy Act
1977	Surface Mining Control and Reclamation Act
	Earthquake Hazards Reduction Act
	Marine Sanctuaries Act
1978	Outer Continental Shelf Act (Amendments)
	Ocean Pollution Research, Development and Monitoring Act
	National Climate Program Act
1979	Fuel Use Act
	Archeological and Historical Preservation Act

Such frustration continued commonplace throughout the 1970s. For example, in 1968, Exxon paid the Federal Government $218 million for offshore leases some 70 km west-southwest of Santa Barbara, California. During 1970, drilling demonstrated that the Hondo seismic prospect contained a "giant" oil and gas field. After three major environmental impact studies, 21 public hearings, 10 major (and many minor) governmental approvals, 12 lawsuits, a county referendum, and the spending of more than $340 million on this one site, the first production from the field finally came in mid-1981. The costly legal approach to offshore drilling still continues; for example, in January, 1981 Exxon sent 150 documents weighing 153 kg to Washington, D.C. as the initial step in getting permission to drill its first exploratory well on North Atlantic Block 133 located on the fish-rich Georges Bank. As an overall consequence of such requirements, oil and gas drilling on federal offshore leases, which had risen steadily from 825 wells in 1950 to over 3,700 wells in 1968, fell back to less than 1,700 wells in 1976 at a time when the politicians were claiming that there was an "energy shortage".[10]

The "Environmental Movement" also acquired considerable political clout in other parts of the world. Even industrial Japan passed a comprehensive group of such laws — but one that industry could smoothly adjust to and still keep the inflation rate below five percent. In Canada, the central government formed the Department of the Environment (also known as Environment Canada) in 1971 and included an Atmospheric Environment Service headed by an Assistant Deputy Minister among its components.[11] Initially, Ottawa cooperated closely with petroleum and mining industries and actually gave subsidies in various forms, including cheap transportation, to assist in opening up the Canadian Arctic and locating commercial deposits of oil and natural gas off Newfoundland and Labrador, as well as in the Canadian archipelago and even in the Arctic Ocean itself. However, by 1971, the dominant political mood in Canada (and in Great Britain and Norway as well) emphasized the creation of elaborate mechanisms for government control and ownership of

---

[10.] During the 1970s, the Rowan Companies alone moved over $200 million worth of drilling equipment out of the United States for lack of proper incentive. Yet there remained some 525 million acres of Federal land offshore that were largely virgin territory for oil and gas exploration.

[11.] Just the year before, this Service of 2,500 personnel and an annual budget of $34.9 million (Canadian) had been relabeled the "Canadian Meteorological Service" and placed in the newly-formed Ministry of Transport.

mineral wealth.[12,13] As a result, the oil industry had to look more and more towards countries interested in becoming net exporters, rather than importers, of energy.[14]

Although the so-called "public interest" groups made industry their chief target, managers in government and academia also felt their impact. Suddenly there was a blossoming of rules, regulations, and court actions relative to equal opportunity, civil rights, affirmative action, bilingualism, and the demands and needs of "minorities".[15] To geophysicists brought up on the idea of "the best man gets the job", this great effort in social engineering seemed a bit out of place. Jews, although rare in exploration geophysics, are among the top leaders of the profession. Blacks are still fewer but not unknown. For example, since 1977, the only geophysicist who rates being on Exxon's Board of Directors is Dr. Randolph W. Bromery, an extremely capable "Black" who started with the U.S. Geological Survey, took a doctorate in geology and oceanography at Johns Hopkins University, and, while a professor and department chairman at the University of Massachusetts (Amherst), became a specialist in mineral geophysics of the African continent. At the Geological Survey, another black geomagnetician, Roland G. Henderson, eventually became one of the world's foremost authorities in applying potential theory to the interpretation of aeromagnetic and gravitational anomalies.[16] Today, the seismologist, Waverly J. Person, serves as the Survey's able spokesman regarding newsworthy earthquakes located by its Branch of Global Seismology.

In the oceanographic field, Edward L. Ridley started near the

---

12. Once Texasgulf, Inc. found that a geophysical prospect near Kidd Creek in the Canadian Shield was one of the world's great copper—zinc—silver mining properties, the governmental Canadian Development Corporation (CDC) acquired 32 percent of the firm's common stock via a special tender offer in 1973. In 1981, the CDC apparently worked behind the scenes and arranged for Elf Aquitaine, an oil company owned by the French Government, to acquire the publicly-owned portion of Texasgulf. CDC then gave about $450 million (Canadian) and its one-third share of Texasgulf, Inc. to Elf and received back clear ownership of all of Texasgulf's Canadian assets.

13. The first environmentalist to become head of state was 41-year Dr. Gro Brundtland, who became Prime Minister of Norway on 4 February 1981 and lasted seven months in the post. An active Socialist, she had served as Minister of Environmental Affairs between 1974 and 1977.

14. During 1981, offshore drilling was extremely active in the Far East, the Middle East, West Africa, North Sea, the Caribbean, and South America. Attendance at the annual Offshore Technology Conference in Houston, Texas, during May, 1981 also set an all-time attendance record of over 90,000 persons.

15. Just what constituted a "minority" proved to be a shifting target. In Canada, the Meteorological Service during 1973 denoted some 300 of its positions as needing bilingual officials under terms of the "Official Language Administrative System".

16. Henderson was the first U.S. government employee to be named an Honorary Member of the SEG (1976).

bottom at the Navy's Hydrographic Office in 1954; years of highly meritorious work resulted in his being named Director of the National Oceanographic Data Center in January, 1981. Dr. John B. Slaughter, who started out in government with the U.S. Naval Electronics Laboratory in San Diego, California, moved even farther ahead by serving as Assistant Director of the National Science Foundation for Astronomical, Atmospheric, Earth and Ocean Sciences, as Director of the Navy-funded, ocean-oriented Applied Physics Laboratory of the University of Washington and finally, as of December, 1980, as Director of the National Science Foundation. In the associated field of global climate modeling, Dr. Warren M. Washington of the National Center for Atmospheric Research, an expert in the response time of the atmosphere to changes in sea surface temperature, joined the National Advisory Committee on Oceans and Atmosphere by Presidential appointment in 1978.

"Hispanics" have also been active in geophysics. The Puerto Rican, Guillermo Medina, for example, spent his entire professional career at the U.S. Navy Hydrographic Office between 1922 and 1968 and became the Office's first Scientific and Technical Director in 1960. The Chilean, Bernardo F. Grossling, took his doctorate in geophysics from the University of London in 1951 and became the ranking "Research Geophysicist" for the U.S. Geological Survey in 1973. By 1980, SEG members lived in 30 Mexican cities; in fact, 85 members had mailing addresses within Mexico City.[17] Some Hindus have also made excellent geophysicists. For example, a graduate of the University of Delhi, Manik Talwani, replaced Professor Ewing as Director of the Lamont-Doherty Geological Observatory in 1972, while Dr. Bimal K. Bhattacharyya, who had earned his B.S. degree at Calcutta University, became an Honorary Member (although posthumously) of the SEG in 1980.

Today, female geophysicists are finding it easier to obtain sea and flight duty, but even in World War II, Lt. Commander Mary Sears, USNR(W), ran a very creditable Oceanographic Branch at the Hydrographic Office between 1943 and 1945.[18] Subsequently, Jean Keen

---

17. The personable Antonio Garcia Rojas, Director of Exploration for Pemex, was designated an Honorary Member of the SEG in 1973 at the same time the Canadian, Dr. J. Tuzo Wilson, was similarly honored. Santos Figueroa H., who had formed the first seismic crew for Pemex in the late 1930s and moved on to be that oil company's Chief Geophysicist in 1950, was likewise named an Honorary Member of the SEG during 1981.

18. Once the war was over, Dr. Sears returned to Woods Hole and served as the founding editor for Pergamon Press's distinguished journal, *Deep Sea Research,* a responsibility which lasted from 1953 to 1973. She also edited *Oceanography: The Past,* which serves as the technical proceedings of the Third International Congress on the History of Oceanography held during September, 1980 (Sears and Merriman, 1981).

at Hydro, Margaret Robinson and June G. Patullo of Scripps, and Betty Bunce of Woods Hole have also been able to achieve national recognition for their oceanographic efforts. In this regard, one should not overlook the role of Ellen (*née* Scripps) Revelle, the wife of Professor Roger Revelle and the grand-niece of Ellen B. Scripps, one of the original sponsors of the Scripps Institution of Oceanography. The oceanographic establishment simply could not have gotten along without Mrs. Revelle, for whenever a visiting scientist came to the Scripps campus, it was almost certain that he would be entertained, often into the wee hours of the night and without much notice, at the Revelles' beautiful home above the beach at La Jolla. Moreover, if a junior oceanographer's wife needed any cajoling in order that her spouse could disappear at sea for months at a time, Ellen was always willing to do so.

In the field of exploration geophysics, it has taken more time for mathematically and field-oriented women to become "doodlebuggers". Even so, by the time of the SEG's annual meeting in 1981, there was a sizeable sprinkling of serious-looking young women carrying brief cases in attendance at the technical sessions, and others had joined the field crews as "jug hustlers", cable repair technicians and surveyors.[19] Nevertheless, the best known woman today in geophysical prospecting is the hard-working public affairs specialist, Bettye Athanasiou. Based in Houston, she had edited three company magazines — *Party Line* (Robert H. Ray, Inc.), *Time Break* (Geo Space Corporation), and *The Phone* (Walker-Hall-Sears) and operated scores of hospitality suites for the thirsty explorationists throughout the world during the past three decades. Graced with near total recall, she knows by face (and generally by name) perhaps some 6,000 geophysicists not only from attendance at innumerable conventions and exhibits, but also by visits to the field, some of which have taken her far north of the Arctic Circle. While on visits to such remote outposts of the geophysical profession, she listens to everyone's hopes and frustrations, takes their picture, and writes down urgent errands she can perform for them once she returns to civilization.[20] For these unique efforts as "Den Mother" to an entire profession, the Society of Exploration Geophysicists made her a

---

[19.] As of mid-1980, SEG's Committee for Women in Geophysics estimated that some 500 women belonged to the society, or less than four percent of the group's total membership of 13,800. Of these 500, nearly 25 percent, had been professional geophysicists for 10 or more years, 30 percent for five to ten years, and 45 percent for four years or less.

[20.] For a memento of the SEG's 50th Anniversary Meeting in November, 1980, NF Industries issued an 84-page album, *Doodlebuggin' with Bettye,* which contained 1,500 photographs she had taken of the geophysical profession during the past quarter century.

"Life Member" in 1978, and the Houston Geophysical Society followed with a similar award a year later.

Besides having to adjust to the environmental movement and to political prioritization as to who should be educated and employed, the geophysical profession experienced definite problems during the 1970s from uncontrolled inflation throughout the world. At the start of the decade, inflation rates in Canada, the U.S., and Great Britain had been about three, six and twelve percent respectively, but ten years later the rates were more like 10, 14 and 21 percent. Noting the dwindling of their purchasing power and the tight world market for petroleum, the Organization of Petroleum Exporting Countries (OPEC) succeeded in jumping their average price for a barrel of crude oil from approximately $3.20 in 1970 to $28.15 by January, 1980, up by 880 percent.[21] An exploration geophysicist working directly for or on contract with an international oil company was thus insulated from increased costs caused by inflation because they were more than covered by the upward jump in the price of crude oil, but the reverse held for geophysical field work in the public and academic sectors. Even as late as 1978, fuel costs were but ten percent of a survey or research vessel's operating cost, but by 1980, fuel costs were approaching 25 percent of the total operational costs.[22] In addition, there was no longer a guaranteed fuel supply for wars, trade embargoes, nationalization actions, and new conservation policies had all contributed to creating market shortages in 1973–1974 and again in early 1979.

Political reaction to these problems was extensive although not necessarily realistic from the long-term point of view. Canada, for example, flirted briefly with a conservative government in 1979, but then turned back to a policy of buying or forcing out foreign-owned oil companies. Great Britain turned conservative in 1979 under the articulate chemist, Mrs. Margaret Thatcher, and continued that way into the early 1980s. The United States waited to see what Lieutenant "Jimmy" Carter, USN (Retired), would do with

[21.] After 1974, oil and gas producers in the United States and Canada found their prices heavily dependent upon political dictates and bearing little relation to the world market price. For example, in 1975 the average price of U.S. oil was $7.50 a barrel and still only $9.00 a barrel in 1978. A comparable situation existed in Canada. In early 1981, price controls on crude oil were lifted in the United States, and the price of high-quality "old oil" jumped to as high as $38 a barrel, while the equivalent price of Canadian oil held at about C$17.75. As a result, the Canadian oil industry began moving a large share of its operations south of the border, essentially "voting with its feet".

[22.] To make up some of this funding deficit, the U.S. National Science Foundation robbed "Peter" to pay "Paul" by deferring maintenance and yard overhaul of its academic research fleet. As a result this fleet may well experience a collapse syndrome during the mid-1980s much like that which faced the "one-hoss shay".

this set of problems upon becoming president in 1977. To assist him, Carter was the first president to select a geophysicist as his science and technology adviser.[23] Looking for someone who knew arms control, the energy field, environmental issues, and last, but not least, minority issues, Carter decided that the 53-year-old Dr. Frank Press, Chairman of MIT's Department of Earth and Planetary Sciences, fitted this profile. After taking his doctorate in 1949 under Professor Maurice Ewing, Press had migrated to CalTech where he eventually succeeded Dr. Beno Gutenberg as Director of the Seismological Laboratory in 1957. Returning across the country in 1965 to be a department chairman at MIT, he subsequently found time to co-author *Earth,* the best introductory college text for geology written in the half past century, as well as to continue to be involved in a multitude of national and international scientific policy matters (Press and Siever, 1974).[24]

Growing up in a mixed section of Brooklyn, New York, Press was quite familiar with minority problems and active in the early 1970s relative to having the American Geological Institute establish a "Minority Participation Program".[25] Even at age 34, he was interested in arms-control issues and participated in 1958 in the first Geneva talks on a nuclear-test ban, followed almost immediately by serving on the U.S. State Department's Panel on Seismic Improvement (the Berkner Panel) for improving the proposed Geneva nuclear-test-ban monitoring system. Urbane and quiet spoken, Press had the knack of moving groups to accomplish something significant, whether it was to start a national program in earthquake prediction or to place geophysical sensors on the moon, on Mars, and in overlooked portions of Mother Earth. In other words, he was most expert in the arcane art of "D.C. seismology — Washington version".

During his four-year tour with President Carter, Press regularly

[23.] After being inaugurated President of the United States in January, 1981, Ronald Reagan reverted to form and selected a nuclear physicist, Dr. George Keyworth, from the Los Alamos Scientific Laboratory to be his science advisor. However, Henry Salvatori, the Founder of the Western Geophysical Company, is reputed to be a member of the President's "Kitchen Cabinet".

[24.] Among noted groups that Dr. Press participated in were the President's Science Advisory Committee (1961—1964), National Science Board (1970—1976), and, beginning on 1 July 1981, a six-year term as President, National Academy of Sciences. In addition, he has authored or co-authored over 160 papers, including the one which announced the discovery of the earth's free oscillations initiated by the great Chilean earthquakes of 1960. He also kept a strong love for the sea; upon being named "California Scientist of the Year" in 1962, he promptly spent the $5,000 award on a "Lapworth-25" sailing vessel.

[25.] This Program now awards some 50 scholarships annually to promising and deserving minority college majors in the geosciences. The SEG makes another 110 scholarships available each year to best qualified young people of all types.

sat in on Cabinet and senior staff meetings and frequently helped create and implement such presidential initiatives as increased federal support for basic research, involvement of science and technology more deeply in regulatory decision-making, provision of better incentives for technological innovation, and strengthening the role of technology and science in international affairs. In the latter instance, Press met with Vice Premier Deng Xiaoping of the People's Republic of China in Beijing during July, 1978 which led to the ultimate signing of thirteen protocols fostering closer technical and scientific cooperation between two of the world's great powers (Press, 1981).

Within the White House establishment, Press directed the Office of Science and Technology Policy. Thus it fell to him to decide whether a unique double light flash observed on 22 September 1979 by a Vela nuclear monitoring satellite was actually the signature of an unannounced four-kiloton nuclear explosion high over the South Atlantic Ocean not far from South Africa. Assisted by a panel of nine scientists chaired by Professor Jack Ruina, now at MIT, Press concluded that the cause of the unique double-flash was "technically indeterminate", and the U.S. State Department subsequently allowed that it was not in a position to accuse anyone of anything.[26] Curiously enough, the Vela satellite detection system again spotted a strange flash in the same general region on 16 December 1980, although this time the unofficial consensus was that the satellite had detected the trail of a meteorite burning up in the atmosphere.

The choosing of Press for the highest scientific advisory post in the United States was an arbitrary, rather than an elective, act. In contrast, when the paleomagnetician, Bruce Babbitt, was elected governor of the State of Arizona during November, 1978, his became the highest elective role ever achieved by a geophysicist within the United States.[27] An undergraduate geology major at Notre Dame University (where he was also student body president), Babbitt graduated *magna cum laude* in 1960 and proceeded to study under dynamic Professor Stanley K. Runcorn at the University of Newcastle where he earned his master's degree in 1963, using as thesis material a collection of rocks taken from his beloved Grand Canyon area. However, Babbitt's interest in the world around him did not stop with deciphering the history of the earth. In 1965, he received

---

[26.] One theory was that the satellite was hit by a small meteoroid which split, leaving a double trail. However, radio-astronomers at the National Radio Astronomy Facility in Arecibo, Puerto Rico, also claim that they tracked a large gravity wave in the ionosphere moving away from the South Atlantic region with a speed of 600 m/sec at about the appropriate time that same night.

[27.] Babbitt was also the youngest Arizona citizen to be elected to this role.

a law degree from Harvard University, and 13 years later found himself in a position to begin "ram-rodding" a "model" ground-water law through an almost hopelessly divided state legislature. Once this was accomplished in 1980, he advised his citizenry that ". . . if we plan wisely, in twenty years Arizona will be the high technology center of the whole nation". He may well be right, for the leadership that this articulate, sensible geo-politician has provided the state to date has been excellent.[28]

Boom Days Once Again for Exploration Geophysics

In 1970, the exploration geophysics industry found itself at the bottom of a slump which, in the United States, had dragged on for a full 18 years (see Fig. 7.1). In 1952, for example, the domestic industry had fielded 670 crews, but by 1970, the demand could be met by only 185 crews. Internationally, the picture was slightly better. Even so, the data indicate that worldwide seismic activity, when expressed in crew-months, fell gradually from 8,800 in 1960 to 6,500 in 1970, followed by a peak of 8,100 in 1974 before dropping back to but 6,000 crew months in 1977. Within these crew months, marine work normally comprised only ten per cent or less although the faster speed at which seismic boats worked produced 60 percent or more of total line miles from 1965 on. Starting in 1977, the world's energy crunch was better understood by the public and many politicians; consequently, the petroleum exploration business began to boom once again. By 1981, the old record set back in 1952 of seismic crews at work was finally broken as the count of known seismograph crews at work outside the Communist World came to 1,084 on the first of September. In fact, the worldwide geophysical exploration industry, which had had its first $2 billion (10^9) year in 1979 followed by a $3 billion year in 1980, expected a $4 billion year during 1981. To be sure, much of this increase came from higher oil prices and the advent of new offshore drilling and production technology which permitted working in more hostile and deeper waters.[29] Yet the fact remains that the quality of detailed seismic surveys had improved so much that sophisticated

---

[28.] In early 1981, Babbitt acquired as a speech-writer Ms. Anna Simons, the daughter of the *Washington Post*'s Managing Editor, Howard Simons, after she had performed a similar function for President Carter.

[29.] In 1980, Shell was producing profitable oil from the Cognac platform in 310 m of water off the Mississippi River delta, exploratory drilling was going on globally in water depths over 1,000 m, and oil company officials were willing to testify before the U.S. Bureau of Land Management that it might be possible to obtain commercial production in water depths of 1,800 m by 1990 should appropriate lease sales be held in the North Atlantic Ocean during 1981.

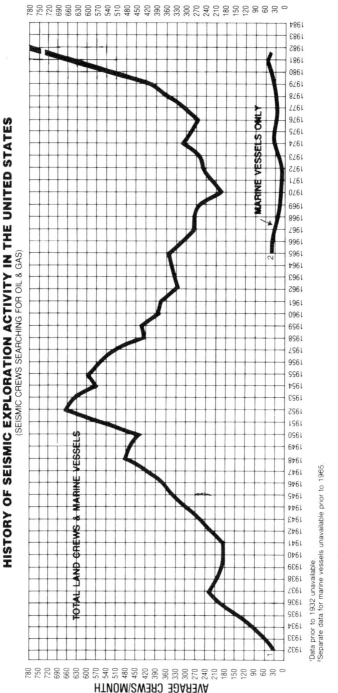

HISTORY OF SEISMIC EXPLORATION ACTIVITY IN THE UNITED STATES
(SEISMIC CREWS SEARCHING FOR OIL & GAS)

TOTAL LAND CREWS & MARINE VESSELS

MARINE VESSELS ONLY

AVERAGE CREWS/MONTH

[1]Data prior to 1932 unavailable
[2]Separate data for marine vessels unavailable prior to 1965

FIG. 7.1. Number of seismic crews searching for oil and gas within the United States, 1932–1981. (Courtesy of the Society of Exploration Geophysicists.)

field work and computer data processing took the place of much exploratory drilling and freed up capital for additional geophysical exploration.

By the early 1970s, fairly standard data-processing steps existed that provided a useful first look at the geologic substructure within a seismically surveyed area. Basically, these steps consisted of: (1) separating the multiplexed recorded data from the several geophone arrays into separate, continuous traces, (2) gathering the traces into CDP groups, (3) applying an initial deconvolution to the individual CDP traces to make the reflections sharper and more uniform, (4) applying static corrections to the CDP traces, (5) performing velocity analyses on the CDP "gathers" at regular intervals along the seismic line, (6) using the resulting velocity curves to correct all the CDP "gathers" for normal moveout, (7) if necessary, migrating reflections to their true spatial positions, (8) stacking the CDP traces, (9) applying numerical filters or additional deconvolution as needed, and (10) finally plotting out the first stacked seismic cross-section. Of these, steps (1), (4), and (5) are manpower intensive, and steps (3), (4), (5), (8), and (10) require the most machine time.

Evidence had also existed for many years, although not usually recognized, that sandstones containing gas could exhibit abnormally low velocities (Savit, 1960). By the mid-1960s, a number of binary-gain recording systems were being used that could record seismic amplitudes from geophones in terms of precise six db (multiples of two) gain steps according to a binary code (Siems and Hefer, 1967). This has made possible the much more accurate preservation of true relative amplitude for a single trace and from one trace to another throughout the entire data processing operation. Amplitudes of weaker reflections arriving later are carefully controlled by correcting for geometric spreading of a downgoing wave instead of applying the usual "brute-force" automatic gain control. As a result, such traces frequently display a "bright spot" over gas-containing sandstone.

When the major offshore Texas lease sale was held in May, 1974, this high reflection amplitude ("bright spot") technique was being used extensively. During May, 1975, Shell's general manager for geophysics even indicated that his co-workers had been "routinely predicting hydrocarbons" for a few years (Flowers, 1976). The success ratio for new-field "wildcat wells" jumped a fourth (from 11 percent in 1972 to over 14 percent in 1973 and 1974) with a gas/oil discovery ratio of 55/45 (Rice, 1976).[30] However, some of this

30. By 1980, the success rate had hit 19 percent for new field wildcat wells and nearly 30 percent for all exploratory wells (Barry, 1981).

increase was too early to have been influenced substantially by use of the bright-spot technique and was likely caused by additional attention to gas exploration as a result of increasing gas prices. Trial and error also proved the hard way that bright spots were not necessarily an "open sesame" to finding commercial production, as many major oil companies learned from their unrewarding $600 million leasing and drilling campaign over the extensive Destin dome seaward from Appalachicola, Florida.[31] Bright spots, in fact, are not the only seismic indicator of the presence of hydrocarbons. A "flat-spot" reflection from a gas—oil or oil—water interface is even a better indicator, as is a decided character change in the reflections from the hydrocarbons relative to the reflections on either side (see Fig. 7.2).

On another tack, more attention began to be paid to seismic techniques capable of locating stratigraphic traps such as reefs, sand bars, bed pinchouts against anticlines, unconformities and faults,

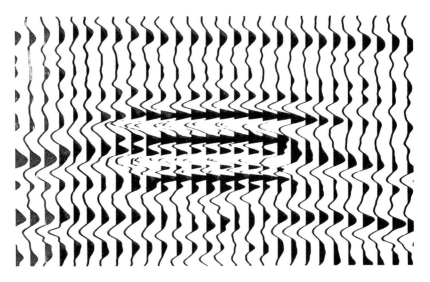

FIG. 7.2. Section from a processed vertical seismic profile showing three indicators of a "bright spot": (1) the signal is stronger than are the nearby reflections, (2) the signal polarity has reversed, and (3) there are perfectly horizontal reflections indicating gas—oil and oil—water contacts (from Savit, 1974). (Courtesy of Western Geophysical Company.)

[31.] Extensive laboratory work showed that a very small percentage of gas in a porous sand could produce almost as low a velocity value as could a high concentration of gas (Domenico, 1974).

and facies changes giving porosity developments within otherwise tight formations. To find consistently such traps required the maximum resolution of reflections from individual strata and of lateral changes in reflection character. A number of companies tried many varied approaches, but the real breakthrough came in 1972 from work by Roy O. Lindseth, president of a small geophysical firm, Teknica Resource Development, headquartered in Calgary, Alberta. His Seislog® technique has since been applied in various areas of the world with good results, particularly after additional processing and interpretation refinements had been added (Lindseth, 1979). Use of the technique has, however, emphasized the need to obtain seismic data of much broader frequency content, perhaps up to 500 hz, instead of the conventional upper limit of 100 hz. The change was not easy to come by for it has meant completely redesigning the recording equipment to pick up the much weaker higher-frequency signals, digitizing more rapidly, and considerably increasing the number of geophone stations to make the spatial sampling rate correspond to the increased time sampling rate. Even so, two geophysical contractors now offer high-resolution seismic crews recording at sampling rates down to 0.25 millisecond and one contractor even records on 1,024 channels in order to provide a three-dimensional picture of areas with steep dips and complex geology, as around and over a salt dome.

As a result of these and other advances, the geophysicist is beginning to offer petroleum and mining geologists a rapidly increasing spectrum of displays in both black-and-white and in color that indicate stratigraphic relationships in a variety of modes: (1) quantitative stratigraphy with anomalies presented on a given plane that show gas—fluid contacts and probable pore fill, (2) qualitative stratigraphy in which specific rock units are analyzed to show lithology, porosity, structural setting, and areal extent, and (3) macro-stratigraphy wherein regional and basinal depositional patterns are shown. These types of analyses are now so far along that Savit has predicted that he will see the day when computers create displays showing where oil and gas occurs throughout a structure even before it is drilled. But there is one "Catch 22" to this — to acquire field data of such an extremely high quality is not only painstaking and expensive, but the amount of electronic computer time required is truly phenomenal. Already the geophysical industry uses more magnetic tape and digital computer power than any other. Let us take one example: high-resolution seismic recording at, say, a sampling rate of 0.5 ms will produce

® is a trademark of Teknica Resource Development, Ltd.

288,000 "numbers" or "words" (5,184,000 "bits") in just one 24-trace, 6-second recording. And the CDP 12-fold stack from such traces will involve about 3.5 million words (62 million bits) of data, to say nothing of the additional capacity needed to do deconvolution filtering and normal-moveout corrections. But then, these one-dimensional operations are very simple compared to the speed and capacity required to perform wave-equation migration in two dimensions, to say nothing of three dimensions. The largest and fastest computer in use in seismic data processing today is Texas Instruments' *A*dvanced *S*cientific *C*omputer (ASC). But the ASC is not powerful enough to do wave-equation migration *before* CDP stacking on three-dimensional data sampled at 4-millisecond intervals, which is badly needed for accuracy, so three-dimensional processing of data sampled at a higher rate will have to await more powerful computers. It is estimated that by 1985, the seismic industry will need computers 50 times larger and 50,000 times faster than the machines of 1980 (Savit, 1980).

Although 95 percent of all geophysical exploration funds spent during 1980 went into oil and gas prospecting, good progress has been made on other applications. With mineral exploration requiring ever deeper and more costly drilling, the mining industry has been acquiring increasingly greater amounts of geophysical and geochemical information. More than 30 commercial variations of the following geophysical techniques are used for this purpose (see Table 7.2). Their multiplicity and sophistication are so great, however, that space limitations keep us from describing the status of each technique. For specific details, you are referred to the review articles in the "Golden Anniversary Issue" of *Geophysics* (November, 1980) and to a subsequent "New Developments" paper published in the same journal eight months later (Rice *et al.*, 1981). Suffice it to note here that the more expensive techniques, such as spectral induced polarization, borehole logging with extensive suites of probes, and high-resolution seismology supplemented by mini-computers and micro-processors at the field site, are all receiving much better acceptance than in the past. Remote sensing has also developed so well that the SEG, sponsored, under the leadership of Dr. Anthony R. Barringer, President of Barringer Resources, Ltd., the Sixth Annual Pecora Symposium on "Remote Sensing and its Application to Exploration" at the U.S. Geological Survey Earth Research Observational Satellite Data Center, Sioux Falls, South Dakota during April, 1980.

These same geophysical techniques have been often used in other applications, particularly in construction, where increasing public

TABLE 7.2.

Geophysical Techniques Available for Mineral Exploration in the 1980s

Technique	Locale				
	Land	Marine	Airborne	Satellite	Drill-Hole
Seismic (reflection)	X	X			
Seismic (refraction)	X	X			
Side-scan sonar		X			
Gravity	X	X	X		X
Magnetic	X	X	X		X
Resistivity	X				X
Self-potential	X				
Electro-magnetic	X		X		X
Combined electro- magnetic/magnetic	X		X		
Very low frequency					X
pulse	X		X		
"Input"			X		
"Afmag"	X		X		
Magnetotelluric	X				
Induced polarization	X				X
Earth current	X				
Radiometric (gamma ray)	X		X		X
Radiometric (alpha track)	X				
Sonic/velocity logging					X
Side-scan radar			X	X	
Infra-red scanning			X	X	
Multi-spectral photography			X	X	

awareness of natural and man-induced hazards from earthquakes, incompetent footings, and permafrost require extensive pre-engineering field surveys. Such applications include the siting of dams, power plants (particularly nuclear ones), pipelines, highways, tunnels, subways, offshore drilling and production platforms, underground fuel storage, and hazardous waste disposal by either storage ponds or underground injection.[32] Heritage preservation laws also require surveying the sea bottom offshore for archeological artifacts before construction activities can go ahead. Ground-water surveys are also quite common, particularly in water-short areas. One of the new growth areas for exploration geophysics may be the discovery and mapping of geothermal resources. Although many techniques have

32. For example, the original engineering design of the 1,280-km-long Trans Alaska Pipeline System from Prudhoe Bay to Valdez, Alaska, primarily called for conventional burial of the line. Upon closer analysis at the urging of the U.S. Geological Survey, it was found that permafrost or other potentially unstable soil conditions occurred over 85 percent of the route, thereby requiring an above-ground support system for 675 km of the line at a cost of $2.8 billion. In addition, decisions had to be made whether to bridge or bury at more than 800 river and stream crossings, as well as to how best to cross the major, highly active Denali Fault zone (an allowance of 1.5 m vertical motion and 6.0 m horizontal motion was eventually decided on).

been tried, including micro-earthquake monitoring, much has still to be learned before such geophysical surveys become routine and economic. In fact, after having spent four years studying the Roosevelt Hot Springs thermal area in Utah, the highly competent Department of Geophysics at the state university disappointedly concluded that detailed seismic reflection and refraction surveys had delineated neither the thermal reservoir nor its heat source (Ward *et al.*, 1978). However, during May–July, 1977, the U.S. Geological Survey operated a 15-element seismometer array encircling the area and recorded 72 teleseismic events from all quadrants. Upon working up the P-wave travel-time residuals, the analysis indicated a small but very anomalous nearby region of low seismic velocity and high signal attenuation, thereby suggesting the presence of abnormally high temperature and perhaps partial rock melt at a depth of about 15 km (Robinson and Iyer, 1981).

Non-profit Geophysics Acquires a Political Hue

Once Richard Nixon became President of the United States on 20 January 1969, political pressure came from many sources to do something tangible about scores of major recommendations included in the just released Stratton Commission report, *Our Nation and the Sea.* On 28 January, a Union Oil Company oil well on Platform A nine km off Santa Barbara, California, blew out and briefly contaminated some of the most beautiful and expensive coastline in the entire nation. This event gave special impetus to addressing the issue of national ocean policy. Nixon took the standard bureaucratic way out — reorganize! After much stirring within the bureaucracy, on 9 July 1970 he announced the establishment of the *N*ational *O*ceanic and *A*tmospheric *A*dministration (NOAA) within the U.S. Department of Commerce, as well as formation of an independent agency, the *E*nvironmental *P*rotection *A*gency (EPA). As a major "umbrella" agency, NOAA's core would be Commerce's Environmental Science Services Administration, supplemented by the Sea Grant Program from the National Science Foundation, the Bureau of Commercial Fisheries from the Department of Interior, the National Ocean Data Buoy Office from the Coast Guard, and the National Oceanographic Data and Instrumentation Centers from the Navy.[33]

33. Despite extensive lobbying to place the 43,000-man Coast Guard (with a budget of $653 million) in NOAA, Secretary of Transportation John Volpe was able to fight off the raid. After all, who wants to lose a *de luxe* executive jet aircraft and uniformed mess boys?

NOAA was headed by the able Dr. Robert White, backed up by a popular Alaskan politician, Howard W. Pollack, who had failed to be re-elected to the Congress. NOAA's initial staff consisted of 8,000 persons and an initial budget (FY 1972) of $330 million, $128 million of which was ocean oriented. In good Washington style, NOAA kept adding missions, responsibilities, and highly paid staff.[34] By the end of the decade, its funding needs had tripled, even though it was forced to spin off its seismological mission, including the National Earthquake Information Center, to the U.S. Geological Survey on 7 September 1973. White's emphasis had been on expanding NOAA's research arm, but once Carter became President in January, 1977, he placed a 37-year-old humanist, Richard A. Frank, in White's slot.[35] A Harvard-trained lawyer, Frank's previous role had been Director, Center for Law and Social Policy, the nation's first multi-purpose (and largest) public-interest law firm. As an added credential, his wife, Margaret, also a lawyer, served as Deputy Secretary for the Presidential Cabinet.[36] Another young lawyer, James P. Walsh, who had served as counsel to the U.S. Senate's National Ocean Policy Study between 1972 and 1977, was then fleeted in as the Deputy Administrator of NOAA, thereby placing "legal eagles" in both top slots of a scientific agency.

By 1980, NOAA was as much a regulatory agency as it was a scientific services organization. Its major new programs included coastal zone management, coastal energy-impact subsidies, offshore fishery management and enforcement, designation of marine sanctuaries, and regulation of the embryonic ocean mining and ocean thermal energy industries. In the summer of 1980, NOAA also acquired the highly useful "LANDSAT" series of earth resource satellites from the Geological Survey despite anguished protests by Senator Harrison Schmitt.[37] As a result of this raid, NOAA relabeled its National Environmental Satellite Service the "National Earth Satellite Service", continued it under the able directorship of Dr. David A. Johnson, and proposed adding an extra $122.4

[34.] In 1978, NOAA utilized 140 executive level and "super-grade" positions, plus six admirals. The Coast Guard, with twice the budget and five times the staff, used only five supergrades and 27 admirals.

[35.] White stayed on the Washington scene, first becoming Executive Director of the National Research Council and, then, in 1980, President of the University Corporation for Atmospheric Research.

[36.] Early in the Carter Administration, the *Washington Post* pointed out that the Franks were one of the new husband-and-wife teams drawing $100,000 a year in joint Federal pay.

[37.] With a Ph.D. in geology from Harvard, Schmitt was the first earth scientist to "pound rocks" while on the moon (Apollo Flight 16 in December, 1972).

million to its FY 1982 budget. This made the proposed annual budget come to $1,050,800,000, thereby exceeding one billion dollars for the first time. NOAA's staff now totaled nearly 13,000, 1,400 of which were assigned to 12 research laboratories managed by an "Environmental Research Laboratory Directorate" in Boulder, Colorado.[38] In addition, the *National Advisory Committee on Oceans and Atmosphere* (NACOA) was headed by an environmental activist, Dr. Evelyn F. Murphy.

But being at the cutting edge of the pending "Decade of Ocean Policy" didn't help either NACOA or NOAA very much when the conservative Reagan Administration took over in January, 1981. Dr. Murphy, to be sure, had already been replaced by a *bona-fide* oceanographer, Dr. John Knauss, Dean of the College of Oceanography, at the University of Rhode Island. By April, 1981, the word was also out that the new president wanted NOAA's budget cut back to $800 million, its staff cut by 900 people, and its new administrator to be a marine geologist, Dr. John Byrne, the Vice President of Research and Graduate Studies at Oregon State University. Thus, this first effort by the "environmentalist movement" to have a public-interest lawyer direct a major federal scientific agency lasted just four years.

For the first time in its 100-year history, the U.S. Geological Survey, NOAA's competitor in such fields as mapping and hydrology, also became involved in presidential politics when the Carter Administration took over. The USGS's Director, 61-year-old Dr. Vincent McKelvey, was an economic geologist who believed in early exploitation of sea-bed minerals. This concept didn't fit with that held by the Department of Interior's new assistant secretaries who, being in their early thirties, came from a different generation of thinking. So, despite some hue and cry from the scientific community, McKelvey was shunted back to being a research geologist, and his place taken by Dr. H. William Menard, one of the top geological oceanographers at Scripps. A veteran of 20 oceanographic expeditions, Menard lasted throughout the Carter Administration, even though he had the effrontery in 1979 to state publicly that the great petroleum-bearing Jurassic reef of Mexico's Reforma Platform in the southern Gulf of Mexico extended northeastwards perhaps as far as the Atlantic continental margin of the United States.[39]

[38.] The engaging, well-organized high-energy physicist, Dr. Wilmot N. Hess, ran the directorate between 1970 and 1980, when he shifted up to the top of the mesa at Boulder to run the National Center for Atmospheric Research. Hess's deputy during the 1970s was no less than Colonel Joseph Fletcher, USAF (Retired), of "Fletcher's Ice Island".

[39.] As part of the politicization of governmental earth science within the United States,

Menard further suggested that the most favorable location for testing the oil potential of this reef, which might contain up to six billion barrels of recoverable oil in the mid-Atlantic region, would be 230 km off the New Jersey coast in water depths of approximately 1,800 m. Fortunately, he got away with such positive thinking because his close friend, Frank Press, had convinced President Carter that the optimal use of the moth-balled deep-sea drilling ship, *Glomar Explorer,* would be deployment in an "Ocean Margin Drilling Program" to cost $495 million during the 1980s (Office of Technology Assessment, 1980).[40] Part of this funding was to come from ten oil companies, and the remainder from the National Science Foundation. However, after putting up $5 million the previous year, the oil companies suddenly canceled out during October, 1981, leaving the long-term fate of the program much in doubt.[41]

Military geophysics also saw its share of administrative turmoil during the 1970s, particularly in the U.S. Navy. During 1978, the Naval Oceanographic Office was merged with the Naval Weather Service into a "Naval Oceanography Command" and shifted to the National Space Technology Laboratory complex in southern Mississippi 1,500 km southwest of Washington, D.C.[42] As part of this consolidation, the category of "aerological officer" was changed to "geophysical officer", "weather centrals" became "oceanography centers", and the Command's realm of responsibility defined as

---

Menard was asked to resign when the Reagan Administration came to power in 1981. His replacement was Dr. Dallas Peck, who moved up from the Chief Geologist's slot at the Geological Survey.

[40.] At the behest of the Central Intelligence Agency, Global Marine built the *Glomar Challenger* during the late 1960s in order to raise secretly a portion of a Soviet submarine sunk in the Pacific Ocean. When refitted with enough drill stem, a riser, and a blowout preventer, she should be able to drill holes six km deep in water depths of four km.

[41.] The participating companies were Atlantic Richfield, Cities Service, Conoco, Exxon, Mobil, Pennzoil, Phillips, Chevron, Sunmark and Union. One apparent excuse for bowing out was that such major firms as Gulf, Shell, and Texaco had never joined the group. University participation via the Joint Oceanographic Institutions consortium would have been Universities of California (Scripps), Hawaii, Miami, Rhode Island, Texas and Washington; Lamont-Doherty Geological Observatory of Columbia University; Oregon State University; Texas A&M University and the Woods Hole Oceanographic Institution.

[42.] Outsiders found it hard to pin-point exactly why this amalgamation took place at this time and in this manner. To be sure, NSTL was drastically under-utilized, the Pentagon had spun off hydrographic charting functions from the Navy and assigned them to the new "Defense Mapping Agency", and Congress, in general, believed there were too many Navy agencies in the Washington region. The individuals who took both the "kudos" and "brick-bats" for this action were Senator John Stennis of Mississippi (Chairman of the Senate's Armed Services Committee), Rear Admiral J. Edward Snyder, Jr., USN, then Oceanographer of the Navy, and Mr. John Stephens, who eventually became Assistant Commander for Management Information, Naval Oceanography Command.

extending from the bottom of the ocean to the upper atmosphere.[43] In addition, the Mississippi complex was boosted by the formation of a "Naval Ocean Research and Development Activity" (NORDA) sponsored by the Office of Naval Research, as well as ONR's maintaining part of its Ocean Science and Technology Division in the same location. Although the Gulf Coast location irritated the academic clientele of the Office of Naval Research, it has made it possible for NORDA to be increasingly alert to the capabilities of the geophysical industry; as a result, NORDA and the SEG now jointly sponsor an annual technical conference which delves into a facet of signal processing of importance both to the industry and to military geophysicists.

Within Canada, governmentally sponsored geophysics was much more stable. To be sure, such effective individuals as Drs. W. E. van Steenburgh, Patrick McTaggart-Cowan, William Ford, and John Hodgson no longer called the shots, but a hard-working group of post-war trained earth scientists was there to provide in dealing with such new legislation as the Clean Air Act, the Environmental Contaminants Act, the Canada Water Act, the Northern Inlands Water Act, the Arctic Water Pollution Act and the Fisheries Act. As a result, the marine science side of government continued to provide a "fast track" to a wide number of diverse posts within the federal service.[44] For example, after Arthur E. Collin took his doctorate degree from McGill University during 1962, he had been named Scientist-in-charge, Arctic Programme, at the Bedford Institute of Oceanography just a year later. By 1965, he was Scientific Advisor to the Director General, Canadian Maritime Forces, and then moved on to become Dominion Hydrographer three years later. During the 1970s, Collin

---

43. Even the famed "Fleet Numerical Weather Facility" with a staff of 60 officers, 128 enlisted personnel, and 104 civilians located in Monterey, California, split into two components, the Naval Environmental Prediction Research Facility and the Fleet Numerical Oceanography Center. The Office of Naval Research was also required to move its Ocean Sciences and Technology Division to the NSTL site amidst large-scale objections by the academic oceanographers who preferred to do "one-stop shopping" while in the Washington, D.C. area on other business. Just the same, U.S. Comptroller General Report CED-79-27 of 25 January 1979 still noted: "Over 200 oceanographic activities or projects are being conducted or managed in separate Navy commands and many academic institutions without any overall coordination or single point of management within the Department of the Navy." This, it also noted, was against the intent of Congress.

44. As one of Canada's earliest physical oceanographers, Dr. William Ford's career included: Marine Physicist, Naval Research Establishment, Halifax (1948–1952); Director of Scientific Services (Navy), Ottawa (1953–1954); Superintendent, Pacific Naval Laboratory, Esquimalt (1955–1958); Scientific Adviser to the Chief of Naval Staff (1959–1963) and Chief of Personnel, Defense Research Board (1963–1965), both in Ottawa; Director, Bedford Institute of Oceanography (1965–1974), and Director General, Atlantic Region of Ocean and Aquatic Sciences, Department of the Environment (1974–1978).

also had three other important assignments — Director General for Fisheries Research, Assistant Deputy Minister for Ocean and Aquatic Sciences, and then, in August, 1977, he took over as Assistant Deputy Administrator for the Atmospheric Environment Service from Mr. J. R. H. Noble. However, his role as an oceanographer in charge of a century-old meteorological service proved relatively short and a bit tumultuous, and by May, 1980 he had moved on to the Associate Deputy Minister in the Department of Energy, Mines, and Resources dealing with scientific matters.

During the 1970s, the Canadian Government continued to expand its Arctic research program and eventually centralized it within the Polar Continental Shelf Project which receives support from both the Department of Energy, Mines, and Resources and from Environment Canada.[45] By 1978, the Project ranged across the entire Canadian Arctic and fielded as many as 140 research parties within a single summer season. One of its major efforts during 1978 was an extensive hydrographic and gravity-measurement program conducted as part of the requisite effort to generate adequate baseline data for the proposed Eastern Arctic Pipeline and an equivalent route for liquefied natural gas tankers. In 1979, as part of the *Lo*monosov *R*idge *Ex*periment (LOREX), the Project operated three ice camps, including one at the North Pole, to survey the oceanographic, geological, and geophysical conditions associated with this submerged ridge which extends from Ellesmere Island to Siberia.[46]

Canadian-based petroleum exploration in Canada's far north was also extremely active during the 1970s, resulting in the estimate by the Geological Survey of Canada that the Arctic Islands and the Mackenzie River Delta—Beaufort Sea region might jointly hold some 55×10^9 barrels of producible oil, or nearly twice the known reserves of the United States.[47] In fact, offshore seismic exploration

---

[45.] The charter of the Polar Continental Shelf Project provides for a long-term study of the Canadian Arctic starting with the earth's mantle below and working up through the crust and sea floor to the waters above, as well as the islands in the archipelago and the straits and sounds between. Started during 1958 with a budget of C$60,000, funding has subsequently grown to C$3.47 during fiscal year 1980. Dr. Fred Roots served as the project's director from 1958 to 1972, at which time he moved over to be Science Advisor to the Minister of Fisheries and the Environment, and his leadership role for the Project assumed by George D. Hobson.

[46.] The LOREX effort is closely coordinated with the U.S. Office of Naval Research's manned ice station program in the Arctic, which, in 1979 and 1980, operated FRAM I and II in the eastern Arctic Ocean north of Greenland with special emphasis on underwater acoustics involving a suspended 24-element hydroacoustic array (Baggeroer, 1981).

[47.] The same projection by the Geological Survey of Canada calls for the possibility of there being 55×10^9 producible barrels of oil in "Iceberg Alley" off the eastern coast of Canada.

succeeded in finding some sixty major geological structures in the southeastern Beaufort Sea alone, causing Dome Petroleum Ltd. to begin using four Arctic drill ships, as well as artificial offshore islands, in determining the exact oil and gas content of some of the more promising structures.[48] In addition, Canadian scientists and engineers in both industry and the federal government (particularly at the Institute of Ocean Sciences, Patricia Bay) have worked actively and jointly during the late 1970s in ascertaining how best to contain any Arctic oil and gas blow-outs, particularly those that might occur within an ice field, in a major technical effort called "The Beaufort Sea Project".

Despite the high priority given to Arctic research, the Research Council of Canada has not neglected "soft-water oceanography". When the Council launched during 1977 a "Strategic Grants Program" to stimulate all types of science, oceanography was one which received special emphasis. Thus, during 1980, 64 ocean grants were allocated in the amount of C\$2.4 million, or roughly 14 percent of the total funding available.

In Great Britain, the hue and cry relative to "Mother Earth" led to the creation of the Department of Environment in 1970. A Natural Environment Research Council also came into being. Assigned to this Council are such major research facilities as the National Institute of Oceanographic Sciences at Wormley, Surrey (the old NIO), with subunits at Taunton, Somerset, and Bidston, Liverpool, the Institute for Marine Environmental Research at Plymouth, Devon, and the London-based Institute for Geological Sciences which has a strong "Division of Geophysics".[49] However, the Ministry of Defence still operates the Meteorological Office (the federal weather service) headquartered at Bracknell, Berkshire, and an Oceanographic Data Centre at the Hydrographic Department, Taunton.

International politics continued to play an important role in the field of geophysics during the 1970s. The long-term issue of a controlled, comprehensive nuclear test ban persisted throughout the

---

48. Other major offshore operators in this difficult area include Esso Resources Ltd. and Gulf Canada Resources, Inc., although neither of these have yet used ice-reinforced drill ships and supporting ice-breakers.

49. The Institute of Oceanographic Sciences is charged with operating the RRS *Discovery* and supporting several other vessels used in geophysical and biological research under the aegis of the Natural Environment Research Council. In addition, the Admiralty's Hydrographic Department continues its keen interest in marine geophysics and all of its ocean survey ships are equipped with magnetometers and gravity meters. The powerful Science Research Council also supports marine technology centers combining postgraduate training and research at London, Edinburgh, Glasgow, Newcastle, Strathclyde, and within the North Western Universities Consortium.

decade. Even the tiny Earth Physics Branch of Canada's Department of Energy, Mines and Resources wrote nearly 50 reports and papers on how to improve the seismological verification of such a test-ban treaty (Whitham, 1972).[50] The Soviet Union and the United States also signed in Moscow on 3 July 1974 a "Threshold Test Ban Treaty" limiting single underground nuclear tests to those with yields under 150 kilotons. This was followed on 28 May 1976 by the dual signing in Moscow and Washington of an "Underground Nuclear Explosions for Peaceful Purposes Treaty", an agreement which requires either party who wishes to conduct such an explosion to notify the other in advance as to site geography and geology, as well as the purpose of the explosion. Despite all this treaty signing, however, the North Atlantic Treaty Organization believed it quite appropriate to hold during September, 1980 a ten-day "Advanced Study Institute" in Oslo, Norway, concerning the "Identification of Seismic Sources: Earthquake or Underground Explosion?"[51]

With respect to international politics and the ocean — and even with a war on — President Lyndon Johnson had gone ahead and told the Congress on 8 March 1968 about one of his upcoming bold, new initiatives, to wit:[52]

> I have instructed the Secretary of State to consult with other nations on the steps that could be taken to launch a historic and unprecedented adventure — an *I*nternational *D*ecade of *O*cean *E*xploration (IDOE) for the 1970's.

Forty-six nations were advised of this concept, and hints were dropped to the press that the United States might spend $5 billion in this effort (Wenk, 1972). However, the "Decade" never did acquire the aura, scientific prestige and international financial backing that the International Geophysics Year had ten years before.[53] To be sure, 36 countries became involved in the American projects, with the United Kingdom participating in nine, France in eight,

---

[50.] Members of the impressive Canadian seismological team included Kenneth Whitham, P. W. Basham, Eric B. Manchee, Peter D. Marshall, F. M. Anglin, R. B. Horner, R. B. Hayman, and D. H. Weichert.

[51.] Conference arrangers were Dr. E. S. Husebye, Norwegian Seismic Array Research, Kjeller, Norway; Dr. H. C. Rodean, Lawrence Livermore Laboratory, University of California; and Mr. William Best, Air Force Office of Scientific Research, Washington, D.C.

[52.] Dr. Edward Wenk, Executive Secretary of the Marine Council, was the chief instigator and "leg-man" for this Johnson–Humphrey initiative. Unfortunately for the Decade, he left town during 1970 and entered academia.

[53.] Major categories of U.S. IDOE projects were: *C*limate: *L*ong *R*ange *I*nvestigation, *M*apping, and *P*rediction (CLIMAP); *M*id-*O*cean *D*ynamics *E*xperiment (MODE); *G*eochemical *O*ceans *Sec*tions *S*tudy (GEOSECS); *C*ontrolled *E*cosystem *P*ollution *E*xperiment (CEPEX); *M*anganese *No*dules *P*roject (MANOP); Eastern Atlantic Continental Margin Study; Plate Tectonics; *I*nternational *S*outhern *O*cean *S*tudies (ISOS); *C*oastal *U*pwelling *E*cosystem *A*nalysis (CUEA); Pollutant Baselines, and Pollutant Transfer.

the Federal Republic of Germany in seven, and Canada in five. Between 1970 and 1978, the IDOE cost the American taxpayer over $183 million. Even so, its sponsor, the National Science Foundation, and the non-profit oceanographic community succeeded in shifting it away from a major national scientific initiative with political and economic overtones into an open-ended research program of the type spelled out in *An Oceanic Quest* (National Academy of Sciences, 1969; Hooper, 1979).[54]

By the mid-1970s, the time was finally deemed appropriate to tidy up the diverse international conventions pertaining to the use of the "global commons". Accordingly, the initial meeting of the United Nations Conference on Law of the Sea opened in Caracas, Venezuela, during August, 1974 and was still sputtering along eight years later.[55] To be sure, some 154 nations send up to 1,400 participants to the semi-annual working sessions. However, there is a marked dichotomy of views concerning how mankind should retrieve and market metallic minerals from the deep sea floor.[56] The "Group of 77" (which numbers more like 119), typically from the "Lesser Developed Nations", have held out for creating, in the words of U.S. Ambassador Elliott L. Richardson, "the first piece of global government". This would come about by forming an international seabed mining "Authority", perhaps to be headquartered in Jamaica, and an associated "International Seabed Tribunal" possibly based in Hamburg, Federal Republic of Germany. The Authority would then establish a mining "Enterprise" which would operate as a commercial entity and in direct competition with any privately-held firms which, very likely, would have to turn over any patented mining technology to the Enterprise on demand with little or no charge. To assist the American firms that have already invested upwards of $100 million in the potential of deep-sea mining, the U.S. Congress did pass the Deep Sea Mining Act of June, 1980 to give some modest degree of protection to private enterprise.[57]

[54.] Feenan Jennings served as IDOE Head within the NSF structure from 1970 to late 1977. As the Decade approached its end, the Ocean Sciences Board of the NAS complex then proposed spending $509 million as a follow-on between 1981 and 1990 (National Academy of Sciences, 1979). The concept didn't fly, however.

[55.] By 1981, general agreement has been reached on issues pertaining to innocent rights of passage through straits, policing of marine pollution, conducting oceanographic research over another nation's continental shelf, and delineating a 200-mile exclusive economic zone off each coastal country.

[56.] The current international interest pertains only to manganese nodules, which offer a non-monopolistic source of manganese and cobalt, plus the possibility of profitable returns from the extraction of copper, nickel, molybdenum, and vanadium by A.D. 2000 (McKelvey, 1980).

[57.] The only profitable deep-sea mining during the 1980s would likely occur under the

On the international scientific scene, the time continued propitious for big science. For example, during 1980, the $5,000 American Association for the Advance of Science—Newcomb Cleveland Prize for the best article in *Science* went to the co-Chief Scientists, Drs. Fred Spiess and Kenneth MacDonald of Scripps, of Project Rise and to 20 co-authors.[58] Academic seismology took a comparable route and launched the *Co*nsortium for *Co*ntinental *R*eflection *P*rofiling (COCORP) based at Cornell University, Ithaca, New York.[59] Under co-Chief Scientists Jack W. Oliver and Sidney Kaufman, the project is funded at approximately $5.6 million a year by the National Science Foundation and targeted at acquiring deep crustal profiles using the latest in geophysical equipment and techniques.[60] The areas selected for these studies are widely distributed, including the Rio Grande rift area of New Mexico, the Wind River uplift (Wyoming), the southern Appalachian Mountains (Georgia—Tennessee), southwestern Oklahoma, northern Texas, and areas of eastern South Carolina and central Michigan.

However, the most expensive program in pure geophysical research during the 1970s has been that associated with the American space program. The six manned landings of the Apollo series onto the moon's surface between 16 July 1969 and 19 December 1972 became massive generators of geophysical data because perceptive earth

---

aegis of Saudi Arabia and the Sudan in the Red Sea's Atlantis II Deep at a depth of 2.2 km. There, metalliferous muds about 20 m thick fill a five by thirteen km basin within the rift valley. Experimental work during 1979 by the West German research vessel, *Valdivia*, and the *Sedco Drillship* 445 suggested that some 800,000 tons of copper and 2,000,000 tons of zinc, plus some silver, might be extracted by pumping this special hot-spring mud to the surface for processing back in West Germany. Additional deposits of extremely rich polymetallic sulfide muds are now being found near vents associated with the deep-sea fracture zones of Oregon, the tip of Baja California, the Galapagos Islands, and Easter Island. Rate of mud deposition appears to be a function of spreading rate, which can be as much as 18 cm/yr in the latter instance.

[58.] Institutional ties for the co-authors were: Scripps Institution of Oceanography — eight; University of California (Santa Barbara) — two; Universidad Nacional Autonoma de México — four; Woods Hole Oceanographic Institution — one; Lamont-Doherty Geological Observatory — one; U.S. Geological Survey — one; Centre Océanologique de Bretagne — one; Université Louis Pasteur — one; and Université de Paris — one. Based on work done at a depth of 2.6 km with the research submersibles *Alvin* and *Cyana*, the authors describe how, among other tasks, they conducted seismic, gravity, magnetic, and electrical sounding experiments at this crustal spreading center.

[59.] COCORP institutions include: Cornell University, Princeton University, and the Universities of Houston, Wisconsin, and Texas (at Austin).

[60.] During 1980, COCORP data were generated under contract with Geosource, Inc. using five vibrators designed to provide 24-fold CDP stacking, an eight to 32 hertz linear upsweep of 30-second duration, and a recording time of 50 seconds. Recording was by a 96-channel MDS-10 system using an eight-millisecond sampling rate and 24 geophones per recording station.

physicists such as Maurice Ewing and Frank Press were able to get their ideas incorporated into the *Apollo Lunar Surface Experiments Package* (ALSEPS). This package was then left behind on the lunar surface by the astronauts to continue sending telemetered data back to earth (see Table 7.3). Although the *National Aeronautics and Space Administration* (NASA) had started flying the "*Polar Orbiting Geophysical Observatory*" (POGO) as early as 1965, a Geophysics Branch was not established until 1974, and then at the Goddard Space Flight Center in near-by Greenbelt, Maryland. Previously, however, geophysicists were well served by Dr. Wilmot N. Hess, the broad-minded Director of Science and Applications at NASA's Manned Spacecraft Center near Houston, Texas, between 1967 and 1969, and by the aerospace geologist, Dr. Paul D. Lowman, Jr., at Goddard who became the Geophysics Branch's first chief. Since the heady days of the lunar landings, a suite of geophysical instruments has also been placed on Mars, and the Pioneer and Voyager space vehicles have carried both gravimeters and magnetometers to record the potential fields in space and near the other planets and their satellites.[61] In addition, NASA has been alert to the geophysical implications and capabilities of other earth satellites, such as SEASAT and MAGSAT, and is planning a combination gravity—magnetic

TABLE 7.3.
Geophysical Experiments During Apollo Flight Series

Type of experiment	ALSEP code	Apollo mission number*				
		12	14	15	16	17
Lunar Surface						
Passive seismic	S-031	X	X	X	X	
Active seismic	S-033		X		X	
Magnetometer	S-034	X		X	X	
Heat flow	S-037			X	**	X
Portable magnetometer	S-198		X		X	
Traverse gravimeter	S-199					X
Lunar seismic profiling	S-203					X
Surface electrical properties	S-204					X
Gravimeter	S-207					X

Source: *Apollo Scientific Experiments Data Handbook,* 1974 (NASA Report TM X-58131).

*Date on which ALSEPS were implanted: #12: 19 November 1969; #14: 5 February 1971; #15: 31 July 1971; #16: 21 April 1972, and #17: 12 December 1972.

**Cable parted during deployment.

[61.] A useful summary of Apollo experiments and the results obtained may be found in *Lunar Science: A Post Apollo View* (Taylor, 1975).

satellite (GRAVSAT/MAGSAT) that may be flown in the late 1980s (Taranik, 1981) (see Fig. 7.3). Recently, the Geophysics Branch has also turned out two impressive publications concerning Planet Earth: *A Geophysical Atlas for Interpretation of Satellite-Derived Data* (Lowman and Frey, 1979) and *A Global Tectonic Activity Map with Orbital Photography Supplement* (Lowman, 1981).[62]

The Good Earth as a Killer

Today, many spokesmen emphasize over and over again the adverse impact of human-induced hazards on *Homo sapiens.* Such hazards, they point out, include over-population, nuclear war, nuclear power, chemically treated foods, improper disposal of chemical wastes, and climatic change. In Canada and the United States, the media have, in fact, often created a "Carcinogen of the Week", and Jeremy Rifkin, in his popular book, *Entropy — A New World View,* writes of the impending death of Planet Earth. Even Pope John Paul II, in his encyclical *Redemptor hominis,* has found occasion to say, "Man today seems always menaced by what he produces. . . . This seems to constitute the principal act of human existence today." On 10 November 1979, John Paul II commented further, "Man must

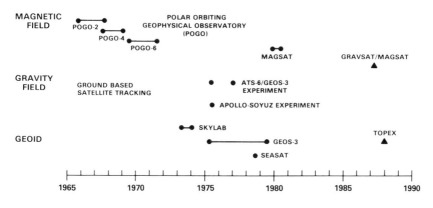

FIG. 7.3. Earth satellites flown (or to be flown) by the National Aeronautics and Space Administration with geophysical implications. (Courtesy of Mark Settle and NASA.)

[62.] Geologists — and to some extent geophysicists — have also made major use of the LANDSAT series of satellites during the 1970s (Halbouty, 1980). As a result, industry established during 1976 the Geosat Committee, Inc. based in San Francisco to serve as "the voice of the exploration geology community for Satellite Remote Sensing for energy and mineral resources".

emerge victorious from this drama, which threatens to degenerate into tragedy, and he must rediscover his authentic kingship over the world in his full dominion over the things he produces." The geophysicists of today agree fully with the Pope's views; however, they also recognize that the "Good Earth" can also be up to no good upon occasion.

For example, the public — and particularly its youth — are perpetually reminded that the atomic firestorms at Hiroshima and Nagasaki during August, 1945 took about 150,000 lives. Yet on 28 July 1976, just one earthquake of Richter magnitude (M_S) 8.0 in northeastern China killed more than 240,000 persons and leveled the industrial city of Tangshan.[63] One must also remember that earthquakes are not the only type of natural phenomena that kill, maim, and destroy, for volcanic eruptions, typhoons and hurricanes, floods, tsunamis ("tidal waves"), and droughts often do so as well. Within the United States, earthquakes are still considered the largest single natural hazard facing the nation for at least the next three decades; after that, some scientists argue that climatic change will take first place (see Table 7.4).

Because of such losses — past, present and future — geophysicists are not averse to using the "gloom and doom" approach when

TABLE 7.4

Estimates of Annual and Sudden Loss Potential from Natural Hazards within the United States (in 10^9 dollars)

Natural hazard	Annual loss	Sudden loss potential
Earthquakes	0.2	50
Tsunami (tidal wave)	Negligible	0.1
Floods	2.5	3.5
Hurricanes and tropical storms	0.2	4.2
Tornadoes	0.5	2.0
Landslides	0.1	0.3
Volcanic activity	Variable	1.0
Drought/Climatic change	17.0*	—

Sources: U.S. Geol. Survey Circular 816 (1979). NOAA Tech. Memo. NWS-NHC 7 (1978) and EDIS Report of 7 Sept. 1980.

*Cost of U.S. heatwave during June—September 1980 estimated at $17 billion.

---

63. The magnitude quoted is that of the U.S. Geological Survey. Unofficial reports have also been circulated that suggest the number of casualties may have been as high as 650,000 killed and 700,000 injured.

seeking more funding by the taxpayers. Thus, in the 1960s when Dr. Robert White, as Administrator of ESSA, went before his Congressional committee to ask for more funds, Chairman John Rooney quipped to his staff, "Well, here come the Calamity Boys!" Since 1974, the National Research Council's Geophysics Research Board has been issuing reports which evaluate the adequacy of today's knowledge to deal with societal problems having a high geophysical content.[64] Moreover, the U.S. Congress has been listening. In 1977, it passed the Earthquake Hazards Reduction Act, thereby permitting the Geological Survey and the National Science Foundation to expand their seismic monitoring and prediction programs, as well as to begin seismic hazard mapping and risk assessments.[65] Similarly, at the instigation of Congressman George E. Brown, Jr. of California, President Carter signed the National Climate Act of 1979 which fosters studies at the state level.

Although the geophysical hazards program is but a few years old within the United States, it worked surprisingly well when Mount St. Helens in southwestern Washington state became volcanically active on 27 March 1980. The contiguous United States had not had a volcanic eruption since Mount Lassen's in 1914. At first blush, the residents of the populous Northwest considered the initial eruptive phase of Mount St. Helens, the "Fujiyama of America", primarily a tourist attraction. Still, two U.S. Geological Survey "hazard geologists", Dwight Crandell and Donal Mullineaux, had carefully studied the area and predicted as early as 1975 that this mountain would be the first of the Cascade Range volcanoes to enter an eruptive phase.[66] Moreover, they believed this would happen by the end of the century even though Mount St. Helens had been dormant for 118 years (Crandell, Mullineaux, and Rubin, 1975). In 1978, they published maps in USGS *Bulletin* 1383-C which

---

[64.] Among the more important reports of this series are: *Geophysical Predictions; The Impact of Technology on Geophysics; Energy and Climate; Climate, Climatic Change, and Water Supply; Continental Tectonics,* and *The Upper Atmosphere and Magnetosphere.* Related reports issued by the Committee on Seismology, a parallel group, include *Seismology: Responsibilities and Requirements of a Growing Science* and *Global Earthquake Monitoring: Its Uses, Potentials and Support Requirements.*

[65.] Earthquake precursors currently under study include: changes in crustal seismic velocities, accelerated release of radon gas, crustal strain and tilt, gaps in regular strain release along major fault lines, unusual animal behavior, variation in ground-water level, and anomalies in electrical fields. Such measurements are normally done within 10 km of the major fault zone in question.

[66.] Cascade Range volcanoes receive their magma from the downward subducted edge of the small Juan de Fuca plate which moves eastward at about three cm a year. This is the plate originally delineated in Vine and Wilson's important paper on continental spreading back in 1965.

showed the likely limits of destructive avalanches, pyroclastic and mud flows, and ash fall-out. Backing up these geological observations was a network of seismographs throughout Oregon and Washington established for the purpose of delineating seismic areas for scientific purposes and the safe engineering design of nuclear separation and power plants, present and future.

Seismic precursors proved to be the best indicators of what Mount St. Helens would do next.[67] On 20 March 1980, an earthquake of magnitude 4.1 occurred at depth of five km or less at the dormant volcano. This was soon followed by a seismic sequence never recorded before in the Pacific Northwest, including a peak of 24 magnitude four earthquakes in just one eight-hour period on 25 March. Two days later, the volcano began a full-scale ash and steam eruption series although it was more spectacular than hazardous. As the weeks wore on, the untrained came to believe that the volcano presented no major hazard to humans. Considerable pressure was therefore exerted on the U.S. Forest Service, which controlled much of the mountain, and Washington State officials to allow the public back into the immediately surrounding area, at least during daylight hours and particularly on weekends. But the mountain's north side was experiencing a steadily growing bulge, visible even to the naked eye, and the Forest Service held the Crandell–Mullineaux hazard plot in too high a regard to permit access to the region by the general public.

At about 15:32:00 GMT (08:32 local time) on Sunday, 18 May 1980, a magnitude 5.1 earthquake occurred at a depth of about 0.3 km below sea level and 1.6 m km north of the summit (Rice and Watson, 1981). Within five seconds, a massive avalanche started down the north face of the mountain, followed by a tremendous lateral explosion 15 seconds later as water in the volcanic rock below flashed to steam. Within 50 seconds, this blast cloud of superheated steam and gas, as well as pulverized rock, was travelling with a speed of about 100 m/s (360 km/hr) towards the east and northeast at an altitude not exceeding six km. At 15:34:50 GMT, a massive new fan-shaped mass of ash was propelled laterally at initial speeds of about 450 m/s towards the north, splitting the initial surge so that the old ash, with surface temperatures of about 225°C, moved both eastwards and westwards at speeds up to 250 m/s and reached as far as 25 km from the source. Shortly after 15:35,

67. The Graduate Program in Geophysics at the University of Washington gradually built up a network of 13 seismic stations in the area and connected them by telephone line to a harried analysis center in the basement of the University Physics Building. Signals from three of these stations were also "hard-wired" to the Federal Emergency Center 60 km southwest of the volcano at Vancouver, Washington.

the volcano began emitting vertically and by 15:42 ejected an ash plume which reached a peak altitude of approximately 30 km.[68]

As in other major volcanic explosions, distant geophysical observatories recorded both atmospheric pressure and Rayleigh seismic waves at the appropriate times. For example, A. R. Ritsema at the Royal Netherlands Meteorological Institute has reported that the Mt. Saint Helens' atmospheric wave was comparable to that observed at De Bilt from Soviet atmospheric nuclear explosions over Novaya Zemlya in 1958–1962 which had blast energy equivalents of about 10 megatons of high explosive (Ritsema, 1980). However, this value appears low, for it is but twice the energy safely released underground at a depth of 1,780 m by the U.S. Atomic Energy Commission's five-megaton Cannikin device on 6 November 1971. Subsequently, however, Donn and Balachandran at Lamont have raised the explosive yield of Mt. St. Helens' main eruption to between 25 and 35 megatons (Donn and Balachandran, 1981). In addition, magnetic field perturbations were reported as far as 3,122 km away (Fougere and Tsacoyeanes, 1980).

Once on-site measurements were possible, they showed that the initial blast and following eruption had removed between 1.5 and 2.0 cubic kilometers from the mountain, lowered its crest by 400 meters, and covered 49 percent of Washington State with visible ash. As of early 1981, monetary losses are still being determined, but they appear to be less than the original estimates of $900 million. For example, crops in eastern Washington during 1980 were about seven percent below normal, thereby suggesting a crop loss of only $100 million (Cook *et al.*, 1981). Loss to the timber crop on the volcano's flanks might come to $200 million but much depends on how fast the downed timber can be harvested.[69] Dredging of the shoaled Columbia River and tributaries has also proved to be an expensive operation, as has the replacement of roads, bridges, homes, and lumbering equipment. Human fatalities apparently numbered 82, many of whom had entered the prohibited "RED" danger

---

[68.] With a consistency of dry Portland cement, the ash does not appear to be dangerous to humans, for it consists primarily of pulverized dacitic and andesitic rock, plus some volcanic glass. The heaviest fall-out, which averaged over 30 kg per square meter of uncompacted ash (a thickness of over five cm), occurred about 300 km east-northeast of the volcano and even caused darkness starting at 1500 hours local time in Pullman, Washington, 390 km distant. For reference purposes, one may note the famed Vesuvius eruption that buried Pompeii in A.D. 79 released about 1 cubic km of ash, while the Krakatau eruption of 1883 generated about 20 cubic km.

[69.] Weyerhauser Company, the major lumber operator in the area, reported a loss of about $66 million.

area illegally.[70] By November, 1980 casualty insurance companies had paid out only $11.7 million in homeowners' claims, even though they had, at the state government's request, classified the event as an "explosion" rather than "volcanic activity" so that the normal insurance policy would apply. So far, the Pacific Northwest has gotten off relatively easy after some frightening moments. Unfortunately, the predictions are that Mount St. Helens will stay active for another 25 years. In addition, there are some indications that more massive volcanoes, such as Mount Hood, 80 km east of Portland, Oregon, and Mount Rainier, 50 km south-east of Tacoma, Washington, need not continue in a dormant phase because the Juan de Fuca plate continues being relentlessly shoved inland and downward.[71]

Is Mankind Modifying the Earth?

During the 1970s, oceanographers — both self-styled and academically trained — were talking about the possibility that entry into the sea of sewage sludge, dredge spoil, chemical and nuclear wastes, spilled petroleum, and non-biodegradable pesticides might ultimately lead to a "dead ocean" as the shallow Baltic Sea was threatening to become.[72] Should you believe Captain Jacques Cousteau's scenario that the sea might become a vast anerobic cesspool, it then follows you should worry about such a dying sea ceasing to absorb carbon dioxide, the waste product of all animal life as well as fuel burning. In such a case, the oxygen-regenerating phytoplankton would disappear, and the sea would no longer be a major contributor in replenishing the atmosphere's oxygen. In addition, the "thermal green house effect" caused by increasing amounts of carbon dioxide would be enhanced. Unfortunately for the Cousteau thesis — but fortunately for mankind — the sea still appears vast enough to tolerate man's insults. By the way of example, the Mississippi River drains over 40 percent of the contiguous United

---

[70.] On 19 October 1980, Mullineaux observed that the tenacious stand of the Forest Service against opening up the "RED Zone" to week-end visitors saved as many as 5,000 lives. Dr. David Johnson, the USGS volcanologist, was the only scientist to lose his life. His last words from a site eight km away were, "Vancouver, Vancouver — this is it!" The assumption which cost his life was that any major blast would be vertical, rather than lateral.

[71.] Geophysicists in the Pacific Northwest may wish to initiate a facetious protest movement and issue T-shirts emblazoned with "Re-unite Gondwanaland", "Stop Continental Drift", and "Fine Mother Nature — She's a Thoughtless Polluter".

[72.] Despite the London Dumping Convention of 1975, ocean dumping is still common practice. In 1978, the United Kingdom dumped 902,000 tons of sewage sludge, Hong Kong 800,000 tons, and the United States 640,000 tons to cite just a few instances.

States and thereby becomes a low-cost sewer for removing millions of tons of agricultural, mining, industrial, and metropolitan waste. Yet the fishing remains good near its mouth and becomes excellent at the thousands of oil and gas platforms that surround its delta.

Meteorologists have been among the most active of the environmental "Cassandras". In the early 1970s, one of their atmospheric chemistry models suggested supersonic commercial aircraft, if introduced in quantity, would eventually diminish the earth's protective ozone layer and cause a notable increase in skin cancer because of increased exposure to ultra-violet radiation. This possibility helped shoot down the American effort to build an "SST" aircraft.[73] Shortly after that, it appeared that the release of chlorofluorcarbons of the Freon type, if continued at the 1977 level, would decrease the level of stratospheric ozone by 16 percent thereby leading to several thousand more cases of melanoma (malignant warts) per year in the United States, many of them fatal, and several hundred thousand more cases of other skin cancers (U.S. Council on Environmental Quality, 1980). Consequently, the use of this chemical is now banned by Sweden, Canada, Norway, and the United States, and the European Economic Community has asked its member nations to reduce its use by at least 30 percent below 1976 levels as of 31 December 1981.

Two new global meteorological issues are being widely publicized, both associated with the increasing use of fossil fuels. The first is build-up in acid rain levels. Apparently fossil fuels, when burned in association with a tall smoke stack, tend to form sulfur dioxide (SO_x) and nitrogen oxide (NO_x) aerosols which may reside in the atmosphere for days before combining with rain or snow to form a dilute acid with a pH of about 4.2 (on a pH acidity scale of zero to seven, wherein numbers near zero are extremely acid and 7.0 is neutral; vinegar, for example, has a value of approximately three). Should this "acid rain" fall onto an area deficient in carbonate rocks, as in a Pre-Cambrian Shield area, the pH of such lake water can fall to five or less from a normal value somewhat above six. As a result, a die-off of fish occurs for they cannot stand the extra aluminum and other metals released from the lake bed by the increased acidity. As of January, 1981 some reports indicated that there were already 280 known "killed lakes" in the Adirondack Mountains of New York

73. The British are far less "flappable" in such matters. For example, Sir John Mason, the Director of their Meteorological Office, noted in his Symons Memorial Lecture of March, 1976 that several hundred SSTs, even if they flew five hours each day, would not reduce the ozone level by an amount distinguishable from that of natural atmospheric fluctuations (Mason, 1976).

State and hundreds more in Ontario and Quebec. Because such a pollutant can form in one country, such as the United States or the Federal Republic of Germany, and be deposited in another, as in Canada or Sweden, we now have a new international problem. Consequently, discussions are going on at both national and international levels to decide just how serious the problem is likely to become and what ought to be done about it.[74]

The second global issue asks whether man has finally reached the stage where he can affect global climate in an adverse manner and not realize what he has done until after an irreversible change sets in regarding the earth's heat budget. Unfortunately (or fortunately if you are looking for an additional source of research funds), this issue boils down to a very complicated search for a "true signal" within large amounts of "random noise" (see Fig. 7.4). On one hand, the earth's climate over the past 3.5×10^9 years (out of a total earth lifetime of 4.5×10^9 years) has been benign enough to sustain

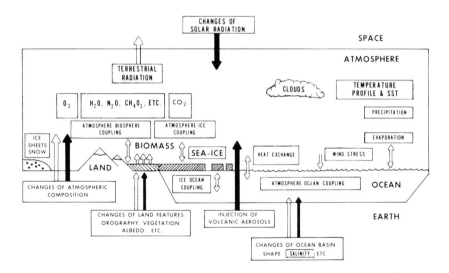

FIG. 7.4. Schematic diagram of the complicated interaction between physical, chemical and biological factors that determines the earth's climate. (Source: *Understanding Climate Change, A Program for Action*, U.S. National Academy of Sciences, 1975.)

74. The acid-rain issue was one of the few substantive problems discussed on President Reagan's first visit to Canada to meet with Premier Pierre Trudeau during March, 1981. Tagging the exact culprit may be more difficult than originally assumed. In working with Antarctic glacial ice samples up to 350 years old, Drs. Paul Mayewski and William Berry Lyons of the University of New Hampshire found mean pH values of 4.8 to 5.0, suggesting that "acid rain" may not be all that new (*Wall Street Journal* of 19 September 1980).

continuously some form of life (EOS, 1980). On the other hand, the geological record suggests repeated warming and cooling, including glacial periods going back into Pre-Cambrian time; in addition, there is evidence that at the end of the Cretaceous Period about 65 million years ago, there may have been the sudden elimination, perhaps within a 500,000-year interval, of more than 70 percent of all living species, including the dinosaurs and much of the marine surface-dwelling life forms. At the moment, mankind appears to be enjoying the peak of a warm interglacial period and, if history repeats itself, pronounced cooling should be expected within the next thousand years or so (Lamb, 1977).

But members of the new "advocacy school of science" have noted that there is a steady and measurable build-up of carbon dioxide within the earth's atmosphere which has been occurring at least since 1958. This build-up rate is now about 1.5 ppm per year. Some calculations have already suggested, based on the increasing amount of fossil fuel projected for burning in the coming century, that there is a real possibility that the atmospheric carbon dioxide level could double over that of pre-industrial times by the mid-21st century (U.S. Council on Environmental Quality, 1980).[75] The question becomes, what happens next? One publicized scenario assumes that a "green-house" effect sets in from the increased amount of carbon dioxide. Then the earth's average temperature jumps by $3°C$ and in polar regions by as much as $7°C$ (see Fig. 7.5). This drastic global

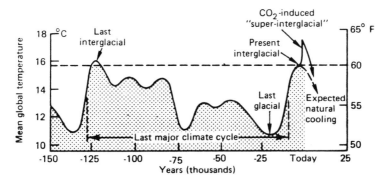

FIG. 7.5. Range of global-scale mean temperature, with and without projected carbon dioxide effect. (Source: U.S. Council on Environmental Quality, 1981.)

[75.] The build-up of atmospheric carbon dioxide to a level of 334 ppm in 1979 represents an increase of 20 ppm in 21 years. This is equivalent to adding 38×10^{12} kg of carbon during a period when fuel burning emitted 71×10^{12} kg of carbon.

warming is followed by major changes in wind, precipitation, and ocean current patterns, probably leading to disappearance of the polar pack ice and the West Antarctic ice sheet. Should the latter event happen, sea level would rise by five to eight m in just a few decades and flood the coastal regions where a sizeable percent of the world population now lives.[76] Similar adverse impacts could also appear in forested areas, perhaps causing disappearance of the tropical rain forest and desertification of extensive agricultural areas, thereby initiating additional food shortages.

In view of such horrific scenarios, the media has lined up for public scrutiny both "climate coolers" and "climate warmers".[77] In 1974, the *B*ritish *B*roadcasting *C*orporation (BBC) featured a television special, *The Weather Machine,* which left the viewer with the impression that cooling might dominate and mankind would likely face a "snow blitz" in the not-too-distant future. Contrary-wise, a White House sponsored effort by a National Academy of Sciences' *ad hoc* Study Group on Carbon Dioxide and Climate in July, 1979 gave some validity to the "climatic warmers". Whether conclusions of the present climatic modelers can be trusted, however, is still debatable. Dr. J. Murray Mitchell, one of the National Weather Service's best climatologists, wonders whether they are correctly filtering out climatic signals from the random noise as, in any given year, he notes that localized climates can be drastically different from the overall normal.

One of the best American long-range weather forecasters, Dr. Jerome Namias, long of the U.S. Weather Bureau and now at Scripps, has repeatedly noted the great control that anomalies in ocean surface temperatures appear to exert on the position of the polar jet stream and associated storm tracks. As of the moment, Namias believes "that the climate effects of CO_2 variations may be much more complex than is often surmised, so it is not wise to draw inferences of regional anomalies from hemispheric means or latitudinal profiles" (Namias, 1980).[78] Dr. Reid Bryson, Director of

---

[76.] Once the full implications of plate tectonics were understood, it became evident that there would be marked changes in global sea level as the continents shifted about. Some of the best work in this area has been published after a research team at Exxon Production Research Company, Houston, Texas, led by Peter R. Vail announced what this meant relative to oil exploration (Vail *et al.,* 1977). In this paper, they point out that sea level has twice gone 100 m above and 100 m below present sea level during the past 28 million years.

[77.] The "cooling school" believes that the earth is, on the average, about 8°C cooler than 150 million years ago when "ice ages" again began to reappear, and that another drop of 10°C is not improbable within the next several million years.

[78.] Namias, who has studied United States weather data as much as anyone, also observes that the mean temperature of the "Lower 48" states is not highly correlated with mean

the Institute of Environmental Studies at the University of Wisconsin, has gone even further towards developing a geophysical model for predicting monthly mean air temperature and precipitation a year in advance by feeding in data keyed to sun—earth geometry, earth and atmospheric tides, volcanism, and carbon dioxide level.[79] He then finds global volcanic activity levels appear to cause climatic fluctuations because of the introduction of excess dust into the high atmosphere and may, in their own right, be related to earth tide perturbations and therefore predictable as well, a concept not agreed to by most of the geophysical profession.[80] Further complicating the issue is that recently a special NASA satellite flown to measure the solar maximum found that the "solar constant" fluctuated by 0.1 percent on several occasions.[81]

In view of these complexities of climatic interpretation and prediction, meteorological bureaucrats in developed countries which use large amounts of fossil fuel are paying less heed to the need of improving weather forecasts up to five days duration and, instead, are increasing the priority on understanding driving mechanisms which affect world climate. To assess the issue, the World Meteorological Organization sponsored a broadly based conference on world climate variations during 1979 that was attended by over 500 people, primarily from developing countries.[82] In addition, an even more extensive conference was organized by the WMO in late 1981 to review in detail the latest knowledge concerning the relationship between carbon dioxide and world climate. Individual countries are also strengthening their climatic research efforts. For example, the United States has an agreed-on federal five-year plan, costing about $120 million a year, to expand climatic knowledge and its societal interplay.

---

temperature of the Northern Hemisphere. Similarly, precipitation over the contiguous U.S. does not appear well correlated with global mean temperature, thereby making a modeler's task more difficult than it appears at first glance.

[79.] Back before climatology was a "sexy" science, Bryson's group worked hard at placing paleoclimatology on a quantitative basis during the 1960s (Bryson and Padoch, 1981). Since then, several universities, such as the University of East Anglia under the leadership of Professor Hubert H. Lamb during 1972, have set up "Climatic Research Units".

[80.] Even now, it remains open season for suggesting new climatic change models and computer programs. For example, the journal *Science* carried a half-dozen papers on this subject during just 1980, and other comparable journals also ran frequent articles of the same type.

[81.] Some calculations suggest that a drop of one percent in the solar constant would reduce the earth's mean temperature by over $1°C$; if the drop were as much as six percent, today's "water planet" would become an "ice planet".

[82.] Typical of papers presented by the industrial sector was one by Dr. Gaskell entitled, "Climate Variation as it Affects Offshore Resources", on behalf of the International Oceanographic Commission.

By what Means will Geophysics Progress into the Future?

Mankind has been on this earth for perhaps 20,000 generations. Continued wise use of science and technology, when combined with prudent political, social and economic systems, should go far towards doubling this number. In the past, geophysicists have repeatedly shown that when faced with difficult problems, they thrive, rather than being of a type which "runs in circles, screams and shouts". In times of crisis, they still conduct their measurements, sort out variables, draw on the accumulated experience of those who went before, and, after proper analysis, generate answers which stand the test of time. In fact, one geophysicist, Dr. King Hubbert, started his own one man prediction campaign back in 1948 for the purpose of alerting his generation to the effect that they would see limits (Hubbert's "pimple") to the ever-expanding production of oil and gas. He did point out, however, that a high-energy industrial civilization could be maintained indefinitely if the human population were stabilized at some reasonable figure, perhaps at something under nine billion (10^9) (Hubbert, 1950).[83]

Hubbert's worry has been further expanded in a recently released study prepared at the express request of President Jimmy Carter entitled, *The Global 2000 Report to the President* (Council on Environmental Quality, 1980). Billed as "the first U.S. Government effort to look at all three issues (population, resources, and the environment) from a long-term global perspective that recognizes their relationships and attempts to make connections among them", the three-volume report has many internal consistencies and places selective emphasis on conclusions that fit the beliefs of the study organizers, who were perturbed environmentalists. As a result, so much emphasis is placed on the dark side of the problem that the Honorable Henry Reuss, the Chairman, Subcommittee on International Economics, Joint Economic Committee, United States Congress, noted: "The *Global 2000* report documents a world a bare twenty years from now that is desolate and dying, the result of the past, present, and prospective follies of its people." Needless to say, accounts in most news media reflected this grim point of view, for after all, bad news attracts more attention than does good news. Yet if one looked carefully at the study, he found that the central staff could only reach such dismal predictions by maintaining that the

---

[83.] Included in this assumption is that mankind would make full use of fossil fuels, nuclear power (both fission and fusion), hydro-electric power, and geo-thermal energy. Major use of extra renewable resources, such as solar, wind power, ocean currents, and bio-energy, would begin to occur in the latter half of the 21st century.

initial projections by the supporting federal agencies were incomplete, inaccurate, or inconsistent.

If today's (and tomorrow's) geophysicist is to deal successfully with so much "challenge and change", he must have a supporting infra-structure which permits him to grow professionally, socially and economically, as well as one which provides him with a sounding board wherein his views can be mingled with those of others in order that many may speak as one. Various types of non-profit groupings, particularly professional societies, have been created for this specific purpose. Over the past 60 years, their growth has been phenomenal. For example, when it first came into being during 1919, the American Geophysical Union had 65 members and was a formal committee of the National Research Council. By 1971, the AGU had 7,000 members, which seemed too large for a committee, and it was spun off on its own. Now, a decade later, its membership totals 13,000 and resides in more than 100 countries (see Table 7.5). The Society of Exploration Geophysicists has experienced a similar growth and even exceeds the AGU in membership by perhaps a thousand. Both own impressive headquarters buildings as well. However, their sister societies, the Seismological Society of America and the Marine Technology Society, are still small, impecunious, and have to operate out of rented space.

Each of these societies have their own style of operating. At the urging of Dr. Lloyd Berkner, the American Geophysical Union made a determined effort to become the scientific society for the burgeoning area of aeronomy and magnetospherics when the Space Age started in the late 1960s. Even so, the Union remains extremely appealing to physical oceanographers and to seismologists. For example, the annual Pacific Coast winter meeting in December, 1980 had 200 papers in seismology. In the realm of exploration geophysics, the SEG has always emphasized its global nature and continues pleased that its founders in 1930 intentionally omitted the prefix, "American", when naming the organization. The SEG's annual meeting, whether in the United States, Canada, or Mexico, is always spectacular. Thus, at the Golden Anniversary meeting in Houston, the celebration involved 105 events, including 36 technical sessions, and required the use of 20 hotels and 80 buses. In contrast, the Seismological Society of America remains as a purist scientific society and has only a modest annual meeting. The oceanographers are still not certain where their interests can best be met, and oceanographic sessions are scheduled by such diverse groups as the American Meteorological Society, the American Geophysical Union, the Society of Exploration Geophysicists, the American

TABLE 7.5

Key Aspects of Major Geophysical Societies Based in the United States

Aspect	$AGU^{(1)}$	$SEG^{(2)}$	$SSA^{(3)}$	$MTS^{(4)}$
Membership				
Total	13,000	15,196	1,767	3,200
Student	1,775	1,214	160	475
Sections/Chapters		22		16
Student Sections/				
Chapters		$54^{(5)}$		
Resident Countries	100 plus	100 plus		
Annual Budget				
(millions of $)	4.3	4.1	0.2	0.09
Estimated Net Assets				
(millions of $)	1.2	2.5	0.1	0.05
Journals published	$13^{(6)}$	1	1	1
		(Geophysics)	*(Bulletin)*	*(Journal)*
Pages printed annually				
(journals and books)	19,811	1,992		
Staff members	65	35	$4^{(7)}$	6
Attendance at Annual	2,500 at			
Meeting	each of two			
	meetings	12,319	338	1,600
Scholarships Awarded		112		

(1) As of 31 December 1980.
(2) As of 30 June 1981.
(3) As of 1 May 1981.
(4) As of 1 May 1981.
(5) Student sections external to the United States: Canada: 5; Australia: 2; England: 1; Japan: 1; Mexico: 1; Spain: 1, and New Zealand: 1.
(6) *Journal of Geophysical Research, Geophysical Research Letters, Reviews of Geophysics and Space Physics, Radio Science, Water Resources Research, EOS* (weekly newsletter), seven Russian geophysical journal translation series, and several book series.
(7) Shares space and staff with the Earthquake Engineering Research Institute.

Society for Limnology and Oceanography, the Institute of Electrical and Electronic Engineers and the Marine Technology Society. The last mentioned society was primarily the brain-child of the first Oceanographer of the U.S. Navy, Rear Admiral Edward C. Stephan, who hoped that finally, as of 1964, a critical mass of people existed who would support a major technical society addressing problems of the sea (see Table 7.6). As of the moment, however, it remains touch-and-go as to whether the MTS will flourish or not.

External to the United States, Canadian practitioners of the geophysical sciences tend to have multiple loyalties in the field of societal membership. For example, there are such groups as the Canadian Society for Meteorology and Oceanography, the Canadian Exploration Geophysical Society, and the Canadian Society of Exploration Geophysicists. Canadians have also participated vigor-

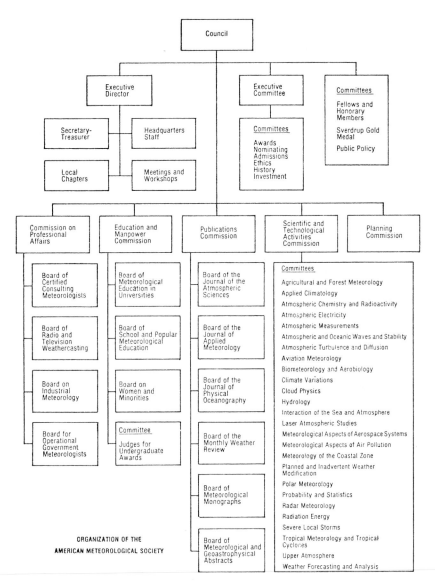

FIG. 7.6. Organization of the American Meteorological Society.

TABLE 7.6
Interest Areas of the Marine Technology Society

Societal Division	Divisional Committees
Ocean and Coastal Engineering	Buoy Technology, Cables and Connectors, Diving, Marine Materials, Power Systems, Salvage and Towing, Ocean Energy, Oceanographic Instrumentation, Offshore Structures, Satellite and Aircraft Remote Sensing, Seafloor Engineering, and Undersea Vehicles
Manpower and Professional Development	Education
Ocean and Coastal Management	Coastal Zone Management, Marine Law and Policy, Ocean Economic Potential, and Technology Exchange
Marine Resources and Environment	Geology and Geophysics, International Marine Food and Drug Resources, Marine Biology, Marine Fisheries, Marine Geodesy, Mineral Resources, Oceanographic Ships, Undersea Physics, Underwater Photography and Sensing, and Water Quality

ously in the larger British and American geophysical societies and provided a number of presidents and other officers to the U.S.-based geophysical societies.[84] In the United Kingdom, the Royal Society, the European Association of Exploration Geophysicists, the Royal Astronomical Society, the European Geophysical Society, and the Society for Underwater Technology provide most of the societal focal points needed by that country's geophysicists (see Table 7.7). In addition, such publication outlets as *Nature,* the *Geophysical Journal* of the Royal Astronomical Society, and *Geophysical Prospecting* are considered to be as prestigious as the comparable American journals.

Even though marine geophysicists and other seagoing scientists have never agreed on whether "oceanography" is a true science suitable for teaching at the undergraduate level, they are in full agreement that they need comprehensive shoreside facilities, sizeable research fleets, and specially dedicated aircraft. Today, the privately operated Woods Hole Oceanographic Institution has two complete campuses (one harbor-side and the other inland), an annual operating budget of $30 million, and six world-ranging

---

[84.] The Canadians Norman J. Christie and Roy O. Lindseth served as President of the SEG during 1963–1964 and 1976–1977, respectively. Dr. J. Tuzo Wilson served as President of the AGU during the 1981–1982 period.

TABLE 7.7

Aspects of Major Geophysically-oriented Societies Based in Great Britain and Canada

Aspect	Royal Astronomical Society (1/81)	Society for Underwater Technology (11/81)*	Canadian Society for Exploration Geophysicists (1/82)
Membership	2,718	2,000	2,000
Annual Budget	£125,000	–	C$75,000
Journals published with geophysical content	Geophysical Journal	Journal	Journal
Geophysical pages published annually (1980)	799	130	70
Full-time staff	7	4	1
Estimated Net Assets	£349,000	–	C$70,000

*Many British oceanographers belong to this London-based group which includes 86 corporate members drawn from various fields of endeavor as London-based oil companies, British universities, Det norske Veritas, the Geological Survey of Ireland, ocean engineering firms, and fish farming companies. During 1975–1976, HRH Prince Philip, Duke of Edinburgh, KG, KT, OM, served as the Society's President.

vessels, including the deep submersible, *Alvin,* and its tender, *Lulu.*[85] Scripps, too, has an equally impressive budget, campus and marine capability. NOAA has acquired a number of impressive laboratory buildings and is now building a $70 million marine facility on the shores of Lake Washington in Seattle. Even the low-key Naval Oceanographic Office has a new building of its own. In Canada, there are now two outstanding Canadian Marine Research Centres, one at Dartmouth, Nova Scotia, near Halifax, and one at Patricia Bay, Vancouver Island. The comparable expansion in British marine-oriented research facilities during the past decade has, of course, already been touched on earlier in this chapter.

Manpower-wise, the field of geophysics has never been in sounder condition. Geophysics is something no longer taught as a solitary course off on the side in another department. In 1980–81, 13 Canadian schools taught geophysics and three oceanography, while in the United States, the comparative numbers were 40 and 44. In fact, some schools have gone so far, as in the case of the prestigious University of Chicago, to establish their own "Department of Geophysical Sciences". Student enrollment in 1980–1981 is also strong. In Canada, 124 students study oceanography at the graduate level,

[85.] The Institution's current Director is Dr. John Steele, a Scottish specialist in fisheries.

while 245 undergraduate and 107 graduate students study geophysics and geophysical engineering. Within the United States, 881 undergraduates are studying oceanography and 1,462 are doing so at the graduate level. U.S. undergraduates studying geophysics and geophysical engineering number 1,108, while 993 are seeking advanced degrees (American Geological Institute, 1981). Salary levels and job openings within the profession are also excellent, particularly on the industrial side. With respect to the philanthropic side of the profession, all are pleased that some of the early day exploration geophysicists have already been able to contribute additional science facilities to selected campuses. These include donations by Cecil (and Ida) Green to MIT, Colorado School of Mines and the Marine Science Institute of the University of Texas, by the late William B. Heroy to Southern Methodist and Syracuse Universities, and by Henry (and Grace) Salvatori to Stanford University. In addition, both the AGU and the SEG fund extensive fellowship programs as well.

To summarize, over the past six decades geophysicists have been highly productive and financially well rewarded. Ahead of the profession are still a vast number of problems pertaining to the physics of the earth that require analysis and solution. In this regard, the pragmatic Athelstan Spilhaus has noted, "Things worth doing are worth doing for money — for that is the popular judgment of 'worth'!" With this truism in mind, our next chapter addresses geophysics as a business.

Transfer ceremony on 13 October 1966 of Central Radio Propagation Laboratory to ESSA. Left to right: C. G. Little (Laboratory Director), J. H. Hollomon, A. V. Astin (Director, Bureau of Standards, the losing agency), R. M. White (ESSA Administrator), and Lloyd V. Berkner. (Courtesy American Meteorological Society.)

Dr. William A. Nierenberg, Director of Scripps Institution of Oceanography. (Courtesy of SIO.)

Gloria Mark I side-looking sonar dissembled. (Courtesy A. S. Laughton.)

Sir George Deacon, founding Director, National Institute of Oceanography. (Reproduced by permission of Institute of Oceanographic Sciences.)

Dr. Anthony Laughton, Director, Institute of Oceanographic Sciences, since 1978. (Reproduced by permission of Institute of Oceanographic Sciences.)

Dr. and Mrs. Richard W. James on 14 April 1961 after receiving a check for $5,400 for pioneering the routing of naval vessels to avoid adverse wave conditions. (U.S. Navy photograph.)

Dr. Lloyd V. Breslau attempting to locate gravel deposits in the Mekong River delta during 1967 using high-frequency echo sounding. (U.S. Navy photograph.)

Remains of USS *Thresher* along route of her fatal dive. (U.S. Navy photograph.)

Hatch opening to messenger buoy compartment of the ill-fated USS *Scorpion.* (U.S. Navy photograph.)

RADM W. Guest, USN, and others inspect hydrogen fusion bomb retrieved off Palomares, Spain by USS *Petrel* and the CURV device. (U.S. Navy photograph.)

General Dynamics' *Star III* submersible. Left to right: John Drewry, Captain Paul Bauer, USNR, Captain Samuel Applegarth, USN and C. Bates.

Signing of marine geophysical survey contract between NAVOCEANO and Texas Instruments during March, 1965. Seated: Hazel Bright (NAVOCEANO contract negotiator) and Robert C. Dunlap, Jr. (TI Vice President). Standing: Charles Converse and Richard A. Arnett (TI) and Larry Walker (Office, Inspector of Naval Materiel.)

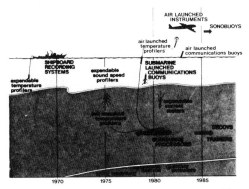

Diagram showing evolution of oceanographic sensors for undersea warfare. (Courtesy Sippican Corporation.)

Tanker *Manhattan* stopped in ice floe for testing of ice properties. (Courtesy Exxon USA.)

President Lyndon Johnson and Dr. William G. Stroud inspect interior of TIROS satellite during 1964. (Courtesy Goddard Space Flight Center.)

NASA Earth Resources Aircraft (a Convair 240-A) showing locations of remote sensors. (Courtesy NASA.)

Professor Randolph W. Bromery. (Courtesy University of Massachusetts, Amherst.)

Roland G. Henderson, Honorary Member of the Society of Exploration Geophysicists. (Courtesy R. G. Henderson.)

Dr. Warren Washington. (Courtesy NCAR.)

Edward Ridley, Director, U.S. National Oceanographic Data Center. (Courtesy E. Ridley.)

Dr. John Slaughter, Director, National Science Foundation. (Courtesy NSF.)

Professor Manik Talwani, second Director (1972—1980) of the Lamont-Doherty Geological Observatory. (Courtesy Lamont-Doherty.)

Linda Benedict, Junior Surveyor on Seismic Party 1741 of Geophysical Service, Inc., at work outside Evanston, Wyoming, during July, 1979. (Courtesy GSI.)

Bettye Athanasiou, Life Member of the Society of Exploration Geophysicists. (Courtesy B. Athanasiou.)

President Jimmy Carter inspects Dr. Frank Press's book, *Earth*. (White House photograph.)

President Jimmy Carter congratulates Congressman George E. Brown, Jr. on passage of the National Climate Act of 1979 (name of staff assistant unknown). (White House photograph.)

Ellen Scripps Revelle dedicating the research ship, *Ellen B. Scripps*, on 1 October 1965. Left to right in background: J. S. Galbraith, Chancellor of University of California; Roger Revelle, past SIO Director, and Jeffrey Frautschy, SIO faculty member. (Courtesy SIO.)

Diving geologists of the early 1950s who later gained international recognition. Left to right: Drs. H. William Menard, Harris B. Stewart and Robert Dietz. (Courtesy H. B. Stewart.)

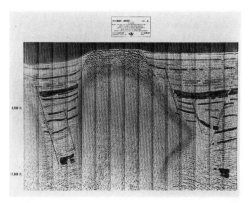

Seismic delineation of the outline of a salt dome as of the mid-1970s. (Courtesy GSI.)

Collection of documents submitted to U.S. government during January, 1981 for permission to drill on Georges Bank. (Courtesy Exxon USA.)

Honorable Bruce Babbitt, Governor of Arizona (1977—present). (Courtesy B. Babbitt.)

Dr. Richard Frank, Administrator of the National Oceanic and Atmospheric Administration (1977–1981). (Courtesy NOAA.)

Dr. John Byrne, the first geological oceanographer to head the National Oceanic and Atmospheric Administration (1981–present). (Courtesy NOAA.)

RADM J. B. (Brad) Mooney, Jr., USN, seventh Oceanographer of the Navy. (Courtesy NAVOCEANO.)

Dr. Arthur E. Collin, Associate Deputy Minister, Department of Energy, Mines and Resources of Canada, since mid-1980. (Courtesy of A. Collin.)

Dome's four drillships moored at their McKinley Bay wintering location on the Tuktoyaktuk Peninsula.

Drillships of Dome Petroleum wintering over at McKinley Bay, southeastern Beaufort Sea. (Courtesy Dome Petroleum Ltd.)

Waverly Person, National Earthquake Information Service, points out U.S. real-time seismological telemetry network. (Courtesy USGS.)

Mount St. Helens on 10 April 1980 during initial eruptive phase. (Courtesy of USGS.)

Mount St. Helens on 18 May 1980 after destroying its top earlier in the day. (Courtesy of USGS.)

Lowering five-megaton Cannikin nuclear device into Amchitka Island during November, 1971. (Courtesy of Las Vegas Office, U.S. Department of Energy.)

George D. Hobson, current Director, Polar Continental Shelf Project. (Courtesy Energy, Mines and Resources, Canada.)

Dr. Wilmot H. Hess, Director of Science and Applications, Johnson Manned Spacecraft Center, during Apollo flights to the moon. (Courtesy W. Hess.)

Dr. Paul W. Lowman, initial Director of the Geophysics Branch, Goddard Space Flight Center. (Courtesy P. Lowman.)

Installation of ALSEP package on the moon during 19 November 1969. Undeployed magnetometer is just to left of astronaut's long shadow. (Courtesy NASA.)

Carl Savit, RADM Odale D. Waters, USN (Oceanographer of the Navy), and Sidney Kaufman at a SEG meeting. (Courtesy Bettye Athanasiou.)

Dr. Jerome Namias, America's leading proponent of oceanic controls on seasonal weather patterns. (Courtesy J. Namias.)

Professor Reid Bryson, Director, Institute of Environmental Studies, University of Wisconsin. (Courtesy R. Bryson.)

Sir John Mason, the cloud-physicist who has been Director-General, Meteorological Office, since 1965. (Crown copyright; reprinted by permission of Her Majesty's Stationery Office.)

Dr. M. King Hubbert, one of the most original thinkers among American geophysicists. (Courtesy M. K. Hubbert.)

Professor Jack Oliver, Co-Director of the COCORP deep crustal seismic profiling project. (Courtesy of J. Oliver.)

Dr. Robert L. Bernstein, Principal Investigator, Satellite Oceanography Facility (the first of its kind), based at Scripps Institution of Oceanography. (Courtesy SIO.)

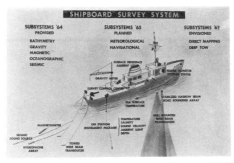

Diagram of a multi-sensor survey ship, the USNS *Silas Bent*. (Courtesy NAVOCEANO.)

NAVOCEANO's new oceanographic facility at Bay St. Louis, Mississippi. (Courtesy NAVOCEANO.)

Cecil and Ida Green before the Green Earth Sciences Building at the Massachusetts Institute of Technology. (Courtesy MIT Historical Collections.)

Sketch of American Geophysical Union headquarters, Washington, D.C. (Courtesy AGU.)

Aerial view of Harbor Campus, Woods Hole Oceanographic Institution. The Marine Biological Laboratory is at the extreme right. (Courtesy WHOI.)

ESSA vessels at the ESSA Marine Center, Seattle, Washington, in late 1970. Ships present are the *Oceanographer, Pathfinder, Surveyor, Davidson, McArthur, Fairweather, Rainier, Miller Freeman* and the *Kelez*. (Courtesy NOAA.)

Facilities of the Institute of Ocean Sciences, the Pacific Geosciences Centre, and the Marine Technology Centre at Patricia Bay, Vancouver Island. (Courtesy Marine Technology Centre.)

CHAPTER 8

Geophysics as a Business—
Then and Now

The successful entrepreneur is driven not so much by greed as by the desire to create. He has much more in common with the artist than with the bureaucrat. His principal satisfaction comes not from his bank balance but from the jobs he provides and the benefits he spreads through his products and services.

Paul Johnson, *Wall Street Journal* of 22 January 1981

This chapter will describe in some detail the nature of today's geophysical exploration industry. We will touch on the backgrounds and traits of some of the individuals involved; note various types of technical services offered and how they have changed with time; provide some interesting details concerning some typical companies, large and small, young and old; point out the relationship of growth in size and in number of firms to the economic and political climate and advancing technology; offer a cross-section of the industry as it is today; and comment briefly on prospects for the future. We hope that the material will constitute a source of satisfaction for a job well done by the older members of our profession, and will lend encouragement to our younger colleagues. We also hope that it will add to the available documentation delineating the all-important role played by the innovator-capitalist operating in a free-enterprise system that has led to a major technology and service industry of great importance to the world's economic well-being. Our discussion here pertains strictly to the profit-making side of exploration geophysics; therefore, no mention is made of geophysical firms or surveys primarily operated or financed by governmental entities. Similarly, no mention is made of industrial meteorology or oceanography for such functions have already been touched on elsewhere in the book.

The Dollar Volume of the Exploration Geophysics Industry

The total annual expenditure of the free-world for geophysical exploration in 1980 was in excess of 3×10^9 dollars (U.S.). Of this, 97 percent was spent for petroleum exploration (up a record 49 percent from the year before), and over 94 percent involved use of the seismic method (see Table 8.1).[1] In mineral exploration, however, 33 different techniques are used. On shore, the most common type of seismic energy source was still dynamite (56 percent), but mechanical vibrators ran a strong second (37 percent). In 1980, marine crews cost, on the average, $400,917 to operate each month, while land crews cost $240,324 monthly. Average costs at sea per seismic line-mile were but $638 compared to $4,612 for a similar mile of profiling on land. Seismic data processing and interpretation costs were estimated to be $650 per line-mile for the Western Hemisphere, where the prospects required more rigorous processing, and $520 per line-mile in the Eastern Hemisphere. In non-seismic surveys for petroleum, gravity survey expenditures, 86 percent of which were in marine areas, were about $55 million. Magnetic surveys, 94 percent airborne, cost about $13.6 million. In the search for minerals, land expenditures for gravity and magnetics were about $2.5 million and $571 thousand respectively, but $7.4 million was spent on airborne magnetics. Mineral exploration still places considerable emphasis on various electromagnetic survey techniques, which cost almost $17 million during the year.

The Role of the Individual Entrepreneur

It will be recalled that the geophysical exploration service industry began during the 1920s as the result of the dedicated striving of such technically oriented entrepreneurs as Lord Cowdray of England, by Fessenden, DeGolyer, Karcher, Eckhardt, Haseman, McCollum and the Petty Brothers in the United States, by Mintrop in Germany,

---

[1.] Exploration data for 1979 and 1980 include a major, unique event. By late 1979, a sixth of the world's geophysical survey fleet was working off the coast of mainland China, an area with the potential of becoming the world's fourth most productive oil and gas province. Eight major oil companies had been assigned exploratory blocks roughly 150 by 350 km in size and could bring in as many as thirty-five partners. By mid-1980, well over 100,000 km of seismic line had been acquired using 96-channel recording gear operated by a *mélange* of surveys crews. Exxon used a survey vessel of its own, supplemented by one from GSI; Mobil also used one of its own crews, plus a contract crew. The block assigned to Chevron and Texaco was shot jointly by GSI and Digicon, Inc. crews. AMOCO used two Chinese deep-water crews, and BP and Total used one Chinese marine crew and one Prakla-Seismos, Ltd. crew.

TABLE 8.1
Free-World Expenditures for Geophysical Surveys in 1980* (in millions of U.S. dollars)

Exploration type	Petroleum	Minerals	Engineering	Geothermal	Groundwater	Oceanography	Research	Total
Land	2,046.8	23.0	3.6	4.1	4.1	—	5.4	2,087.0
Marine	841.2	—	10.5	—	—	5.3	4.0	861.0
Airborne	18.2	27.4	0.3	0.3	0.1	0.2	0.4	46.9
Drill hole	0.8	3.3	0.6	1.9	0.3	—	0.4	7.3
Total	2,907.0	53.7	15.0	6.3	4.5	5.5	10.2	3,002.2
Exploration Region								
USA	1,300.3	8.4	9.2	3.5	0.7	1.1	5.0	1,328.2
Canada	313.7	15.2	0.4	0.7	—	1.1	0.2	331.3
Mexico	71.5	0.4	1.0	0.1	—	—	—	73.0
S. America	235.4	0.7	0.2	—	0.1	2.3	0.1	238.8
Europe	257.0	12.3	2.5	0.7	0.2	0.3	0.3	273.3
Africa	292.3	9.0	0.5	1.3	3.5	0.4	0.1	307.1
Middle East	151.5	0.4	0.2	—	—	—	—	152.1
Far East	203.8	0.8	0.8	—	—	0.2	1.0	206.6
Australia/ New Zealand	56.9	6.5	0.2	—	—	—	—	63.6
International areas	24.6	—	—	—	—	0.1	3.5	28.2
Total	2,907.0	53.7	15.0	6.3	4.5	5.5	10.2	3,002.2

Petroleum Exploration by Major Method and Region

	Land seismic	Marine seismic	Gravity/ Magnetic	Airborne surveys
USA	873.1	142.6	22.9	10.4
Canada	230.1	27.4	0.8	0.1
Mexico	28.9	14.7	1.6	1.6
S. America	115.3	42.7	9.5	—
Europe	120.0	60.1	13.0	0.7
Africa	202.4	29.9	6.5	2.1
Middle East	120.5	7.9	4.7	1.3
Far East	92.5	47.7	10.1	2.0
Australia/ New Zealand	23.6	18.0	—	—
International areas	—	24.6	—	—
Total	1,806.4	415.6	69.1**	18.2***

*Source: SEG Geophysical Activity Report for 1980 based on submissions by about 500 entities (Senti, 1981). The first two tables include both data processing and acquisition costs, while the third shows only data acquisition cost.

**All other gravity/magnetic application expenditures totaled $18.2 million.

***All other airborne survey expenditures totaled $19.5 million.

and by the Schlumberger brothers in France. The industry then expanded so rapidly that, by the 1930s, any enterprising physicist, electrical engineer, or geologist with ingenuity and line of financial credit, coupled with considerable fortitude (or folly), could start a business by assembling or buying a set of equipment (seismic or otherwise) for a moderate to sizeable investment. And if his personal contacts provided an initial contract or two in a promising seismic area and he subsequently turned in several attractive drilling prospects, the firm typically prospered, and he could pay out his initial investment within a relatively short time. Today, start-up costs for entering the data-acquisition or data-processing business in petroleum exploration are in the upper hundreds of thousands of dollars or more.[2] Nevertheless, today more geophysicists than ever before are starting up their own consulting, field-crew, or shothole drilling businesses, initiating new processing centers, computer software development, and data storage—data brokerage services, or creating manufacturing concerns to produce innovative equipment.[3] Frequently, the financial backing is obtained from larger service, manufacturing, or client companies who stand to benefit in some way from the new venture.

In the early days, the new industry's entrepreneurs were rather young, often in their late twenties or early thirties. Today, they are of all ages. Typically, though, they have obtained prior training and experience with large petroleum or contracting companies. Some have preferred to limit their efforts to the management-worker role in small companies providing a specialized or unique service to some segment of the industry.[4] Others have ridden their firm's growth into large enterprises which, in many cases, merged with or were acquired by other large companies or conglomerates. Of the 49 men who have served as President of the SEG since its founding in 1930, 45 percent of these leaders were independent consultants or had their own firms sometime during their career. An additional 18 percent were

[2.] In 1981, the cost of equipping a crew with the most advanced 1,024-channel seismic acquisition-processing system plus four vibrator-type energy sources and other vehicles was about $3.25 million.

[3.] Shop-talk has it that an entrepreneur can make as much money contracting the shothole drilling for seismic crews as in acquiring and processing the seismic data.

[4.] For example, in April, 1981, M. Turhan Taner announced that he had formed the firm of Seismic Research Corporation in Houston, Texas, to develop specialized data-processing techniques. Taner had been a co-founder and chairman of Seismic Computing Corporation, which in itself became Seiscom Delta, Inc. in the late 1960s, resulting in his being Vice President of Research for that new firm. He could obviously afford to strike out on his own at long last. Public records indicate that in a seven-month period during 1980, he sold 13,400 shares of Seiscom Delta common stock (old shares) for approximately a quarter of a million dollars while still retaining 50,875 shares worth about $1.25 million.

key officers of large contracting companies and thus in the forefront of managing these cost-conscious enterprises, while another 12 percent were also employed by such firms, although not in a top managerial role. Overall, 75 percent have been involved with independent geophysical firms, and 25 percent have worked only for oil companies.

Also of interest is that 51 percent of the SEG presidents received their academic training in physics, or in physics and mathematics combined. The remainder's training is about evenly distributed beween the disciplines of electrical engineering, geology, and geophysics. About one-third of the group had achieved the Ph.D. degree, although an advanced degree is certainly no requirement for a successful entrepreneurship in exploration geophysics. Only a half dozen of the SEG presidents had any solid academic background in geophysics before they entered this field. Most of them, in fact, heard about geophysics by chance, and in the 1920s and 1930s were faced with a scarcity of employment opportunities for physicists and mathematicians. Even in the 1940s, other industries had few job openings for mathematicians or physicists, so entering geophysics appeared to present an exciting opportunity for getting in on the "ground floor". In fact, the early pioneers tended to have a minimal knowledge about the earth itself, thereby requiring much on-the-job and self-training.

Work on a geophysical crew is still an exciting way to see the world for petroleum and other minerals are not only ubiquitous but exist in some of the most difficult geographic regions of the world.[5] Work assignments involving fighting through swamps and jungles, contending with the extreme heat of the Middle East and the cold of the Arctic, and keeping the data coming in while the survey craft bounces in the rough seaway of a North Sea or a Gulf of Alaska, lose their glamor quickly despite today's excellent field equipment and feeding arrangements. So can makeshift living arrangements, particularly if one is accompanied by his family. Hence, it is not surprising that a few geophysical firms were founded by field personnel who had become tired of being nomads. This is not to say that geophysicists avoid going into the field, for any earth scientist worth his "salt" needs to get outside and contend with Mother Nature. One thing is certain, however. The wide variety of shared

---

5. As a memento of the SEG's Fiftieth Anniversary, the Houston Geophysical Society compiled comments of the spouses of many "old-timers" into a book, *Reflections* (Houston Geophysical Society, 1980). A few of these memories were also published in an article, "Reminiscences", in *Geophysics* (Darden, 1980).

experiences and mutual dependencies among "doodlebuggers" down through the years has produced an unexcelled camaraderie.

Growth and Proliferation of Geophysical Firms

We related in Chapter 2 how the first seismic data-acquisition contractor, the Geological Engineering Company, quickly became a financial failure by 1921. But we also saw how one of its principals, Karcher, was hired by DeGolyer four years later to head the new Geophysical Research Corporation as part of the overall Amerada Corporation. This time around, the geophysical firm became a highly profitable operation within three short years. However, once DeGolyer proceeded on the sly to break his own employer's monopoly, the sky seemed to be the limit in establishing new seismic explorations firms, as exemplified by the "Amerada tree" (see Fig. 8.1). Out of the original 30 "Amerada tree" companies (not counting the parent GRC), at least 17 still existed in 1955. As time passed and geophysical technology expanded, more and more companies emerged to fill the increasing needs for specialized equipment, supplies and service, and even outside corporations became increasingly involved in supplying technical support. By late 1981, about 1,000 companies (not including oil companies and individual consultants) had a significant involvement in the geophysical exploration industry. Of these, at least 172 firms were fielding seismic crews (five firms with over 90 crews, 17 firms with eight to 40 crews, and some 150 firms with one to seven crews) on a global basis.

During October, 1955 in Denver, Colorado, the SEG celebrated its 25th Anniversary by holding the first annual meeting separate from the American Association of Petroleum Geologists and the Society of Economic Paleontologists and Mineralogists. Thus, this date provides an excellent bench-mark for examining further the growth of the geophysics business. Figure 8.2 shows both the total number of firms (upper curve) that have exhibited yearly since 1955 at the SEG Annual Meeting and the number of first-time exhibitors during any given year (lower dashed curve). While a new exhibitor may or may not have been a newly-formed company, it definitely had to be at the stage in which it wished to expand via public advertising.[6] The SEG's Annual Meeting is unexcelled for this purpose because of the large percentage of members in attendance and the opportunity for one-on-one discussions regarding new

---

[6]. To maintain the non-profit nature of the SEG's annual exhibit for new techniques and equipment, sales are not permitted to be made on the exhibition's floor.

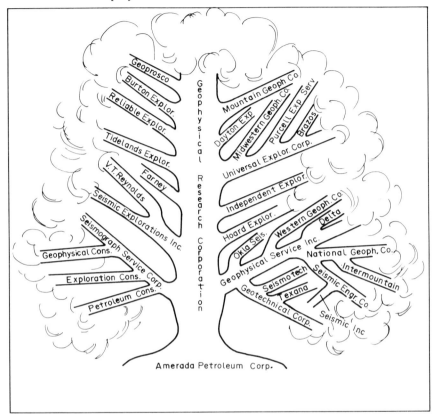

FIG. 8.1. The Amerada "Tree" illustrating the proliferation of firms between 1930 and 1950 stemming from the Geophysical Research Corporation. (Original drawing by H. M. Houghton; courtesy of K. M. Lawrence.)

developments. One interesting aspect of Fig. 8.2 is that it indicates a new peak in exhibitor displays each time the annual meeting is held in Houston, Texas. Such peaking is a normal phenomenon for Houston holds the largest concentration of geophysical contractors and practicing geophysicists to be found anywhere in the world. The 50th Anniversary Meeting in Houston during 1980 had more first-time exhibitors — 78 to be exact — than the total number exhibiting anywhere prior to 1971. The trend lines of Fig. 8.2 also confirm other important events impacting on the industry. For example, the industry was still vibrant in 1955 despite the steady decline in U.S. seismic crew activity from its all-time 1952 peak. This was because of the introduction of magnetic recording and processing. Once the downtrend in seismic activity finally stopped in

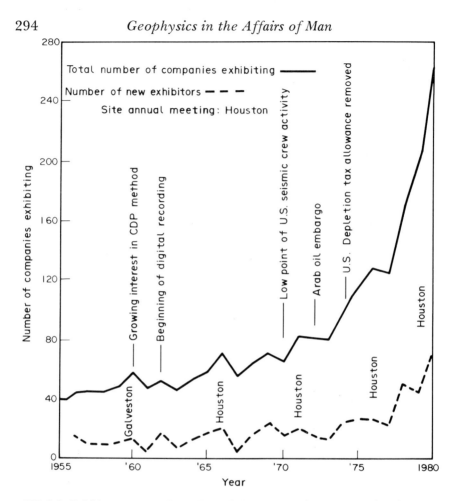

FIG. 8.2. Exhibitors at annual meetings of the Society of Exploration Geophysicists (1955—1980).

1970, interest in exhibiting began to pick up and then, after the Arab oil export embargo of 1973—1974, drastically accelerated. This acceleration continues even today despite petroleum price controls and unfavorable tax legislation in many parts of the world.[7]

Back in 1955, the global cost of petroleum geophysics had been $350 to $400 million, followed by a gradual increase to approximately $450 million in 1964. By 1969, despite a decline in total land crews and marine crew-months worked, the annual expenditure rate nearly doubled to $847 million because of the increased

7. Crude oil prices within the United States were finally decontrolled by President Ronald Reagan in January, 1981.

use of CDP shooting and its associated additional data-processing costs. This type of additional expense was, however, well worth it because of marked improvement in data quality. After 1970, sophisticated and costly data acquisition and processing techniques were being profitably employed in most seismic surveys, so that the controlling factor in the notable jump in cost of worldwide surveys was actually caused by the turnaround in the amount of seismic activity taking place within the United States (see Fig. 8.3) (Senti, 1981).

Geophysical exploration for minerals has contributed modestly to the industry's overall growth since 1955. Thus, total world-wide expenditures for this purpose increased from about $7 million in 1955 to $41 million in 1968, followed by a low of $27 million in 1974 and a high of $52 million in 1977. The complexity and potential of this field is also indicated by the fact that research and development expenses, as a ratio of total survey cost, has been greater for mineral exploration than that for locating oil and gas.

As the industry grew, it faced a number of problems. Accordingly, the *I*nternational *A*ssociation of *G*eophysical *C*ontractors (IAGC) was founded during May, 1971 with 20 charter members and a prime goal of fostering the best interests of the geophysical contracting group and its clients, primarily the geophysical departments of integrated petroleum companies. At first, the IAGC grew slowly, but in 1976, the Association hired a dynamic full-time President, Charles F. Darden. Since then, growth has been much more rapid and, as of early 1981, some 260 firms belonged. The Association is performing a very valuable service, especially in the more recent years when growing government regulations of the types cited in Table 7.1 have increasingly hindered exploration activities. Because of such environmental issues, practically all companies involved in field data gathering belong to the group. In addition, most data processors are members. The Association has also been active in assessing future trends and problems within the industry, and recently polled its membership to update its data bank and projections.

Several other important information sources exist relative to the services and products offered, the most comprehensive of which is the annual *Geophysical Directory*. Starting out in 1946 as a four-page mimeographed compilation by Sidney Schafer and Associates of Houston, Texas (a geophysical consulting firm as well), the 1981 *Directory* contained 584 pages and 31 categories of geophysical services. The Directory also contains a master index of company names and key officials, including names of important personnel at branch offices, as well as an index of oil companies known to use geophysical survey techniques. Finally, advertisements are included

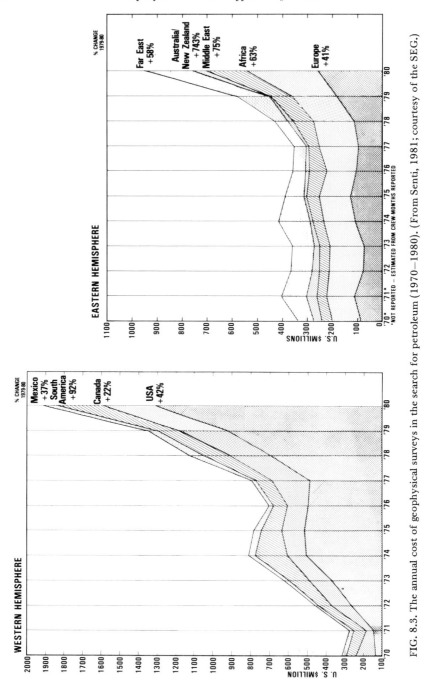

FIG. 8.3. The annual cost of geophysical surveys in the search for petroleum (1970–1980). (From Senti, 1981; courtesy of the SEG.)

along with an index of advertisers making it easy for one to pick up information on new products and services. In 1976, the *Directory* already contained 177 advertisers representing a good cross-section of the industry, but, by 1981, that number had almost doubled to 345 advertisers.[8]

Another source of information about new developments and products is the occasional article written by one or more company researchers or spokesmen. Such an article may appear in one of the more popular oil industry trade journals, but is morely likely to appear in the two professional journals, *Geophysics,* now a monthly publication, and *Geophysical Prospecting,* a newer, bi-monthly publication issued by the European Association of Exploration Geophysicists (EAEG), the only other international society of professional exploration geophysicists. Because of proprietary reasons, lesser technological advances that still may add up to significant progress are generally not reported.

Finally, the advertising carried in *Geophysics* and *Geophysical Prospecting* are important data sources. These advertisements, which include everything from descriptions and claims concerning new products and services to equipment photographs and comparisons of old-style and new-style data, are very much a tell-tale barometer of the industry's progress and which firms lead in introducing new concepts and services. Sheriff, in fact, has written a brief history of geophysical exploration based entirely on advertisements which have appeared in *Geophysics* down through the years (Sheriff, 1980). There is one qualifier in this regard, however. Often the marketing people may be just a little ahead of the field operations people, and, in truth, despite positive, multi-color advertisements for a new technique, the client may soon learn that the "bugs" are far from having been worked out. Both cited journals carry a good mix of advertisements for both European and North American firms, although the latter predominate simply because they are more numerous.[9]

A rather complete picture of the European-based geophysical exploration industry can be obtained from an analysis of firms which exhibited at the largest EAEG meeting to date — the Silver Anniversary Meeting held in The Hague during June, 1976. The exhibition was honored by being opened during its first morning

---

[8.] Prior to 1981, the *Directory* carried, for historical purposes, an all-inclusive list of client oil companies. Today, only the names of active oil corporations are included.

[9.] For example, the December, 1980 issue of *Geophysical Prospecting* carried sixteen, six, and two advertisements for firms headquartered in the United States, Europe, and Canada respectively.

by His Royal Highness, Prince Claus of the Netherlands, who presented a knowledgeable address (in excellent English) regarding historical and modern aspects of geophysics.[10] Fifty firms had exhibition booths. Of these, slightly over half were sponsored by companies based in Europe. Between them, England and France accounted for about 30 percent of the total number of exhibitors, and there were several each from Belgium, Federal Republic of Germany, the Netherlands, and Norway. Canada also accounted for about ten percent, and the United States the rest.

The degree of interaction between American and European firms and individuals is quite high. On the occasion of the EAEG's Silver Anniversary, Rice found that, as of that time, 54 percent of all EAEG members (some 1,400 individuals) were also members of the SEG and, conversely, about 15 percent of all SEG members belong also to the EAEG (Rice, 1976). In the best capitalistic tradition, the competition between the major European and American geophysical exploration firms throughout the world continues as keen as that existing within the American domestic industry. Each of the major contractors has managed to find one or more ways to excel over its fellows at any given time and thus to stay among the world's leaders.

Corporate Profiles — General

It would take volumes to describe the individuals and corporations that have made significant contributions, both through technical innovation and business acumen, to the development of exploration geophysics, even in just the past decade. Fortunately, details of the evolutionary history of exploration geophysics, both within the contractors and major oil companies, have been compiled to a considerable degree (although in a somewhat disjointed manner) by one of the early "doodlebuggers", George Elliott Sweet (Sweet, 1966, 1969).[11] Our treatment here will address eight major groupings: (1) the large, full-service geophysical exploration contractors, (2) smaller seismic data acquisition firms, (3) firms specializing in seismic data processing, (4) other types of service companies, (5) equipment manufacturing and supply companies, (6) consultants, (7) gravity and magnetic surveying companies, and (8) the mineral exploration

---

[10.] Prince Claus's address may be found in the September, 1976 issue of *Geophysical Prospecting.*

[11.] George E. Sweet, with his brother, Reginald, were participants in the founding of the Independent Exploration Company in 1932, followed by founding (on their own) the American Seismograph Corporation a year later.

business. The treatment and the choice of material are, of necessity, biased towards those facets of the industry which we have been closest to over the past 35 years, namely, the areas of data acquisition, processing, and interpretation. Thus, unfortunately, equipment manufacturing, one of the larger segments of the industry, will be short-changed to the extent that only major developments and companies that affected downstream activities in a very significant manner will be touched on. To stay within the space and resources available, we have used a high degree of selectivity. For this, we offer our sincere apologies to those many individuals whose important efforts have been omitted, even though they, too, have been indispensable in bringing this vital industry to such an advanced state of technical maturity.

The Nature of Large, Full-Service Contractors

The importance of large full-service contractors to the growth and current well-being of the worldwide exploration geophysics industry should not be underestimated. Through good times and bad, they, along with the research groups of the major oil companies, have led the industry through successive eras of rapidly improving technology. They have risked tens of millions of dollars in pioneering efficient, economical, global marine surveying techniques, and in shooting tens of thousands of kilometers of speculation surveys in hopes of selling the data to clients at bargain rates. They have made land crews and data-processing centers available worldwide, frequently at great economic risk, under the most difficult operating conditions or in very unstable political environs. They have continuously sought more efficient and more economical ways to process data and to make it more accurate and meaningful for the interpreter. In so doing, they have developed a small but powerful computer industry and have spurred on the manufacturers of the largest computers and auxiliary devices to new achievements.

Of the 658,474 line-miles of petroleum-survey, marine seismic data acquired in 1980, most of the effort can be attributed to a score of companies. Of these, perhaps half a dozen were by non-government-owned major oil companies which operated their own geophysical survey vessels. The remainder were large seismic contractors, of which nine were headquartered in the United States, one in Canada, and four in Europe. The international nature of the competition is indicated by the fact that even in U.S. waters as of early 1981, one Canadian and one European contractor were competing directly with six domestic firms. Since the wholesale adoption of digital data recording and processing in the 1960s, full-service seismic exploration contracting and speculation shooting, particularly off-shore, have become big business, with the *potential* for above-average profit. But very large capital investments are required. For example, the cost of obtaining and equipping a global seismic surveying ship is now at least seven million dollars, including elaborate onboard positioning and computer equipment. And this is only a part of the required investment, because the voluminous data obtained have to be further processed on land on the finest digital computing equipment available in order to obtain optimum results and to keep processing costs and turnaround times to a minimum.[12]

These large costs and sizeable potential profits have fostered frequent mergers within the industry itself and also the acquisition of many of the major contractors by large conglomerates. Fortunately, when such buy-outs occur, most of the contractors still operate as fairly independent subsidiaries of their parent companies, so that their identities and individual characteristics have remained largely intact over the past decade or so. As a result, we have compiled in Table 8.2 some historical and current facts about the eight largest full-service contractors (in alphabetical order) and are also able to present below considerable information on how they got to where they are today.

Compagnie Générale de Géophysique (CGG)

CGG's roots go back to 1919 when Conrad and Marcel Schlumberger signed an agreement with their father whereby he promised to pay them "a sum of 500,000 francs, necessary for conducting research experiments to find ways of determining the nature of the subsurface by means of electrical currents" as an aid to geologists. In return, the sons promised to devote all of their time to the research and not to let financial interest take precedence over scientific interest. Their steadfast efforts resulted in the first discovery in 1922 by the resistivity method — a salt dome near Aricesti, Romania, which produced gas, and in the mapping in 1926 of five anticlines and a salt dome on the Alsatian plain of France. But the Société de Prospection Electrique, nicknamed "Pros", still had no steady source of income. So in September 1926, they expanded operations to the United States, where contracts were obtained with Gulf,

---

12. Marine processing and initial interpretation costs in the Western Hemisphere in 1980 were estimated at $650 per line-mile (Fig. 8.3). This represents only about one hour of processing time on a large computer, yet the amount of data handling and computation involved in processing one mile of data is staggering.

TABLE 8.2
Some Pertinent Data on the Eight Largest Geophysical Contractors

Company (parent firm)	Year founded	Founder(s)	Acquisitions or Affiliates	Chief officer(s)	Headquarters	No. of U.S. cities	No. of countries outside U.S.
				1981		Offices	
Compagnie Générale de Géophysique	1931	R. Maillet, Conrad & Marcel Schlumberger	Geodigit, Georex cos., Geoterrex, Helidrill, Sercel cos. S.P.S.	R. Desaint, Chairman & Pres.	Paris	4	26
Digicon Inc.	1965	D. L. Brown, G. A. Cloudy, P. H. Poe, E. R. Prince, D. R. Steetle, E. A. Robinson	Alaskan Geophys., Digisonics, Ramco Labs., Houston Drilling, Ocean Marine Services Information Products Systems	E. R. Prince, Chm. & Pres.	Houston	3	2
Geophysical Service, Inc. (Texas Instruments)	1930	E. DeGolyer, J. C. Karcher, E. McDermott	*See* text	GSI: C. H. Green, Hon. Chm., G. A. Dove, Chm., D. K. McDaniel, Pres. TI: M. Shepherd, Jr., Chm. J. F. Bucy, Pres.	Dallas & Houston	7	25
Petty-Ray (Geosource)	1925	Dabney & Scott Petty		D. B. Sheffield, Pres.	Houston	6	22
Seiscom Delta United (Seiscom Delta, Inc.)	1964)	M. T. Taner, Scott Kelso	Delta Exploration (1968) United Geophysical (1981)	G. D. Reese, Chmn; T. B. Portwood, Jr., Pres.	Houston	8	
Seismograph Service Corp. (Raytheon)	1931	J. H. Green, W. G. Green, C. H. Frost, G. F. Martin	Cie. Francaise de Prospection Séismique	R. C. Anderson, Pres.	Tulsa	6	14
Teledyne Exploration (Teledyne Inc.)	1936	R. F. Beers	Geotech, IX, Nat'l, Continental	J. H. Frasher, Pres.	Houston	4	2
Western Geophysical (Litton Industries)	1933	H. Salvatori	Aero Service	B. B. Strange, Chm., M. H. Dingman, Pres.	Houston (originally Los Angeles)	8	17

Note: Prakla-Seismos GmbH is excluded from this list because it is government-owned. It was founded by the German Reich (in 1937) and is still owned by the Federal Republic of Germany. It bought out Mintrop's old firm, Seismos Ltd. (founded in 1921) in 1963.

Shell, and Standard. By 1929, the total number of employees had climbed to 95 and the profits to 1,400,000 francs. Then the great depression hit, and all sources of business dried up. In order to survive, "Pros" was obliged to merge with the Société Géophysique de Recherches Minières (SGRM), whose land seismic and magnetic prospecting operations were complementary to the "Pros" electrical work and whose tools, licenses, and five million francs of financial backing constituted an irresistible combination. Thus, CGG was formed in March, 1931, with R. Maillet, founder of SGRM, as Managing Director.[13] A third company, the Société de Prospection Géophysique, which specialized in gravity surveys, was acquired in 1937, thereby completing a full scope of services offered.

Although it was consistently a leader in electrical work and also carried out important gravity surveys for British Petroleum and others, CGG's most notable early success was with the seismic method, mainly refraction, in the Saharan Desert. Here, during the period 1952–1963, using up to thirty crews, they solved difficult prospecting problems and discovered the major oil fields in Algeria (Pieuchot and Richard, 1958; Layat *et al.,* 1961). Success in *reflection* seismograph work was not fully realized until after a major foothold had been established in the North American market. CGG's first technical representative for the United States was the *bon vivant* but capable Richard F. Hagemann, hired in 1956. They exhibited at the Annual SEG Meeting for the first time in 1959, and have prepared increasingly elaborate exhibits every year since, with major emphasis on seismic techniques. Major expansion came with the introduction in 1973 of heliportable seismic operations in frontier areas of the Rocky Mountains. Today, 25 percent of CGG's operations — 20 land crews and one marine crew with an associated positioning network — are in the United States and Canada. The company is among the three or four largest geophysical contractors in the world, and is probably unexcelled for diversity of operations, as indicated in Table 8.2 by the number of subsidiaries and the number of countries in which they have offices.

The present corporate structure is too complex to explain in detail as generally there is a CGG branch office and/or subsidiary to handle each type of activity in each country or area. For example, in the United States and Canada, Geodigit is the direct data-processing subsidiary. But all the crew activity in the United States, including drilling equipment manufacturing (Georex Industries), shot-

---

[13.] The Schlumberger brothers preferred to manage the well-logging business, which was kept separate and became the multibillion-dollar Schlumberger well-logging company mentioned in Chapter 2.

hole drilling (Helidrill), positioning (Survey Positioning Systems), and recording equipment construction (Sercel Industries), is now carried out by subsidiaries of Consolidated Georex Geophysics, Inc. (a new corporate entity as of January 1981), which in turn is a subsidiary of Georex, Inc., a CGG subsidiary headquartered in Paris. Sercel Industries also manufactures positioning equipment and the well-known Sercel plotter. But all Sercel equipment is sold in the United States through Sercel, Inc., a subsidiary of Sercel S.A. of France, which in turn is the subsidiary of CGG that handles both manufacturing and sales of Sercel equipment outside the United States. Geoterrex is CGG's electrical contractor world-wide, and Georex Data is their data exchange company in the USA.

CGG has prided itself on its innovation, which has been fostered by an effective research and development group under the continuing leadership of D. Michon, Vice-President of Research and Technique, and J. M. Fourmann, Manager, Scientific Dept. Their work has covered a broad spectrum of subjects from important theoretical formulations to computer processing and interpretation techniques to major engineering developments. Scientists such as V. Baranov and G. Kunetz (now both retired), H. Naudy (still in charge of gravity and magnetics research), and D. Paturet (manager of software development) have achieved world-wide recognition.

An example of CGG's aggressiveness in innovation and development is the Vaporchoc® marine energy source. This was developed in the late 1960s to greatly reduce troublesome bubble effects on marine seismic records. An underwater bubble is formed from a standard explosive source which travels to the surface to burst and cause a secondary source of considerable strength. Its effects must be negated through the use of a properly designed array of sources or by means of "debubble" computer processing.

CGG's logical approach was to effect an *im*plosion, rather than explosion. This is done by introducing superheated steam into the water which rapidly cools and collapses to produce the primary energy source but very little bubble. This required building shooting vessels with extra large and costly steam generators and associated equipment. But CGG pushed ahead, made many improvements, and is still using the technique, apparently with good results.

CGG's current capabilities embrace all the techniques which form the leading edge of today's seismic technology. Furthermore, they remain strong in gravity and magnetics, and are effectively using their decades of experience in multidiscipline approaches to

® Registered trademark of CGG.

mineral exploration and engineering problems. Finally, their long experience in electrical methods is being applied at the forefront of the wave of renewed interest in direct detection of shallower petroleum deposits by use of transient electromagnetic techniques.

In addition to its fine technical capabilities, CGG does a very good job of selling these capabilities to the profession. Their scientists continue to give talks on a great variety of topics (six papers were presented at the 1980 Annual SEG Meeting). In addition, they offer state-of-the-art speculation and group surveys in exploration areas of current interest, and their continuously updated literature is excellent. Hence, CGG should continue to maintain its position as one of the very largest and most versatile full-service geophysical contractors.

Digicon, Inc.

Digicon is the only one of the eight companies cited in Table 8.2 which started completely from scratch after the advent of the seismic digital revolution in the 1960s.[14] Beginning with practically no financial base, the company did interpretation work, which requires no capital investment, for about a year.[15] Then they leased their first processing computer, and soon afterward leased their first seismic field equipment. But what the group lacked in capital was compensated for in depth of experience (see Table 8.3). The first Chairman and President, David Brown, had been a crew supervisor for GSI for many years. Rudy Prince, who became chief executive officer in 1970 when Brown decided to cut out so much traveling, had been one of GSI's Area Geophysicists, a talented group who worked directly with clients in pioneering the applications of GSI's new digital technology to exploration problems. Then there was the already famous Enders Robinson (see Chapter 5), who served as a Vice President until 1970. He was succeeded by the very talented John Sherwood, who became Digicon's Director of Research in 1978.[16]

By mid-1968, the company's revenues and net income for the fiscal year had climbed to $2.4 million and $304 thousand ($0.21 per share) respectively. Later the same year, Alaskan Geophysical

---

[14.] Seiscom was founded in 1964, but it had to acquire the much older Delta Exploration (founded in 1947) in order to become a full-fledged geophysical exploration firm.

[15.] Dave Brown recalls that the company's capital was $2,338.

[16.] During 1962–1963, Dr. Sherwood, then an employee of Chevron's research arm, spent several months with the Tabor Pluto group at Blacknest as part of the overall cooperative effort between the Tabor Pluto and Vela Uniform programs.

TABLE 8.3
Financial History of Digicon Since Going Public*

	1969	1970	1971	1972	1973	1974	1975	1976	1977	1978	1979	1980	1981
Total Revenues (1,000's of U.S. $)	6,051	8,672	9,194	10,036	9,554	20,046	26,097	26,244	23,646	29,057	38,728	53,681	107,983
Net income (loss) (1,000's of U.S. $)	780	(4,101)	(2,183)	857	1,417	2,279	2,425	175	(119)	267	457	2,340	5,969
Net income per share	47¢	($2.08)	($1.12)	44c	71c	$1.13	$1.20	9c	(6c)	13c	22c	$1.01	$2.07
Trend of U.S. work	Up	Down	Down	—	Up	Up	Up	Down	Down	Down	Up	Up	Up
Trend of work outside U.S.	Up	—	—	Up	Down	Up	Down	Up	Down	Up	Up	Up	Up
Research & Development expenditures (1,000's of U.S. $)								(— — — Average 809 — — — —)				1,420	4,057
Business backlog (1,000's of U.S. $)												32,400	91,600

Breakdown of employees as of early 1981:

Executives	29
Administrative and accounting	115
Research and development	64
Clerical and other office personnel	57
Marine seismic operations	173
Land seismic operations	116
Geophysical data processing	815
Compressor packaging	136
Tug-supply operations	68
Scientific manufacturing & distribution	127
Total	1,700

*Digicon's fiscal year ends 31 July.

was acquired from J. A. Riendl, who became a Vice President, through an exchange of shares. But the opportunities for both land and marine surveying were increasing rapidly. So the company decided to go public and raised almost $6 million through the sale of shares at $22. This proved to be a very profitable move, for the following year, revenues and net income were $6.05 and $0.47 per share respectively. By this time, the company had four land seismic crews (three operating on the North Slope of Alaska) and four marine crews (two working in Alaskan waters and two off the eastern United States) and were processing data in Houston and Singapore from many land and marine areas of the world. From then on, expansion and purchase (rather than leasing) of seismic data collection and processing equipment along with increased sales and operations outside the United States were effected as rapidly as world prospecting opportunities permitted.

A careful scrutiny of Digicon's financial history shows the impact of several important external factors which seriously affect the entire full-service geophysical contracting industry—and particularly the immediate fortune of any new firm just entering the field. The first of these factors involves the difficult problem of deriving an optimum mix between land and marine crews, between operations in various parts of the world, and between resources assigned to data acquisition in the field and to data processing at a computer center. The operation of land crews can be advantageous in that equipment and operating costs are less, making down-time much less costly, and the effort is not quite so subject to the vagaries of governments. On the other hand, marine work has the potential of yielding higher profits, but down-time costs and equipment losses can eat up these profits in a short time. This accounts for Digicon's poor net-income showing in 1970, when low activity forced the derigging of two boats, and again in 1971, when three more vessels were derigged — at considerable expense. From 1973 to 1975, primarily as the result of the Arab embargo, United States offshore leasing was greatly accelerated and Digicon obtained a substantial share of the evaluation work. By 1976, the United States' political climate for the oil industry began deteriorating again, with Congressional talk of divestiture of integrated operations (never passed) and of much more time-consuming and costly rules governing federal offshore leasing (passed in 1978). As a result, the entire United States leasing schedule was essentially abandoned and the marine geophysical industry suffered, with Digicon derigging their West Coast marine crew in 1977. Then, mechanical difficulties with two of their remaining three ships added significantly to Digicon's down-time costs during fiscal 1978. Finally the outlook for the United States' offshore began to improve in 1978 with the drilling of the first exploratory wells offshore the East Coast and the announcement in 1979 of a new five-year lease schedule. This renewed activity coupled with a surge in exploration offshore Mexico and in the China Sea have greatly improved the profitability of marine operations in recent times.

Digicon's history has demonstrated the value of using worldwide operations to help balance depressed activity in major areas such as the U.S. In Table 8.3, although the "up" or "down" indications of fluctuations in seismic activity in and outside the United States are purely qualitative, they do show that Digicon's revenues and profits were probably helped significantly through their international operations in 1972, 1973, 1975, 1976, and 1978, or nearly half the time. This is particularly true for marine data acquisition. Data-processing operations in different areas have counterbalanced each other some, but have provided the most consistent revenues and profits. The reason is that there is always much data processing and reprocessing to be done to generate new prospects which is not tied to specific offshore leasing schedules. In fact, continuous improvements in seismic data-processing techniques have rendered previous processing results obsolete about every five years, which has made the reprocessing business a steadily growing industry.

The second important factor revealed by Digicon's financial record concerns the role of competition, which determines what proportion of the total available geophysical exploration work throughout the world is carried out by a single contracting company. Obviously, marketing effectiveness is one factor in this determination. But there are two much more important factors in the geophysical exploration business. First and foremost is the ability to obtain high-quality data which permit the accurate interpretation of subtle stratigraphic as well as structural traps plus as many lithologic parameters as possible. Cost continues to be important, but only if individual clients or groups are convinced that the contractor can deliver state-of-the-art data quality. It is evident that Digicon has done very well both technically and in terms of competitive bidding, for starting from scratch, it obtained 0.7 percent of the world-wide petroleum geophysical data acquisition and processing business ($847.3 million) in 1969 and 1.67 percent of a greatly increased volume ($1942.2 million) in 1979.

Two examples will illustrate Digicon's innovation capabilities. The first has to do with "wave-equation migration", which was barely mentioned in Chapter 7. The wave equation mathematically describes how an acoustic wavefront which is generated by a seismic source on or near the surface radiates outward and downward through the earth in the same manner that a pebble dropped at the edge of a pond causes an ever-broadening ripple to travel out into the pond. But when the wavefront encounters a boundary (for example a rock in the pond), a new reflected wave is generated for each point on the boundary which radiates back toward the surface (the edge of the pond). Now the sum total of all these reflected waves which are recorded at the surface of the earth is a function of all the reflection boundaries, including faults, in the subsurface, and the wave equation can be used to reconstruct accurately the location of all these boundaries. "Ray-path" techniques used exclusively until recent years only deal with one point on each of the radiating and reflected wavefronts and in many cases do not produce an accurate picture of the subsurface. Professor Jon Claerbout of Stanford University was the first person to develop approaches that would facilitate the implementation of wave-equation methods on digital computers.[17] In 1974 he expanded this effort under the Stanford Research Program with the support of most of the major oil companies in the United States and some of the large contractors. Digicon joined this support group at the start and the same year became the first contractor to reduce wave-equation migration to commercial practice (Loewenthal *et al.*, 1976).

Another recent Digicon innovation, which took 30 man-years to develop, is a digital seismic streamer (hydrophone cable) for marine work, which was installed in 1980 on the first ship that Digicon owned rather than leased. This cable provides 240 channels of seismic data (as opposed to the usual 48 or 96), gives more accurate data than analog cables, and is much more reliable because it contains 95 percent fewer wires.

Since the advent of the digital revolution in seismic data processing, there has been a gradually accelerating market for data-processing hardware—software systems, both for smaller field-oriented systems and powerful central processing facilities. Digicon offers the same two systems that it uses in-house: (1) a smaller field-oriented system based on the Texas Instruments TIMAP computer and an expanded and improved version of software originally obtained from Systems Corporation of America; and (2) a fairly new system with large throughput capacity built around a Digital Equipment Corporation VAX 11/780 computer and Digicon software. Digicon

---

17. Claerbout received the SEG Reginald Fessenden Award in 1973 for his pioneering work.

began switching from Xerox Sigma computers to the VAX system for its major processing centers late in 1978. By January 1981, it had also sold 14 of these systems to clients, and a backlog of orders was continuing to build. These sales are now contributing substantially to the company's revenues.

Some rather unusual diversification activities have been undertaken by Digicon during its relatively brief life. The first of these was initiated in 1974 through the acquisition of a 50-percent interest (later increased) in Ocean Marine Services and the development of joint plans to build eight ocean tug-supply ships to service the offshore drilling industry. Also, by 1975, research by its subsidiary, Digisonics, was well along on improving the application of seismic techniques to obtaining static and dynamic cross-section photographs of the human heart. Another medical subsidiary, Ramco Laboratories, was formed in 1976 to market a kit for differentiating iron deficiency from other types of anemia without the trauma associated with the usual bone-marrow-extraction technique. Both medical endeavors have been substantially profitable since 1978. Ocean Marine Services contributed 10 percent of 1978 revenues, but has suffered since from an overcrowded market. In January, 1981, Digicon purchased Bluewater Maintenance, Inc. which designs, fabricates, assembles, delivers, and installs compressor modules and other oilfield production equipment. This is a profitable enterprise which had 1980 revenues of about $69 million. All in all, Digicon should be able to maintain or even increase its share of today's very favorable geophysical exploration market.

Geophysical Service, Inc. and its Holding Company, Texas Instruments, Incorporated

Our chronicle has already provided glimpses into the history of *Texas Instruments* (TI) and its initiating company, *Geophysical Service, Inc.* (GSI). The evolution of this unique enterprise is one of America's greatest success stories in combining advanced concepts of technology and managerial skills. We have therefore decided to provide additional details in three separate ways — as a vignette by one of the firm's early owners (see vignette by Cecil H. Green); as a narrative which immediately follows and describes how a small cadre of talented men put together, in just 50 years, a multinational corporation employing nearly 90,000 people in 50 countries; and as a listing of significant corporate events carried as an appendix.

Between 1930 and 1945, GSI was a typical geophysical contracting firm. However, by 1945, the "art" of seismic prospecting had become fairly well stabilized and even appeared to be reaching the point of diminishing returns. As a result, the owners could see trouble ahead as there would be increasing difficulty in acquiring useful data for their clients in new and more complicated areas. Fortunately, Erik Jonsson had made contact with some impressive young engineers on the Washington scene who were anxious to use their talents in the commercial world once the war was over. In this regard, Jonsson had become reasonably well acquainted with Lt. Patrick Haggerty, USNR, the Bureau of Aeronautics' desk officer for airborne magnetometer acquisition. Haggerty tended to believe, as did Jonsson and McDermott, that there would soon be a definite need for small, smart companies capable of successfully blending electrical and mechanical know-how in the design and manufacture of military electronics. So Jonsson invited Haggerty to Dallas in September 1945 to look over the GSI set-up and to meet the other two partners. Peacock developed some doubts from this visit about diversification, but Jonsson prevailed and Haggerty reported to work on 1 November to become the General Manager of a new Laboratory and Manufacturing Division. "Pat" then proceeded to get Robert Olson on board from the Bureau of Aeronautics for engineering and Carl J. Thomsen for accounting. He further argued that he needed a 37,500-square foot plant for manufacturing (to cost $150,000), plus 17 people from geophysics, to initiate a three-pronged program: (1) manufacture geophysical equipment for both GSI and the open market; (2) manufacture appropriate non-military electronic items; and (3) manufacture military electronics as a "second source" of revenue and eventually having them become a "prime source". When the "Lemmon Avenue plant", dubbed the "dream castle", was completed in 1947, the cost had doubled, and the GSI banker was limiting the firm's credit.

Haggerty's group won their first major military contract in 1947 and another in 1949. Overall sales climbed from $2.25 million in 1946 to $5.8 million in 1949, and the number of employees from 554 to 792. However, geophysical services to the oil industry still generated 75 percent of the company's revenues. When the first contract was obtained from the Navy to build military radars, Jonsson and Haggerty believed the time had come to force the issue as to the exact nature of long-term goals for the corporation. By early 1952, Peacock decided to sell his 25 percent share of company stock (50,030 shares subsequently split 8.8 for 1 in May, 1953). In spite of the company's profit-sharing plan initiated in 1942, ten years later there were still only 28 shareholders, all company officials or directors. But when Peacock sold his stock at approximately $46.20 per share to his associates, an additional 19 key staff and line employees were added to the number of shareholders. This action ultimately created a sizeable number of new millionaires, for Peacock's stock, sold for slightly over $2 million, grew over the years via splits and a stock dividend to 2,201,360 shares, which was worth more than $330 million at 1980's peak price. At the end of 1953, GSI had 47 crews in the field, but still their efforts brought in only 32 percent of the total revenues of $27 million for the year. Property, plant, and equipment were valued at $6.1 million, including $1.6 million in geophysical equipment, $1.0 million in automobiles and trucks, $1.2 million in machinery and furniture, and $1.6 million in buildings. Long-term debt was very low, being but $2.4 million as compared to total assets of $14.9 million.

TI common stock was first traded publicly on the floor of the New York Stock Exchange on 1 October 1953, with President Jonsson buying the first 100 shares for $5.25 a share. As of that time, 2,987,000 shares of common stock were outstanding, annual sales came to $27 million, and the net income was $1.27 million. During the next 17 years, the company grew largely from within, and by September, 1980 there were over 23 million shares of common stock in existence because of stock splits, conversion of preferred stock, and the exercise of stock options. Moreover, during the same period, annual sales had grown to over four billion dollars and net profit to about $200 million, for a total growth of more than two orders of magnitude. Haggerty was very much at the center of this bootstrapping operation for three decades, serving as either company President or Chairman of the Board between 1958 and 1976.[18]

Haggerty's dominant method of operating was by using the "O–S–T" approach, a technique that had much in common with standard military doctrine. As he explained it, "O" stood for carefully selecting the company's objectives for the long haul, "S" for agreed-on strategies needed to reach the objectives, and "T" for the supporting tactics needed to carry out each strategy.[19] The first major company strategy picked by this technique came in 1951 when it was decided to develop, manufacture, and market semiconductor devices (Haggerty, 1965). This field was chosen because the vacuum-tube market was saturated. In addition, the military was asking for miniaturization and for the ability to withstand severe vibrations in such vehicles as intercontinental ballistic missiles. TI's tactics in this instance were: (1) seek a patent license from the

---

[18.] Haggerty retired at age 62 in accordance with company policy for top TI executives.

[19.] By 1976, TI had 11 objectives, 65 strategies, and 422 supporting tactics (Haggerty, 1977).

Bell System for manufacturing transistors, (2) set up an in-house Project Engineering Group to develop solid-state devices, (3) manufacture and market the end products, and (4) establish a supporting research laboratory emphasizing semiconductor materials and devices. As a result, the transistor license was acquired in 1952, the research laboratory started in 1953, and the first commercial pocket transistor radio introduced into the mass market in October 1954, about two years ahead of the competition.

The results of TI's decision to be revolutionary, rather than evolutionary, are indicated in Fig. 8.4, which shows an annual growth rate of over 23 percent for both net sales billed and net income for three decades. Note that the goals set in 1949 for revenues and net income of $200 million and $10 million respectively by about 1959 were right on target. Furthermore, a goal for 1980 of 3×10^9 worth of net sales billed was exceeded by 33 percent. However, it remains to be seen whether the firm can continue its breathtaking growth rate. As of mid-1981, profits were sharply down (by 58 percent) because of weak markets for semiconductors and associated products, forcing the total number of employees to be reduced by three percent. However, electronic matériel sales to government and GSI's exploration contracting business remained strong. Be as it may, the firm's current goal is to achieve $15 billion in annual sales some time in the late 1980s, or a rate nearly four times that of the 1980 sales level.

It is also quite appropriate to provide some detail on how the GSI—TI team was able to effect the technical breakthrough in digital

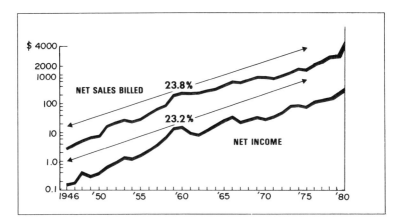

FIG. 8.4. Plot of net sales billed and net income of Texas Instruments (in millions of dollars) between 1946 and 1980.

recording and processing of seismic data, undoubtedly the single most important accomplishment in the history of exploration geophysics once the seismic reflection technique was found to work back in 1921. While their competitors were placing attention and resources on how to upgrade analog recording and processing equipment — for example, providing a better set of amplifiers, during the early 1950s GSI assembled a group composed of Haggerty, McDermott, Burg, Olson, and Hal Jones to determine whether it was possible to improve the signal-to-noise ratio of recorded and processed data by a truly significant amount. It was eventually decided that the first step should be to study possible applications of signal-to-noise ratio improvement methods developed by Professor Norbert Weiner of MIT and Claude Shannon of the Bell Telephone Research Laboratories and called "statistical communications theory". Following a presentation by "Ken" Burg to GSI's Board of Directors in late 1952, the first research budget of $25,000 was authorized to study a portion of the digital data-processing theory for improving the detection and resolution of overlapping reflections. This approach was simultaneously being studied, it will be recalled from Chapter 5, by the MIT Geophysical Analysis Group which had just been organized after two years of promising preliminary work by Professor George Wadsworth and the doctoral degree candidate, Enders Robinson. Relative to the GAG effort, it is significant to note that GSI–TI was one of only two geophysical contractors (the other being United Geophysical) willing to contribute about $8,000 for such advanced research. This support, of course, was a direct derivative of Cecil Green's effort in having started the already mentioned "Student Cooperative Plan" under MIT's good auspices and the devoted efforts of Professor Shrock during the summer of 1951. In any event, through their firm's research effort paralleling that of the GAG team, GSI–TI was able to maximize benefits from the industry-sponsored program at MIT.

By 1954, TI's Central Research Laboratories were able to issue a report addressing in some detail the digital processing of digitized analog seismic data, including digital filtering, cross-trace correlation, automatic record picking, and signal plotting. Then a hybrid machine, called seisMAC, was designed and built under the direction of Fred Bucy as project engineer that could actually perform some digital processing on analog magnetic recordings.[20] Two of these devices were later put into commercial service — one in Dallas and one in London. To back up this effort, "Ken" Burg then hired and orga-

---

20. Bucy became TI's president in 1976.

nized a research group of MIT graduates headed by Mark Smith and including Robert Bowman, Milo Backus, Peter Embree, and Lawrence Strickland.[21] This team launched a full-scale research effort, initially testing their digital processing methods on an IBM 650 computer at GSI's Dallas facilities. Next came an experimental vacuum-tube computer for digital processing (DARC for *D*ata *A*nalysis and *R*eduction *C*omputer) and corresponding field digital-recording equipment during 1958. Finally, work began on the "DFS-9000" field recording system and a new Fred Bucy development, the all-transistor "TIAC" computer.

But there was still the major hurdle of convincing the petroleum industry that they should accept and use this new technology. Here, a masterful marketing move was effected by Edward Vetter of the Industrial Products group who, in 1961, obtained the financial and technical help of two of the largest oil companies, Mobil and Texaco, in exchange for two years lead on the rest of the industry. The result, as mentioned earlier in Chapter 6, was that Texaco fielded the first digital seismic recording crews in 1962. Initial results obtained by both Texaco and Mobil were encouraging, and other oil companies were eager to hire GSI digital crews on an experimental basis when they became available in 1964. Typically, a crew would be asked to record a small amount of data in each of a number of areas known to be difficult, the data would be processed by GSI, and then a conference would be arranged between the client's representatives and one or more of GSI's Area Geophysicists. The results were generally mixed but encouraging, and the conferences were of great educational value both to the client and to GSI. Curiously, one of the bones of contention was that GSI wanted to use the new technology to locate stratigraphic traps using their new deconvolution methods. But this objective frequently resulted in noisier, higher-frequency seismic sections than the client was used to or desired, for he was usually interested mostly in obtaining more accurate structural interpretation. So in effect, GSI was about a decade ahead of the industry in their thinking.

Clients who gave this new technology a fair trial knew for certain that a revolution in seismic prospecting was under way. A large share of the credit for selling this revolution must go to James Toomey, who arranged client seminars in Dallas and Europe, and to GSI's Area Geophysicists of that era. The earliest and best known of this group were: Robert Graebner (who rose to President of GSI in

---

[21.] Smith and Bowman were both members of the GAG team. Smith served as GSI's President between 1966 and 1969.

1973), Peter Embree (now Chief Area Geophysicist), Clem Blum (left to join Digicon in 1975), Melvin Carter (founded his own firm, Energy Analysts, in 1974), Ben Giles (after becoming GSI's Chief Geophysicist, Western Hemisphere, he resigned in 1977 to become Digicon's Chief Geophysicist), Richard Matthews (now GSI's Processing Manager), and Emir Tavella (now GSI's manager in Buenos Aires). The knowledge and quiet confidence displayed by these and other GSI representatives through the years have had a great deal to do with GSI's historic successes.

Once Cecil Green became Vice-President with total responsibility for employees, field operations, and foreign business, GSI placed unusually strong emphasis on employee training. Much of this training has been under the auspices of "Ed Stulken's College of G-Function Knowledge", with appreciable help from the likes of Ben Giles and "Bob" Graebner. These programs include formal class instruction and extensive GSI correspondence courses, plus special schools for drillers and observers conducted by Jack McManus. Fortunately for the entire geophysical profession, past and present, Green's consuming interest in education could not be contained within the sphere of GSI and its clients, nor even within the realm of geophysics.

Having no children of their own, Cecil and his wife, Ida, have oriented their entire life during their later years towards helping young people make their way in the worlds of science, engineering, medicine, and the arts. As MIT's President, Dr. Jerome Wiesner, has noted, the Greens have proved to be "pollinators" for a wide variety of disciplines ranging from oceanography to elementary school teaching. At an elaborate dinner, complete with string orchestra, hosted by MIT at the U.S. National Academy of Sciences on 9 November 1978, the distinguished head table estimated that financial gifts by the Greens had, although they were still in their mid-70s, directly affected 200,000 students and a million university alumni. Just how much they have given away is known only to the two of them, but a listing of some of their better known actions is not out of order (see Table 8.4). Dr. Frank Press, a former occupant of MIT's Cecil and Ida Green Building for Earth Sciences and then Science Advisor to President Jimmy Carter, said it all when he observed: "One can make a difference — and they have. They've made the world a little wiser and a better place in which to live."

GSI has been a strong supporter of the geophysics profession through the SEG, and, to some extent, the European Association of Exploration Geophysicists, both in terms of technical presentations at professional meetings and in allowing their very capable leaders to

TABLE 8.4

A Selection of Contributions by Cecil and Ida Green to their Fellow Man

Location	Type of contribution
Global	20 endowed scholarships, largely in sciences and engineering, at various universities
Massachusetts Institute of Technology	Cecil and Ida Green Building for Earth Sciences
Colorado School of Mines	Cecil and Ida Green Professional Center, plus Cecil Green Geophysical Observatory
Oxford University	The Green College to accommodate post-graduate medical students in clinical medicine
La Jolla, California	Cecil and Ida Green Hospital to supplement other facilities at the famed Scripps Clinic Partial support of the Institute of Geophysics, University of California at San Diego (Scripps Campus)
Palo Alto, California	Cecil Green Library Addition to main library
Vancouver, British Columbia	Cecil Green Park which provides quarters for the University's Alumni Association
Dallas, Texas	Cecil Green Center which houses School of Management and Administration Science Building at St. Mark's School Endowed "Master Teacher Chair" at St. Mark's School
Austin, Texas	Ida Green Communication Center at Austin College
Galveston, Texas	275-ton ocean-going research vessel, the *Ida Green*, for use by the Institute of Marine Science, University of Texas
Northern Texas	Educational television system operated by Association for Graduate Education and Research for North Texas
Global	Project "IDA" which allows "International Deployment of Accelerometers" at some 20 points to study world-wide effects of earth oscillations, with data center at Scripps Institution of Oceanography, La Jolla, California
National	Several "Green Fellowships" at universities to encourage women to work in the fields of science and engineering, particularly in biological science

serve as officers of the SEG. GSI has supplied five SEG Presidents — McDermott (1933–1934), Karcher (1937–1938), Peacock (1941–1942), Green (1947–1948), and Dunlap (1955–1956).[22] Backus retired from GSI before becoming President in 1979. No other contracting company has had more than two (GRC), and only one oil company (Humble–Carter–Exxon) had as many SEG presidents as did GSI. GSI–TI technical presentations at professional meetings

22. DeGolyer, Karcher, Green, McDermott, Peacock, and Burg have all been made Honorary Members of the SEG. Green was also the first person to receive the prestigious Kauffman Gold Medal Award in 1966 for "outstanding contributions to the advancement of geophysical exploration". External to the world of science and engineering, McDermott served as Mayor of Dallas, one of America's largest cities, between 1962 and 1964.

have been known for their excellence year after year. Many of these papers have been published and have added significantly to the permanent literature. In the Silver Anniversary Issue of *Geophysics* (February, 1960), two out of the 35 seismic papers chosen as "Classic Papers of Geophysics" were by GSI authors (Green and Backus).

What of GSI's future? It is still a very strong and profitable organization. However, about five years ago, it lost its place as the world's largest geophysical contractor to Western Geophysical, a Litton company.[23] As a result, it is generally believed to rank second along with CGG and Petty-Ray Geophysical, a Geosource company. Even so, GSI operates fifteen well-equipped marine survey vessels and dozens of land crews throughout the world. It currently remains the leader in the emerging field of three-dimensional seismic technology, which has a very bright future for reservoir delineation and development.

No other geophysical contractor has research and development support of the scope that the massive TI complex can offer GSI, a situation at least partially attributable to the fact that all TI Presidents to date have had extensive experience in geophysics. This combination to date has been of tremendous value to geophysical exploration, to the economic well-being of many nations, and to mankind in general.[24]

Petty-Ray Geophysical Operations

Although Petty-Ray ranks among the top four geophysical field contractors, its present corporate structure dates back only to 1973 (see Fig. 8.5). Nevertheless, Petty-Ray is an amalgamation of some of the oldest and best-known firms in the geophysical industry. The story of how it became a major operating division of Geosource, Inc., a firm founded in 1973 with the financial backing of The Aetna Casualty and Surety Company and the Rockwell International Corporation, is complicated but intriguing.[25] The core firm was

23. TI Presidents Erik Jonsson (1951–1958) and Fred Bucy (1976–present) began their professional careers with GSI. "Pat" Haggerty (1958–1966) began his career with Badger Carton Company, Milwaukee, Wisconsin, before joining the Navy and then GSI. Mark Shepherd (1967–1976) joined GSI–TI in 1948 after having worked for General Electric Company and Farnsworth Radio and Television Corporation.

24. For example, TI pioneered in making possible the mass production of cheap but reliable solid-state radios, watches, and calculators subsequently sold in many countries of the world.

25. As of early 1980, Rockwell held 11.5 percent of Geosource's common stock. However, Rockwell then sold its holding to the general public during March, 1980. Aetna, on the other hand, continued to acquire additional shares and owned over 29 percent of Geosource by November, 1981.

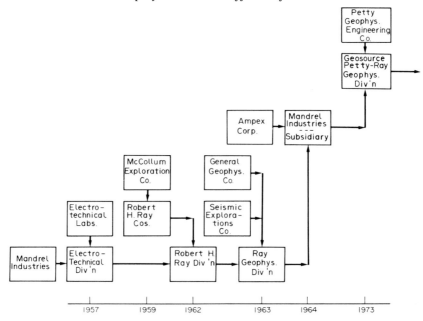

FIG. 8.5. Principal events in the formation of Geosource's Petty-Ray Geophysical Division.

Mandrel Industries, which first got into geophysics by acquiring Electro-Technical Laboratories of Houston, Texas, one of the oldest manufacturers of a complete line of seismic equipment. "Electro-Tech" was basically the creation of the dynamic, talented Harold (Hal) A. Sears, who served as its founder and President until he started another firm, Hall-Sears, Inc., with Ernest O. Hall in 1957 to specialize in geophone manufacture.[26]

Mandrel's next major acquisition came in 1962 when it picked up the Robert H. Ray Companies. This group had been founded back in 1939 by Ray to specialize in gravity prospecting and by the start of World War II had become the largest gravity contractor in the world. By acquiring McCollum Exploration Co. in 1959, plus the acquisition with CGG of a joint license to use McCollum's weight-dropping "Geograph" seismic technique in most of Europe and the French territories, the Ray firm became an important entity in the

---

[26.] By 1954, Electro-Tech had furnished the recording equipment for some 90 percent of the 700 seismic crews operating in the U.S. and Canada, as well as over 115,000 seismometers. In 1963, Hall-Sears merged with the Geo Space Corporation. A geophone introduced by Geo Space in 1970 was the most popular seismic detector ever made, with more than two million sold by 1980.

seismic business as well.[27] A year later, Mandrel acquired both the General Geophysical Company and the Seismic Explorations Company and labeled its total geophysical effort the Ray Geophysical Division.[28]

Mandrel's buy-out of General was an extremely important decision. The General Geophysical Company had been started back in 1935 by Earl W. Johnson after he served as one of GSI's first party chiefs back in 1930 and helped Henry Salvatori start up the Western Geophysical Company in 1933. In Johnson's case, his sponsoring firm was, in large part, The Ohio Oil Company (now Marathon). Even so, during the war years, pickings were slim, and Johnson was one of Dr. Merle Tuve's closest collaborators in developing the proximity fuse at the Applied Physics Laboratory of Johns Hopkins University. After that, Johnson returned to General but died much too soon, age 50, in 1953. However, General had two very capable vice presidents — Chester Sappington, who became Chairman of the Board, and Thomas O. Hall, who became President.[29] The addition of General gave the new Ray Geophysical Division not only a large number of experienced seismic land crews but also an excellent research and development organization directed by the very experienced and innovative Dr. Lewis Mott-Smith who took on a similar assignment in the new firm.

Corporate-wise, Mandrel Industries was still venturesome, and in March, 1964 it became a subsidiary of a large West Coast tape-recorder manufacturing firm, the Ampex Corporation, through an exchange of ten Mandrel shares for nine Ampex shares. Fortunately, the name, "Ray Geophysical Division — Mandrel Industries", was preserved. Robert H. Ray continued as President of the subsidiary and was then Chairman of the Board from 1966 until his death in 1968. However, this corporate marriage was not totally successful, for the top executives on the West Coast over-expanded the firm,

---

[27.] After selling out to Ray at the age of 79, Dr. Burton McCollum established McCollum Laboratories to continue his beloved research in geophysics. Death came to this pioneer of the geophysical industry at the end of a normal working day when he was 84 years of age.

[28.] Seismic Explorations was one of the "Amerada tree" companies, having been founded in 1932 by the confident Fisher F. Reynolds, one of the SEG's founding members who lived to attend the society's golden anniversary in November, 1980. After selling out to Mandrel, Reynolds continued at the new Division, as did John D. Marr, who served as Vice-President in both geophysical organizations. Marr is particularly well known for an excellent series of articles on "Seismic Stratigraphic Exploration" (Marr, 1971).

[29.] When General was bought out, Sappington became Vice President (Domestic Operations), while Hall became Vice President (Foreign Operations). Hall also served as the SEG's President during 1959—1960 and was instrumental in opening up the dialog between the "Oil Patch" and the Pentagon relative to the geophysical industry's heavy participation in the Vela Uniform program.

causing it to get into financial difficulties. Hence, the Texas-oriented geophysicists of Ray Geophysical were pleased when the Geosource founders acquired them in 1973 along with the family-owned Petty Geophysical Engineering Company of San Antonio, Texas.[30] Petty, in particular, brought considerable prestige and royalties, for it still owned the Harry Mayne "CDP" patent issued back in 1955 that had resulted in such a boost to the firm during the late 1950s and all of the 1960s. In fact, at one time, Petty claimed to own more geophysical patents than any other geophysical contractor or oil company in the world, and their star performer, Mayne, had generated 25 patents of his own by late 1980 (Petty, 1976).

This new corporate marriage worked out far better (see Table 8.5). Back in 1973, the incorporators of Geosource, Inc. stated the firm's goal would be:

> To be the world leader in exploration for depletable natural resources; inventory and management of renewable natural resources; environmental monitoring and impact assessment; and development planning and management services.

Certainly the geophysical side of Geosource, labeled the "Petroleum Exploration Group" and directed by D. B. Sheffield, has contributed its share during the past decade in meeting these goals. For example:

1. As of 1980, the Group contributed half of the firm's revenues and twice as many earnings as the other three segments of the company combined.[31]

2. On 2 July 1980, the People's Republic of China awarded a $34.2 million, three-year contract to Petty-Ray Geophysical to survey seismically the Tsaidam Basin of western China, as well as to supervise Chinese crews working in the same area. This was the first U.S. geophysical contractor to be invited to work in mainland China.

3. In May, 1981, Petty-Ray Geophysical fielded its 100th land

---

30. O. Scott Petty, the youngest of the three Petty brothers, served as President and Board Chairman from 1925 to 1952, followed by service as Chairman from 1952 to 1973. His son, Scott Petty, Jr., after receiving bachelor's and master's degrees in petroleum engineering from the University of Texas, joined the company in 1961, served as President from 1967 to 1973, and remained active in Geosource management even after the merger. The other founding brother, Dabney E. Petty, served as Vice President until 1954 and as a Director until 1960, three years prior to his death. The eldest of the original brothers, Van A. Petty, Jr., a lawyer and independent oil operator who formulated the company's first charter and negotiated its first client contracts, served as Vice-President, Chief Counsel, and Director until his death. His son, Van A. Petty, III, after taking B.S. and M.A. degrees in geology at the University of Texas, also served as one of the firm's vice presidents and directors between 1942 and 1955, when he resigned to do independent consulting.

31. It is noteworthy that between 1976 and 1979, Geosource's petroleum exploration revenues increased 65.7 percent, almost exactly the same as the increase in total worldwide petroleum exploration expenditures (from $1.185 billion to $1.955 billion).

TABLE 8.5
Five-year Financial Record of Geosource

Sales & Service Revenues (millions)	1980	1979	1978	1977	1976
Overall Geosource	$441.3	$346.1	$263.7	$213.8	$204.5
Petroleum Exploration Group*	$231.9	$164.4	$124.6	$ 97.9	$ 99.2
Percent of Total	53	48	47	46	49
Operating Earnings					
Overall Geosource	$ 65.9	$ 44.2	$ 36.7	$ 30.0	$ 29.4
Petroleum Exploration Group	$ 45.7	$ 26.1	$ 21.2	$ 15.1	$ 12.5
Percent of Total	69	59	58	50	43
Net Income (millions)	$ 32.7	$ 21.1	$ 17.3	$ 13.5	$ 10.8
Earnings per share	$ 2.77	$ 1.83	$ 1.57	$ 1.36	$ 1.13
Total Assets (millions)	$304.8	$236.7	$182.8	$142.7	$127.4
Shares Outstanding (millions)	11.8	11.4	10.7	10.3	9.2
Return on Assets (percent)	10.7	8.9	9.5	9.5	8.5
Employees (1,000's)	6.8	6.2	5.4	4.8	4.7

*The firm's other Operating Groups are: Petroleum Development, Petroleum Processing and Distribution, and Electronics.

geophysical crew, thereby becoming the largest land exploration contractor in the world.

4. A new unit, GeoSurvey Operations, was formed to utilize a combination of surface and down-hole seismic data augmented with well log information for solving problems of oilfield development and detecting the presence of hydrocarbon-bearing horizons not penetrated by the borehole.

5. The Group claims to be the only manufacturer making and selling all of the equipment needed to outfit a seismic field crew. In addition, it markets the TEMPUS seismic data processor package and a very fine laser plotter, the LASER-DOT®, which provides resolution of graphical data up to 500 lines per inch.

6. Because of pioneering work in high-resolution seismic field recording equipment, Geosource is able to obtain higher frequency data useful in making detailed interpretations of stratigraphy and lithology, thereby providing a technique which identifies small, thin oil and gas reservoirs overlooked by older survey techniques.[32]

® Registered trademark of Geosource, Inc.

32. Another very important factor in high-resolution recording is the use of much denser surface sampling by means of more seismic detectors and recording channels. The Geophysical Systems Corporation also has a system for handling this factor (see the section, Other Data Acquisition Companies).

Looking ahead, company management expects the Petroleum Exploration Group to double in size by 1985. Within this frame of reference, Petty-Ray's principal weakness may be in the area of marine prospecting, for it operates only four offshore crews. On the other hand, there is certainly no end in sight for land exploration, and Petty-Ray may be wise in concentrating on remaining the leading seismic contractor onshore.

Seiscom Delta, Inc.

On 28 July 1981, Seiscom Delta became the world's largest independent publicly-owned company engaged in the collection and processing of geophysical data. As such, its annual revenues will likely be in excess of $200 million. Once again, the firm's name is less than a decade old, but the corporate roots go back to the mid-1930s. In this instance, the surviving firm actually consists of three large, well-known firms: Seismic Computing Corporation, Delta Exploration Company, and United Geophysical Company. The present corporate name came about in the following way. Delta Exploration had been a second-tier field contractor since its founding in 1947, but lacked standing and competence in the data-processing area. This deficiency was eliminated during March, 1968 by exchanging its stock with that of the Seismic Computing Corporation (Seiscom). Because "Seiscom" was publicly owned and Delta was not, Seismic Computing Corporation became the surviving company, with Delta Exploration being a wholly-owned subsidiary. Four years later, however, it was decided to give both capabilities equal billing and the name of the firm, which is listed on the American Stock Exchange, was changed to Seiscom Delta, Inc.

Delta Exploration was formed as a five-way partnership in Jackson, Mississippi, between the following ex-GSI employees: Ewin Gaby, John J. Babb, George Augustat, Etoyle Conly, and Audio G. Harvey. Gaby had been an Area Supervisor, while Babb, Augustat, and Harvey had been party chiefs under him, and Conly had been a supervising surveyor. In 1952, a sixth partner, Phil Gaby, a brother of Ewin, was added. Until his death in 1961, Ewin Gaby served as the firm's President, when he was succeeded by John Babb who had been Director, Field Support Activities.

During the critical start-up months in early 1948, Delta Exploration was able to field two crews for the California Company in the swampy region southeast of New Orleans even though the firm had but one set of field equipment. This was possible because the crews had to live in and work out of quarterboats; thus, it was perfectly feasible to employ a work schedule of seven days on and seven days off, permitting both crews to use the same set of recording equipment until a back-up set became available in March, 1948. Before long, the Union Producing Company also hired two crews — one a quarterboat crew for swamp work and one for conventional work on dry land. The California Company's parent, the Standard Oil Company of California, then contracted for two Canadian crews, one in 1949 and one in 1950. Late in 1951, Delta sent a seismic crew to Trinidad, and by a year later had a total of twelve crews working not only domestically but also in Bahrain, Turkey, Chile, Sumatra, and the Philippines. This proved to be the peak year for U.S. seismic activity and for Delta's crew count as well, since by 1967 the firm was operating only three field parties. But, as the saying goes, "When times are tough, the 'Tough' get going!" Delta accordingly opened a Houston office and by mid-1968 had three marine crews operating. In addition, the decision was made to merge with Seiscom, the firm doing Delta's digital data processing.

Seiscom dated back only to late 1964, having been conceived earlier that year during a luncheon discussion between M. Turhan ("Tury") Taner and Professor Fulton Koehler of the University of Minnesota's Mathematics Department, where Taner had taken graduate work in applied mathematics.[33] The discussion indicated that there might well be a need for a company dedicated to performing advanced digital data processing for the petroleum-exploration industry. Taner then reviewed the concept with A. Scott Kelso, a colleague of his at Scientific Computers in Houston. When it was learned that Scientific Computers did not believe the market potential was great enough, Kelso and Taner formed the Seismic Computing Corporation with some financial backing from Jacob Greenberg. Seiscom bought back Greenberg's stock a few years later. Taner was Chairman of the Board and Director of Research, Kelso President, and Koehler Vice-President. The company was prepared to spend two years without income to develop commercial seismic data-processing techniques. Initially, computer time was rented from Control Data Corporation, who bought out the Houston branch of Scientific Computers.

By the summer of 1965, Taner and Koehler had developed their own version of deconvolution which gave results competitive with GSI's.[34] Then on the basis of a remark by Ray Sanders of Amoco in February 1966 about GSI's "Dynamic Correlation Analysis"

[33.] Taner was born in Turkey and obtained an engineering degree at the Technical University of Istanbul before coming to the United States.

[34.] Ray Geophysical was one of the first clients to use Seiscom's deconvolution program until they could develop their own.

method for obtaining velocity information from CDP data, Taner and Koehler developed over a weekend the elements of their "Semblance Coherence" technique for computing "velocity spectra" (Schneider and Backus, 1968). The Taner—Koehler method remains the standard of the industry even today (Taner and Koehler, 1969; Neidell and Taner, 1971).[35] These spectra not only provide the velocity information as a function of record time needed to effect an optimum stack of CDP traces but also give the seismic interpreter important information about data quality and change in rock lithology.

After the 1968 merger, Seiscom was the first contractor to incorporate the tracing of seismic ray-paths with layer-by-layer velocity analysis to obtain more accurate results in complex geological situations (Taner *et al.*, 1969). Other notable achievements of Taner's research team include: (1) a unique automatic way of computing static time corrections for near-surface velocity variations (Taner *et al.*, 1974), (2) early use (about 1970) of color to represent various parameters within seismic sections, and (3) the computation and use of seismic amplitude, phase and instantaneous frequency in stratigraphic interpretation of seismic sections (Taner *et al.*, 1979). Mention should be made that these last two developments were partially due to the skilful efforts of the noted English geophysicist, Nigel Anstey, a Seiscom executive in London who used an in-house laser plotter to carry out research studies from 1969 on.

During the 1970s, the lead managerial role at Seiscom Delta was assumed by D. Gale Reese, a lawyer by training who had worked his way through college by serving as a roughneck on various oil rigs. When he joined the firm in 1971, Reese was named Senior Vice-President and Directorate of Corporate Affairs and Finance. Within two years he was named company President, and Chief Executive Officer in 1974.[36] During his tenure as President, Reese has seen his organization have three losing years — 1973, 1974, and 1976 (see Table 8.6). The 1973 losses were caused in large part by unusually adverse weather conditions impacting their U.S. land-crew operations and by the inability to sell extensive proprietary data from Peru because of a reversal in the Peruvian political climate. Losses during the following year derived largely from continued weak sales of proprietary seismic data, resulting in a large write-off

---

[35.] Norman S. Neidell, a Ph.D. in geophysics from Cambridge University (England) in 1964 and formerly with Gulf, joined Seiscom in 1968 and made important contributions for several years before starting his own consulting firm.

[36.] Reese's competitors elected him to the distinguished post of Chairman, International Association of Geophysical Contractors, for the period, 1977—1978.

TABLE 8.6
Financial History of Seiscom Delta

	1968	1969	1970	1971	1972	1973	1974	1975	1976	1977	1978	1979	1980
Total income (1,000's of U.S. $)	8,156	12,316	16,143	22,059	25,784	22,557	30,284	29,089	24,972	32,076	37,595	41,348	63,221
Net income (loss) (1,000's)	557	742	286	893	1,411	(1,324)	(696)	555	(1,371)	825	1,594	1,362	3,233
Net income per share ($)	0.53	0.66	0.25	0.75	1.13	(1.12)	(0.58)	0.47	(1.15)	0.69	1.29	1.07	1.60
Assets (1,000's)	6,416	13,045	18,326	19,332	22,879	21,715	19,483	17,598	15,445	15,584	20,381	26,026	47,529

of data-acquisition costs. The 1976 loss came about when there was a loss of interest by the United States Government in fostering offshore land sales; as a result, Seiscom Delta was forced to cut its marine capabilities by 40 percent. In comparing the loss years with those of its competitor, Digicon (see Table 8.3), one finds that the two firms had only one bad year in common — that of 1976. In fact, one may generalize that an exploration contractor's fortunes rise and fall with how well its operations mesh with unforeseen circumstances. In the absence of any diversification outside the geophysical industry, company managements need to be lucky, unusually gifted at prophecy, or willing, like farmers, to average results over extended periods.

Starting in 1977, Seiscom Delta began to surge ahead as the international oil industry reacted to the escalating prices induced by the OPEC nations and by gradual lifting of controls on the price of oil and natural gas. By 1980, annual revenues had grown to $63 million, or nearly double that of 1977. Even more important was that net income had quadrupled to $3.2 million per year, while net assets had tripled to $47.5 million. Moreover, Reese and his cohorts had also funded most of this growth with internally-generated funds; as a result, their firm's long-term debt was only $11.2 million as of 30 September 1980.[37] This permitted them to capitalize on an opportunity of a lifetime when they learned in early 1981 that the Bendix Corporation, in an attempt to raise $900 million by divestments, was willing to sell its subsidiary, the United Geophysical Corporation of Pasadena, California. United was doing $80 million of business a year, and Bendix was willing to sell for exactly that amount.[38] A few trips to the First City National Bank of Houston, Texas soon provided Seiscom Delta with the extra money they needed, for the bankers were willing to provide a financing package of $86 million, of which $51 million was a term loan and $35 million was in the form of revolving credit. The final deal with Bendix was struck on 28 July 1981, and Seiscom Delta doubled in size overnight. The corporate name remained Seiscom Delta, Inc. but the geophysical exploration division became Seiscom Delta United.

The acquisition of United offers considerable economies of scale, marketing, and geographic positioning, for United was one of the

---

[37.] Seiscom Delta's comparable competition, Digicon, had 1980 revenues of $53.7 million and $24.3 million of long-term debt because it had started to diversify into supplying service vessels to offshore drillers and manufacturing geophysical and medical equipment.

[38.] At the time of purchase, United was operating 55 seismic land crews and had a net asset value of $43,646,000. United's gross profit margin as a percentage of sales was about 27 percent in contrast to Seiscom-Delta's gross margin of 22 percent.

older and most respected geophysical firms in spite of a somewhat checkered history. It came into being during 1935 when Grant W. Corby, a distinguished petroleum geologist and geophysicist on the West Coast, told his friend and fellow "ham" radio operator, Herbert Hoover, Jr., about the exciting possibilities of exploration geophysics (Sweet, 1969).[39] They persuaded Dr. Harold W. Washburn, the laboratory director for Western Geophysical Company in nearby Los Angeles, to join them and build their first set of seismic instruments. Their first field crew came on line during 1936. Corby soon sold his interest in the firm to Hoover, and Hoover continued as President until 1952 and then as Chairman of the Board for another year. However, back in 1949, Hoover's research man, Dr. Raymond A. Peterson, had interested Hoover in experimenting with mineral-detection technology (Mansfield, 1980). Because southern Arizona was both mineral rich and relatively close to Pasadena, a regional reconnaissance program got under way in mid-1950. Luckily, as a starter, they chose the old Twin Buttes mining district about 25 km southwest of Tucson. Beginning near an old mine, they worked south across a gravelly area with electrical, magnetic, and gravity sensors and soon found an anomaly associated with favorable geology. They then brought in a standard United truck-mounted shot-hole drilling rig and soon began coring ore after penetrating bed-rock (Heinrichs and Thurmond, 1956).[40] By the time the drill reached a depth of 180 m, the cores were showing five percent ore (chalcopyrite plus molybdenite). Once shafts were sunk, this proved to be the rich Pima ore body. As a result of this and other nearby discoveries which soon followed, the district revived to become the most active mining area in Arizona with four major operators (Anamax, Asarco, Cyprus Pima, and Duval) extracting about 200,000 tons of ore daily by the late 1970s. Nevertheless, Hoover's interests were elsewhere, and in 1952 United was sold to one of its best customers, Union Oil of California.[41]

Union soon learned that other oil companies did not wish to

[39.] The son of the 31st President of the United States, Hoover had acquired a bachelor's degree in mining engineering from Stanford University (as had his father) and a master's degree in business administration from Harvard University. After working as a communications engineer for TWA Airlines between 1929 and 1934, he had become a teaching fellow at CalTech when he and Corby decided to go into the geophysical exploration business.

[40.] The principal in Heinrichs Geoexploration was Walter E. Heinrichs, Jr., a very early graduate of the Colorado School of Mines.

[41.] Hoover left California in 1954 to serve a three-year tour as Under Secretary of the U.S. Department of State. He then returned to California to serve on the Board of Directors of a half-dozen corporations and to direct the affairs of the Hoover Foundation before dying at age 66 in 1969.

hire a geophysical crew from a competitor, so United was sold to its employees in 1956. By that time, the firm had extensive world-wide operations and considerable experience in Arctic operations, for it operated under contract to the Navy's Petroleum Reserve Number Four activity as far back as 1944. United also developed an excellent research program under the able leadership of Dr. Raymond A. Peterson. "Pete" had joined United in 1938 and acquired the title, Director of Research, in 1943. He was the first to demonstrate how continuous velocity logs could be used to construct synthetic seismograms, one of the most important seismic interpretation aids still in use today (Peterson *et al.*, 1955).[42]

In 1965, United was sold by its employees to the Bendix Corporation, an automotive and aerospace conglomerate based in far-away Southfield, Michigan. Why the Pasadena group wished to acquire a distant master is not general information, but it may well have been one way to acquire the extra capital needed to go "digital". In any case, United was in extremely capable hands at the time with L. A. Martin as President and Samuel J. Allen as Executive Vice-President and Operations Manager. In fact, it would have been hard to find a better operations executive than the modest and highly principled Allen. In order to avoid missing time in the office or the field, it was Sam's practice to work long daytime hours and then catch a "red-eye" flight overnight to the locale of the next day's activities. In addition to this, he also did a yeoman's work for the benefit of the geophysical profession at large.[43]

United had the services of another great geophysicist during the 1960s after Canadian-born Milton B. Dobrin joined the firm in 1961 as Chief Geophysicist and Head, Technical Services Department.[44] In addition to writing the most-used modern text in geophysical prospecting, Dobrin worked with scientists of the small Conductron Corporation of Ann Arbor, Michigan, to perfect a commercial laser-

---

[42]. Peterson has also become widely known for his work on a unique series of monographs, distributed free by United, on various technical aspects of seismography. He continued in this interest even after the demise of United's research organization and, though semi-retired, still carries the title of Director of Research. Some of the more intriguing titles in the series are: "The Writing of the Earth Waves" (1970), "Through the Kaleidoscope — A Doodlebugger in Wonderland" (1974), and "Seismic Imaging Atlas" (1976, 1977, 1978).

[43]. Allen's many services include President of the SEG (1972—1973), SEG Distinguished Lecturer (1975), and Chairman of the SEG Geophysical Activity Committee (1968—1969). He single-handedly fought to get an acceptable registration law passed for geophysicists in the state of California. He was made an Honorary Member of the SEG in 1980 and received the Kauffman Gold Medal Award in 1973.

[44]. Dobrin was Editor of the SEG from 1953 to 1955, President (1969—1970), and was made an Honorary Member in 1978. His last 11 years were spent in teaching academic and industry courses at the University of Houston and in consulting all over the world.

scanning technique for removing unwanted events and frequencies on variable-density seismic sections (Dobrin *et al.*, 1965).[45] Although this specialized technique was ultimately doomed to be superseded by improved digital processing techniques, the method did see extensive industrial use for a number of years.

In 1969, Bendix further enlarged its operations by acquiring GeoProspectors Corporation, an Oklahoma-based seismic data acquisition, processing, and interpretation firm formed in 1955 and owned by Hugh M. Thralls, Theodore Green, and Waynard B. Perry.[46] Upon selling their firm, Green became a consultant, Perry continued as Vice-President with Bendix—Geoprospectors for several years, and Thralls became United's Chairman of the Board for four years.[47] In the mid-1970s, Bendix overseers drastically and surprisingly changed the nature of this old-line geophysical firm by eliminating all research and technological development and buying any new technology needed for its field crews on the open market. In fact, even the capability to process seismic data was dropped. In the short run, this made United a fierce competitor for new exploration contracts for its overhead was low and its profitability high. However, such a favorable position could not have been maintained for long, and the sell-out to Seiscom Delta undoubtedly came at an optimum moment.

Just how the combined efforts of Seiscom Delta and United will work out in the future remains to be seen. The company is now in the 100-land-crew class along with Petty-Ray and Western, and can effect many economies. On the other hand, Seiscom Delta has lost a number of its top entrepreneurs and technical people during the past few years. For example, in 1977 the firm lost R. Hugh Ervin, W. Peyton Weems, Wayne L. Underwood, Willis T. Brown, and George E. Pawley, all important executives, when they left to form a competing firm, Seis Pros Inc.[48] Even worse, however, is the loss of the genius of Turhan Taner, who resigned in April, 1981 to found a consulting firm, Seismic Research Corporation. The new firm will do some consulting for a competitor, the Grant Geophysical Cor-

---

[45.] First optical processing of seismic data was done about 1961 by Philip L. Jackson at the Acoustics and Seismics Lab. of the University of Michigan under a Vela Uniform contract, initially as part of the effort to detect atomic explosions from seismograms.

[46.] An associated partnership, called Thralls, Green and Perry, was entirely a consulting firm and charged GeoProspectors for interpretation work.

[47.] Thralls was President of the SEG in 1964—1965 and was awarded Life Membership in 1980 in recognition of his extensive services to the Society.

[48.] The founders of Seis Pros ranged in age from 32 to 50. As of March, 1981, Ervin, Weems, and Underwood were being paid $140,000 per year and the two junior founders, Brown and Pawley, were earning $112,000 per year.

poration, on matters related to enhanced data processing techniques. To many clients, "Tury" was Seiscom. He will be missed by Seiscom customers for his unique, tautological solutions to problems, and for his ability to build client confidence through open and lucid presentations of complex issues. However, the entire profession undoubtedly will continue to benefit from Tury's frequent technical papers and short courses.[49] Backing up "Tury" was another gifted research geophysicist and Senior Vice-President, Dr. Robert E. Sheriff.[50] He, too, took off in early 1981 to become Professor of Geophysics at the University of Houston to help fill the gap created when Dr. Dobrin died suddenly at age 65 while jogging on 22 May 1980.[51] Thus there is no question but that D. Gale Reese and United's past president, Thomas B. Portwood, Jr., who became President of Seiscom Delta, have a challenge ahead of them.

Seismograph Service Corporation

The Seismograph Service Corporation, better known as "SSC", is another of the old-line geophysical firms that grew out of the Geophysical Research Corporation and the Geophysical Services, Inc. evolution nearly 50 years ago and then got sucked up by a major corporate conglomerate (Raytheon) in the mid-1960s. In this case, William G. Green, an electrical engineering student at the University of Oklahoma, used his "ham" radio experience and associated radio operator's license to land an observer's job during 1929 on a GRC refraction seismograph crew. He switched over to GSI within about a year of its founding in 1930. However, his father, J. H. Green, had even greater ideas for his son. Why not start a geophysical contracting firm of one's own? The father raised $4,000 for this, with the funds coming primarily from George F. Martin and C. E. Frost, an oil-lease broker who also, through acquaintances, arranged a start-up contract with Mid-Continent Petroleum Company. Young William then quit GSI and became the first President of SSC when it was incorporated on 3 November 1931.

[49.] Taner was a Distinguished Lecturer for the American Association of Petroleum Geologists in 1976 and was made an Honorary Member of the SEG in 1978.

[50.] Sheriff has authored two modern textbooks, *Geophysical Exploration and Interpretation,* and *Seismic Stratigraphy,* as well as the SEG's *Encyclopedic Dictionary of Exploration Geophysics.*

[51.] Dobrin had jogged in 70 different countries. In his memorial in the December, 1980 issue of *Geophysics,* the son, David, poignantly notes:

> "There was just so much to do. No one, if he knew, would have been more irritated, frustrated, or angered by his own death than my father. He died with the rest of a short course to give, lectures in Norway to prepare, a report unfinished on his desk."

By mid-1932, the younger Green had an operable, although crude, set of equipment patterned after that of GRC. It was immediately put to work for Mid-Continent, and two more crews were added during the next 12 months (Sweet, 1969).

In the fall of 1933, Gerald H. Westby, the Chief Geophysicist of the Cities Service Company, was persuaded to gamble and became SSC's initial Vice President. Westby was actually quite a catch; although a geologist, he had taken additional courses in geophysics at the Colorado School of Mines and was well versed in mathematics. As a result, he was equally conversant with both petroleum geologists and geophysicists.[52] Once on the job, he was alert to the fact that a number of oil companies were interested in starting up geophysical departments. Accordingly, he and Green agreed to offer oil companies the opportunity to buy "skeleton crews", that is, a complete set of field equipment and a bare minimum of operating personnel. The first crew of this type was sold to Cities Service, which assigned Robert L. Kidd as Party Chief.[53]

By 1935, SSC had become large enough to be split in two. The research and manufacturing components were spun off as the Engineering Laboratories, Inc., with William Green as President.[54] Westby then became President, a post he would continue in for the next 31 years. He was then elevated to Chairman and Chief Executive Officer, an assignment he did not relinquish until 1971 at age 73. In fact, stability of top management has been one of SSC's strong-points. For example, Westby's successor as President in 1966 was Elmer D. Wilson, a graduate of the nearby University of Tulsa, who started in the accounting end of SSC after World War II and continued as President and Chief Executive Officer until an early death at age 59 during 1981. On the operations side, the able, likeable Hugh Thralls started with SSC in 1936, became Chief Geophysicist in 1948, and moved up to Vice President and Director, Domestic Exploration in 1954 before retiring and subsequently starting his own firm, GeoProspectors, a year later. Another key employee was the German-born Dr. Albert J. Barthelmes, a Shell Oil man before he joined SSC in 1935 as a party chief. Barthelmes steadily worked his way up, becoming Vice President and Director of Exploration in 1947 and

---

52. As a founding member of the SEG, Westby was one of those honored during the 50th Anniversary of the SEG in 1980. During the intervening years, he had served the SEG as a three-term Vice President (1931–1932, 1932–1933, and 1936–1937) and as its Secretary-Treasurer (1935–1936).

53. In 1957, Kidd was named President of the Cities Service Company and became Chairman of the Board in 1961.

54. Engineering Labs. became Engineering Products later. Green stayed with this organization until 1958, when all manufacturing was placed under the Seiscor Division of SSC.

Executive Vice President (Exploration) during 1955 before retiring in 1972 to do consulting.

Technical management has been equally stable and skilful. Hard-nosed James E. Hawkins, an instructor at the Colorado School of Mines, joined SSC in 1941 and became Vice President and Director of Research, Manufacturing and Supply four years later. As of 1966, he became Executive Vice President, a post he held until retiring in 1975, when he, too, became an active consultant. After starting with SSC in the field back in 1942, S. W. ("Shell") Schoellhorn moved into the research area during 1953 and was named Director of Research and Development in 1961 and Vice President, Research, in 1963, a position he still holds as of 1981.[55]

Even with long tenures in a given office, SSC's managers stayed alert and took advantage of many opportunities.[56] Despite the difficulties of the war years, by later 1946 SSC was operating 16 domestic crews, 11 in Venezuela, and nine elsewhere. In mid-1947, it started the already-mentioned ill-fated negotiations with the Hastings Instruments Company to acquire a workable electronic positioning system for offshore surveys, and by the early 1950s was offering its own location service via the Lorac Service Corporation. In 1954, SSC became the first contractor to offer a continuous velocity logging service. Three years later, it acquired the Birdwell Logging Company and installed the company's founder, J. M. Bird, as Vice President of the Birdwell Division. By that time, SSC had run about half of the 1,482 velocity logs that existed within the United States. A quarter of a century later, Birdwell continues to offer velocity surveys, except now they are digitally recorded and use the Vibroseis® technique as the energy source. In the case of Vibroseis®, SSC was the first licensee of this unique method when it became available during 1961 from the Continental Oil Company.

After becoming a wholly-owned subsidiary of the much larger high technology firm, the Raytheon Company, based in Lexington, Massachusetts, in 1966, SSC pushed ahead in developing combination hardware—software suitable for processing seismic data digitally. By 1970, they claimed to be the first to introduce a complete floating-point processing system capable of preserving the dynamic range associated with binary-gain digital recording. SSC's series of

---

[55.] Schoellhorn served as the SEG's First Vice President during 1979—1980.

[56.] A letter of commendation attests to SSC's performance in the field: "In some 39 years of observing field parties, I have never seen a crew operate in such difficult terrain recording straight lines and still getting reasonable production. Supervisory personnel must be doing an excellent job. . . ."

® Vibroseis is a registered trademark of Continental Oil Company.

Phoenix seismic processing systems utilizing Raytheon computer hardware have kept pace with the state-of-the-art and sold well for both field and central processing applications. By 1981, SSC was operating two marine crews, a number of "transition zone" shallow-water parties, and many land crews. To do this took 3,100 employees, 1,200 of which were U.S. based. With continued emphasis on stable management and high-caliber, experienced personnel, SSC should be able to maintain its position as one of the world's top four full-service geophysical contractors.

Teledyne Geotech and Teledyne Exploration

The story of these two companies is, again, a reasonable duplicate of several of the other mainline geophysical contracting firms — initially a small tightly-held firm started in the 1930s that sold out to a major technical conglomerate (Teledyne) in the mid-1960s. In this case, the founder of Geotech, Dr. Roland F. Beers, after doing a seven-year stint with such firms as Western Electric, Raytheon, and the Submarine Signal Corporation, joined the pioneering group of geophysicists at the Geophysical Research Corporation in 1928 as one of their party chiefs. In June, 1930 he became GSI's first party chief and moved up to Vice President during 1934. By 1936, he felt confident enough to found his own firm, Geotechnical Corp., in Dallas, Texas. However, with meager financial resources, his firm had trouble in surviving the war years, and after the war he brought in Dr. William Heroy as a major partner and Executive Vice President.[57] In addition, the two formed the partnership of Beers and Heroy, Inc., to handle consulting work in geology, geophysics and engineering, primarily for the Federal Government. By early 1948, the partnership was asked by Headquarters, U.S. Air Force (AF-OAT-1), to investigate the various ways in which geophysics could be applied within the Federal Government, particularly in the area of nuclear-test detection, even though the first nuclear test by the Soviet Union was still a year away.

From 1949 on, Geotech and the partnership devoted more and more time to Air Force work and increasingly less to exploration, for the Air Force asked them to build a seismic detection network. Heroy took on the assignment of locating seismically quiet detection sites, while Beers began developing and mass-producing 15

[57.] Prior to World War II, Heroy had been Vice President and Chief Geologist, Sinclair Consolidated Oil Corporation, as well as President, Pilgrim Exploration Company. During the war, he was a key executive in the U.S. Office of Petroleum Administration for War (1942–1946).

vertical and 30 horizontal seismometers patterned after the models created by the master craftsman, Professor Hugo Benioff.[58] Even more difficult was the problem of designing a suitable, standardized recording galvanometer device capable of recording seismic wave trains in the one-hertz range. Nevertheless, the gear was built and on line by the time the United States started a nuclear testing series in Nevada and at Eniwetok Atoll in January, 1951. The following year, Heroy became Geotech's President, thereby permitting Beers to spend more and more time as Professor of Geophysics at his *alma mater,* Rennsalaer Polytechnic Institute in Troy, New York.

By 1959, the advent of cheap foreign oil had dried up most of Geotech's oil-exploration work.[59] Total revenues for the year were about $1.6 million, of which only $40,494 were commercial sales, the remainder primarily being seismic detection work for the Air Force. Net income had dropped to close to $22,000, and the difference between current assets (less inventories on hand) and current liabilities was but $180,000. However, with the advent of Vela Uniform, Geotech's backlog for work in this program grew from $300,000 in 1959 to $4.3 million by mid-1961. Beers, however, preferred to concentrate on his consulting work for the U.S. Atomic Energy Commission, so in 1962, Heroy took on the role of Geotech's Chairman, while his son, William, Junior became the firm's President. By June, 1963, current assets exceeded current liabilities by more than 3:1, the long-term debt of approximately $331,000 was exceeded by cash on hand, and the firm had $12.3 million of unfilled orders. This was exactly the kind of company that Teledyne, Inc. of Los Angeles, California, an aggressively run firm with a heavy debt structure, was looking for. Dr. Henry E. Singleton, Teledyne's President, had already snapped up United Geophysical Corporation's government-contracting subsidiary, United Electrodynamics, in 1964. Hence, in 1965 he proposed to the senior Heroy, then 81 years old, that they exchange common stock between the two firms.[60] This was accomplished by late 1965 and later, in 1968, Richard Arnett was hired back from Texas Instruments to be President of Teledyne Geotech. As a forward-looking manager, Arnett was anxious to

[58.] Two youthful hires for this overall effort, Carl F. Romney and Richard Arnett, eventually became leaders in their own right within this unique technical area.

[59.] Major financing to support this rapid expansion began to be supplied during 1960 by the American Research and Development Corporation. By 1962, this firm owned 25 percent of Geotech via the exercise of stock options valued at $300,000 in 1960.

[60.] At the time of the exchange, Heroy Senior's Teledyne stock was worth about $3 million, while Heroy Junior's stock was valued at about $500,000. By 1968, Heroy Senior had given about $1.5 million of common stock to both Southern Methodist and Syracuse University for new buildings to house their earth science departments.

diversify the technical base of Geotech. However, this took capital, an action that Singleton did not concur in, so Arnett left in 1974 to become Business Manager, Southern Methodist University where William Heroy, Jr. had been Vice President for Financial Affairs since 1966.

Shortly after the Geotech acquisition, Teledyne moved to build up Geotech's latent geophysical exploration capabilities. The following leading geophysical contractors were all acquired during 1966: *I*ndependent *E*xploration Company (IX) of Houston, Continental Geophysical Company of Fort Worth, and the National Geophysical Company of Dallas.[61,62,63] The following year all of Teledyne's geophysical exploration capabilities were consolidated into one subsidiary, Teledyne Exploration, based in Houston, Texas, under J. H. ("Jerry") Frasher. By 1968, Teledyne was advertising single-boat marine geophysical system for simultaneously recording seismic, gravity, and magnetic data, a valuable practice now quite common. When U.S. seismic activity hit a 30-year low in the early 1970s, Teledyne turned its attention to acquiring non-exclusive data which could be sold in the open market place. By 1973, its clients could choose from a library of 47,000 miles of seismic data in the Gulf of Mexico, 65,000 miles of seismic data from other parts of the world, and 52,000 miles of bottom-read gravity measurements from the Gulf of Mexico and Gulf coastal waterways. The same year, it also offered a "bright spot" data-processing technique.

In more recent years, emphasis has been placed on the firm's capability to offer a total geophysical survey capability and to acquire

---

[61.] IX, an "Amerada tree" company, was founded in 1932 by Dr. E. Eugene Rosaire, who had been GRC's first party chief and discovered the first salt dome in the United States (Moss Bluff dome in Liberty County, Texas) using refraction techniques. During 1934—1935, Rosaire served as the SEG's President. In 1937, he left IX to form a new company, Subterrex, the first American firm to contract for doing geochemical surveys. In 1962, IX became aggressive in its own right, acquiring Empire Exploration (founded by Howard E. Itten in 1952), followed by three Canadian firms in 1965 — Farney Exploration (founded in 1948 by Harold L. Farney), Nance Exploration (founded by A. G. Nance in 1948), and Exploration Consultants, Inc. (founded in 1950 by Norman Christie, who served as SEG President during 1963—1964). Christie's firm was funded in part by Thomas A. Manhart, who also started two other geophysical firms in Tulsa — Geophysical Consultants and Petroleum Consultants, plus his own firm of Manhart, Millison, and Beebe). Finally, in 1966, IX acquired Tidelands Exploration Company (founded in 1946 by U. E. Neese and John L. Bible as a gravity surveying company).

[62.] Continental was founded in 1952 by A. E. ("Sandy") McKay, a graduate in geology from the University of Oklahoma back in 1928.

[63.] National was founded in 1935 by William Salvatori, brother of Western's Henry Salvatori, and by John A. Gillin. Salvatori soon resigned to devote all his time to the stock market, and Gillin became President. By 1963, Gillin was Chairman of the Board, and Walter R. Mitchell assumed the presidency until the buy-out by Teledyne when both Gillin and Mitchell retired.

and sell additional non-exclusive geophysical data within the main producing trends of the Gulf Coast of the United States. The Teledyne organization has been fortunate in retaining the services of two outstanding data processors — Kevin M. Barry, Vice President and Manager of the Data Processing Division, Teledyne Exploration, and Dr. Robert Van Nostrand, Vice President of Teledyne Geotech, who directs the 50-man Vela Seismic Data and Analysis Center in Alexandria, Virginia, under contract to the Air Force Technical Applications Center.[64],[65] Corporate-wise, we surmise that Singleton considers his two geophysical companies to be "average subsidiaries" within a billion-dollar, highly-diversified growth conglomerate that has been very profitable for long-time stockholders via the capital-gains route.

The Western Geophysical Company of America[66]

Western Geophysical, although the last of this group to be discussed, is certainly not the least, for as of 1981, it is the largest geophysical contractor in the world. The driving force behind Western since its inception has been Henry Salvatori, born in Rome, Italy, in 1901 and brought to the United States at the age of four by his parents. During the mid-1920s, he earned an electrical engineering degree from that generator of early exploration geophysicists — the University of Pennsylvania — followed by a master's degree from Columbia University. These credentials were good enough to have Dr. Karcher hire him to be the Party Chief of GRC's Party Number Four in 1926. He, too, joined Karcher and his associates when they left in early 1930 to set up GSI; in fact, Salvatori was the chief of GSI's third field party. However, Henry saw no chance of moving higher from the ever traveling position of Party Chief, so he resigned in August, 1933 to set up Western Geophysical in Los Angeles (Sweet, 1969). His business partner was Dr. Harold Washburn, who had earned his doctorate in electrical engineering at CalTech in 1932 before moving on to be an instructor at MIT.

Washburn's chief task was instrument design and manufacture.

---

64. Barry served as the SEG's president during 1980–1981, while Van Nostrand, who served as the Editor of *Geophysics* during 1965–1967, was named an Honorary Member of the SEG in 1979.

65. Teledyne Geotech, now under the direction of an outsider, Gordon Breland, has five major operating divisions: Geophysical Instruments, Alexandria Laboratories, Meteorological Systems, Industrial Systems, and the Manufacturing Division. It remains the world's leader in manufacturing seismic monitoring equipment and, during 1980, was installing an earthquake-detection system in Greece, Bulgaria, and Romania under UNESCO sponsorship.

66. As of 1960, Western became a wholly-owned subsidiary of Litton Industries, Inc.

His first set of instruments were ready for field use in September, 1933, quickly followed by enough additional instruments to allow Western to have ten crews working in California, Colorado, Kansas, Louisiana, and West Texas by the end of the following year. In fact, by 1936, Western was the second largest seismic contracting firm. More of Western's crews might have been concentrated in California but for the fact that exploration managers of California-based oil companies thought it a waste of time trying to do reflection work in such a geologically complex area; thus, a typical contract ran for a week or less. From 1937 to about 1950, the design of Western's instruments was a joint undertaking of Western and of the Stanolind Oil and Gas Company (the forerunner of Pan American Petroleum Corporation) in Tulsa under the supervision of the talented and enthusiastic Dr. Daniel Silverman, who had started his geophysical career at Western.

Salvatori proved to be a devoted scientist and an excellent executive. During the era of analog data recording and processing, he was a prolific inventor, yet was always willing to chat with any employee as long as his door was open. During the early 1950s, he initiated a generous employee profit-sharing plan, and the relationship between Henry and his co-workers was one of mutual respect and loyalty.[67] Certainly one of Salvatori's wisest acts was hiring in 1948 the astute mathematician-physicist, Carl H. Savit, who later would prove to be one of the most articulate public spokesmen for Western and for the entire geophysical profession. Except for three years as a junior officer in the U.S. Air Force, Carl had been a student continuously at CalTech from 1938 to 1948. With most of the Ph.D. degree requirements out of the way, he decided it was time to get a job. CalTech's Placement Office advised that two local firms had openings for mathematicians — one at North American Aviation, Inc., and one at a small outfit called Western Geophysical. Carl had interviews with both, including Henry Salvatori himself. He knew that North American was too dependent on government contracts. On the other hand, he didn't really know what geophysics was all about, although he had heard Dr. Beno Gutenberg give a guest lecture about "seesmeec eechoes".

[67.] Once Salvatori became wealthy in the 1960s, he funded the Henry Salvatori Laboratory of Geophysics at Stanford University and made generous donations to the University of Southern California, Claremont College, and the University of Pennsylvania. Additional fiscal support has also been provided to many other organizations via the Henry and Grace Ford Salvatori Foundation. Grace Ford, by the way, was a beautiful movie starlet from Tulsa, Oklahoma, who consented to become his wife back in 1937. And, as has been true of many who came of immigrant stock, Henry's patriotism has been outstanding. In 1966, he was named "American of the Year" by the Americanism Educational League.

Henry liked the fact that Carl had been doing radio signal reflection research off the ionosphere and offered him a job at $400 a month, which proved to be notably less than an offer from North American. Salvatori then sweetened his offer by $50 a month, and Carl was soon sweeping the floor in a dusty crew office in tiny Taft, an oilfield community in the San Joaquin Valley of California.[68] After six months on the crew, Carl was called back to Los Angeles to become, along with a draftsman, the "Research Department" of Western Geophysical.

During the 1950s, Western began showing a knack for making better than usual profits. Using a mathematical approximation developed by Savit and several of his associates, it proved possible to devise equipment which could correct analog records for the shooting geometry ("normal moveout") at 24 traces a pass on the spinning magnetic drums, while their competition could only correct one trace at a time using a more exact formula which gave no better results. Western's Vice President, Booth Strange, also came up with a method for preparing composite record sections from individual paper records with slightly different time scales by variable stretching of the records when they were wet. Using just these two techniques, Western could underbid the competition and still make a sizeable profit.

Another of Strange's better ideas was to bid marine jobs at so much per line-mile rather than on a monthly basis. This freed the crew from client restrictions such as no overtime. Because ship costs were the dominant expense factor, whether in port or at sea, crews actually could work overtime at even double-time wages and still maximize company revenues with a nominal increase in operating cost. Savit also introduced another geophysical first when he used in-house research funds to experiment with multi-channel seismic recording over the Puerto Rico Trench to the east of the Bahamas during 1961. The results obtained were far superior to those obtained by the academicians using single hydrophones. However, when Savit published the results (Savit, 1962), the geophysicists at the Lamont Geological Observatory were incensed about a commercial firm "trespassing on their domain". None the less, as soon as "Doc" Ewing saw the actual seismic traces, he, too, was quick to shift over to multi-channel recording.

---

68. This was about the going rate for a Ph.D. following several years of relatively rapid wage increases after the end of World War II. Co-author Rice, with almost exactly the same near-Ph.D. training, started to work for Phillips Petroleum Co. in 1945 in the Gasoline Dept. for $250 per month, a level raised to $285 when he transferred to geophysical research a year later. By 1948, he was making $432 a month.

Western's entry into digital recording and processing in 1964 is another example of its attention to profits. By adapting available digital technology and computers to its needs, it greatly reduced the capital investments and time required to enter and begin profiting from the new market. It was still not cheap. In fact, when Salvatori gave Savit the digital go-ahead on a Friday afternoon at 3:30, Carl committed half a million dollars for equipment in two hours and another $2.5 million by the end of the following week. But GSI–TI must have spent many times that amount to develop comparable technology from scratch. From the point of view of the industry, Western initiated binary-gain recording very early, with the first gear going on a marine crew operating off Alaska during June, 1966. Also, Western's decision that all computer software be written in "floating point" gave Western a monopoly on the creation of true relative amplitude seismic cross-sections for nine months in the 1972–1973 period when finally the petroleum industry became excited about finding "bright spots".[69] Western's cost for generating these sections was only somewhat more than for automatic gain control sections, so its extra charge for preserving signal amplitude was mostly profit. As a result, this one technique made a multimillion dollar profit for Western during the first year it was marketed.

In 1960, Salvatori arranged for Western to become a subsidiary of a fast rising California technology firm, Litton Industries, through an exchange of stock. Litton happened to be the brain-child of Charles V. Litton, an outstanding electrical engineer (see vignette by Cecil H. Green). Fortunately, Western's corporate identity remained intact as did its management, although Salvatori did leave to pursue his private interests in the world of politics and philanthropy. Dean Walling, one of the original GRC–GSI group who had helped Salvatori start Western back in 1933, moved up to the role of President. In 1962, Litton Industries was able to acquire the Aero Service Corporation of Philadelphia, Pennsylvania, also through an exchange of stock. The largest and oldest of its kind, Aero Service had started back in 1919 as a magnetometer survey firm and remained solely in that business until the 1950s. Under the guidance of Virgil Kauffman, the firm then expanded into gravity and radioactive surveying on land, aerial photography, mapping, and surveying of all kinds for both exploration and engineering, digital base map construction from any type of data, marine gravity and magnetic

---

69. It was common then to write some scientific programs in floating point in order to avoid round-off error in IBM and other machines with a word length of 32 bits or less. But floating-point computations required more processing time than fixed-point and so were avoided by most programmers whenever round-off error was negligible or could be tolerated.

processing and interpretation, digital processing and interpretation of LANDSAT satellite imagery, and radar surveys and interpretation. Homer Jensen, the former project engineer for the development of the airborne magnetometer at the Naval Ordnance Laboratory during World War II, joined Aero Service shortly thereafter to become the firm's first Chief, Magnetometer Division. Today he is Vice President (Development) in charge of synthetic aperture radar applications. Other executives have come and gone, including James Whitaker, who left to become the Under Secretary of Interior during 1961–1964, Michael S. Reford (Chief Geophysicist, 1961–1966), and Roger H. Pemberton (Director of Exploration and Resources Development, 1961–1966).[70]

In 1965, Booth Strange moved up to become Western's President; one of his first actions was to move Western's Headquarters from Los Angeles to Houston, Texas. This promotion permitted Savit to slide up from Chief Mathematician to Vice President of Systems Research and Development, further enhancing his stature as a budding spokesman for the geophysical industry. As far back as the 1950s, Carl had become involved with the question of how many fish were being killed by Western's use of nitramine explosive charges while conducting seismic surveys off southern California. This issue quickly came into very sharp focus when inadvertent shooting in a school of mackerel caused several hundred thousand dead fish to wash ashore along the famous beaches of Santa Barbara, California. Because of this, Dr. Carl Hubbs and Andreas Rechnitzer at the Scripps Institution of Oceanography were asked to determine the feasibility of substituting a slow-burning, low-energy source, black powder, and Savit was drafted to present the technical results to the State of California. He proved so adept at getting the facts across without agitating the opposition that later he was asked to represent both the seismic industry and the sports fishermen at the public hearings held to review new state regulations on the use of explosives in offshore seismic surveying.[71]

During the 1960s and 1970s, it continued to be Western's policy to free Savit from many external tasks. As a result, he is the only person to have served as President, SEG (1971–1972), President, International Association of Geophysical Contractors (1973–1974), and Chairman of the Board of Directors, National Ocean Industries

---

[70] Pemberton wrote an excellent article on airborne electromagnetic techniques in 1961 (Pemberton, 1962) and Reford a similar article on the 1964 status of aeromagnetics (Reford and Sumner, 1964).

[71] Savit gives much credit to his devoted wife, Sandra ("Sandy"), a trained social worker, for expert advice on how to get issues resolved by playing a low-key, conciliatory role.

Association (1975–1976).[72] In addition, he was honored by being asked to serve a 14-month stint as the Special Assistant for Earth, Sea, and Air Sciences to Dr. Lee Dubridge the Science Advisor to the President, during 1970–71. Among Savit's more important accomplishments on the Washington, D.C. scene has been his involvement in the passage of Public Law 89-99 which freed geophysical exploration ships from cargo vessel regulations, particularly with respect to indicating that geophysical personnel are neither passengers nor crew. In more recent years, he has been deeply involved in what seems to be a never-ending effort to protect the confidentiality of exploration data acquired over the outer continental shelf at private cost but, as required by law, subsequently made available to the U.S. Government.

In 1974, Western advertised that it had become the largest geophysical contractor in the world. By then the Aero Service Corporation had been moved to Houston and later made into a major division of Western.[73] In fact, Western had advertised as early as 1961 that each year between 1954 and 1960, its marine crews conducted more offshore seismic exploration than all the other geophysical contractors combined. Since that time, it has continued to introduce new marine techniques, including new energy sources (Aquapulse® in 1967 and the most powerful Maxipulse® in 1971), the first high-pressure air-gun to operate reliably at over 300 atmospheres, and a 500-channel digital recording cable to increase signal resolution (Savit and Siems, 1977). During 1980, Western added eight marine crews, ten land crews, and doubled its data-processing capabilities, an action it could well afford to take for its world-wide sales, with a profit factor of about 30 percent, totaled $350 million, or better than 17 percent of the total world market for geophysical services.[74] It will not be easy for any of Western's major competitors to pass it. The Vice President of Research and Development under Savit is Dr. Kenneth L. Larner, one of the top seismic-data-processing experts in the world. In fact, he is a leading researcher in one of the most difficult problems remaining in seismic data processing and

---

® Registered trademarks of Western Geophysical Company.

[72.] Despite this heavy external work load, Savit has 29 patents to his credit and to the benefit of Western Geophysical.

[73.] The Division's new president was Dr. Emil Mateker, Jr., who left the teaching field at St. Louis University in 1969 to understudy Savit at Western. By 1971, he was Western's Vice President of Research and Development as Savit moved up to a new position, Senior Vice President (Technology).

[74.] As of 1979, Booth Strange became Chairman of the Board, while M. Howard Dingman, who had started out on one of Western's crews, had moved up to President. As of mid-1981, Western was operating some 50 geophysical survey vessels, or about two-thirds as many vessels as were in the entire U.S. non-profit oceanographic research fleet.

interpretation, that of proper spatial positioning ("imaging") in time and depth of reflections in areas with significant lateral variations in geology, as in the very active and economically attractive Overthrust Belt of the Rocky Mountains.

Litton is also in good shape relative to the design and manufacture of reliable field equipment. In 1970, the conglomerate acquired Digital Data Systems, a seismic digital field recording equipment manufacturer who had essentially repackaged Redcor instruments which Western used from the beginning of their digital work. By 1975, the name had been changed to *Litton Resources Systems* (LRS) and the product line had been expanded to include a complete line of seismic source, recording, processing, navigation, and cable equipment. Today, new LRS products include a high-resolution, 960-channel recording system (with a one-half millisecond sampling rate) and a data-processing system which is claimed to have twice the capacity of any other mini-system on the market.

To close out this section on the main-line contractors, it is interesting to speculate a bit on how the geophysical subsidiaries, Western and GSI, will continue to fare under the wings of their much larger holding companies, Litton Industries and Texas Instruments. As of the end of 1980, both of these giants had assets of over two $\times$ 10^9 U.S. dollars, and Texas Instruments' earnings of $9.22 per share were only about one-third larger than Litton's. However, TI is mainly in the semiconductor and electronics business which is undergoing a major slump in sales and profitability, while Litton has concentrated on defense and marine systems where sales are turning up, not down. Nevertheless, TI management controls their firm's fate, while Litton's alumni, Henry E. Singleton, controls 27 percent of Litton's stock via his conglomerate, Teledyne, Inc. About all we can hope for is that as the "Big Boys" compete with each other, the geophysical subsidiaries are aggressively allowed to do "their thing", which is finding more oil, natural gas, and mineral resources for the benefit of us all by the global use of highly motivated, well-equipped field crews.

Smaller Seismic Data Acquisition Firms

The *Geophysical Directory for 1981* lists 143 seismograph contractors (counting Seiscom Delta and United Geophysical Corporation as one). Of these firms, 15 were engaged in at least one major activity outside the field of seismic data acquisition. The remaining 128 companies have headquarters scattered all over the globe, namely: 95 in the United States (35 in Houston, nine in Denver, and 51 in 29 other cities), 20 in Canada, four in Australia, two each in England and France, and one each in the Federal Republic of Germany, Mexico, Norway, Poland, and Venezuela. In addition to the eight main-line contractors just reviewed in the preceding section, 15 firms are large enough to have permanent offices in more than one country (see Table 8.7).

By 1981, the *Geophysical Directory* listed 65 more seismic contractors, or nearly twice as many as it did in 1976. Of these, 46 firms were the same in 1981 as in 1976. However, in going back to 1966, one finds there were 119 seismograph contractors, of which only

TABLE 8.7
Countries in which Smaller Seismic Data-acquisition Firms have Permanent Offices (as of 1981)

Firm and headquarters location	Countries with permanent offices
Century Geophysical Corporation, Tulsa, Oklahoma	Canada, United States
Comap Geosurveys, Ltd., Manchester, England	England, Saudi Arabia, Spain, United States (2)
Fairfield Industries, Houston, Texas	England, United States
Geophysical Company of Norway, Hovik, Norway	China, Norway (2), Singapore, United States, Venezuela
Geosignal Exploration, Ltd., Calgary, Alberta	Canada, United States
Globe Universal Sciences, El Paso, Texas	Italy, United States
Grant Geophysical Corporation, Houston, Texas	Canada, United States (3)
Horizon Exploration, Ltd., Swanley, England	Argentina, Australia, Brazil, England
Hunting Surveys and Consultants, Ltd., New York City, New York	Australia, England, Greece, Kenya, Madagascar, New Zealand, Nigeria, Oman, Portugal, Saudi Arabia, United States, Zambia
Kenting Exploration Services, Ltd., Calgary, Alberta	Canada, United States
NORPAC Exploration Services, Denver, Colorado	Canada, England, United States (8)
Prakla Seismos GmbH, Hannover, Federal Republic of Germany	Australia, Austria, Brazil, Egypt, Federal Republic of Germany, France, Indonesia, Italy, Libya, Netherlands, Peru, Singapore, Spain, Turkey, United States
Quest Exploration, Ltd., Calgary, Alberta	Canada, United States
Servicios Tecnologicos Hidro Petrol, Caracas, Venezuela	Venezuela, United States
Sigma Explorations, Ltd., Calgary, Alberta	Canada, United States

20 survived to 1981 (this does not count the seven firms acquired by or merged into Petty, Seiscom, Teledyne, and United Geophysical). If the seven main-line firms that survived intact (CGG, GSI, Petty, Prakla-Seismos, SSC, Teledyne and Western) are eliminated, we are still left with the following 13 companies that are to be congratulated for their staying power: Beaver Geophysical Services (Calgary), Bradley Exploration, Inc. (Wichita Falls, Texas), Central Exploration Company (Oklahoma City), Century Geophysical Corporation (Tulsa), Dawson Geophysical Company (Midland, Texas), Globe Universal Sciences (El Paso, Texas), Heinrichs Geoexploration Company (Tucson, Arizona), Hunting Surveys and Consultants (New York), Newton Exploration Company (Billings, Montana), M. L. Randall Explorations, Inc. (Houston), Rogers Explorations, Inc. (Houston), Seismotech 64, Ltd. (Calgary), and States Geophysical Corporation (Houston). Of this group, Dawson Geophysical Company was selected for detailed treatment.

An Exceptional Small Firm — the Dawson Geophysical Company

Dawson is an excellent example of a company founded to do quality seismic data-acquisition work that has concentrated on this single objective through good times and bad for almost 30 years. Two of its advertisements back in 1970 are still pertinent today:

"At Dawson, we concentrate on getting better, not bigger!"

"At Dawson, we are small enough so that our more competent people are still doing geophysical work!"

L. Decker Dawson, Jr., founded the firm in 1952. A graduate in civil engineering in 1941 from Oklahoma State University, he was able to work for a year on one of Magnolia Petroleum Company's seismic crews before spending the rest of World War II as an engineer in both the Army and the Navy.[75] The war over, he again joined a seismic crew, this time with Republic Exploration Company, a new Tulsa firm founded by Frank E. Brown. Within six years, he was the Area Supervisor at Midland, Texas. Republic was now using 12 seismic

---

[75.] Because Dawson, as was true of many other embryonic "doodlebuggers", had never received any academic training in geophysics, he studied Dr. C. A. Heiland's text, *Geophysical Prospecting*, for two months prior to graduating. "Papa" Heiland was born and educated in Germany. He served as Professor of Geophysics at the Colorado School of Mines from 1926 to 1948 and in 1927 gave the first formal course in geophysical exploration. In 1934, he was the co-founder of Heiland Research Corporation and served as its President and of its successor, the Heiland Division of Minneapolis—Honeywell, until his death in 1956.

crews, and the demand for more crews in West Texas was still not being met. Dawson decided to strike out on his own and was joined by one of Republic's observers, Calvin J. Clements, an excellent electronics man. To start his firm, "Decker" floated a loan and soon owned a set of field equipment worth about $50,000. He was able to field a second crew in 1954, followed by two more in 1955.

The firm initially concentrated on working in West Texas because of its familiarity with the region and the opportunity to improve data quality by paying close attention to significant details and using 36-hole shot patterns and large geophone arrays.[76] Fortunately, some of the early prospects that the firm located proved to be major oilfields for Forest and Ralph Lowe, a pair of Midland-based independent operators. Five years after its founding in 1952, Dawson incorporated his firm, a year after all his crews had been converted to analog magnetic tape recording. Playback of these records was accomplished, however, at outside playback centers. Dawson was one of the first to go to 48-channel analog recording, thereby permitting the use of longer spreads without sacrificing the close geophone station spacing needed for good surface sampling. Shooting with the CDP technique was initiated in 1959, although it took time to learn how to make optimal use of the concept.

In spite of being known for good work, the company was impacted by the sharp cut-back in domestic oil exploration during the 1960s and for a while operated only one to four crews.[77] Nevertheless, to maintain its reputation of technical excellence, Dawson started converting to digital recording during 1968. Because of the great expense in doing so, the conversion took seven years to complete. Going digital also introduced an additional problem — that of delay in getting the records processed. Farming out this function to other companies did not work, for processing West Texas seismic data required very careful line-by-line custom processing which high-volume processing centers were not accustomed to handling. In 1976, Dawson learned that Howell W. Pardue, an expert data processor for the Ray Geophysical Company, did not wish to accept a transfer from Midland to Houston, so he was hired to set up Dawson's first data-processing center in Midland. SSC's Phoenix data-processing equipment was used to outfit the Midland Center, and a little later, another unit was purchased for a second center in Denver, Colorado.

---

76. Much of West Texas and eastern New Mexico is very poor shooting country because of the deep water table and, in many areas, the presence of a hard surface rock called "caliche".

77. As of early 1981, Dawson was operating six land crews.

A major part of Dawson's business comes from repeat customers.[78] As a result, although many new contracting firms have come on the scene recently, Dawson's growth has not slowed. Today, Dawson's operations extend to most Mid-Continent areas of the United States, and he is hopeful that the resource-oriented Reagan Administration will facilitate further geophysical exploration of federally-owned lands. To finance further expansion, the decision was made to go public in 1981 with an offering of 685,000 shares of common stock at $13 per share.[79] Of the nearly six million dollars the firm received from the April, 1981 stock sale, $2.7 million went for additional equipment. Of this $1.2 million was used to retire a short-term loan incurred in the purchase of a revolutionary 1,024-channel GEOCOR IV seismic recording system which, coupled with four vibrator energy sources and other vehicles, cost a total of $3.25 million. Early results obtained with this new equipment have been quite pleasing. Overall, the new Dawson stockholders should fare quite well, because revenues for nine months of 1981 of $11.4 million were almost 70 percent greater than for the corresponding period in 1980, and the net income of $2.15 million was 150 percent higher.

A Unique Firm — Geophysical Systems Corporation (Geosystems)

Geosystems is the only new geophysical company in at least 25 years to be founded on a revolutionary concept, namely, the improvement of seismic resolution by greatly increasing the number of surface sampling points. The new approach uses 1,024 channels of seismic information and is equally applicable to surveying in two or three dimensions. The development is the result of the perseverance of Samuel J. Allen, former United Geophysical Company Executive Vice President, and of Lincoln A. ("Linc") Martin, who had retired as United's President five years before.[80] Early in 1972, the

---

[78.] The respect with which Dawson is held in the geophysical community was reflected in his having been elected Chairman of the Board of the International Association of Geophysical Contractors for 1980–1981.

[79.] Before going public, Dawson owned over 55 percent of the firm, Clements 13 percent, and Pardue six percent. They each offered 10 to 15 percent of their holdings to the public and thereby realized, after commissions, approximately $1.6 million, $440,000, and $253,00 respectively. Compared to the rest of the geophysical industry, executive salaries were relatively modest, with Dawson earning $97,300 as President and Clements and Pardue each $58,600 as vice presidents.

[80.] Back in the summer of 1946, Allen, as a Lieutenant (j.g.) in the Naval Reserve, had the challenge and excitement of initiating the Navy's Microseismic Research Project on the island of Guam just prior to the start of the typhoon season.

two had discussions regarding the lack of fundamental progress in seismic data acquisition. Although there had been tremendous advances in seismic data processing and electronic technology, the industry was still using 24 channels of seismic data as standard field practice just as it had for the past 30 years. To be sure, there had been a drastic increase in the size of geophone and shothole arrays to reduce seismic noise, but this approach increased the spatial smearing of individual reflections and severely limited the improvement that could be effected in signal resolution and timing accuracy during data processing.[81]

Allen and Martin decided that these problems could be alleviated or overcome by developing a seismic data-acquisition system capable of recording simultaneously on 1,000 or more independent channels. Also, it was highly desirable to have the capability of processing the data in the field and in "real-time" as far as possible in order to allow constant monitoring of equipment operation, optimum field choice of geophysical parameters, and timely pursuit of promising exploration leads. Finally, to be competitive, they knew that the equipment must be cost-effective and no bulkier, heavier, or harder to operate than conventional seismic systems. In spite of the obvious difficulty of the objectives, Allen and Martin decided to give it a try, so they formed the Geophysical Systems Corporation in Pasadena, California, during March, 1972.

For the first two years, all of their effort was in research and development, without income of any type. It soon became evident that extending conventional recording technology to 1,000 channels could require recording equipment many times larger than standard equipment. In addition, the cables between the geophone subarrays and the recording truck would have to be capable of transmitting 1,000 data channels, even if the analog rate at the geophones were converted to 16-bit digital format for multiplexing down the cable to the recording unit. As a result, Allen and Martin almost gave up until they began studying in detail the degradation of analog signals along a cable. Others had already reported that because of this degradation, signals reaching the recording truck did not represent true signal to an accuracy of more than four to eight bits. Hence, was it possible that the technology could be greatly simplified by using a single-bit (just a plus or minus sign) measurement at the geophone, provided the signal accuracy were boosted in other ways?

---

81. Lack of resolution affects the accuracy of time corrections for near-surface variations and of CDP velocity determinations as well as the delineation of reflections from individual subsurface beds of interest to the explorationist.

For example, could a coded signal source such as Vibroseis® coupled with signal correlation and stacking techniques translate a single-bit measurement into 16-bit resolution of reflection amplitudes in a seismic cross-section? Mathematical studies and computer simulations indicated this to be true, and field tests using vibratory sources soon gave results under most conditions as good as those obtained with conventional systems. Moreover, surprisingly better results were obtained under certain conditions, such as when noise bursts occurred.

From then on, this pioneering effort accelerated. In early 1974, Geosystems fielded its first crew using the first sign-bit-only system, GEOCOR II, with a 256-channel capacity, and a second crew was started up later that year. Once several patent applications had been filed, it was time to invite in personnel from major petroleum company research laboratories one at a time to witness the system's operation.[82] Generally the visitors reacted in shock, finding the results hard to believe even after running comparable tests in their own laboratories. Extensive field tests soon showed that 256 channels were not sufficient to provide the resolution of signal desired in many instances. Accordingly, between early 1976 and early 1978, emphasis was placed on designing and building the 1,024-channel GEOCOR IV system. The first of these advanced systems went into the field in early 1978, the second in 1979, and four more during 1980. Approximately 16 additional systems are projected for 1981. The technique is more than living up to its promise of providing much greater resolving power when searching for indications of subtle stratigraphic traps and for providing a better means for interpreting regions of complex geology. As a result, good agreement currently exists that Allen's and Martin's dream of nine years ago is well on its way to inaugurating a new era in seismic data acquisition.

In 1978, Geosystems received $3,251,207 for seismic field services. However, as the demand for additional field crews and for marketing GEOCOR IV systems built up, there was a severe shortage of funds with which to finance this expansion. As a consequence, in 1980 Geosystems sold 1,479,840 common shares of stock to two brothers, Stuart and Sherman Hunt of Dallas, Texas (not the H. L. Hunt clan), at $1.96 per share for 45 percent of the firm. By 1981, the outlook for future business was still so good that it appeared that capital requirements would approximate $14 million more than could

---

® is a registered trademark of the Continental Oil Company.

82. As of May, 1981 Geosystems had 17 U.S. patents covering the principal elements of this proprietary system. Associated with Allen and Martin in the beginning was J. Robert Fort, who has since left the firm. This group also received major help from part-time consultants J. A. Westphal and the renowned Professor C. Hewitt Dix of CalTech.

be generated internally. To raise the additional financing, Geosystems made a public offering of one million shares during June, 1981 at $20.00 a share, of which $13,950,000 went to the firm and $4,650,000 to selling shareholders. The firm's proceeds from this sale went for building eight GEOCOR IV systems for sale during 1981 and towards equipping an additional six seismic field crews for its own account.[83] Today a contract for one of these field crews, which utilizes approximately 30 employees and 15 vehicles (including three or four ground vibrators), costs about $250,000 per month. The crews can presently perform on-site processing overnight. However, more sophisticated processing and analyses are obtainable at the firm's central processing centers in Pasadena and Houston which utilize high-speed computers and proprietary software. For the time being, these centers are almost exclusively dedicated to processing GEOCOR data, either from Geosystem's own crews or from companies with GEOCOR systems. Sales for the first quarter of 1981 were about $7 million, which could indicate an annual rate approaching $30 million. The firm's five largest customers accounted for 84 percent of its revenues. This indicates considerable repeat business and apparent satisfaction with the results being obtained. In terms of both industry acceptance and cash flow, Geosystems appears to have emerged from its difficult formative period with flying colors.[84]

Some New Firms on the Scene

Among the large number of new seismic contractors now on the scene, several are growing rapidly and appear to have bright futures. Five such firms are briefly described here to provide an overview of how such firms operate during a boom period.

Fairfield Industries, Inc.

In the late 1960s a firm called Aquatronics began developing the

[83.] The basic GEOCOR-IV field system currently sells for about $1.75 million, including formal training of client representatives.

[84.] Allen and Martin own about 10.8 and 10.4 percent of the outstanding common stock, which now totals 4,015,856 shares. The Hunt Brothers also retained 33.8 percent of the firm. As of 12 November 1981, the bid price for the stock in the over-the-counter market was $34.00. Prior to the public stock offering, only one officer of the firm received over $50,000 for an annual salary during 1980, namely, Allen who was paid $78,746 as President and Director. Once the operation became successful, Martin returned to retirement and growing oranges in the San Joaquin Valley of California, although, as a new challenge, he did decide to learn how to fly helicopters.

Telseis® radio telemetry recording system wherein the seismic detectors were linked to a central recording unit by radio, rather than by wire.[85] By the mid-1970s, this technique had become the most efficient method of acquiring seismic data in swamps, shallow coastal embayments, jungles, extremely mountainous regions and other areas where it was extremely difficult to move equipment. In 1975 to 1976, however, Aquatronics was bought out in two different stages by a privately-owned New York firm, Fairfield Maxwell. The various activities were then consolidated under the name Fairfield Industries, Inc., and L. Tom Nicol hired away from Petty-Ray Geophysical Operations to be President. By 1981, Fairfield was operating three standard marine crews and six Telseis crews while simultaneously phasing out the marine engineering survey business because that specialty was being fought over by a number of even smaller firms. Fairfield is also actively marketing Telseis systems. For example, CGG has six systems under lease, while Western Geophysical has bought three systems outright.

Geophysical Company of Norway (GECO)

Founded in 1972 as GECO-State A/S, GECO is now a subsidiary of Det Norske Veritas (a leading ship inspection and rating bureau) and Kongsberg Vapenfabrikk. However, Anders Farestveit remains as Managing Director and O. Elmstad as Technical and Assistant Managing Director. As of 1981, the firm offers a complete spectrum of land and marine seismic data acquisition, processing, and interpretation services. With eight seismic vessels now in operation, the company claims to be the third largest marine seismic contractor. During the last two years, the firm's operations have expanded far beyond Europe, and offices are now maintained in the United States, Singapore, China, and Venezuela. With strong financial and technical backing from its holding companies, the firm is in a good position to continue to expand, particularly in the marine area.

Grant Geophysical Corporation

The Grant Geophysical Corporation is another of the third-generation contracting firms. Founded in 1977 by Henry L. Grant, formerly President and Chief Executive Office of Petty-Ray Geophysical, who had the financial backing of the family-owned Hillman Company

---

[85.] Although based in Houston, Aquatronics was founded by Troy V. Post, a Dallas insurance executive and financier, and Curtis A. Baker, who became the firm's President.
® Registered trademark of Fairfield Industries, Inc.

of Pittsburgh, Pennsylvania, the new firm was able to begin acquiring older firms during 1979. In that year, Grant acquired the Petroleum Geophysical Company of Denver headed by Orley E. Prather, President, and Phil M. Hurd, Vice President, as well as the Dresser Olympic Division of Dresser Industries, a full-service land and marine contractor.[86] In turn, during 1980 Grant Geophysical was purchased by Petrolane, Inc. for stock worth about $27.5 million.[87] However, expansion has continued with the recent acquisition of another geophysical firm, Loxco, Inc. (formerly Louisiana Oil Exploration, Inc.), which dates back to the early 1950s and was headed for many years by B. H. Treybig, Jr. Soon after, Grant also picked up GUS Manufacturing Company of El Paso, Texas, a firm with close ties to the Shell Development Company and a world leader in the manufacture of seismic telemetry systems. As of 1981, Grant operates 32 land crews in Canada and the United States and is making preparations to operate offshore and outside North America. No less noteworthy, however, was the announcement made in mid-1981 that the esteemed Turhan Taner would become Grant's lead consultant for enhanced data processing.

NF Industries, Inc.

NF Industries is an example of how quickly an earth-resources corporation can grow during boom times. The firm was founded in December, 1979 by two already successful executives, L. Tom Nicol, the former President and Chairman of the Board for Fairfield Industries, and by J. E. (Bud) Franklin, the former President and Chairman of the Board for Geosource, Inc. The initial contracts were for conducting seismic exploration work in Argentina. By late 1981, however, they had also built up a domestic capability by acquiring a variety of companies (seismic exploration, helicopters, fixed-wing aircraft, engineering support, and trucking) involved in the petroleum service industry. As a result, in just a year and half, the firm's employees grew in number from four to 800 and annual sales approached the $60 million level. If all goes well during 1982, ten additional seismic crews will have been added, there will be nearly 1,500 employees on the payroll, and revenues should come to nearly $100 million.

[86.] Upon being bought out, Prather resigned to start another company, while Hurd remained as a Vice President.

[87.] As of 1980, Petrolane had revenues of $1.4 billion and assets of about $707 million. About three-fourths of its income was being derived from oilfield services and the sale of liquefied petroleum gas.

NORPAC Exploration Services, Inc.

Although a new name within the industry, NORPAC is actually a strong, expanding amalgamation of older geophysical contracting firms. In this instance, the oldest firm is the North American Exploration Company of Denver founded by G. L. Scott in 1967 and dedicated to seismic data storage and resale, speculation surveys, and consulting. In 1974, Scott helped Harold F. Murphree finance the launching of Pacific West Exploration Company of Denver for seismic data-acquisition purposes. At about the same time, essentially the same group founded Consolidated Geotechnics of Denver, with Bill A. Rosser as President and Murphree as Vice President. Another firm, Mile-Hi Exploration Company, was formed in 1977 for the purpose of offering data acquisition in regions so difficult of access that the seismic equipment had to be transported by air. By this time, they needed their own seismic data-processing center in Denver, so Jimmie A. Jarrell was hired away from Western Geophysical to set up North American Processing, Inc.[88] Finally, a sixth firm, NORPAC International, was set up in 1980 to provide global seismic services out of Denver and London. In 1981, it became appropriate to consolidate all these companies, less Consolidated Geotechnics, into one firm, NORPAC Exploration Services, Inc. and make a stock offering to the public at $23 a share.[89] By then, 1980 revenues came to over $63 million, or a nine-fold increase over 1976, while the net income was $4.4 million ($1.25 a share), or 18 times that of 1976. Business has continued to be good, and by the fall of 1981, 37 seismic crews (14 shothole, 12 helicopter transportable, and 11 vibrator) were in the field, triple the 1979 number.

Firms Specializing in Seismic Data Processing

The *Geophysical Directory* for 1981 lists 119 service companies engaged in analog or digital processing of geophysical data, including the eight full-service contractors. Of these firms, 35 were common to the 72 companies listed in the *Geophysical Directory* of five years prior. But if one goes back to the early days of the digital processing era in 1966, he finds that of the 49 early data processors, only the

---

[88.] In 1978, this was Denver's seventh seismic data-processing center. In early 1980, Jarrell left to set up his own center, Seismic Services, Inc.; by then, it was the 21st processing center in the city. As of mid-1981, there were 43 processing centers in this booming metropolis.

[89.] After the stock sale, Scott and his immediate family held 12.6 percent of the 4,446,428 shares of common stock outstanding, while Murphree held 6.9 percent.

majors plus the GTS Corporation survive.[90] This finding does not imply that most of the data processors have failed financially, but it does reflect the extremely dynamic character of the industry in which mergers, sellouts, and shifts in type of service offered are commonplace.

In 1981, a breakdown of the various types of data processing companies indicates that 21 are primarily seismic data-acquisition contractors or full-service çompanies, 19 prefer to be known as "consultants", five specialize in gravity, magnetic or electrical prospecting, five are best known as data brokerage or exchange services, three are data-processing equipment or instrument manu-facturers, one is a software developer, another is a positioning survey firm, and one is basically a helicopter service company. This leaves 63 companies primarily dedicated to data processing. Of these, 24 are headquartered in Houston, 12 in the Denver area, 13 in other United States cities, six in Canada, two each in England and Australia, and one each in Czechoslovakia, Germany, India, and Venezuela. Eight firms large enough to have processing centers outside of their home country are: Dimirsu and Associates (two in the United States and one in Turkey), Denver Processing Center (two in the United States and one in Argentina), Digital Resources Company (United States, Syria, Algeria, and China); Digitech Ltd. (Canada, United States), Interra Exploration (United States, India, and Venezuela), Oil Data Processing Company (United States and Canada), Riley's Datashare International, Ltd. (Canada and the United States), and Sefel Geophysical (Canada, England, and the United States).

Typically, a pure data-processing firm is owned by one or more individuals who operate it using a few professional geophysicists and a much larger number of assistants. The one or more computer hardware—software systems used within such a center are frequently leased.[91] While the firm's software may be modified or augmented to adjust it to the client's particular job needs, sales pitches are most likely to be based on quality of results, cost and turnaround time than on uniqueness of the software—hardware system. Two quite unique companies will be described here, one Canadian and one U.S.-based. Both prefer to be known as consulting organizations, but their major contributions to the industry have been in data processing.

[90.] The GTS Corporation (also known as the Geoscience Technology Services Corpora-tion) has played a unique role for the industry by specializing in the digitization and proces-sing of old paper seismic records and well logs. Dr. Sam G. Taylor, as Chairman of the Board, directs their New Orleans office and Mr. Phillip L. Work, as President, the Houston office.

[91.] In addition to reducing capital funds needs, leasing facilitates updating of hardware and software.

Teknica Resource Development, Ltd.

This firm, which pioneered in deriving and marketing stratigraphic information from seismic records, is the result of the genius and dedication of Canadian-born Roy O. Lindseth. Roy began his geophysical career in 1944 working on United Geophysical crews in the U.S., Canada, and South America. In 1951 he joined Richmond Exploration Co. (now Chevron Oil Co.) in Venezuela and was put in charge of the Eastern Division. This job gave him more time at home with his young Colombian wife, Lucy, whom he met while she was working as an airline stewardess. After considerable soul searching, in 1954 he decided to strike out on his own, both in seismic consulting and in renting heavy equipment for oilfield servicing and for clearing seismic lines and moving gear through the bush and jungle. The latter was in partnership with Lucy's brother. Both activities went well, and Lindseth also joined the newly-formed geological and engineering firm of Martin, Sykes, and Associates as the only geophysicist. This company became a vehicle for Sun Oil Co.'s entry into Venezuela and resulted in Sun's acquisition of a very prolific lease in Lake Maracaibo.

Then a revolution toppled the pro-development government, and all new exploration activities by foreign firms were terminated. The number of seismic crews in Venezuela dropped from 50 to three, and drilling equipment had to be stacked, resulting in forced liquidation of construction and service companies. Lindseth quickly saw his million-dollar business become worthless. Only by selling a car he had bought for Lucy as a Christmas present and prevailing upon the Creole Oil Company (Esso) manager to let him haul abandoned equipment out of the jungle and store it in the Creole yards near Lake Maracaibo was Roy able to pay off his debts and return to Canada.

While waiting for a visa that would permit him to work in the United States, Lindseth went to computer school and also took employment with George F. Coote's firm, Accurate Exploration, Ltd. One of Coote's clients was the H. L. Hunt Company, then pioneering in oil exploration in Canada's far north. Because the complex geology of the Canadian Archipelago is rather like that of Venezuela and far different from the "layer-cake" geology typical of the Albertan plains, Roy's services came into demand. Hence, he canceled plans to move south across the border, and in 1964, he and Coote formed the first Albertan digital computing firm, Engineering Data Processors (EDP). By purchasing a digitizer and a plotter, plus using some Lindseth software on a rented computer at the IBM

Data Center in Calgary, they were able to generate synthetic seismograms, contour maps, and other data needed for making seismic interpretations. Their original product, the "Sonigram", is still widely used. Once the digital processing of seismic data began accelerating, a number of Calgary geophysical firms bought out EDP in 1967 and organized Computer Data Processors (CDP).

This new firm acquired a Control Data Corporation 3300 computer, an automated analog-to-digital converter for seismic magnetic tapes, and a seismic plotter. While waiting on special hardware for speeding up time-domain convolution (filtering) operations, Roy developed fast routines for frequency-domain filtering using the new "Fast Fourier Transform" method. For this effort, he was a recipient of a "Canadian Industry Development Award" worth $13,000.

At the start of the 1970s, CDP had to struggle to survive because the Federal Government in Ottawa and the provincial government of Alberta began squabbling over Ottawa's take of proceeds from the sale of Albertan oil. The upshot was a substantial reduction in oil company profits at the same time that the oil embargo by OPEC countries caused sharp increases in exploration activity elsewhere. As a result, there was a wholesale exodus of petroleum personnel and equipment from Canada.[92] Because of this drop in business and his own growing interest in "inverting" raw seismic traces into synthetic continuous velocity logs for stratigraphic mapping, Lindseth left CDP at the end of 1972 to form Teknica Resource Development, Ltd.[93]

To get his new firm under way, Roy rented computer time from CDP and did just enough consulting to meet running expenses. He had already given a paper at the Annual Meeting of the SEG in November, 1972 on a crude version of his Seislog® inversion process. Since then, either he or his right-hand man, A. Verne Street, have given additional papers at every annual SEG meeting (with the exception of 1975) which delineate further a new application or an improved aspect of this technique. In the early years, this took courage and self-confidence, for many were skeptical that an inversion process could ever be relied upon for stratigraphic interpretation. In fact, it must be still considered remarkable that under favorable conditions, the technique can reliably reveal subtle features such as porosity

---

[92.] As noted in Chapter 7, another large exodus began in 1980 when the federal government in Ottawa initiated a new aggressive policy to increase domestic ownership in the Canadian petroleum industry.

[93.] Going from field seismic traces to synthetic velocity logs is called an "inversion" process because it strives to remove or "divide-out" the filtering effects of the shot wavelet and recording equipment.

® Seislog is a registered trademark of Teknica Resource Development.

development within a bed of sandstone or limestone but 8 to 15 m thick at a considerable depth below the surface (Lindseth, 1979).[94] Teknica continues to lead the contracting industry in this type of processing and interpretation not only because it is their specialty but also because they have steadily pressed for improvements and new applications, particularly in the realm of developing and producing petroleum reservoirs. Today, in-house Seislog processing is done by some oil companies who have purchased the entire Seislog software package or hired Lindseth to come in and give special seminars on the technique. Roy has been a noted lecturer on data processing for many years. From the late 1960s through 1980, he gave an SEG-sponsored short course on seismic data processing which became the most popular ever offered by the society and earned him the society's Kauffman Medal in 1970.[95]

GeoQuest International, Inc.

This Houston firm was started in the early 1970s to offer special services to the industry not commercially available elsewhere, particularly in the area of advanced seismic computer modeling techniques and the associated consultation needed to apply such approaches to real-life situations. Prior to 1977, its impressive stable of experts included Dr. Norman S. Neidell and the late Dr. Milton B. Dobrin. In 1977, John B. Butler, Jr., was elected Chairman of the Board, while Robert N. Hodgson became President. Geoquest's extensive modeling system, termed the *Advanced Interpretive Modeling System* (AIMS), has led the industry for almost a decade. Its primary use has been to try to duplicate a portion of an actual seismic cross-section by duplicating in the model the exact field shooting geometry and as much information about subsurface geology and geophysical parameters as can be obtained from the field data and any available well information. If, by adjusting these parameters,

---

[94.] The thickness of such a strata is only a fraction of the wavelength of typical seismic wavelets. In fact, any seismic record, good or bad, lacks the high-frequency components needed to determine the general trend of velocity versus depth. Because these missing components have been filtered out by a combination of the recording system, the source wavelet, and the earth itself, none of which can be defined with much certainty, it requires very clever "bootstrapping" to restore the missing components accurately enough to yield useful velocity-versus-depth logs.

[95.] This course has been given over 30 times in many different countries; the scope of attendance is indicated by the fact that over 8,000 copies of the course notes have been handed out. Lindseth also has many additional honors to his credit, including President, Canadian Geoscience Council (1973), President of the SEG (1976–1977), President, Association of Professional Engineers, Geologists, and Geophysicists of Alberta (1978–1979), and recipient of the J. Tuzo Wilson Medal of the Canadian Geophysical Union (1979).

a fairly good match can be obtained between the seismic "signature" obtained from the model and the field results, it is highly probable that the model then represents the correct geological or geophysical interpretation of the field data.[96] Originally, seismic models could not be made very realistic in terms of the more complex earth-filtering effects and velocity variations. But with the power of modern computers and mathematical techniques, state-of-the-art modeling packages such as AIMS can be made to include most of the factors that significantly influence seismic signatures (Rice *et al.*, 1981).[97]

GeoQuest has also been one of the leaders in the field of "wave-let processing", particularly in correcting recorded reflection wave-lets for recording equipment distortions, and in shortening and shaping long wavelets from marine energy sources. Special attention has been paid to the effects of these factors on important seismic anomalies such as "bright spots". In addition, the firm has moved into the apparently insatiable and profitable worldwide market for three- to five-day training courses for geophysicists, geologists, and petroleum and design engineers. Their 1981 course schedule, for example, includes 31 seminars in Houston, Denver, Calgary, London, and Singapore on the subjects shown in Table 8.8. Tuition for these "open" seminars ranges from $575 to $1,200 per person. In addition, GeoQuest conducts private seminars for individual companies with course location, duration, theme, and material tailored to specific needs.

TABLE 8.8
Geophysical Courses Offered by GeoQuest During 1981

Topic	Topic
Basic Seismic Acquisition	Modern Seismic Data Processing
Basic Geophysics for Technical Assistants	Wave Equation Migration
Geophysics for Geologists	Interpretive View of Wave Migration
Geophysics for Engineers	Applied "AIMS" Modeling
Well Log Analysis for Geophysicists	Applied Seismic Stratigraphic Interpretation
Electrical Prospecting for Oil	Applied Well Log Interpretation

96. This is called the "forward" approach to resolving and interpreting complex seismic reflection signatures because one starts with the earth model and computes the seismic response. "Inversion" starts with the actual seismic response and strives to compute the earth model that produced the response.

97. Geoquest currently offers the use of AIMS and other advanced seismic software by running computer programs for a client, by letting the client run his own data using Geoquest's software, or by selling the software outright.

GeoQuest continues to be an evolving organization. In late 1980, it entered the seismic data-acquisition business itself by forming a subsidiary, GeoQuest Exploration, Inc., and fielding several seismic crews. In 1981, a second subsidiary, Geoquest Data Services, was also organized to specialize in speculative and group seismic surveys and in seismic data brokering. As a result, Geoquest is rapidly becoming a full-service contractor with great technical depth in seismic modeling, data processing, and interpretation.

Other Service Companies

Besides the full-service contractors and those service companies specializing in seismic data acquisition or data processing, there are several hundred companies specializing in other types of data acquisition or auxiliary services (see Table 8.9).

The seismic data exchange and storage business has particularly mushroomed during the past decade for a number of reasons. First, increasing data-acquisition costs due to both inflation and greater multiplicity of coverage have encouraged the recovery of some of these investments through the selling of proprietary data that has out-lived its competitive advantage for a given company. These recovered funds can then be applied to additional data acquisition. Thus, total exploration costs for the entire industry are spread among many different exploration entities, large and small, many of whom would have to forgo the benefits of geophysical data if

TABLE 8.9

Number of Specialized Service Companies with Principal Listings in the
Geophysical Directory

	Year		
Type of company	1981	1976	1966
Aircraft services	49	41	19
Data exchange	53	36	5
Data storage	10	6	—
Shot-hole drilling contractors	84	49	138
Surveying contractors	51	34	25
Velocity data contractors	8	5	9
Core-drilling contractors	19	15	75
Core-hole logging contractors	12	17	30
Electrical & EM contractors	14	22	13
Gravity meter contractors	17	19	33
Magnetometer contractors	21	21	44
Radioactive surveys	3	7	28
Soil analysis contractors	11	15	14

they were forced to pay full data-acquisition costs. The result is higher drilling success ratios because more explorationists are given the opportunity to apply scientific methods as well as their own individual ideas and skills. A second reason for the emergence of a large data-exchange and storage business is that exploration companies cannot justify the cost of the building space, with carefully controlled temperature and humidity, and the special staff required to store, maintain, retrieve, and market an ever-increasing volume of data. On the other hand, such a business conducted efficiently on a large scale can return good profits to a specialty company. One major problem has emerged, however. If the data exchange company fulfills requests by shipping original tapes to a client for copying instead of doing the transcription in-house, another client may have to wait weeks for the same tape(s) and may miss a critical exploration deadline.[98] A number of data-exchange companies now offer very complete services, including data transcription, processing, field crew contracting, and consulting.

Equipment Manufacturing and Supply Companies

Mention has already been made that the full-service contractors market large amounts of equipment for seismic data acquisition and processing. But there are also many other companies, old and new, who share this market to a varying degree. In addition, a large number of firms supply a wide variety of supporting products that range from electronic parts to drilling bits and explosives (see Table 8.10).

Most of these companies are primarily concerned with the seismic exploration industry. One exception is the few oceanographic instrument and service companies that specialize in precision bathymetry, bottom sampling, underwater scanning for possible archeological artifacts, and a variety of ocean engineering applications. One company, InterOcean Systems, actually offers a "sniffer system" for measuring and locating minute concentrations of hydrocarbons emanating from small oil or gas seeps in the ocean floor. In fact, it claims an 85 percent success rate in being able to distinguish between oil and gas provinces. The second major exception is large to giant corporations who offer services to the geophysical industry as only one of many product lines. In the data-processing field, these include such manufacturers as General Electric, *I*nternational *B*usiness *M*achines (IBM), Sperry Univac, *Ca*lifornia *Co*mputer *P*roducts

---

[98.] One simple way of avoiding such delays in the case of "group seismic shoots" is to have a rule which says that the data-exchange center will tell the company who is waiting for overdue tape the name of the company causing the delay.

TABLE 8.10

Number of Equipment and Supply Companies with Principal Listings in the Geophysical Directory

Type of company	Year		
	1981	1976	1966
Automotive & auxiliary equipment	22	14	50
Boats	31	55	61
Data-processing equipment	44	48	–
Drilling bits	30	32	66
Drilling-rig builders	34	37	23
Electronic supplies (including Cables & location instruments)	96	116	102
Explosives	37	36(1977)	–
Field instruments	106	100	155
Oceanographic instruments	57	65	–
Marsh buggies & tracked vehicles	23	33	37
Non-dynamite energy source equipment	8	10	–
Shot-hole casing	11	10	14
Miscellaneous supplies	59	72	31

(CALCOMP), Digital Equipment Corporation and Floating Point Systems. Such companies are being pressed hard by the competition from producers within the industry of increasingly powerful mini-computers.[99] Dominant suppliers of explosives are E. I. DuPont, Hercules, Atlas Powder Company, and the Austin Powder Company, while Magnavox and Motorola are the major suppliers of positioning equipment.[100] Also, as might be expected, the Hughes Tool Company supplies many of the drill bits needed by the geophysical industry, while Eastman Kodak and E. I. DuPont supply much of the photographic film and paper, more and more of which is being utilized in color rather than black-and-white.

Within the geophysical industry, 17 firms produce most of the equipment needed by seismic field crews (see Table 8.11). As noted in the tabulation, eight of these firms have already been described in some detail earlier in this chapter. Highlights concerning the remaining nine are given in Table 8.12.

99. As of today, the major petroleum companies continue to use very large general-purpose systems as well as mini-computers in their central seismic data-processing centers. This practice continues even though the choice of equipment is no longer partially dictated by geophysics divisions having to share computer facilities with accounting or other corporate divisions. In the earlier digital processing days, this sharing was a real deterrent to program development and to minimizing turnaround times.

100. In the early days, the largest number of cocktail-room hosts at geophysical conventions were dynamite companies, and such cocktail suites are still known as "powder rooms". However, with the advent of a variety of non-dynamite energy sources, both large and small, during the past two decades, the "powder business" is not what it used to be.

TABLE 8.11
Major Geophysical Equipment Companies by Product Lines

Company name	Vibrators	Cable and/or detectors	Data-processing mini-computers	Field recording instruments	Auxiliary equipment
			Type of product		
ARDCO Industries					X
George E. Failing	X				X
Fairfield Industries*				X	X
Geo Space		X	X		X
Geosource*		X	X	X	X
GUS Manufacturing				X	X
Input/Output					X
Litton Resources Systems*	X	X	X	X	X
Mark Products		X			
Mertz	X				X
Seiscom Delta*			X		
Sercel of CGG*		X	X	X	X
SIE				X	X
SSC of Raytheon*			X		
Tesco Cable		X			
Texas Instruments*	X	X	X	X	X
Vector Cable of Schlumberger		X			

*Firm described earlier in this chapter.

The Consulting Business

The 1981 *Geophysical Directory* lists 392 firms primarily engaged in consulting, as compared to 272 in 1976 and 151 in 1966. And there are a great many more, primarily individuals, who have not asked to be listed. Of the 392, about 60 percent are operated by one professional person or one family. The consulting company plays a very important role in the geophysical industry for two reasons. First, the consultants are the primary data interpreters outside of the petroleum or mining companies. Hence they are the ones depended on for turning seismic or other data into maps for the delineation of drillable prospects. Secondly, they are the prime source of inter-pretation assistance needed to fulfill an exploration company's fluctuating needs. Consulting also offers a mutually profitable way of retaining the help and skills of older geophysicists who are retired from large exploration companies.

During bad times, the number of younger consultants tends to decrease as many seek more stable employment with large com-

TABLE 8.12
Cogent Points Concerning Certain Major Geophysical Equipment Manufacturers

Company name	Founder(s) and date	Home office	Remarks
ARDCO Industries	David L. Crowell, 1954	Houston, Texas	Initially a major international supplier of specialized off-the-road vehicles, it also manufactures high-pressure air-guns and portable drills
George E. Failing Company	George E. Failing, 1931	Enid, Oklahoma	Oldest and best-known manufacturer of shot-hole drilling rigs, as well as a major manufacturer of seismic vibrators. Became subsidiary of Westinghouse Air Brake in 1965 and then of Azcon Corporation in the early 1970s
GeoSpace Corporation	Louis B. McManis and Fred W. Hefer, 1961	Houston, Texas	Initially designed and manufactured magnetic tape compositing equipment for use with the weight-dropping technique. Merged with Hall-Sears in 1963, and started a Gravity Observatory Division in Boston, Massachusetts, during 1964 involving Dr. Lloyd G. D. Thompson, Dr. William B. Agocs, and Sam P. Worden to do pioneering work in airborne gravity measurements. In 1981, became subsidiary of AMF Inc.; however, still produces some of the best in geophones, with over two million of their GS-20D detectors sold in the past decade
Input/Output, Inc.	Aubra and Dorothy Tilley, 1968	Houston, Texas	Develops supplemental seismic equipment such as precise seismic source synchronizers and multiple air-gun control systems. Also sole producer of recording and processing equipment for repetitive surface sources such as the MINI-SOSIE(TM) gasoline-powered earth tamper. Purchased by Kidde, Inc. in 1980

Company	Founder/Year	Location	Description
Mark Products, Inc.	Eugene F. Florian, 1965	Houston, Texas	A former employee of Electro-Technical Laboratories, Florian's objective, with the aid of W. P. Johnson, was to design and manufacture a line of dependable geophones, hydrophones, and cable, a tradition still followed closely today
Mertz, Inc.	Roy Mertz, 1927	Ponca City, Oklahoma	Originally the Mertz Iron and Machine Works, purchased by the grandsons, Donald and Forrest, in 1948. Incorporated in 1968 and began manufacturing vibrators in 1969. In 1980, Mertz introduced the first Universal Vibrator(TM) capable of being switched in field from compressional to shear wave generation
SIE, Inc.	Richard H. Parker and Keith R. Beeman, 1945	Houston, Texas	Leading independent firm committed exclusively to geophysical instrumentation. However, it was sold to Dresser Industries in 1955 before becoming independent again in 1973. Current products include a 240-trace digital recorder and three-dimensional well geophone.
Tesco Cable Company	Sheldon Miller, 1946	Tulsa, Oklahoma	Major producer of land and marine seismic cable which has introduced a number of improvements in cable performance. Acquired by a newly-formed company, Tescorp, Inc., in 1980
Vector Cable Company	D. Y. Gorman, "Spec" Harrison, and Finley Robideaux, 1945	Houston, Texas	A subsidiary of Schlumberger since 1966, Vector produces the widest and most diverse line of seismographic detector cables and connector systems in the industry

panies — if such employment can be found in a diminishing job market. In good times, many are tempted by the freedom and higher pay to go on their own. Most, of course, are located near the principal petroleum centers, such as Houston and Denver in the U.S. But some manage to find plenty of work in other places of their own choosing. Of course, many large petroleum companies are not located near major petroleum centers. But if the demand and incentives are sufficient, this problem can be overcome. For example, during 1979 and 1980, contracts to acquire and interpret huge quantities of seismic data from the China Sea were accepted by a number of large companies.[101] This placed an unusually heavy demand on the already very busy U.S. seismic consulting pool. One of these companies was Occidental Petroleum Corp., whose exploration headquarters are in Bakersfield, California, far removed from the major supply sources for consultants. So in order to obtain the extra help needed, Occidental quickly offered consultants living in other areas free housing in Bakersfield and regular weekend trips home coupled with a salary of about $70,000 per year.

The Business End of Gravity and Magnetic Surveys

Although the use of gravity and magnetic surveys to locate mineral resources far predates the use of seismological surveys for commercial purposes, these earlier tools have been treated as step-children within the geophysical family. Nevertheless, a few dedicated and enterprising geophysicists outside the petroleum companies and the major geophysical contractors have been able to earn a good living most of the time by specializing in these secondary techniques. Certainly the methods warrant specialization, for they have practically nothing in common with the seismic technique. Field instrumentation and measurement techniques are drastically different, as is the consequent data interpretation based on the mathematics of potential fields, rather than on wave propagation and convolution. About the only common ground is that both use high-speed computers to reduce and analyze the field data, and that modeling has to be frequently used by both groups to determine the optimal geological and geophysical configuration that fits a given set of field observations. It is unfortunate that the exploration industry devotes only about two percent of its effort to using gravity and magnetics. The advent of highly accurate measurement tech-

---

101. Fully one-sixth of the world's offshore geophysical survey fleet was involved in shooting well over one million km of marine seismic lines off the east coast of China.

niques coupled with sophisticated processing and interpretation procedures justify increased use of these methods in non-reconnaissance applications, either in their own right or as a check on seismic measurement programs (Hammer and Anzoleaga, 1975; Hammer, 1975). However, because of low demand for such measurements, there is a lack of research and applications experience to prove this point, and the self-limiting cycle continues.

Airborne profiling completely dominates the magnetic surveying business. In 1980, for example, the airborne magnetometer accounted for $12.8 million in petroleum-exploration expenditures compared to $835,000 for land and marine magnetic operations combined. In mineral exploration, $7.4 million was spent for airborne magnetic data, while $571,000 was spent on land surveying. As far as is known, the Aero Service Division of Western Geophysical Company still accounts for more airborne magnetic than any other contractor. If one scrutinizes the list of 66 magnetometer contractors in the *Geophysical Directory* for 1981, he finds that only 21 are primarily engaged in magnetometer work. Of the remaining 45 companies, 18 are primarily seismograph contractors, 11 are mainly in gravity-meter work, and the rest prefer to be known for work in other categories. Looking backwards, one finds the *Directory* reporting only 21 primary magnetometer contractors in 1976, while the 1966 number was 44. Moreover, the only such firm besides Aero Service to survive the 15-year period intact was Kellogg Exploration Company of Los Angeles, which had been founded in 1959 by William C. Kellogg, the former Chief Geophysicist of the major aerial mapping firm, Fairchild Aerial Surveys.

The gravity survey business in 1980 was four times as large as the magnetics business, accounting for expenditures of $55.4 million in the search for petroleum and $2.4 million in the search for minerals. According to the *Geophysical Directory* for 1981, 52 gravity-meter operators were active, 17 of which were primarily engaged in gravity work, or two less than in 1976. Back in 1966, there were 33 such firms, eight of which still existed in 1981. These firms with staying power are listed here along with their headquarters sites: Thomas J. Bevan (Tulsa), Consolidated Geophysical Surveys (Tulsa), Exploration Surveys (Dallas), Neese Exploration Company (Houston), Photogravity Company (a division of Berry Industries) (Houston), Pohly Exploration Company (Fredericksburg, Texas), Sweet Geophysical Company (Santa Monica, California), and the Wongela Geophysical Party, Ltd. (Sydney, Australia). Four of these offer magnetic surveys as well, and there are still another 12 in both the gravity or magnetics business which consider one or the other to be their principal line of business. Three of these 16 firms are based in Australia (indicating the keen interest in mineral prospecting there), one in Poland, two in Canada, and the rest in six different American cities. From the above it can be seen that the fraction of work not done by the large, full-service geophysical contractors is spread out through a number of relatively small companies. As might be guessed, the "big name professionals" in gravity surveying tend to be old-timers who started before reflection seismology dominated geophysical exploration. Four of the most highly respected men of this type are Craig Ferris, E. V. McCollum, Lewis L. Nettleton, and Sigmund Hammer, and their efforts are cited here in modest detail to provide a fuller picture of geophysical exploration.[102]

The Craig Ferris—E. V. McCollum Partnership

As of 1981, Craig Ferris had spent 43 years in the gravity business, 37 of which were as a partner with E. V. McCollum in the firm E. V. McCollum and Company, of Tulsa.[103] After graduating from Friends University in Wichita, Kansas, at the height of the depression in 1934, Ferris began graduate work in physics at the University of Oklahoma. While at the university, he happened to meet Elliott Sweet, one of the owners of the American Seismograph Company (the other owners were Reginald Sweet and J. L. Copeland). Sweet eventually offered him a job, and in September, 1936 he went to work as a "jug hustler" for four dollars a day on a seismic crew working out of Bay City, Texas. A year later, while the crew was working in Illinois, Ferris observed Gulf's Hoyt gravimeter in operation.[104] The following spring between work assignments, he learned

---

[102.] Three of these men served as SEG's Presidents, namely, McCollum (1958—1959), Nettleton (1948—1949), and Hammer (1951—1952).

[103.] E. V. McCollum & Company was dissolved in 1980. Ferris then formed a new firm, Gravimetrics, Inc., also Tulsa-based.

[104.] Other early gravity meters were the Hart—Brown, Humble, LaCoste—Romberg, World-Wide, and three instruments (the Frost, Stanolind, and North American) designed and built by Reginald Sweet. The LaCoste—Romberg steel-spring gravimeter was conceived by the Sweet brothers and J. L. Copeland after reading an article, "A New Type Long Period Vertical Seismograph" by a young physicist, Lucien J. B. LaCoste, Jr., then an Assistant Professor of Physics at the University of Texas (LaCoste, 1934). Copeland went to the Chairman of the Physics Department there concerning the possibility of Copeland financing the development of a new gravity meter based on this concept of a long-period seismograph. From there, he was referred to Professor Arnold Romberg, who thought the idea was a good one if it could be implemented. Copeland then talked to LaCoste, and arrangements were made to finance the project. The result was the zero-length spring gravity meter used first by the Sun Oil Company. Shortly thereafter, the Sweets and Copeland withdrew from the enterprise and settled out of court with LaCoste and Romberg, who set up their own firm. Forty-five years later, LaCoste still runs this company which develops and produces the best instruments of their kind.

more about this new device which was rapidly replacing the torsion balance. As a result, he decided to join the rapidly expanding gravity-meter contractor, the Mott-Smith Corporation of Houston, which used its own type of instrument.[105] Ferris' instructor on the use of the instrument was Sam P. Worden, one of Mott-Smith's former students.[106] His indoctrination completed, Craig then spent between 1938 and 1943 working on a Mott-Smith gravity party ranging from the Gulf Coast all the way north into Canada.

While working for the firm, Ferris became well acquainted with E. V. McCollum, the corporation's Chief Geophysicist as of 1939. After having received a master's degree from the University of Oklahoma back in 1926, McCollum worked for the old Marland Oil Company for three years and then moved on to its successor, the Continental Oil Company. Once at Continental, he persuaded the firm to purchase three Mott-Smith quartz torsion fibre "Series A" gravi-meters. Because the response curve was non-linear, McCollum then had to develop calibration curves which were subsequently used by all field crews operating this particular Mott-Smith meter. In 1939, McCollum shifted over to become Chief Geophysicist of the Mott-Smith Corporation. While in this position, he began his pioneering work on the concept of establishing a global network of gravity benchmarks using gravimeters instead of the time-consuming geo-detic pendulums (McCollum and Brown, 1943).

By September, 1943, McCollum, Ferris, and their wives decided to found the partnership, E. V. McCollum and Company, in Tulsa, the city that liked to call itself the "Oil Capital of the World". Their goal was simple: "Provide a better gravity service to the petroleum industry!" Over its lifetime, the company claimed it conducted more crew-months of high-quality gravity surveys than any other con-tractor, accomplishing perhaps an eighth of all the gravity surveys made in the United States during the good years, such as in 1955 when the total countrywide effort was 1,074 crew-months. World activity peaked a year later at 2,224 crew-months. During the next decade, however, world activity dropped by 60 percent and U.S. surveying by 80 percent. The firm's concept for survival included: (1) conducting its own proprietary surveys, (2) opening a gravity

---

[105.] This corporation was owned by Lewis M. Mott-Smith, then Professor of Physics at Rice University, his wife, and V. J. Meyer, a former employee of the Humble Oil and Refining Company.

[106.] In the late 1940s, Worden founded Houston Technical Laboratories and developed the famed Worden® quartz spring gravimeter. More Worden gravimeters have been built than all other makes combined. Worden, too, sold out; in this case, the purchase, which occurred in 1953, was by Texas Instruments.

® Registered trademark of Texas Instruments, Inc.

data-bank service, and (3) entering the foreign gravity survey market in South America, Europe, and Africa.[107] As a result, the partnership realized a profit in 34 years out of 37, even though the true price of U.S. oil (when expressed in 1967 dollars) continued to dwindle from $3.66 a barrel in 1957 to $2.79 a barrel in 1972 before starting to come back up to a profitable exploration level.

The Gravity Meter Exploration Company (GMX)

During the tail-end of the boom during the late 1940s and early 1950s which used gravity surveys to locate petroleum prospects, the corporate name of GMX, when coupled with that of Dr. Lewis L. Nettleton, indicated an unsurpassable wealth of talent, experience and equipment in this special field. Nettleton's familiarity with gravity surveys dated back to 1928 when he went to work for Gulf Oil's newly formed research department housed in the Mellon Research Institute. Among his first work assignments were writing reports of surveys made with the Eötvös torsion balance, a tedious task at best. Early on, he came to know Dr. A. E. Eckhardt of the abortive Geological Engineering Company who had moved west to work with the Marland Oil Company at Ponca City, Oklahoma, until it, too, fell apart at the seams. Eckhardt also joined Gulf in 1928 and brought along certain ideas for building a better gravity-measuring pendulum.

Not too much happened until 1935 when Gulf finally deployed Eckhardt's pendulum in the field, even though it took a truck load of equipment and two full days to establish a single gravity station with a probable error of two to three milligals. At about this time, Gulf learned of the gravity meter developed by O. H. Truman, a former Humble employee who made a million dollars by building the first successful instrument of this type. Gulf, however, soon replaced the Truman device with its own gravity meter designed and built in-house during a nine-month period by young Archer Hoyt with the help of Dr. R. D. Wyckoff.[108] This sensor could occupy up to 400 stations per month and, by early 1937, was providing Gulf with the means for rapidly and cheaply detecting many of the salt domes existing along America's Gulf Coast.

[107.] In 1956, E. V. McCollum and Company reached its peak field effort of 120 crew-months. By 1974, business hit an all-time low of ten crew-months, followed by a gradual build-up back to 30 crew-months in 1978.

[108.] Wyckoff had been with Eckhardt at the U.S. Bureau of Standards and then at the Marland Oil Company. Wyckoff became Chief, Physics Division, in 1942 when Eckhardt moved up to the role of Vice President, Gulf Research and Development Company.

Early in this evolutionary period, Nettleton began teaching exploration geophysics at the University of Pittsburgh. Originally directed at geologists who might go to work for Gulf, the course gradually grew in complexity. As the mimeographed notes increased in scope, they were converted into the early, basic textbook, *Geophysical Prospecting for Oil,* published in 1940.[109] By 1946, Nettleton felt it was time to strike out on his own, so he became a partner along with W. G. Saville and A. C. Pagan in the Torsion Balance Exploration Company in Houston, Texas.[110] This name was technologically quite out of date, so in 1948, the partners relabeled it the *Gravity Meter Exploration Company* or GMX, as its advertising had it. Nettleton, who tends to be low-key and scholarly by nature, found the rough-and-tumble world of geophysical consulting to be far more demanding than the research department of a great oil company. There was far more travel, particularly to foreign countries like Venezuela, and there were sizeable fluctuations in the level of work, although new work seemed always to arrive just before the firm got into serious financial difficulty.

Some of GMX's and Nettleton's best years were spent developing improved methods for measuring gravity in offshore coastal regions.[111] By working with LaCoste and Romberg, they expanded the reliability and depth capability of underwater gravimeters until they could be read by remote control while sited on the sea floor 900 m below the surface. In fact, GMX took all of the first dozen marine gravimeters that LaCoste and Romberg built and found gainful employment for each of them. The first shipboard gravimeter was basically a pendulum allowed to swing with the vessel's motion. Even though the correction for such swings might be as much as several hundred milligals, Nettleton and his associates obtained moderately successful readings. However, the big break-through came when LaCoste and Romberg placed a specially-designed gravimeter on a gyro-stabilized platform during the early 1960s, thereby permitting the Navy, for

---

109. Nettleton's rapidly growing stature within the geophysical community was indicated by his being elected Editor of *Geophysics* for 1945–1947 and SEG President in 1948–1949, as well as being named an Honorary Member of the SEG during 1956.

110. The migration of well-trained engineers and scientists from noted universities in the northeastern and central part of the United States to the hot, steamy Gulf Coast during the late 1940s was caused in large part by the financial and technical opportunities existing there. However, the advent of air-conditioned homes, automobiles, and offices about that time accelerated the process.

111. The first underwater gravity meter was developed by the Mott-Smith Corporation during 1940. The device primarily consisted of a standard Mott-Smith "Model C" meter housed in a submersible chamber and capable of being read by remote control (McCollum, 1941).

one, to substitute large merchant ships as the sensor platform in place of deep-running submarines.[112]

During the 1950s, GMX experienced major managerial changes. Early in the decade, M. W. Baynes and Dr. Nelson C. Steenland, a former student of Professor Maurice Ewing, were brought in as partners. A highly perceptive geophysicist, Steenland was outspoken and yet so highly respected that he served as Editor of *Geophysics* between 1959 and 1961. In the meantime, one of GMX's founding partners, W. G. Saville, died during 1954, and was followed in death during 1958 by another of the founding partners, A. C. Pagan. To broaden its base, GMX combined with Geophysical Associates, a seismic exploration firm owned by Joe Shimek and Hart Brown, during 1957 to form a new parent company, Geophysical Associates International (GAI). In 1964, the energetic Dr. Thomas R. LaFehr was named a partner and Vice President, Technical Development, but that did not keep him from resigning five years later to join the faculty at his *alma mater,* the Colorado School of Mines.[113] This resignation actually came two years after GAI had sold out to the high technology firm of Edgerton, Germeshausen, and Grier whose basic specialty was the triggering and subsequent high-speed photography of nuclear explosions. Although Dr. Nettleton remained a ranking official in the GAI–GMX Division, top management in Boston did not truly understand the highly competitive business of geophysical surveying. As a result, contracts began to dwindle in number, and finally GAI–GMX totally disappeared from the scene during 1978.

Sigmund Hammer

Although he turned 80 in 1981, Dr. Sigmund Hammer continues to be one of America's leading consultants in gravity interpretation. As did Dr. Nettleton, Dr. Hammer, or "Sig" for short, began his geophysical career with Gulf's Research Department in 1929. One of Sig's first assignments was developing new interpretation formulae

---

112. As of 1980, LaCoste–Romberg had sold about 100 of these unique instruments. By then, it had been found possible to obtain useful readings with the device aboard a moving helicopter; some 37,000 line-miles of gravity data were taken during that year alone in the United States and Mexico by this new technique.

113. In the early 1970s, LaFehr (even though he was teaching full-time) and several associates started Exploration Data Consultants (EDCON) in Denver. During 1975, he then took Adjunct Professor status and became EDCON's full-time President. After that, the company grew rapidly into one of the world's leading contractors for gravity surveys and moved into electrical prospecting and borehole geophysics (particularly borehole gravity). LaFehr served as Editor of *Geophysics* during 1971–1973 and, in 1979, became the youngest Honorary Member of the SEG at age 45.

for the torsion balance, which were used to analyze the gravity data which led to Gulf's first discovery of an oilfield (the Hankamer field near Beaumont, Texas). From then on, his reputation as a gravity expert built rapidly. Sig's formula for determining the mass of an ore body from its gravity anomaly became of fundamental importance to mining geophysicists.[114] Unlike Nettleton, Hammer chose to stay with Gulf until the normal retirement age of 65.

Upon retirement in 1967, Sig began a five-year stint as Professor of Geophysics at the University of Wisconsin and started an active consulting career on the side. After retiring from the University of Wisconsin, Hammer became an "Advisor in Gravimetry" for the United Nations and worked with the national oil companies of Bolivia and Turkey for six years. He then taught gravity and magnetics in Spanish at the University of Mexico in Mexico City. Moreover, during this period, he found time to publish 15 articles, thereby bringing his career total to 50.

But in Hammer's case, it appears that the past is but a prelude, for he now thinks that some of his most exciting work is just getting under way during the 1980s. As Interpretation Consultant for Carson Geoscience, Inc. of Perkasie, Pennsylvania, he is deeply involved in the long-awaited development of a practical airborne gravity survey system. In fact, Sig's paper for the 1981 Annual Meeting of the SEG in Los Angeles was entitled "Airborne Gravity Is Here!" In this presentation, he postulated that the new ability to obtain one-milligal accuracy during airborne gravity exploration now provides the exploration industry with a powerful tool for reconnoitering the 400 sedimentary basins of the earth which are only partially or totally unexplored for oil and gas.

The Mineral Exploration Business

Mineral exploration and petroleum exploration have much in common, but still there are important fundamental differences. Both are subject to inflation, price fluctuations of the produced raw product, and vagaries of government policies relative to land access, regulations, and taxation. Both face the problem of searching for increasingly deeper and more elusive targets. But the costs and risks involved in exploiting a mineral discovery are much greater, relative to initial finding costs, than for a petroleum prospect. In petroleum reservoir evaluation, open-flow well-testing procedures which are now

---

114. Hammer served as President of the SEG (1951–1952) and was also named an Honorary Member of the Society in 1962.

increasingly coupled with improved seismic methods permit adequate commercial determination of the size and boundaries of a reservoir with a minimum number of carefully-placed exploratory wells. On the other hand, evaluation of a mineral discovery is subject to much uncertainty until costly physical access is achieved. Potential profits versus finite risks must be continuously reappraised during the entire development of a mining project (Peters, 1969). Thus, the mineral explorationist, in conjunction with his geophysicist, has the multiple task of finding geophysical and geological anomalies suggesting ore bodies, delineating those worthy of initial mining commitments, and of helping quantify the risk versus anticipated ore-body factors before and during the specific development stages requiring large capital investments.

The use of geophysics to spot anomalies indicative of important ore bodies dates back over three centuries, but only became sophisticated during the 1920s and 1930s when the Schlumberger brothers of France and a number of others, including Hans Lundberg and Sherwin F. Kelly, fielded a number of geophysical parties utilizing magnetometers, earth resistivity spreads, and self-potential meters for this purpose. As noted earlier in Tables 7.2 and 8.1, geophysical prospecting for minerals has now become a $50-million per year business employing a score of different techniques in a variety of physical environments. Moreover, these techniques are becoming increasingly sophisticated and interfacing ever more closely with geochemical methods (Rice *et al.*, 1981). An excellent example of how this is being presently done is the case history of Barringer Resources, Inc., which is dually based in Toronto, Ontario, and Golden, Colorado, near Denver.

Barringer Resources Inc.

The story of Barringer Resources Inc. is actually the story of Dr. Anthony ("Tony") R. Barringer. Born in Bognor Regis, England, during 1925, Tony was fascinated by geology ever since childhood. When in his early teens, he decided to emigrate to Canada after reading intriguing accounts of the discovery of gold in Ontario long before. To ready himself, he studied mining geology at the Imperial College of Science and Technology in London and earned "First Class Honours" upon graduating in 1951 after a tour in the British Army where he became, at age 21, the youngest captain in the 6th Airborne Division. By 1954, he had earned a Ph.D. in economic geology. Selco Exploration Company, Ltd. of Toronto then paid his way to Canada and placed him on their payroll. He soon came to the conclusion that there was no satisfactory way to locate mineral deposits beneath overburden and came up with the idea of an airborne electromagnetic (EM) system. By 1958, Tony had talked Selco's Chairman into providing $70,000 (Canadian) and a small laboratory building. The upshot was the development of the highly useful INPUT® (*IN*duced *PU*lse *T*ransient) system and the formation, in 1961, of Barringer Research, Ltd. which took over much of Selco's Airborne and Technical Services Division. Barringer became President and the firm's major stockholder, while Selco put up $50,000 (Canadian) and got 40 percent of the profits. This proved to have been an excellent decision, for by 1979, the INPUT system was being used by licensees in 70 percent of the world's commercial airborne EM surveys and was credited with having been responsible for discovering 16 major ore deposits with a combined value of over 10×10^9.

Dr. Barringer's drive, persuasiveness, and technical know-how, plus a staff of 140, including 52 scientists and 12 engineers, is moving the firm along quite rapidly.[115] By mid-1981, Barringer Resources (there had been a name change in 1980) held 91 issued and pending geophysical patents. These methods include COTRAN®, an improved INPUT system using digital processing and pattern recognition, and Radiophase® and E-Phase® airborne systems which employ a spectrum of selected frequencies from radio broadcasting stations to measure variations in the conductivity of the earth's crust. The firm's inventory of geochemical methods currently involves some 260 issued and pending patents, including Surtrace®, Airtrace®, and Vapourtrace®. These systems permit chemical analyses at the surface or while airborne of key organic compounds that may be associated with mineral deposits, hydrocarbon accumulations and geothermal sources.[116]

To raise funds for expansion, Barringer's firm sold stock to the public as early as 1968.[117] By 1979, revenues were about $5 million (Canadian) per year, and net earnings $460,000 per annum, or about $0.25 per share. In spite of great technical success and diversi-

---

[115.] As of June, 1981, Dr. Barringer still owned 40 percent of his firm, and Executive Vice President, Dr. D. Richard Clews, owned 10 percent.

® indicates registered trademarks of Barringer Resources, Inc.

[116.] In recognition of his work to date, Barringer was designated the 15th recipient of the SEG's Kauffman Gold Medal Award during 1980.

[117.] As might be expected, the price of a share of Barringer Resources' common stock is quite volatile. For example, in the first ten months of 1981, the bid price of a common share on the over-the-counter market has ranged from $5.00 to $12.50 (U.S.). Even so, such a share of stock could have been bought for as little as $0.75 a share back in the first quarter of 1979.

fication, Barringer became dissatisfied with the financial return from the firm's work, which remained closely tied to the highly cyclical mineral exploration business and to government contracting. Even as far back as 1976, he had set up a branch research office in Golden, Colorado in order to increase the use of Barringer technology in petroleum exploration. Tony then based himself at this new office and soon had carried out interdisciplinary studies concerning the use of seismic, LANDSAT imagery, magnetics, geochemistry and active EM sounding in the search for petroleum. By 1979, Canadian clients were, in fact, achieving an 80-percent success ratio in exploratory drilling based on the integration of Barringer reconnaissance techniques with seismic surveys and sub-surface geology. A year later, Petro-Canada, the national oil company of Canada, awarded Barringer Resources a three-million dollar contract for further development of an airborne detection system (Tivac$^{(TM)}$) potentially capable of detecting hydrocarbon leakage from oil and gas fields by sophisticated remote sensing methods. Thus, should there be an important breakthrough during the coming decade in the direct detection of hydrocarbon deposits, it is likely that some of the extensive suite of Barringer's multi-disciplinary survey and analytical techniques will prove to have been involved.[118]

The Industry's Prospects for the Future

As noted at the beginning of this chapter, the geophysical exploration industry is now well into its third "boom period" since its founding in the early 1920s. Within the United States, there is much optimism that the present strong uptrend in exploration primarily created by increasing "real" prices for oil and natural gas represents a lasting movement back towards a freer market economy. However, this domestic optimism can probably only continue if there is steady progress towards reducing the rate of inflation, balancing the federal budget, and providing relief from the great burden of "excess profit" and other types of anti-investment taxes. A somewhat comparable situation exists on the international scene, where continued growth will depend on free movement of trade and international investments and a slowing of the monopolistic trend towards state-owned companies. Such companies totally dominate all internal decisions relative to future oil and gas supplies within their respective countries

---

[118.] Barringer's methods make extensive use of the following fields: physics, chemistry, electronics, mathematics, optics, mechanics, geology, geophysics, geochemistry, botany, microbiology, meteorology, fluidics, and computer science.

and often avoid using multiple approaches to exploring a given area. Be as it may, let us now examine future prospects from three points of view — that of the industry, that of manpower supply, and that of potential advances in technology.

The Situation as Viewed by Members of the International Association of Geophysical Contractors (IAGC)

During early 1981, the 260-odd members of the IAGC were polled regarding several facets of their industry, including its present composition and problems, and its outlook regarding the future. Fifty-four companies provided detailed responses.[119] Forty-eight percent of the firms had relatively small staffs (10 to 100 employees), while 41 percent had over 100 regular employees.[120] The percentage of professional employees within a given firm varied widely, 11 companies reporting that less than 10 percent of their staff were professionals, while 14 companies reported that professionals made up 30 or more percent of their employees.

Nearly 78 percent of the reporting firms were privately held, with almost one-third being owned by a single individual. Another 30 percent had two or three principals, and about 20 percent were owned by four or five persons. In 19 cases, the owner(s) had come from a single major contractor, and in seven cases, from a single major oil company. In nearly 70 percent of the firms, the founder(s) had started out with one or more ideas regarding a new or improved service, product or technology, and in nearly three-fourths of these cases, such ideas were the single most important factor that led to creation of the company. Relative to the degree of profitability, 44 percent of the respondents believed that the geophysical industry had been more profitable over the years than United States industry as a whole, while another 40 percent believed their industry to be less profitable. About 56 percent saw a definite correlation between their own business profitability and the peaks and valleys of United States petroleum exploration as indicated by seismic activity and exploratory drilling.

With respect to current problems facing the industry, 18 out of the 54 companies reporting indicated that the impact of increasing

---

[119.] Their combined single or multiple activities were as follows: data acquisition — 34; data processing — 18; data interpretation — 14; data brokerage and/or storage — 11; equipment development — 12; software development — 6; consulting — 10; geophysical drilling — 10; and miscellaneous — 7.

[120.] Firms with less than 10 employees fell into the categories of software development, consulting, data brokerage, or miscellaneous.

government regulations was the most important current problem.[121] In addition, another eight companies listed the cost and difficulty of obtaining land or marine exploration or leasing permits as the biggest problem, a situation frequently tied to the issue of government regulations. Two other firms cited uncertain government policies, while another two cited inflation as their major problem. Counting multiple listings by some, 25 out of 54 firms listed problems involving too much government interference. The major problem cited by 31 companies was the shortage of qualified personnel, which will be taken up in the next subsection. Other identified problems also included the reluctance of exploration managements to accept new ideas, the continuing technical need to locate hydrocarbon accumulations directly, the poor public image of the oil industry, and difficulties in communicating with landowners and clients. Nevertheless, almost 80 percent of the respondents believed that the future of the geophysical industry for the next 20 years is "more promising than ever". Only two firms considered the future to be discouraging, although even one of these indicated that he thought the next dozen years would be good ones.

In view of the industry's current optimism about the future and its continuing 22 percent per annum growth rate which started back in April, 1976, what are the chances for meeting both the current and future demand for well-equipped and well-manned geophysical crews? Equipment-wise, there is every indication that such a continuing demand can be met. A 1979 IAGC study indicated the industry's capacity for manufacturing shothole drills was more than adequate (National Petroleum Council, 1979). However, the study also indicated that the steadily increasing popularity of non-dynamite energy sources would require an increase in the 1979 production capacity of the three major vibrator producers from 350 to 500 units per year by 1985.

In the area of seismic recording equipment, the obsolescence rate had become as high as 50 percent per annum by 1977.[122] As a result, by the early 1980s virtually all recording instruments worldwide would be updated, and it was possible that there would be a surplus production capacity by 1981 of 180 units over the 1979 capability of 300 recording units.[123] One of the unknown factors

121. The effect of increasing government regulations on profitability was cited as "great" by 16 firms, "moderate" by 22 firms, and "slight" by 15 companies.

122. This obsolescence was caused principally by the rapid trend to a larger number of recording channels (Carlile, 1981).

123. This statement was based on a poll of Geosource, Litton Resources Systems, Sercel (CGG), SIE, and Texas Instruments, the principal producers of field equipment.

in estimating the total market is the magnitude of equipment purchases by the Soviet Union and the People's Republic of China. For example, by 1979 the latter had bought about 85 sets of recording instruments and more than 200 vibrators, but still their future requirements remain unknown.

Manpower Issues

The 1981 poll of IAGC members indicated that nearly two-thirds of the firms considered that their most serious current problem was the shortage of qualified personnel.[124] Such individuals are hard to obtain and perhaps even harder to keep. On a long-term basis, much of this shortage may be blamed on the boom-and-bust cycles which take place in petroleum and mineral exploration. According to a global survey made in 1976 by the SEG of its 8,000 Active and Associate members, the employers of these people were as follows (smoothed to the nearest percentage point): oil companies — 51 percent; geophysical contractors — 15 percent; other service companies — two percent; manufacturers and supply companies — two percent; government — seven percent; academic institutions — five percent; mining companies — three percent; consultants — 12 percent; and miscellaneous — three percent. In terms of job function, the same personnel were primarily occupied in: interpretation — 35 percent; management — 29 percent; research — 11 percent; data processing — 10 percent; data acquisition — six percent; teaching — three percent; instrumentation and equipment design — two percent; sales — one percent; and other — three percent.

In the case of the contractors, experience suggests that they experience about a six percent annual attrition rate for the more technical jobs, such as instrument observer, and rates of as high as 32 percent in the case of less-skilled or more broadly trained personnel, such as surveyors who are frequently lost to other industries. Although the contractors do use SEG and commercial "short courses", most of their technician training is conducted in-house by their own staffs. This is a very sizeable effort, and is projected to increase about seven percent per year. In fact, the IAGC study revealed that the effort would reach about one million man-hours per year by 1983, or the equivalent of 500 man-years, 70 percent of which would be by formal course work and the remainder by on-the-job

---

124. The scope of the recruiting problem is demonstrated by the fact that just one geophysical contractor, Digicon Inc., increased its staff from 850 people in July, 1980 to nearly 2,000 in September, 1981 to handle internal growth.

training. The conclusion was that the contractors' training plans would be quite adequate for meeting projected needs, including ample allowance for attrition.

The same cannot be said for the geophysical staffs of the large petroleum companies. Because much of this staff is highly trained and has wide experience, such people frequently are being hired away by a rapidly increasing number of small exploration firms who "offer a piece of the action" as well as other attractive fringe benefits and record salary levels. The only optimistic aspect for the United States — but a sad one for Canada — is that Canadian geophysicists are moving south across the border because they see little hope for a quick return to a positive Canadian federal government policy towards private exploration. Traditionally, the petroleum companies have sought "new blood" on the campuses of colleges and universities, and hired over 300 geophysics graduates in 1978. However, the IAGC study indicated that, starting in 1981, the number of geophysics graduates is expected to start falling. Hence, relative to company needs, there might be a short-fall of about 80 graduates per year by 1985. In the past, oil companies have made up deficiencies of this type by hiring additional mathematicians, geologists, physicists, and computer scientists and giving them both formal and on-the-job training in geophysics. This blend of backgrounds shows up strongly in the 1976 SEG survey, which found that its members had college majors as follows: geology — 30 percent; geophysics — 28 percent; engineering — 15 percent; physics — 13 percent; mathematics — eight percent; computer science — one percent; and miscellaneous — five percent. Unfortunately for the companies, when geophysicists are in great demand, so are geologists. Computer science graduates are meeting only a fraction of the current demand, and the Labor Department predicts that demand for computer programmers will double by 1990. Fortunately, the supply of mathematicians and physicists is adequate and probably an appreciable number of them will opt to shift from pure to applied science during the coming decade.

While most major petroleum companies conduct systematic in-house training programs for new non-geophysicists and for updating their existing employees, they also make extensive use of short courses offered by several universities, by the SEG and its Local Sections, and by individuals and firms specializing in professional training. The American Association of Petroleum Geologists also operates an excellent continuing education program, including a very popular course, "Stratigraphic Interpretation of Seismic Data". In the autumn of 1980, a cooperative industry-wide effort was also started to create a course in "Basic Exploration Geophysics" that would provide 43 hours of personal instruction via a computer-based, interactive, self-paced instructional system. Training of geophysicists is currently so lucrative a business that it is worth noting briefly here three other firms that compete with GeoQuest. In 1981, IED Exploration of Tulsa, Oklahoma, offered three courses in geophysics, while another Tulsa firm, Oil and Gas Consultants International, Inc. offered four geophysics courses (two especially for geologists) and one on geology for geophysicists. A Boston, Massachusetts firm, *I*nternational *H*uman *R*esources *D*evelopment *C*orporation (IHRDC), employs such noted geophysicists as Dr. Robert E. Sheriff of the University of Houston and the lucid Briton, Nigel A. Anstey, for instructors, Both IED Exploration and IHRDC offer course manuals for sale in book form, and IHRDC courses are also sold in a video tape format. As a consequence, the explorationist can obtain much of this course material even if physically far removed from the training centers.

The Potential for Continued Technical Advances and Discoveries of New Natural Resources

Rapid and extensive technical innovation has been the lifeblood of the geophysical industry during the first six decades of its existence. In fact, the digital revolution invigorated both the seismic business and petroleum exploration long before the price of oil and natural gas rose enough to create today's boom conditions. What are the chances for another boost comparable to that brought about by the digital revolution, particularly with regard to enhanced direct indications of the presence of commercial accumulations of oil, natural gas, and ore bodies? The current answer seems to be that there is no pending *single* development of equivalent importance, but that a combination of interactive advances now in the making may prove to be just as significant.

In seismic prospecting, the recent order-of-magnitude increase in the number of data channels that can be recorded, either in two or three dimensions, is setting a new standard for detailed sampling of the subsurface. This coupled with much more sophisticated processing and modeling techniques to make optimum use of such data will provide the increasingly-skilled interpreter with the resolution required for accurate delineation of reservoir boundaries, rock type, porosity, and fluid content (see Graebner *et al.*, 1980). Combined use of shear wave and compressional wave recordings has the potential of much more positive identification of important lithologic changes, and the quantity as well as just the presence of gas (see Meissner and Hegazy, 1981; Rice *et al.*, 1981).

As a further aid to the positive identification of hydrocarbons, the active electromagnetic sounding method is expected to become an increasingly important adjunct to seismic prospecting as understanding improves concerning the direct and indirect hydrocarbon resistivity associations being measured (Rice *et al.*, 1981). In addition, after four decades of offering technical promise but little economic success, the art of geochemical prospecting may even now be coming into its own. Finally, it is important to note that there is no foreseeable letup in advances in electronic computer and equipment technology which continuously renders solutions to yesterday's unsolvable problems feasible today and cost-effective tomorrow.

Thus, exploration technology itself has a bright future. Furthermore, the continued scarcity of and demand for hydrocarbons and minerals will insure the basic need for technical advances for many years to come. Unfortunately, however, the commercial realization of these advances strongly depends on the financial health of the entire exploration industry. The industry's prosperity as of 1981 is very fragile, for it is based on a small but real improvement in petroleum-exploration incentives and on much expectation for the future. Hopefully, the political leaders of the world and their peoples will not destroy what may very well be their last opportunity to effect a smooth transition to the new non-petroleum age of the next century. As Michel T. Halbouty, one of Texas' most erudite and energetic "wildcatters", advised a thousand geophysicists at the 51st Annual International Meeting of the SEG during late 1981, there is good geological evidence that only half of the world's oil and natural gas has yet been discovered. To be sure, that which remains will not be easy to locate and to produce, for much of it lies in stratigraphic traps, in the polar regions and offshore (see Fig. 8.6). Yet it is this very "challenge and change" which makes the profession of geophysics such an exciting one as its practitioners continue to explore new frontiers on earth and in space.

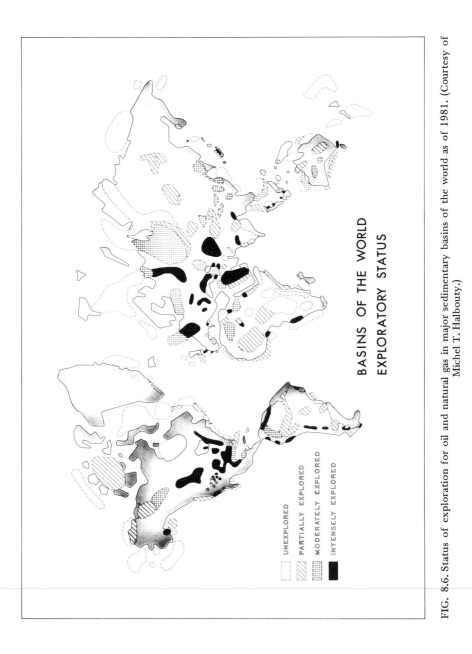

BASINS OF THE WORLD
EXPLORATORY STATUS

UNEXPLORED
PARTIALLY EXPLORED
MODERATELY EXPLORED
INTENSELY EXPLORED

FIG. 8.6. Status of exploration for oil and natural gas in major sedimentary basins of the world as of 1981. (Courtesy of Michel T. Halbouty.)

CHAPTER 9

Geophysics As They Saw It

Think, in this battered caravanserai
Whose portals are but alternate night and day,
How Sultan after Sultan with his pomp
Abode his destined hour and then went on his way.

From *The Rubaiyat of Omar Khayyam*

In Retrospect

At the book's beginning, we stated our goal would be to describe the realm of geophysics in a socio-economic manner as well as in a scientific sense. In working towards this goal, geophysics proved to be, as predicted, an intriguing blend of the theoretical and the practical, the non-profit effort and the profit making venture, and the making of war and the solving of problems that arise during peace. Moreover, the lot of the pioneer often was lonely and unprofitable, while those who followed could sometimes take the same discovery and make it into a highly useful and lucrative technology. As the book progressed, we also found that the total story would have to be streamlined. Hence, to ensure that there was a sizeable perspective on the massive task of exploring, understanding, and predicting the physical processes and properties of the earth, several score geophysicists were invited to list their most noteworthy actions. The response to this request was excellent. Even the 90-year-old dean of British geophysicists, Sir Harold Jeffreys, was kind enough to contribute. In addition, several first-person vignettes became available written by such noted geophysicists as Professor Maurice Ewing, Cecil H. Green, and W. Harry Mayne. These, too, are included in this chapter for they delineate in extra detail the special flavor and excitement that permeates the geophysical profession.

Some Major Personal Achievements

Thirty-eight geophysicists were kind enough to respond to the call

Headquarters of Society of Exploration Geophysicists in Tulsa, Oklahoma. (Courtesy SEG.)

Luncheon gathering on 26 October 1976 of SEG's Past Presidents. Left to right, front row: Roy L. Lay (53–54), H. B. Peacock (41–42), Cecil H. Green (47–48), Lewis L. Nettleton (48–49), W. Ted Born (40–41), Sigmund Hammer (51–52), Paul L. Lyons (54–55); second row: W. Harry Mayne (68–69), W. B. Robinson (70–71), Roy O. Lindseth (76–77), Roy F. Bennett (56–57), John C. Hollister (62–63), Hugh M. Thralls (64–65); back row: E. V. McCollum (58–59), Milton B. Dobrin (69–70), Norman J. Christie (63–64), Robert B. Rice (75–76), J. Edward White (67–68), J. Dan Skelton (74–75) and H. James Kidder (73–74).

"Vin d'Honneur" for His Royal Highness, Prince Claus of the Netherlands, following his address during June, 1976 to the European Association of Exploration Geophysicists. Left to right: A. W. Smit (EAEG Secretary-Treasurer), H. E. R. F. M. Lubbers (Dutch Minister of Economic Affairs), HRH Prince Claus, F. Walter, F. G. L. L. Schols (Burgomaster of The Hague), and A. van Weelden (EAEG founder and its first Honorary Member).

Reception at Silver Anniversary Meeting of the European Association of Exploration Geophysicists (June, 1976).

Key officials of the International Association of Geophysical Contractors at 1981 Annual Meeting. Left to right: Charles F. Darden, President; Thomas B. Portwood (Seiscom Delta), incoming Chairman, and L. Decker Dawson (Dawson Geophysical Company), outgoing Chairman. (Courtesy IAGC.)

The new owners of Geophysical Service, Inc. on 6 December 1941 — Erik Jonsson, H. Bates Peacock, Eugene McDermott and Cecil H. Green. (Courtesy of GSI.)

Patrick Haggerty, catalytic entrepreneur at Texas Instruments between 1946 and 1976. (Courtesy Texas Instruments.)

Henry Salvatori, founder of Western Geophysical Company. (Courtesy of H. Salvatori.)

Hal Sears, designer of the first miniature geophone. (Courtesy Bettye Athanasiou.)

Quorum of Geotechnical Corporation's Board of Directors on 9 May 1964. Left to right: Dr. W. Heroy (Board Chairman), Dr. W. Heroy, Jr. (President), Jack Hamilton (Vice President), and B. R. Ellis, Carey Croneis and Grover Ellis (Directors). (Courtesy W. Heroy, Jr.)

International headquarters of the Western Geophysical Company, Houston, Texas. (Courtesy Western Geophysical Company.)

The Heroy Building for Earth Sciences at Southern Methodist University, Dallas, Texas. (Courtesy W. Heroy, Jr.)

Three interlinked "Betsy" Electric Seisguns being operated by Party X, Seismograph Service Corporation, in Kansas during August, 1980. (Courtesy of MAPCO Inc., Tulsa.)

Turhan Taner, founder of Seiscom, Inc. (Courtesy of T. Taner.)

Roy O. Lindseth, pioneer in the field of seismic stratigraphy. (Courtesy R. Lindseth.)

Professor (Emeritus) C. Hewitt Dix, who taught exploration geophysics between 1948 and 1973 at the California Institute of Technology. (Courtesy C. H. Dix.)

Dr. Anthony Barringer, founder of Barringer Resources, Ltd. (Courtesy A. Barringer.)

for listing actions which had given them the greatest personal satisfaction during their lifetimes. A review of these responses finds that about half of these events involved pure science, not necessarily in the field of pure geophysics. The other half involved applied science, typically involving group interaction to a considerable degree. To avoid any question as to how one can best rank peers, the submissions have been carried in a sequence depending upon the alphabetical order of the contributor's last name and his particular geophysical specialty. The dates cited for each memorable effort typically relate to the initial start of the effort, its initial reporting in the scientific literature, or, in a few cases, to the spread of years within which the special activity was at its peak. Finally, we caution the reader to remember that the actions cited here are but a modest part of the total picture created by these contributors and by their thousands of co-workers, sung and unsung, past and present.

Exploration Geophysics

Dr. Milo Backus (Professor of Geophysics, University of Texas, Austin, Texas):
   (1) Serving on the GSI team which developed and introduced digital processing of seismic data under John Burg and Kenneth Burg (1958).
   (2) Investigating means for directly detecting oil and natural gas by seismic means (1975).
   (3) Working on the early stages of three-dimensional seismic systems (1976).

Anthony R. Barringer (President, Barringer Resources, Toronto, Ontario, and Golden, Colorado):
   (1) Developing the INPUT® system of aeromagnetic surveying for minerals by using a pulsed transient signal transmitted from a towed "bird" (1958).
   (2) Developing rapid means, including the AIRTRACE® and SURTRACE® systems, for conducting rapid geochemical surveys (1970s).
   (3) Developing the digitized COTRAN® system of aeromagnetic surveying (1979).

F. Gilbert Blake (Special Assistant for Resource Applications, U.S. Department of Energy, Washington, D.C.):
   (1) Providing the first rational explanation for the nucleation threshold of water cavitation in presence of intense sound

fields, thereby explaining the physical limit of energy placed in the ocean by a sonar (1948).

(2) Assembling and managing a first-rate research team in exploration geophysics at the La Habra Laboratory of the Chevron Oil Field Research Company (1950).

(3) Inventing a complex method of seismic data collection, processing and interpretation (misnamed "shot steering") that provides greatly improved data for geological regions that are structurally complex (1959).

C. Hewitt Dix (Professor Emeritus of Geological Sciences, California Institute of Technology, Pasadena, California):

(1) Demonstrated value of interval velocity measures from seismic reflections for estimating subsurface lithology (1941).

(2) Introduction of three-dimensional interpretation of seismic records (1947).

(3) Inventing with J. A. Westphal and J. R. Fort a one-bit seismic recording system for 1,024 independent channels of data (1973).

Ralph C. Holmer (Professor of Geophysics, Colorado School of Mines):

(1) Making gravity surveys with the Mott-Smith meter (Mark I weighing 60 kg) in the Near East from the Sinai Peninsula to Iraq and eventually detecting some anomalies near Basra that proved to be productive salt domes (1939–1941).

(2) As an instructor at the Colorado School of Mines, compiling the first state gravity map of Colorado using some 1,700 observations (1947–1954).

(3) Becoming first Chief Geophysicist, Kennecott Copper Corporation, and then moving up to be Vice President, Kennecott Exploration, Inc. and directing four Divisions (Geophysical Operations, Geophysical Research, Geological Research, and Geochemical Research) with a staff of over 100 persons.

Sidney Kaufman (Professor of Geophysics, Cornell University, Ithaca, New York):

(1) First party-chief to take a geophysical crew into the open ocean (1937).

(2) Designation as a professor at a major university after a full career in commercial geophysics at Shell Development Company (1974).

(3) Full-scale utilization of exploration geophysical techniques and equipment within an academic study of the deep crust (1978).

Roy O. Lindseth (President, Teknica Resource Development, Ltd., Calgary, Alberta):
(1) Developing and patenting the process for synthesizing sonic logs from seismic traces (1972).
(2) Derivation and commercial utilization of stratigraphic data from seismic data (1973).
(3) From a personal point of view, marrying the woman he did and changing over to geophysics as his lifetime career (1944).

Robert E. Sheriff (Professor of Geophysics, University of Houston, Houston, Texas):
(1) Writing *Glossary of Geophysical Terms* and the *Encyclopedic Dictionary of Exploration Geophysics* (1968).
(2) Teaching at the University of Houston (1968).
(3) Achieving improvements in geophysical education via SEG courses and tutorial seminars, including the writing of *Geophysical Exploration and Interpretation* and of *Seismic Stratigraphy*.

Earth Physics

Bruce A. Bolt (Director, Seismographic Station Network, University of California, Berkeley, California):
(1) Bringing to light, via new seismic observations and inferences, new knowledge regarding the detailed structure of the Earth's interior, particularly the core (1962).
(2) Pioneering use of long regional telemetry interactions, thereby providing, via standard programs, prompt locations of both local earthquakes and distant teleseisms, singly and in groups (1963).
(3) Emphasizing via lectures, writing, and consulting, the importance of the use and study of strong-motion instruments and recordings to extend basic knowledge of large earthquake sources and near-field seismic shaking, as well as to assist earthquake engineers (1967).

Albert P. Crary (formerly Director, Division of Earth Sciences, National Science Foundation, Washington, D.C.):
(1) Changing over from foreign geophysical work to involvement in the Air Force geophysical program, eventually becoming Chief Scientist for the Ice Island T-3 project (1946).
(2) Fostering use of drifting ice islands as observational platforms for the International Geophysical Year (1955).

(3) Serving as Chief Scientist of the NSF's Antarctic Research Program (1959).

Anton L. Hales (Professor of Geophysics, University of Texas at Dallas, Dallas, Texas):
(1) Publishing (in conjunction with his student, Kenneth Graham) paleomagnetic evidence from the Karoo dolerites indicating that Australia had drifted in relation to the Eurasian continent (1956).
(2) Demonstrating for both the United States and Australia that travel times in stable continental regions show systematic deviations from the Jeffreys—Bullen travel time curves and that as many as nine distinguishable discontinuities occur in the upper mantle (1968).
(3) Demonstrating that relatively high seismic velocities can exist at a depth of 60 to 70 km below an ocean basin such as the Gulf of Mexico (1970).

James R. Heirtzler (Chairman, Department of Geology and Geophysics, Woods Hole Oceanographic Institution, Woods Hole, Massachusetts):
(1) Playing a major role in developing the science of marine geomagnetism and making it a useful tool to study the earth's geologic history (1965).
(2) Director, Hudson Laboratories of Columbia University (1967—1969).
(3) Serving as the American Co-Chief Scientist on the combined French—American Project Famous targeted towards study of a mid-ocean spreading center using deep-diving submersibles (1973).

Wilmot N. Hess (Director, National Center for Atmospheric Research, Boulder, Colorado):
(1) Conducting the first measurement of neutrons in space (1959).
(2) Conducting the first experiment employing an electron accelerator in outer space for the purpose of creating aurorae (1967).
(3) Developing a broad-based ocean-pollution-research program, including environmental assessment of ocean dumping and oil spills (1976).

J. A. Jacobs (Professor of Geophysics, Cambridge University):
(1) Providing a physical explanation of how the Earth could have evolved leaving a solid inner core and a fluid outer core (1953).
(2) Having the opportunity to expand the earth science field by

serving as the initial Director, Institute of Earth Sciences, University of British Columbia, Vancouver, Canada (1961).
(3) Serving as the founding Director, Institute of Earth and Planetary Physics, University of Alberta, Edmonton, Canada (1970).

Sir Harold Jeffreys (Retired Plumian Professor of Astronomy, Cambridge University):
(1) Developing a theory of the thermal history of the earth with major geological implications (1916 onwards).
(2) Developing a "Theory of Scientific Method" which explains how scientific laws can acquire high probabilities and applying this concept to many branches of science (1919 onwards).
(3) Stating, contrary to general belief, that the outer planets are at liquid air temperatures (1923).
(4) Making the first explicit statement that the earth's core is liquid (1926).
(5) Reconstructing the theory of the earth's gravitational field (1938).
(6) Improving the travel times (in association with Gagna, Shimshoni, and Heddon) for earthquakes in Europe, Central Asia, and Pacific regions.
(7) Construction (with K. C. Bullen) of the Jeffreys–Bullen travel-time tables of earthquake waves (1939).

Helmut E. Landsberg (Formerly Director, Institute for Fluid Dynamics and Applied Mathematics, University of Maryland, University Park, Maryland):
(1) Introducing the science of geophysics (both seismology and meteorology) to Pennsylvania State University, leading later to creation of two new academic departments there – Geophysics and Meteorology (1934).
(2) Establishing first federal plan for geophysical research while Executive Director, Committee on Geophysics and Geography, Defense Research and Development Board (1947).
(3) Initiating concept of geophysical reviews by initiating and serving as first editor of *Advances in Geophysics* (Academic Press) (1952).
(4) Starting and editing the projected 15-volume series, *World Survey of Climatology* (1964).

Jack Oliver (Chairman, Department of Geological Sciences, Cornell University, Ithaca, New York):
(1) Identification of the variety of different surface wave types that occur within the seismic wave train (1957).

(2) Discovery (with Bryan Isacks) of downgoing lithospheric slab in island areas (1967), followed by their collaboration with Lynn Sykes to postulate the "new global tectonics", now known as "plate tectonics" (1968).

(3) Initiation of the COCORP effort to explore the continental basement (1978).

Louis C. Pakiser (Past Chief, Office of Earthquake Research and Crustal Studies, U.S. Geological Survey, Golden, Colorado):

(1) Fostering since 1940 increased minority participation in the earth sciences.

(2) Organizing USGS's Branch of Crustal Studies (1961).

(3) Organizing USGS's Office of Earthquake Research and Crustal Studies (1966).

Frank Press (President, U.S. National Academy of Sciences, Washington, D.C.):

(1) Major involvement (separate but simultaneously with Slichter of UCLA, Ewing of Lamont, and Benioff of CalTech) in the discovery of the earth's free oscillations after the great Chilean earthquake of 1960 (1960).

(2) Playing a major role in initiating a national earthquake-prediction program (1965).

(3) Providing major impetus, as Director, Office of Science and Technology Policy, for what will hopefully be a major "Ocean Margin Drilling Program" (1977).

S. Keith Runcorn (Head, School of Physics, University of Newcastle-upon-Tyne):

(1) Initiating, along with P. M. S. Blackett and in collaboration with J. Hospers, E. Irving, K. M. Greer, D. W. Collinson, and N. D. Opdyke, development of the palaeomagnetic survey of the geological column, including obtaining evidence for reversals of magnetic polarity and the first quantitative data for continental drift (1950 onwards).

(2) Being an early advocate (1950) of the relevance of solid-state physics to understanding the earth's mantle, especially with respect to convection through creep processes serving as the mechanism for continental drift, and being the first to suggest that the non-hydrostatic figure of the Moon and the geoid are the result of solid-state convection (1962 and 1964, respectively).

(3) Showing the geophysical importance of counts in the growth increments of corals and other fossils (1964).

(4) Chairing a Science Research Council Committee which initiated the participation of British laboratories in the Apollo lunar sample analysis program (1965). This led to my discovering jointly with D. W. Collinson and A. Stephenson indications of lunar palaeomagnetism, followed by our explanation calling for an early lunar magnetic field generated by a dynamo process in an iron core (1969).

H. I. S. Thirlaway (Superintendent, Seismological Detection and Verification Group, British Atomic Weapons Research Establishment):
(1) Demonstrating in association with Alan Cook of the Cavendish Laboratory, Cambridge University, the true nature of the Worcestershire basin and its boundary fault (1948).
(2) Establishing the Pakistani Seismological Network headquartered in Quetta in association with Moiduddin Ahmed of the University of Ohio and the late Mansur Choudbury, formerly of the Institut du Physique du Globe, Strasbourg (1951/1957).
(3) Demonstrating in association with Frank Whiteway the value and use of phased seismometer arrays in both earthquake and forensic seismology (1960–1965).

J. Tuzo Wilson (Director-General, Ontario Science Centre, Toronto, Ontario):
(1) Recognition (simultaneously but independent of J. E. Gill) of the major age provinces of the Canadian Shield and their boundaries (1946).
(2) Realization that isolated islands such as Hawaii might be caused by plumes rising from the earth's mantle, thus causing "hot spots" and island trails that show plate movements (1963).
(3) Recognition of "transform faults" between earth's tectonic plates (1965).
(4) Elaboration of the concept that cycles of continental drift have occurred more than once, thereby creating and closing oceanic basins (1968).

J. Lamar Worzel (Past Director, Geophysics Laboratory, University of Texas Marine Science Institute, Galveston, Texas):
(1) Development of long-range underwater acoustic monitoring systems (SOFAR and RAFOS) (1945).
(2) Development of first underwater television (1942).
(3) Development of technique for measuring gravity from surface ships under way (1958).

Oceanography

Robert B. Abel (Vice President, New Jersey Marine Sciences Consortium, Fort Hancock, New Jersey):
(1) Serving as Executive Secretary of the Interagency Committee on Oceanography from inception until replacement by the National Council on Marine Resources and Engineering Development (1960–1967).
(2) Directing National Sea Grant Program for its first decade (1967–1977).
(3) Designing "Cooperative Marine Technology Program" for the Middle East, thereby involving the Israelis and the Egyptians in common projects for the first time in 5,000 years (1977).

Wayne V. Burt (Former Associate Dean for Oceanography, Oregon State University, Corvallis, Oregon):
(1) Creation of Department of Oceanography, Oregon State University (1959).
(2) Fostering the creation of a highly successful Sea Grant College advisory system that placed "oceanographers in hip boots" at fishing ports (1968).

William L. Ford (Formerly Director, Bedford Institute of Oceanography, Halifax, Nova Scotia):
(1) Confirmation of oceanographic forecast that the enormous load of radioactive contamination from underwater shot Baker would disperse quickly to give safe operational levels in Bikini Atoll (1946).
(2) Fostering healthy development and growth (of two orders of magnitude) of marine sciences in Canada in post-World War II period (1953 on).
(3) As first Director of the Bedford Institute of Oceanography, creating a management scheme in which researchers were imbued with a satisfying concept of "multiple loyalty" both to the Institute and to their funding Ministry (1965).

Dale F. Leipper (Formerly Chairman, Department of Oceanography, U.S. Naval Postgraduate School, Monterey, California):
(1) Serving as initial chairman of two major academic departments of oceanography (Texas A&M University, 1949–1964, and USN Postgraduate School, 1968–1979).
(2) Directing the academic effort which turned out approximately one-third of all the American graduates in oceanography in the decade following 1953.
(3) Never losing a man at sea while serving as Department Chairman.

Gordon G. Lill (Formerly Deputy Director, National Ocean Survey, Washington, D.C.):

(1) Fostering a broad-ranging approach to naval support of ocean research, including the formation of modern oceanographic institutions at Texas A&M University, Columbia University, Oregon State University, and the Universities of Miami, Rhode Island, and Washington (1949 on).

(2) Launching the deep-sea drilling project at the National Science Foundation — an effort that led to extensive offshore drilling in deep water by both industry and academia (1959).

(3) Interesting Lockheed Aircraft Corporation in entering the deep-sea technology field and in setting up an oceanographic research laboratory (1961).

Waldo L. Lyon (Chief Research Scientist for Submarine Arctic Technology, Naval Undersea Systems Center, San Diego, California):

(1) Agitating for and personally conducting the pioneering developments necessary before submarines could navigate safely below the Polar Ice Pack (1946).

(2) Making 23 informative Arctic submarine patrols (as of 1979).

Arthur E. Maxwell (Director, Institute of Geophysics, University of Texas at Austin):

(1) Collaborating with Roger Revelle and Edward Bullard to obtain first measurements of heat flow through the ocean floor (1952).

(2) Chairing the Technical Advisory Committee for the successful search for the remains of the USS *Thresher* (1963).

(3) Serving as Chief Scientist on Leg 3 of *Glomar Challenger* cruise series that developed clear evidence of sea-floor spreading in South Atlantic Ocean (1969).

H. William Menard (Formerly Chief, U.S. Geological Survey, Washington, D.C.):

(1) Joining the U.S. Navy Electronics Laboratory as a marine geologist, followed by switching to Scripps, and, while at these two locations, making over a thousand aqualung dives for geological studies (1949).

(2) Spending three years at sea as a member of 20 different oceanographic expeditions (1950 on).

(3) Working up scientific results of his cruises in over 100 scientific papers (1949—present).

David S. Potter (Vice President for Government Affairs, General Motors Corporation, Detroit, Michigan; formerly Under Secretary of the Navy):

(1) Developing the torpedo-tracking range of the Applied Physics Laboratory, University of Washington, at Dabob Bay (1952).

(2) Co-authoring (with S. R. Murphy and G. R. Garrison) the definitive paper on relaxation frequency and absorption co-efficient of underwater sound in sea water at 50 to 500 hertz (1958).

(3) Co-authoring (with S. Daubin) a basic paper on design of deep-sea moorings (1963).

Roger R. Revelle (Professor of Science and Public Policy, University of California, San Diego):

(1) Conceiving and directing the oceanographic surveys at the Bikini Atomic Bomb Tests and the Bikini Re-survey a year later (1946).

(2) Creation and/or promotion of oceanographic entities, including the Intergovernmental Oceanographic Commission, the Scientific Committee for *Oceanic Research* (SCOR), and the First International Oceanographic Congress (1959).

(3) Pushing the Scripps Institution of Oceanography into extended and intensive surveys of the world ocean (1960).

Rear Admiral Edward C. Stephan, USN (Ret.), Washington, D.C.:

(1) Serving as first Oceanographer of the U.S. Navy (1960).

(2) Chairing the Navy's Deep Submergence Review Group to determine ways in which survivors can be rescued from submarines lost at depth (1963).

(3) Serving as Founding President of the Marine Technology Society (1964).

Harris B. Stewart (Director, Center for Marine Sciences, Old Dominion University, Norfolk, Virginia):

(1) Convincing the U.S. Coast and Geodetic Survey to permit Mason and Raff to tow their magnetometer behind the C&GS *Pioneer*, thereby revealing magnetic striping for the first time (1955).

(2) Converting an old-line Federal agency, the Coast and Geodetic Survey, into one which conducts truly modern marine science (1957).

(3) Establishing the facility now known as the NOAA Atlantic Oceanographic and Meteorological Laboratories, Key Biscayne, Florida (1965).

James H. Wakelin (Formerly Assistant Secretary for Research and Technology of both the Departments of Navy and of Commerce), Washington, D.C.:

(1) Converting the U.S. Naval Hydrographic Office into a "Naval

Oceanographic Office" and an "Office of the Oceanographer of the Navy" (1960).

(2) Chairing the Interagency Committee on Oceanography during the halcyon days of this science (1960–1964).

(3) As Assistant Secretary of the Navy (R&D), keeping the task force looking for (and finding) the remains of the USS *Thresher* after President Kennedy and Secretary of the Navy Korth passed the word to "Knock it off!" (1963).

Edward Wenk, Jr. (Professor of Engineering and Public Affairs, University of Washington, Seattle, Washington):

(1) Developing concepts for the design of deep-diving submarines, culminating in the design of the Aluminaut (1958).

(2) Developing (with Richard Carpenter) the basic concepts of technology assessment as utilized by the U.S. House of Representatives (1965).

(3) Initiating the concept and obtaining presidential endorsement for the International Decade of the Oceans (1968).

(4) Publishing *The Politics of the Ocean* and *Margins for Survival* to explain how governmental decisions are made in the sphere of science and technology and to inquire why the political apparatus turns a deaf ear to signals suggesting major future societal problems (1972 and 1979).

Oceanography/Meteorology

Patrick D. McTaggart-Cowan (Formerly Executive Director, Canadian Science Council):

(1) Serving as the meteorologist-in-charge at Botwood, Newfoundland, for the first successful commercial trans-Atlantic flights (Pan-American's Captain Gray, Botwood to Foynes, and Imperial Airways' Captain Wilcockson, Foynes to Botwood) (1937).

(2) As the founding President, directing the construction, staffing, and opening within an 18-month period of a full-fledged university — Simon Fraser — at the cost of C$18 million and with a student body of 2,500 (1964–1965).

(3) Directing the major oil-spill clean-up operation resulting from the sinking of the tanker *Arrow* in 30 m of water within Chedabucto Bay, Nova Scotia. By using an extensive task force including 20 Canadian Navy divers and the U.S. Navy salvage tug, USS *Curb*, it proved possible to do the impossible and pump out three million gallons of extremely viscous Bunker-C

oil from the tanker's stern section, thereby avoiding creation of a long-term source of pollution.

Vignettes on the Teaching of Geophysics

Comments on the Early Years of the Lamont Geological Observatory — Professor Maurice Ewing[1]

Our program of work at sea started in 1935, when the late Dr. Bowie of the Coast and Geodetic Survey and Professor Richard Field of Princeton came and pointed out to me the importance of making some measurements of the depth to basement rocks off the east coast of the United States. I was then an instructor in physics at Lehigh University and had practically no knowledge of geology. . . . These men (indicated that there was) the possibility of getting a grant of $2,000 from the Geological Society of America to support the field work of this project. The grant was obtained, and we made our first tests on the ship *Oceanographer* of the U.S. Coast and Geodetic Survey. Captain Nicholas H. Heck of the Coast Survey was a keen supporter of this project, and actually made one of the early trips with us. After these first tests, we did further work on the ship *Atlantis* of the Woods Hole Oceanographic Institution in the autumn of the same year. We were able to trace the basement rock from the outcrop near Petersburg, Virginia almost to the edge of the continental shelf.

After this Professor Field and Dr. Bowie urged us to undertake work in the deep ocean basin. At this time the group consisted of myself and one or two graduate students, and our ship facilities consisted of two weeks each year as guests of the Woods Hole Oceanographic Institution aboard the *Atlantis*. It requires much development of new techniques to get good measurements of the sediment thickness in the deep ocean basins. I was able to devote full time to it for two years by virtue of a grant from the John Simmon Guggenheim Foundation, which enabled me to take leave of absence from Lehigh University. I have always felt that this grant marked the real turning point in my career. . . . At about this time we also initiated work in photography of the ocean bottom. On a small grant of funds from the

---

[1.] Extract from an unpublished report, "A Short History of Lamont Geological Observatory", dictated on 19 April 1955 by Professor Ewing and submitted to Mr. Richard Stevens, Resident Representative for the Office of Naval Research. The commentary has been edited to some degree by rearranging the location of several paragraphs to provide improved continuity. The document provides an interesting overview of what a great marine geophysicist thought was particularly important just a quarter of a century ago.

National Geographic Society we built a few cameras which were very successful, and to this day almost all of the photographs of the ocean bottom in depths greater than a few tens of fathoms have been taken by us or with our cameras.

An incident which occurred in the early days of this work is a good illustration of the difficulty we have had in raising the funds necessary to keep the program going. After the first work in 1935 I realized that I had found the kind of work which I would like to do the rest of my life, so I organized a prospectus with a view to submitting it to a number of people in the oil business and getting a small annual grant from each of several companies to support a modest program of research. This proposal received no support whatever, and I was told that work out in the ocean could not possibly be of interest to the shareholder and could not rightfully receive one nickel of the shareholders' money. This was the year 1935 and these were the tidelands which have since become so important in the petroleum business. I can safely say that no project to explore any part of the earth can possibly look more remote from commercial application today than the investigation of the tidelands looked to these leaders in the petroleum industry in 1935.

Just before the outbreak of World War II the National Defense Research Committee was formed to promote research on projects important for national defense. I took leave of absence from Lehigh and went to Woods Hole to work on one of these contracts. Two of my Lehigh students, A. C. Vine and J. L. Worzel, went with me, and this formed the nucleus of the present team. Our wartime work consisted in the application of underwater sound and photography to problems of submarine warfare. We started out on an anti-submarine project but inevitably were led into some developments which were useful for pro-submarine applications. Toward the end of the war we developed a long-range system of underwater sound signalling for which the Navy made the name SOFAR. It had important possibilities for air—sea rescue purposes, but has always got bogged down in bureaucracy to the extent that no appreciable use has ever been made of it. We were quite fortunate professionally in our wartime experiences in that they involved the use of equipment and methods which were very close to the line of work we had followed before the war and after it.

Toward the end of the war I was offered a chair at Columbia University to initiate instruction and research in geophysics. It turned out that a number of my colleagues there were disappointed when it developed that the bulk of my effort would be devoted to investigations of the oceans. They had rather hoped for a program of work which would apply more directly to the mining industry. But the program has flourished well enough that there has certainly been no opposition to it within the University, and it now receives strong encouragement from the administration and from our Department of Geology. This encouragement does not include any assistance with the financial problems. The University was operating in the red at the time we started our program, and had insisted from the start that it could not assume any additional financial responsibilities.

The move to Columbia marked the beginning of a marvelous growth in our program. Mr. Worzel had come to Columbia as a graduate student at the end of the war, and continued to serve as the nucleus of the team while obtaining his Ph.D. degree. . . . The team has been very largely built up of our own graduate students and, as a result, is remarkably coherent and cooperative. We succeeded in measuring the thickness of sediment in the ocean basin at a number of places and found it to be only about 2,000 feet. Unexpectedly we were able to measure the thickness of the basaltic layer beneath the sediment and show that it was only about two miles in thickness. This constitutes the entire crust of the earth under the ocean basin, and the contrast between this thin crust and the crust twenty miles thick or so which underlies the continents, amounts to proof of the permanence of the ocean basins. The transition zone between continent and ocean is the difficult part to investigate, and we are now working on that by several different methods and believe that by understanding it we will understand the mechanism by which continents have grown from nothing, or from small nuclei to their present size.

The demonstration that the total accumulation of sediment in the ocean basins is only about 2,000 feet and that this sediment is unconsolidated sets up a very challenging project which promises results of the greatest importance in almost every field of science. This is the project for penetrating the 2,000 feet of sediment beneath the three miles of water and sampling it from the ocean floor to the underlying basaltic rock. This should contain a very complete record of the evolution of life with very conspicuous changes in sediment character corresponding to the first development of animals which could concentrate calcium carbonate shells, the first development of plants which would produce free oxygen in the atmosphere, and give clues about the most primitive forms of life through their chemical and biochemical activity. Beneath this, at the very bottom of this sediment, should be clear evidence of the kind of earth we had at the time when water first began to accumulate in the oceans. To date this question has been the subject of much high-level mathematical study. Very divergent conclusions have been reached, and the consequences of definite information about this point would affect fields as far apart as astronomy and biology.

The project is entirely feasible. To date we have developed a sampling apparatus which penetrates as much as 60 feet into these unconsolidated sediments and has brought together the greatest collection of submarine sediment samples in the world — a total of about 1,500 long cores of sediment from the deep sea. The research team which we have got together now includes slightly over a hundred people. A number of these are part-time students, and a number of them are clerical, technical, and mechanical technicians, but it requires a large team like this to undertake the large problems of the character of that just mentioned.

Our work in measurements of gravity at sea has gone on primarily under the leadership of Dr. Worzel, and if we get continued support in this line, we hope to have, within the next decade, enough of these measurements to permit a new determination of the figure of the earth, in addition to definitive information about many important geological features such as the edges of continents and the volcanic island arcs.

The next big turning point in our affairs came in 1949, when the Lamont Geological Observatory was founded through a gift of property by the late Mrs. Thomas W. Lamont. At that time we were provided with an operating fund of $400,000 which was to constitute our assured support for the first six-year period. We have just completed that six years, and have produced a very gratifying series of scientific reports on work accomplished during the time.

At about the time the Observatory was established, we felt it necessary to begin the study of earthquake seismology, and since then have built up what we consider to be the best seismological observatory in the world. We have concentrated on study of the propagation of surface waves along continental paths, and the contrast between these and the waves which have traveled over oceanic paths. From this study we have been able to demonstrate conclusively that the results about the oceanic crust found in the relatively small area which we could investigate by actual voyages are typical of the oceans of the world. Our work in seismology has received wide recognition. It started primarily as a project between Frank Press, then a graduate student, and myself, but like the other things, it has grown.

A considerable group of our people is now working in what we call model seismology. Laboratory-size scale models of various geological formations and ultimately of the whole earth are put together, and the propagation of vibrations through these is used as a guide to what happens in earthquake- and explosion-generated waves.

We are finding wonderful results in our study of the sediment cores from the deep sea. There is a clear record here of the end of the last glaciation, and a demonstration that it was an abrupt event, as far as the oceans were concerned. The opinion formed by all of those who have studied evidences of glaciation on land was that it was a transitional event, probably still in progress. The discovery that it was abrupt, as far as the ocean is concerned, will certainly prove a most valuable clue to the eventual discovery of the reason for these glaciations. At present we have only speculation, much of it of the wildest possible type, about the cause of those enormous changes in climate.

We have added a group under the leadership of Professor J. L. Kulp which works on the general problems of geochemistry. In addition to a considerable program of measuring the ages of rocks from various radioactive analyses, this group is, in my opinion, the leading one in the world for definitive studies on the changes in the level of radioactivity due to experiments in fission carried out throughout the world, and I am certain that the leadership of this group will be well recognized when the expected change in classification of information makes it possible for them to publish their results.

We have been studying changes in the earth's magnetic field over the oceans by towing behind our research vessel a magnetometer which measures the total strength of the earth's magnetic field. This is an adaptation of the airborne magnetometer which has been of importance in exploration for minerals in the past few years. It is my belief that we will produce truly revolutionary results from these studies within the next two years. We (also) have a fine program of work in submarine topography which has led to the discovery of abyssal plains and an explanation of their origin. The North Atlantic has been divided into new physiographic provinces, ordinary submarine canyons shown to extend to depths greater than two miles, and a new type of mid-ocean canyon discovered on it.

Under a grant recently awarded by the Rockefeller Foundation, we are about to undertake a study of the productivity of food materials in the ocean. Dr. Gifford Pinchot, now at Yale University, plans to come here and head up this work. We believe this program of work will expand and will constitute a positive action in response to the prolonged outcry about the necessity for developing additional sources of food if the population of the earth is to continue its growth. In addition, the study of the distribution of small organisms in the present-day ocean will be of great advantage to us in interpreting the climatic changes of the past as recorded in the fossil remains of these organisms. . . .

I believe that we have built up here a unique team of scientists, unique in the diversity of techniques which it can bring to bear on problems and in the fundamental importance of the problems in which the group is interested. It can truly be said that practically every project in this laboratory — certainly every major project — has a direct and important bearing on all the others, and that there is the fullest cooperation between people working on the various projects. In fact, there is much overlapping and interlocking of the personnel, so that the group can work almost as efficiently as a single individual could do. . . .

I believe that this integrated group of scientists, this group of facilities which includes the ship, the chemical laboratory, the collection of sediment cores, and the great seismograph station constitute a facility comparable with the greatest cyclotron or the greatest telescope, and it is unique. It is as though there was just one cyclotron in the world and we had control of it, or just one big telescope in the world and we had control of it. It is true that there are other groups in the general field, but none has a comparable integration of personnel and techniques, or such concentration on the major problems.*

(End of Report)

Learning Geophysics from Professor George P. Woollard — Ned A. Ostenso[2]

George Woollard was a refreshingly irreverent person. His vast reservoir of sanctity for the human spirit and human feelings did not spill over to human foibles and institutions.[3] I first met George quite by accident in the winter of 1952, and our association has been a vignette of hundreds of others who shared the rare opportunity to be associated with him. It all began with my need for a final elective three-credit course within the Department of Geology to meet the minimum requirements (which were my maximum requirements) for a baccalaureate degree in what I fondly hoped to be my terminal semester of college. A necessary requirement for my selection was that the course not start at 7:45 a.m. The selection was rather limited because geology professors tend to be early starters, and I had taken

*Additional details on the accomplishments of Ewing and his group may be found in the book by Wertenbaker (1974) and in Dr. Wallace Broecker's response upon receiving the Maurice Ewing Medal of the American Geophysical Union and the U.S. Navy (Broecker, 1979).

2. Remarks of Dr. Ostenso at a Memorial Service for Professor Woollard on 19 April 1979. Dr. Ostenso is presently Director, Sea Grant Program Office, National Oceanic and Atmospheric Administration.

3. Woollard's special compassion for children is indicated by he and his wife, Eleanore, having four natural children, adopting five others, and supporting 18 foster children.

all the easy courses. Seemingly, the least distasteful choice was a relatively new course called "Geophysics" and taught by a guy called George Woollard. Although his personal class attendance can be most charitably described as spotty, he completely transformed my views of science, the earth, and myself. Rather than a dry compilation of facts, textbooks became frameworks for structured questioning. The printed word did not square with what he and Brackett Hersey observed a couple of years ago off the Carolina Coast. If Statements "A" and "B" are both true, then how can we possibly have "C?" For example, the Charleston, North Carolina, earthquake, which was one of the greatest ever recorded in North America, occurred thousands of miles from areas of normal earthquake activity. Why? "We've just got a proposal funded, Bill Bonini and some of the boys are going to try to find the answer, and we're looking for help." Parenthetically, I might add that some months later, while sitting in a Columbia, South Carolina jail, Bill, Don Plouff, Jack Mack, and I had time to discuss the pluses and minuses of providing George with "help"!

The sepia-and-celluloid-collared scientists that stared bloodlessly from the textbook pages were replaced by the exploits of the Ewing Brothers, Dick Geyer, "Teddy" Bullard, "Bert" Crary, and Walter Munk — to mention but a few of the happy and hapless adventurers who filled our lives through George Woollard's special style of "non-teaching". We were encouraged to wonder about all the wondrous things for which we had never given a single thought. Why are there continents and ocean basins? Why have all continents been seas at various times in their history, whereas there is no evidence that an ocean has ever been dry land? Why does the Earth have an internal magnetic field and why does this field exhibit certain dynamic properties? What is the internal constitution of the Earth? Why earthquakes? And so on. We were given all the scanty and conflicting evidence and challenged to evaluate it, to propose experiments to test it, and most exciting of all, tantalized by the prospects of participating in the grand adventure.

Half-way through the semester, I was committed to go on to graduate school and approached George on the subject. Let me be explicit here in stating that at this point in time my academic credentials were less than stellar. My first semester as a freshman at Eau Claire State Teachers College was spent marking time awaiting my appointment to Annapolis. The social life was straight A's! My second semester was spent brooding over having failed the Academy physical examination and trying to punish my parents for making me stay in college when I wanted to join the Marines. That semester I even flunked social life. The following summer I continued my higher education working in the local canning factory. The hours were long and the pay 75¢ per hour. The niceties of time-and-a-half for overtime and double-time for Sundays and holidays had not yet been invented. The work was wet and miserable for us all — except for the guys who walked around with clipboards. They weren't wet and haggard. They looked pretty well fed, rested, and seemed to be actually enjoying themselves.

Right then, I decided that the clipboard job was the job for me — enough of stuffing lids into canning machines — and I fancied that a college degree was the surest and shortest way to a clipboard job. It was with nothing more than a clipboard in mind that I set my academic sights as I moved on to the Madison campus and began in earnest the delicate balancing act of selecting the easiest courses that did not start too early in the morning. The ensemble of this highly selective process led me *de facto* towards a degree in geology and the afore-mentioned encounter with George Woollard. My singular lack of collegiate performance to date did not bother George one whit. I was willing, I had the blush of new-found youthful enthusiasm, and he figured I would work cheap. That was good enough for George, and the niceties of getting me admitted to graduate school and accepted by the Department were, like sex and religion, something gentlemen didn't discuss.

Through some mysterious combination of his awesome gall and guile, George got me into graduate school, and I found myself crammed into his outer office in the basement of Science Hall. My first encounter was a mothy vintage bearskin suspended from the ceiling and the greater part of two-thirds of a dried-out rattlesnake. These proved to be the most benign denizens of that room now so hallowed in many of our memories. There were also Bob Meyer, John Rose, Bill Black, Rodger Chapman, Bill Heinze, Phoebe Pierce, Ed Thiel, and John Behrendt, to name but a few. This rich zoological mix had certain experiences in common which was a strong uniting force: none of us ever took a course from George that we didn't end up teaching ourselves; we all wondered where George was; we all became infected with George's working habit which was to keep going till you dropped — and then start over again when you came to; and most importantly, the exhilarating sense that we were inventing as we were learning.

My first "official" act as a new graduate student was to enter George's inner office where I was astonished to see him dancing about the floor strewn with bits of paper. This ritual, I was informed, had nothing to do with geophysics but rather was a ceremony in honor of the University's business office. It was called "aging" receipts. Of all the things that have and could be said of George Woollard, he has never been accused of being organized. The orderly collection and retention of receipts from his multitudinous travels represented management acumen to which he never aspired. His system of doing travel accounts had, like the common pyre, the elegance of ultimate simplicity. He knew how much he left with and he knew how much he came back with — the difference is what he spent. To satisfy the less elegantly-minded accountants, at the end of each trip he would pass out a motley array of scraps of paper to his students who had to produce two receipts each — one with the right hand and one with the left hand. We were graded on our originality! These receipts were then aged under foot.

In addition to serving as George's sanctum sanctorum, we found his inner office a convenient place to store dynamite left over from or in anticipation of seismic experiments. The only bowing I have seen George do to higher authority was when he started stashing the explosives under his desk and some of the more nervous of the custodial force complained to buildings and grounds management. Looking back now, those were primitive days without the headlines of the new global tectonics, earthquake prediction, satellites, upper-mantle projects, and so forth. Nevertheless, they were truly exciting days — I think mainly for three reasons.

First, there was a sense of self-sufficiency. What we had was pretty much what we made. There were no Academy committees or international consortia to generate research ideas. There were few companies to supply the instruments we needed even if we could afford "store-bought" stuff. There was no support staff of technicians and data processors. And there were few research grants. The grant that supported my Master's thesis research was the generous loan of George's personal gasoline credit card — I suspect Eleanore (his wife) actually paid for it.

Secondly and somewhat contrastingly, there was a strong sense of togetherness in George's basement room plus a feeling of common purpose and kinship that drew from the strength of George's character. This was an aura that was shared with Woods Hole and Lamont by virtue of his association with these institutions. The fortunate amongst us remember the annual (in those days one didn't have to specify Spring) AGU meetings in Washington when we would all descend on the old Gordon Hotel. The Gordon was a masterful statement of what hotels were meant to be — cavernous rooms that could be rented for $8.00 as a double and into which an additional six sleeping bags would easily fit. The management dignified the then unknown expression — "benign neglect". Frank Press, George Sutton, Jack Nafe, Paul Pomeroy, John Ewing, Chuck Drake, Joe Worzel, and others would join us there. We drank in the good fellowship and shared sense of discovery along with the more tangible spirits of the Hay-Adams Hotel Bar. We couldn't afford to stay at the "Hay" but we liked to drink in style — the ol' Woollard élan.

Thirdly, George shared with his students the rich array of his personal friends: Sam Worden, Nelson Steenland, Cecil Green, Louis Slichter, Louie Nettleton, Lucien La Coste, Teddy Bullard, and others. These were truly pioneers. We stood in their awe and were privileged to be in their presence. Wherever we traveled throughout the world our identification with George Woollard opened doors and hearts.

My story could go on for hours, and thousands more could be told by others whose lives have been touched by George. But I cannot stop without a final comment on the indelibility of that touch. Once one became a member of the Woollard "family", it was a lifetime affair. For instance, while doing my Army service in Alaska, it seemed perfectly natural that I should spend my weekends and leave-time doing a gravity survey of the then "Territory". Somehow, he got me one of Sam Worden's meters, and I "gumshoed" the jeeps, planes, and boats to do the job. Later, while gainfully employed with Geophysical Service Inc., looking for oil in the Gulf of Mexico, George volunteered my services for a year-and-a-half tour in the interior of Antarctica. It never occurred to me that I shouldn't go. The only surprise was when I learned, two days before we sailed, that I was actually going to get paid for going — $1,800 a year yet!

More significantly, George set the standards against which I have measured my judgment and performance since we first met over a quarter of a century ago. The lessons he taught were the lessons of life, not just the relative significance of a Bouguer versus a free air anomaly. That such a great heart has stopped beating is more than pause for reverence — it is also pause for concern. We must take stock of a world without George Woollard and wonder what a poorer place it must be unless someone else picks up the trace. We must look to ourselves. To achieve George's stature most surely exceeds our reach. But if we strive, the many of us who have known and been influenced by him and come even half way to the mark can, together, perpetuate his greatness for another generation — and maybe others to come. This is our challenge.

Vignettes on Geophysical Prospecting

Reflections on the Field of Exploration Geophysics
— Cecil H. Green[4]

My long career in exploration geophysics and electronics has entailed only a faint line of separation between work and pleasure. At this later time in life, I have reached the important conclusion that success in life is mainly a function of one's state of mind — or in a single word — "happiness". This, in turn, is tied to a second important word, "accomplishment". Just how and why did I get started in exploration geophysics in the first place? After all, I had grown up in far-away British Columbia and had then graduated as an electrical engineer at MIT in New England. However, it turned out that, in 1930, I was associated in Palo Alto, California, as a production engineer with Charles V. Litton in the electronics shop of the Federal Telegraph Company, a subsidiary of the *I*nternational *T*elephone and *T*elegraph Company (ITT). I soon admired Litton as one of the smartest young men I had ever met.[5] He could go into the machine shop, take over from the best toolmaker, proceed to show him how to do a job better, and then return to his own office to perform a mathematical analysis of some other problem on the blackboard. At the same time, while designing and building 20-kilowatt transmitting tubes for ITT's overseas radio system, Charlie had become one of the most capable glass-to-metal technologists in the United States.

I mention this detail because this is where a new company, named Geophysical Service, Inc. (or GSI for short), first entered the picture. Charlie's preferred daily work regime with molten glass and metal was from late afternoon until long after midnight. I became greatly intrigued and understudied him after my regular work schedule, with the result that I found myself putting in far too many 12—15-hour days. Quite naturally, Ida, my wife, got just a little discouraged with her loneliness and so expressed herself in correspondence with Helen, the wife of Roland F. Beers, with whom I had worked on a previous research assignment with Raytheon back in Cambridge, Massachusetts. When I had left for the West Coast, Roland transferred first to Submarine Signal Corp. and then to Geophysical *R*esearch *C*orp. (GRC), a subsidiary of Amerada Petroleum Corporation. Then, in early 1930, when J. C. Karcher and Eugene McDermott elected to leave GRC to

---

[4.] Presented on 3 November 1980 to the Dallas Society of Exploration Geophysicists, Mr. Green's remarks are based on having spent 50 years in petroleum geophysics dating back to 1930.

[5.] Litton went on to found the firm which eventually became the conglomerate, Litton Industries, and the owner of the Western Geophysical Company.

launch GSI as a completely independent seismic exploration service company, several other GRC associates, such as Roland Beers, H. Bates Peacock, Ken Burg, Henry Salvatori, and Chalmers Pittman, left as well to become party chiefs for the new firm based in Dallas, Texas. In spite of the great depression, prevailing at that time, business was so good that additional party chiefs had to be recruited. Because of our previous association in the Boston area with Raytheon, coupled with his knowledge via our two wives that I had become a workaholic out in Palo Alto, Roland purposely mailed me a glowing account of the exciting life of a petroleum explorationist. Ida was, of course, quite amenable and so I succumbed to an attractive job offer from Eugene McDermott — a letter which I still have in my memorabilia file.

I included Charlie Litton in my notice of resignation from Federal Telegraph and I still hear his friendly, but harsh, criticism and his reference to the utter folly of looking for new oil because all storage tanks were already overflowing. But the die was cast! Ida and I drove our Oldsmobile from Palo Alto to Seminole, Oklahoma, where I understudied Roland Beers as Party Chief on his crew under contract to Pure Oil. His visiting geologist, Ira Cram, was most cooperative. Then, on 1 October 1930, I became party chief of GSI's tenth field party, headquartered 12 miles away in the big town of Maud, Oklahoma, population 275 and under contract to Twin States Oil Company, a subsidiary of Sun Oil Company. My observer was Bill McDermott and computer, Chester Donnally. I experienced many interesting adventures during the next 12 months as we moved at frequent intervals to such Oklahoma towns as Seminole, Prague, Stillwater, Perry and Guthrie. Seismic energy was, of course, provided by $2\frac{1}{2}''$ diameter sticks of Hercules 60% nitro-glycerine sticks of dynamite.

Since this was before the advent of truck-mounted drill rigs, shot holes were obtained with bucket augers built by the Hall Company. You real oldtimers will recall this hole-digging process as "Bear down and turn to the right!"[6] This was accomplished by three-men labor crews — two men to turn the auger handle, while the third sat on the handle to provide necessary weight as successive 4-foot pipe extensions were added until the hole reached its desired depth of 19 feet. Six closely spaced holes had to be drilled at each shot point, designated

---

[6.] One of GSI's newer employees, Edward Fain, was promptly assigned to hand auger duty. In the process of grumbling about it, he asked a partner on the device what he had been hired as. The answer was simple — "President", for the individual was no less than J. C. Karcher! Fain became a topflight observer, but still was noted for his complaining. Wherever his crew happened to be was the world's worst place, while the last place they had been was usually termed the world's best.

ahead of time by the crew surveyor with a stake decorated with yellow flagging. As many as six holes per shot point were necessary, since the amplifiers of that early day did not feature automatic volume control, so successive charges of dynamite, of varying size per hole, had to be detonated in order to obtain the optimum reflection amplitude.

Water from a truck-mounted tank was used for tamping each charge. A telephone line over the length of the spread of 1,500 to 3,000 feet between the layout of six seismometers and shot point provided communication between the observer and shooter. The shooter would advise by telephone when each charge had been pushed to the bottom of its hole and tamped with water. The observer would then start hand cranking the four-inch record through the camera as he told the shooter to "fire" — the shooter then quickly pushed down the handle of his old-fashioned blasting machine to detonate the charges. The instant of detonation was transmitted automatically from the blaster via the telephone line to be recorded as a "time-break" at the beginning of the reflection record. The surveyor's distance from shot point to each of the seismometers at the recorder's set-up was then checked by means of a blastphone — the time taken for the air wave generated by a very small charge, detonated at the surface, to travel at an air velocity of 1,100 feet per second.

The small number of seismometers was offset by their size and weight. They were our so-called S-2 "jugs" — cylindrical, 6 inches diameter, 15 inches tall, 30 lbs weight and comprising a stationary coil and moving magnet mounted on leaf springs — such heavy moving system being properly damped by the right weight of oil. Each of these cylindrical jugs had to be buried in hand auger holes in order to minimize wind disturbance. At the same time, set-ups near trees were avoided in order to avoid wind noise transmitted by nearby roots.

The essential unit in the recording camera was the so-called "harp" — comprising five 0.00065-inch-diameter gold wires, stretched between two ivory bridges. In the bright light of a properly focused optical system, each of these five small wires, whether stationary, or moving in response to an amplified signal from its particular seismometer, produced a so-called trace on 4"-wide photographic paper as it was hand-cranked through the camera by the observer. The camera also included a miniature size hand-started synchronous motor driven by an electrically actuated tuning fork — the ten tines on its small fly wheel casting timing lines at 0.01-second intervals across the recording paper, so that the reflection arrivals on each of the five traces could be timed to the nearest 0.001 second by the party chief in his field office — that is, after the record had been developed, fixed and dried by the observer's helper as he caught the exposed record with his hand in the developer box located beneath the camera.

Interpretation in the field office by the party chief, with an assistant as a computer, was a process of identifying reflection patterns and correlating same throughout the particular area, or prospect. Corrections for reflection travel time through the weathering, near-surface material, was measured at each seismometer by the observer recording shallow refraction travel time using very small dynamite charges detonated in shallow auger holes beyond each end of the seismometer array. After thus correcting for weathering, as well as the surveyor's surface elevation data, reflection arrival times were then converted to depths utilizing a chart based on time vs. depth information from nearby velocity profiles or from direct measurements in a nearby dry oil well in the general area. In the case of our central Oklahoma work, the prime reflection horizon was an Ordovician limestone known as the Viola. I still recall the wonderment of our crew's wives and their speculations that, since their men talked so much about Viola, she must obviously be quite a girl.

In correlating the reflection patterns in a new prospect, the initial reconnaissance objective was, of course, to discover any reversal from normal regional dip for the prime reflection horizon. Unless such reversal was discovered fairly early, the prospect would be abandoned. On the other hand, an indication of reverse dip would then warrant additional or fill-in reflection profiles to outline a hoped-for anomaly for a subsequent test well. One can't help remembering the aggressive acts of field scouts paid by oil companies other than one's client firm. Indeed, it became very much a game to do one's best in leading scouts astray in order that they would not be aware that interest was intensifying in a given prospect as the need arose to do fill-in profiling.

I would make a personal observation at this point. After starting my professional career as an electrical engineer in both research and production, I had come to the conclusion that exploration seismology gave me the greatest measure of pleasure and satisfaction for the field comprised a wonderful combination of technology and people. I was also impressed that people in exploration rate highest in both personality and integrity. I learned also that "Mother Nature" can be coy and inconsistent so that whenever you feel you have her completely discerned, she is then ready to "throw you a curve". So — in line with this experience, I finally concluded that whenever you meet a geologist, or a geophysicist, who is an egotist, you can be quite sure that you are listening to a beginner. Back in those days of the 1930s, before there were rigid controls on the number of hours worked, our prime objective was to satisfy the client company with a daily level of good information. If breakdowns occurred or trucks were mired overly long on muddy roads, a crew might then find itself working in the field by moonlight, but this could be offset by returning to town in the middle of the afternoon on other days if everything had worked perfectly.

That was also back in the days of "Prohibition", so alcoholic stills were very prevalent in the wooded hills of central Oklahoma. One afternoon I elected to visit my field crew, hoping that as I walked from my car through the woods I would not be shot as a possible "revenooer" by some trigger-happy "moonshiner". I found the crew eventually, but instead of working, the observer and his men were asleep in the shade of the recording truck. On awakening them with a few kicks, I then listened to a plausible story. Thus, while the hand auger crew was drilling its six 19-foot holes at the surveyor's stake with its yellow flagging in a creek bed beneath a large cottonwood tree, a farmer came over the brow of the nearby hill. He was Mr. Snow, the owner of that particular land, and wanted to know why holes were being dug there and, of course, the answer was "We are fixing to set off a little dynamite in order to find some oil." His quick response was — "Dynamite, Hell, I have six kegs of whiskey buried right there. Could you move up or down the creek?" Since the crew felt that the surveyor was probably not "exacting", they very willingly moved the shot point about 100 feet up the creek. In return for their ready cooperation, my crew explained to me that Mr. Snow then insisted that they walk over the hill to enjoy his hospitality, which consisted of "corn squeezings" fresh from the spigot of his still. "Mr. Green, since we are on his land, how could we refuse his kind hospitality?" I agreed, especially since our client representative had not happened on this sleepy situation before I did.

As I was beginning to regard Oklahoma as my newly adopted home, the East Texas oil field suddenly came on stream in mid-1931, resulting in a panic price drop in oil to as low as 10—15¢ per barrel, which resulted in an equally sudden cancellation of seismic exploration. Karcher and McDermott urged as many of us as possible to please get off the payroll for awhile by either returning to some previous job or to graduate studies. I elected to help out by communicating with my old friend, Charlie Litton. "Sure, come on back to your old job, now that you have probably learned your lesson." Very soon after I had returned to Palo Alto, ITT announced that its Federal Telegraph operation would be moved back to Newark, New Jersey. Charlie elected to resign rather than move back East with the result I had no other choice than to take his place as manager of the Electronics Lab back in Newark. This seemed like retrogression to me, but Eugene McDermott came to my rescue 12 months later when in July, 1932 he telephoned that the depressing effect of the East Texas field was beginning to be absorbed. "Would I please come back to my job as party chief, especially since a small anomaly I had mapped just south of Perry, Oklahoma was now a new oil discovery?" I said "yes" and then tackled the task of telling Mr. Clements, Chief Engineer of Federal Telegraph, that I planned to resign. His immediate response was — "Cecil, if this is your strategy for getting a raise in pay, the answer is no!" He was flabbergasted when I told him that I was completely serious about leaving. So his next response was, "Where can you possibly go for another job in this great depression?", followed by giving Ida and myself a very pleasant farewell dinner.

We were soon on our way to Corning, New York, after loading 6-S2 seismometers in my car trunk obtained from Erik Jonsson, then in charge of a small GSI shop in nearby Elizabeth, New Jersey. My first job upon returning to GSI was shooting a small prospect at nearby Campbell, New York, with W. W. "Ike" Newton as observer, J. W. Thomas, computer, and Phil Gaby, driller, on a portable spudder rig. The client was a small independent operator who drilled a test well on our resultant anomaly — then the following year he brought Karcher and me back into a Brooklyn Court because his well turned out to be a dry hole instead of a gas producer. The judge soon dismissed the case since seismic exploration is certainly not a direct oil-finding method, nor could the absence of geologic structure be verified by one lonely test well.

The next four years were full of interesting experiences in many small towns throughout Kansas, Oklahoma, Texas, and Northern Louisiana. Thus, while working for Seaboard Oil Company, I recall mapping a structure at Indian Gap, Hamilton County, Texas. A test well revealed the presence of gas rather than oil for which I received a friendly "bawling-out" by Chief Geologist, Ray Stehr — his criticism being — "What good is gas when we want oil?" Another structure was mapped for Seaboard, south of Minden, North Louisiana, which was drilled to the Woodbine sand — a dry hole! I had no trouble getting permission to run a velocity survey before the well was abandoned. Schlumberger cables were not then available so it meant taping about 5,000 feet of lamp cord to a simple acid bottle line. Fortunately, we made our velocity measurement as we lowered the seismometer to the bottom of the well, because as we started to raise the seismometer through the heavy mud, the small steel line broke. So we obtained our velocity information, but at the cost of our geophone. Ray Stehr laughingly stated that we had done a good job of helping to fill a useless well! A year later, Stehr changed his mind completely. Seaboard had decided to resume drilling, and he worried about the possible difficulty of drilling through our lost geophone. He also noted in a friendly way that we might be sued in case of blockage. As might be expected, the geophone was merely pushed into the sidewall as the well was carried on down to the Pettit zone where it became a distillate producer.

On being promoted to the position of supervisor, life became even more exciting because of continued adventures but on a worldwide basis. An episode in Southern California is worth mentioning. While headquartered in Los Angeles, I tried to add to our client customers by calling repeatedly on the Chief Geophysicist of the Superior Oil Company. I finally caught up with him at a social gathering. When he admitted that he had been avoiding me at his office, I told him that I was anxious to talk to him about a possible contract with GSI. His response was, "Over my dead body!" Naturally, I asked him why. His answer was straightforward, "When I was about to graduate from 'O.U.' ten years ago, I wrote your dear Dr. Karcher asking for a job. I am still waiting for an answer." I have remained forever mindful of that important lesson in people relations.

Accompanying George M. Cunningham, the Chief Geologist for Standard Oil Company (California), around the world during 1939–1940 provided more experiences. GSI was then conducting our first seismic marine survey in the Persian Gulf — one off Bahrein Island. This was much before the advent of seismic streamers. Fortunately, the water was only 20 feet deep — also crystal clear because of the bottom material being a smooth white marl. We were able to use a vertical spread comprising four Type-S10 seismometers on either side of the shot point. Each seismometer was dropped overboard from a small Arab dhow at marked intervals. They generally settled upright on the bottom because of being surmounted by a small metal parachute. Before shooting began, it was necessary to check visually from the surface to make certain each geophone was in an upright position. In case of any being on their side, one of the Arab helpers with pearl-diving experience would dive down to it, for which he was glad to receive a one rupee bonus. We did not know enough to detonate charges in the water, so shot holes drilled beneath the bottom, with casing reaching to the surface for easy loading, were provided by a barge-mounted rotary drill rig. Surveying was by sight triangulation with towers on frequent islets and coral reefs.

One night, when I was on the headquarters barge, a strong seasonal gale from the north (a "shamal") not only caused us to drag our anchors over the smooth marl bottom but, more seriously, the Arab dhow carrying our supply of dynamite disappeared. Knowing the direction of the previous night's wind, it was easy for our observer and a retired British Navy officer to find the dynamite dhow on a beach in the nearby sheikdom of Dhofar. Our men were very relieved to find the dynamite intact in the dhow. After apologizing to some of the local Bedouins for defiling their beach with the uninvited dhow and its unsavory cargo, our men promised to return the next day with a larger craft to remove both the dhow and its cargo. This announced plan was, of course, a real mistake, because the local bedouins very quickly surmised that anything worth coming back for must be valuable in spite of the fact that, on tasting a slice from one of the dynamite sticks, it did not seem a palatable food. So when our men returned the next day, the dhow was still on the beach, but its cargo had disappeared. Without mentioning the word, dynamite, our men tried to scare the local bedouins by describing the cargo as very unhealthy and therefore undesirable to retain in their country. By that time the local sheik had taken full possession and, on being pro-positioned, he demanded a preposterous ransom equivalent to $25,000. There was only one thing left to do — that was to seek the counsel and aid of the British Political Agents of Bahrein and Dhofar. The Bahrein agent agreed to help by first stating that marine law would apply wherein anyone who salvages a ship's cargo is entitled to only half its market value. Our estimate was $5,000, so in quick time the two British agents obtained the release of the dynamite with our payment of $2,500.

At that time, the *C*alifornia–*A*rabian *S*tandard *O*il *C*ompany (CASOC), the predecessor to the *A*rabian–*A*merican *O*il *C*ompany (ARAMCO), was very unhappy that a test well drilled on seismic data known as Abu Hadrya Number One in the northeastern part of Saudi Arabia near the Neutral Zone was giving indications that it would be a dry hole. The well had already penetrated without result all of the known Jurassic producing zones identified in the Damman Dome to the south at Dhahran, but it also had reached the unheard of depth (for foreign areas) of 2,900 m. Everyone in CASOC's Produc-ing Department wanted to abandon the well and, at the same time, use this as a good reason for getting rid of the misleading geophysicists. However, the drilling superintendent had a peculiar feeling, based on his imagined smell of sulphur in the returning mud, that he might strike some new and deeper oil-producing zone. So it was agreed that he continue drilling until he either struck basement or ran out of drill stem or casing.

This was the unhappy situation when George Cunningham and I left for Karachi, India. While we were together visiting a seismic party an overnight train ride up the Indus River Valley, George received two cablegrams — one from San Francisco and the other from Dhahran. He decided to open the San Francisco message first because of it probably being from his boss, Clark Gester, the SOC Vice President for Production. Sure enough, it was, and the cryptic instruction — "Please return to Saudi Arabia before going on to Indonesia for the express purpose of making a post-mortem study *re* the loss of $1.5 million on the dry hole at Abu Hadrya." While still moaning about such a disappointing message, I suggested that he read the second cablegram. It was also a cryptic message signed by Chief Geologist, Max Steineke at Dhahran, that Abu Hadrya had just stuck oil at a depth of 10,200 feet and at a rate of 15,000 barrels per day. My next remark was to suggest that this was a very appropriate time to order up a cold beer!

Soon after World War II, I found myself visiting Richmond Petroleum headquarters in Bogota, Colombia, before taking off on an unusual travel excursion to visit a seismic crew in the upper Magdalena River Valley. I was accompanied by the GSI Party Chief, Martin Linwell, by "Chuck" Ramsden, a field geologist for Richmond, and by a Colombian medical student, known as a "practicanti", who was carrying a thermos bottle of yellow fever serum to inoculate our crew because of the reported threat of a yellow fever outbreak. We first traveled by DC-3 aircraft from Bogota to Bucaramonga in the coffee-bean country, then by a small rail car powered by a "Kalamazoo" one-cylinder gasoline engine downgrade to Puerto Wilches on the Magdalena River, followed by moving on down the river by outboard-engine-powered canoe to Bocas del Rosario. Horses had been scheduled to meet us there for a ride via jungle trail to our houseboat on a nearby tributary of the Magdalena. While waiting in the town plaza for our horses, we decided to have a beer in a small grocery store. While the store owner was expressing his regrets for not having any "Cervesa" (a beer), "Chuck" opened the door of the refrigerator and found it fully stocked with beer as well as Coca Cola. The manager was full of apologies as he pointed to the heads of several peons who were watching us through the half open front door. He pleaded with us to accept Coca Cola instead of Cervesa because if he sold beer to us then he would have to sell beer to the watchful peons. Since their pockets were full of money, it being pay-day, they would drink too much, get terribly drunk, and wreck his store. We, of course, were willing to cooperate until we walked over to the waiting horses and found our "practicanti" missing. Very soon we saw him approaching us, carrying his thermos bottle plus a brown paper sack, and wearing a big smile. When asked as to the contents of the sack, he produced four cold bottles of Cervesa. We asked, of course,

"Where did you get them?" His reply "In the store you have just left." Naturally we asked, "How come, when we were turned down?" "Ah — but Senors, you offered money, while I asked if he would like an inoculation as protection against a bad yellow fever epidemic that is about to break out?" Naturally, our regard for our young Colombian friend increased tremendously in those few moments.

In 1955, I had occasion to visit a seismic party near Exmouth Gulf in Western Australia. I had heard a year or so previously that CALTEX, in partnership with *Western Australia Petroleum* Company (WAPET), had acquired a new concession in that area. In response to my inquiry concerning seismic service, I was informed that the prime feature of the concession was a very prominent surface anticline — about 30 miles long, two or three miles wide, with 500 feet of relief and known as "Rough Range". So all they really had to do was drill. As expected, oil was struck in the first well. At this juncture the market price of WAPET stock down in Perth escalated overnight from 10 shillings to 10 pounds per share. A north offset location was then drilled and it was dry, clearly a case of advancing in the wrong direction. Obviously a south offset was prescribed, but horrors, it was dry also! Then, much in desperation, offset locations both east and west were drilled and tragically were also dry. By this time, the common stock of WAPET had descended back "into the basement". Our seismic task became twofold. The first step was to resolve this enigma. In this instance, the seismic data seemed to indicate a situation of complicated faulting with no apparent relation between the subsurface structure and the prominent surface anticline. The other portion of our assignment was for me, as a so-called "outside expert", to go down to Perth and explain to news reporters why this peculiar and expensive operation was not just a mean maneuver by WAPET and CALTEX to "fleece" the gullible public who had put their savings into what turned out to be an unfortunate gamble for everyone concerned.

In closing this musing, I would like to express my admiration and high esteem for the job that the Society of Exploration Geophysicists has done during the first half-century of its existence. During my term as the Society's President in 1947—1948, we were able to convert the Tulsa headquarters run as an informal office for several years by the devoted efforts of Elizabeth Stiles into a formal business one directed by Colin C. Campbell. Moreover, the concept of "Local Sections" came into being then. This concept not only has stood the test of time but has grown to include sections in Australia, Canada, and Peru. As a result, our professional group is more than living up to the hopes of its founders that it be truly THE Society of Exploration

Geophysicists and know no national boundaries when it comes to improving the economic and social welfare of mankind.

Conception of the Common-depth-point (CDP) Method of Seismic Surveying
— W. Harry Mayne[7]

In late 1949 or early 1950, I made a field trip from the San Antonio office of Petty Geophysical Engineering, to an area somewhere in South Texas to assist in the startup of a new crew. As I recall, the area turned out to have a massive caliche surface which produced severe surface noise (ground roll) of unusually high velocity. Analysis of the noise indicated that extremely long (600 to 1000 ft) arrays would be required to satisfactorily attenuate this noise. Unfortunately, this solution was not considered permissible because the objective of the survey was a search for low-relief structure accompanied by small faulting. The "smearing" effects of such long arrays were thought to be so serious that the desired definition could not be obtained.

Consequently a marginally acceptable choice of array length was made, and the survey proceeded with data of dubious quality. I was far from satisfied with the results obtained with this expedient solution, and continued to think about the problem after returning to San Antonio. How might such long wavelength noise be attenuated without "smearing" the subsurface? The search for an answer to the question led to the basic concept of CRP (also CDP). To do this, an array was formed by summation of the recordings from a series of source-receiver pairs with progressively greater source-to-receiver distances, but with all having a common midpoint. With the concept in hand, a patent was drafted and filed in July, 1950.

At this time, however, there were two major deterrents to routine application of the technique. First, no field-worthy system for making electrically reproducible recordings was available at the time. Second, no acceptable way of applying the normal moveout corrections to the electrical signals prior to summation existed. As a matter of fact, the only way in which I visualized the method could be implemented was by purely graphical means. In 1950 this represented the only viable solution, and the method would probably have remained a scientific curiosity had not other parallel developments taken place.

The Patent Office processed applications in leisurely fashion in those days, and issuance of the patent was not insured until late 1955

7. Courtesy of Petty-Ray Geophysical Operations, Inc., Houston, Texas, a Geosource, Inc., subsidiary.

with final publication in January 1956. By this time, of course, magnetic tape recording had become established as a fieldworthy recording system, but was limited to routine filtering and trace-mixing playback options. The only missing link was the lack of transcription correction capability.* This remaining deficiency began occupying more and more of my attention by mid-1955 as it became increasingly apparent that significant patent coverage would probably be obtained.

All of the geophysical instrument manufacturers were approached with an outline of the requirements for processing equipment which could perform the necessary correcting, transcribing, and summing operations. With the exception of one company, the Texas Division of Brush Electronics, no one could understand the need for transcribing the corrected data to another tape and refused to consider making such a machine. After considerable negotiation, Brush prepared preliminary specifications and submitted a price quotation on a suitable machine, and Petty entered into a development contract in late 1955. Brush was a newcomer in the geophysical instrument business, and not fully aware of the high standards required by the industry. I suspect that their education was a rather painful engineering and financial process. In any case, after two years' elapsed time, considerable engineering trauma, and a price escalation to 2½ times the original price quotation, they developed quite a competent machine.

Unfortunately, the geophysical business had suffered a severe recession in the interim, and Petty as a relatively small contractor was dubious about the business wisdom of risking $250,000 (the final purchase price) on an untried instrument designed to perform a process which had not yet been accepted by the industry. It was a classic case of the chicken and the egg. We could not demonstrate effectiveness of the CRP method without the machine, and we didn't want the machine unless the method proved to be economically viable. All this against the backdrop of a widespread "It will be too expensive even if it works" expression of opinion from a large segment of the industry.

Consequently it was with some relief that we received an inquiry from a major oil company requesting us to relinquish our order for the machine to them. Negotiations were completed between the three parties at the SEG Annual Meeting in Dallas in 1957, and they assumed our order and subsequently took delivery on the first machine of its type. It should be gratifying to all concerned that this particular machine continued to perform creditably until displaced by digital processing in the mid-sixties. Then once again it was back to the drawing board!

*GSI's magnetic disc system did not have enough capacity to handle the CDP requirements.

To regress a bit, we had realized shortly after conception of the idea that it had the capability of significantly attenuating "multiple reflections"* as well as general improvement in the signal-to-noise ratios. Since this could not be accomplished by any other means at the time, we decided to emphasize this advantage in trying to gain industry acceptance. It also happened that at least two oil companies, Phillips Petroleum Company and Shell Oil, had independently conceived the same idea prior to publication of my patent and had been separately developing it on a proprietary basis. Quite logically then, they became the first licensees under the patent, and this acceptance helped to strengthen our commitment to the idea. Shell, in fact, later bought several data-processing units patterned after the prototype which was originally designed and built for us, but delivered to another oil company.

Thus it was that as late as early 1958, Petty was still unable to implement the method and was desperately searching for an alternative data-processing unit which we thought we could afford. About then just such a unit was introduced by Techno Instrument Company in the form of the "Dekatrak"® auxiliary unit for their mechanical moveout corrector. The first of these units, manufactured under license from Continental Oil Company, was delivered to us in mid-1958, and we were finally in a position to fully implement the CRP method. General industry acceptance was still some time away, however, and the first clients who supported this early work were the Texasgulf Producing, Pure Oil, and El Paso Natural Gas companies. Through the interests of these companies we were able to develop a considerable body of experience in a variety of areas, and confirm our hopes that the method offered significant potential for data-quality enhancement in most problem areas. This information was gradually percolating through the industry, and a great deal of interest was apparent by late 1960 at the SEG Annual Meeting in Galveston. As evidenced by licensing activity, and the proliferation of specialized data-processing equipment designed for the method, interest increased rapidly, from 1960 through 1963, and the last major remaining skeptics had finally accepted the method by the end of the 1960s.

Paralleling a part of this period, another technical development was gaining momentum, namely the digital recording and processing revolution. This was also a fortuitous circumstance because the CDP

---

*These are "secondary" reflections which have been reflected more than once, either between the surface and one or more subsurface horizons, or between two subsurface horizons. They are frequently mistaken for bona fide "primary" reflections and hence make accurate interpretations difficult.

®is a registered trademark of the Techno Instrument Company.

method and digital processing were mutually synergistic. The widespread use of the CDP concept also provided the masses of data required to make digital processing attractive, and digital processing soon developed the capabilities of routine velocity analysis and automatic determination of static corrections. Both of these latter capabilities enhanced both the effectiveness and the convenience of the CDP method. One other synergism between apparently unrelated technical developments can be cited between the CDP technique and the development of non-dynamite sources, both land and marine. The development of these inexpensive and effective new sources also undoubtedly impacted acceptance of the CDP method by notably improving its cost-effectiveness.

The Way It Really Happened — The Invention of the VIBROSEIS® System[8]
— John M. Crawford

Bill Doty had just returned from a seminar at Massachusetts Institute of Technology and was reporting to me concerning the new mathematical concepts on "Information Theory" which had been presented. He laid the June, 1950 copy of the journal *Electronics* on my desk. It was open to page 86 where an article by Lee & Wiesner discussed the principles of "Auto-Correlation" and "Cross-Correlation". Bill said something like this, "Here is a principle which we should be able to use in our business, but I don't yet know how."

After he left my office, I scanned the article briefly and we got together again. We agreed that this principle would allow accurate measure of the travel time of a seismic wave through acoustic material *if* we could provide a signal which was "unique" (did not repeat itself) and *if* we had an accurate copy of the signal as transmitted.

The article taught a "white noise" signal as usable but we knew of no available equipment for transmitting such a signal with sufficient power to be of much interest. Again Bill left me to my own devices for a short time. I then recalled an earlier conversation with Bill about his experience as an officer on a minesweeper during World War II. He had described the "Bumblebee" signal which the minesweeper transmitted as a seismic wave through the water in an attempt to detonate "acoustic" mines which were sensitive to certain sound frequencies. The signal varied continuously in frequency from high

---

8. Reprinted from the company journal, *Mertz World*, of March, 1982 by courtesy of Mr. Don Mertz, President.
®Registered trademark of the Continental Oil Company.

to low to high, etc., hopefully hitting the triggering frequency in the process.

It occurred to me that here was our "unique" signal. That is, it would match itself only at one time-phase relationship during a single frequency excursion or "sweep". Also, I remembered an article in the *Oil and Gas Journal* from several years back which had described an old principle for making a mechanical vibrator. Counter-rotating, off-center weights were geared together in such a way that horizontal forces were cancelled and vertical forces added in combination. To make a "sweep" signal — all we needed to do was to continually speed up or slow down the driving engine or motor.

When Bill returned, he agreed that we had a system with possibilities. From these and other conversations between Bill and me the ideas flowed thick and fast. With Frank Clynch's help we came up with a mechanical vibrator, driven by a borrowed gasoline engine and held in contact with the earth by driving a loaded water truck (also borrowed) up on some pads rigidly attached to the vibrator. A seismic detector fastened to the vibrator furnished the sample of the "sweep signal" as transmitted.

With this vibrator and a bunch of electronic gear assembled from some very unlikely sources by some highly effective electrical engineers, we were able to get our first modest field results, and the rest is history. (Thus) the basic invention was made in a single working day by two people — each contributing vital building blocks from mostly unrelated fields. The principle building blocks were not original with either Bill or me, but the combination and adaptations were ours by means of a cooperative effort. It is quite probable that neither of us, working by ourselves, would have come up with the system.

Vignettes on Unusual Oceanographic Expeditions

The Unforeseen Demise of Operation Skijump II[9]
— John Holmes[10]

At 5 a.m. on Tuesday, 25 March 1952, we took off in the R4D to fly 600 miles north, land, meet the covering P2V with two reporters on board, get refueled, and come home. We got out about 500 miles

---

[9.] The general intent of Operation Skijump is described on page 147. Skijump II had already had its share of problems by 25 March for it had been found that its gravity meter went off scale at those latitudes and geophones froze up at temperatures of $-40°C$.

[10.] Technical Director of Project Skijump. The narrative is extracted from Holmes' letter report of 1 April 1952 to the U.S. Naval Hydrographic Office. As of 1981, Holmes was serving as an oceanographic engineer with the Ocean Data Buoy Project, National Oceanic and Atmospheric Administration, Bay St. Louis, Mississippi.

(beyond our point of no return with our heavily-loaded aircraft as far as gasoline supply goes), and then the trailing P2V reported it had a bad oil leak in one engine. We got a little worried. Then the P2V said it was returning to Barrow. So we went ahead and landed with no cover aircraft. Our R4D had about 300 gallons of gas on board which would either fly us 300 miles or keep us going for three days on the ice, since we burn 100 gallons a day when sitting on the ice doing oceanography. So Val and I took our time getting a good oceanographic station while we all wondered about the P2V. It truly was having trouble. About 300 miles out of Barrow it lost the engine and had to proceed on just the other. About 200 miles out, they had to lighten ship by jettisoning the bomb bay tanks. When they got to Barrow and started the landing, one side gear wouldn't come down and they had to take a "wave-off" on one "fan" with gear and flaps down. This is a marginal maneuver. The side-mount then came down and the landing was successful. That was enough for the reporters. They took off the next day! The engine had to be changed, of course.

At 10.08 on Wednesday the other P2V showed up overhead and wanted to push on north. I wasn't in favor of it. The P2V couldn't land on the ice we were on since it was only 41 inches thick. So we took off and went about 200 miles north until our gas was very low and we had to land and let the P2V refuel us. But there weren't any satisfactory places. We chose a large floe of old ice and set the R4D down. She came to a screeching halt since Ed didn't have the skis down. The ice was 7½ feet thick and the snow about 15 inches thick. It was quite rough. The P2V came in and used reverse pitch to stop. One engine didn't reverse; the other wouldn't come out of reverse and also caught on fire. Everybody was out of the P2V with fire bottles and checked the fire. It was the reverse pitch relay. The "mechs" had to take the one off the good engine and put it on the bad one to get the prop out of reverse. The right ski was broken just aft of the wheel. Meanwhile, the R4D was stuck and we had to dig the snow from under the skis to get the skis down. We got the R4D going and made a taxi-run to see if it would get off the snow. It seemed that it would, so we went to the P2V and got a load of gas. Val and I made another oceanographic station during the evening and planned for a 9 a.m. take-off for Barrow.

The P2V got off with no trouble. Next day, we all got in and Ed made a take-off run. It didn't want to fly — too much drag of wheels in the snow. We tried it again. Then the left landing gear collapsed. The "prop" dug into the snow breaking the front housing off the engine and next went 300 feet down the runway in front of the R4D. When we stopped, someone yelled, "Bail out!" We tried to open the door but it was jammed with gear. Someone pushed one of the "mechs" out the escape hatch head first and with a fire bottle at his heels. As soon as the door was opened, we began throwing food and survival gear out on the snow. After the immediate danger of fire had passed, we tried to get the "put-put" going to get a radio message out. But the "put-put" battery had been outside all night and was too low to start the "put-put"; so we had to get the plane batteries out to start the "put-put" to get the radio going. The only station we could raise was Thule who sent it to Westover to Fairbanks to Barrow. We set up a tent and put the useless stuff in it. A watch list was made out to keep gas in the "put-put" and the heater. The rest of Thursday we got the ship cleaned up and waited for word from Cdr. Coley. We got it. He wanted a runway out on the ice and would come Sunday.

The following two days were cold, $-30°F$ during the day and down to a reported $-50°F$ during the night. The weather was clear. We had one red shovel and two axes to make 3,800 feet of runway, so we divided into two shifts to work the one shovel the maximum number of hours. We had plenty of food and heat. Snow was melted for drinking water which became too hot. I made a sleeping bag on the deck, but sometimes I was too hot to sleep inside the bag. By the third day the boys were going outside in their shirt sleeves to get snow to cool the drinking water.

The nights were very light — the sun set about 9.30 p.m. and rose about 3 a.m. There wasn't a bit of movement of the ice that I could see. While on watch, I would sit in the cockpit and listen to the radio. We had wonderful reception — Seattle, Chicago, Radio Moscow, Radio Ankara, BBC, AFRS. The Armed Services Radio Service has much too much sport on it. Almost every other 15 minutes, they pound on sports about like Radio Moscow pounds on bacteriological warfare. Some mornings we could see fine mirages, with the ice behind the horizon looming high in the air. While using the axe on the runway I found some ice worm holes, the first I have seen. This floe was about three years old (a guess based on nothing in particular), and the salt cells had melted vertical holes in the old pressure ridges that resemble worm holes.

Meantime, Cdr. Coley had flown to Kodiak to have the broken ski repaired. He took three wave-offs at Kodiak because of "williwaws"; each time he made an approach one of the skis would be blown in the vertical position. I am told it was rather hairy. The ground crew fixed the broken ski fast and Cdr. Coley came back to Barrow Saturday night. He tried to reverse pitch on the landing and again only one reversed so he ran over one of the runway lights when the brakes failed to work. Sunday morning his P2V took off from Barrow to come out for us. Also on Sunday the snow began to blow and we had grey weather. We were told that we would have only 2,000 lb. total weight limitation; for 9 men it meant that we would have 220 pounds per man, including body weight, clothing, and scientific records. When the P2V was two hours out, our radio transmitter started to act up. Moreover, we had no standby radio until they would be within VHF (line-of-sight) range. Fortunately, the transmitter continued to work after a fashion. The P2V circled about 5 times, then came in for a landing. It was a very smooth landing except that he didn't use reverse pitch so he ran 300 feet beyond the flags at the end of the runway. Everybody jumped out and began to guide the P2V back over the two-foot pressure ridges. It took about 15 minutes and full power to get the P2V out of the boondocks. Also, the other ski had been broken on the landing in exactly the same weak point as the first ski!

After our pictures were taken, the survival gear was loaded aboard. Just three hours after the P2V had landed, we were all on board and counting the major bumps in our makeshift runway as we took off. There were four of these and we clearly felt all four before the P2V lumbered into the air. Thus we were back to Barrow about 8.35 p.m. on 30 March 1952 after flying 800 miles to get home. But my main conclusion still is:

TRANSPORTATION IS STILL THE MAJOR DRAWBACK TO
ARCTIC OCEANOGRAPHY!

Glossary of Jargon:
R4D: Ski-equipped DC-3 (or C-47 in U.S. Air Force parlance).
P2V: Navy Neptune long-range patrol aircraft built by Lockheed Aircraft Company. This was a piston-driven aircraft with the longest flying range ever built. As a commander, the future Chief of Naval Development, RADM Thomas D. Davies, USN, flew the "Truculent Turtle" (a P2V-1) non-stop and without mid-air refueling from Perth, Australia to Port Columbus, Ohio, on 29 September—1 October 1946 for a distance of some 17,975 km.
"Ed": Commander Edward Ward, USN, project officer for Skijump and pilot of the R4D.
"Fan": Propeller.
"Fire bottles": Bottles of pressurized carbon dioxide that would put out oil and gasoline fires.
"Mechs": Aircraft mechanics.
"Prop": Propeller.
"Put-put": A small, noisy gasoline-powered electric generator.
"Heater": A Herman-Nelson heater burning gasoline and extensively used in the Arctic for blowing hot air wherever one needed it.
"Williwaws": Violent, persistent wind that poured in gusts down the mountainsides surrounding the airstrip in Kodiak. This particular airstrip is an interesting one to land on for it ends abruptly against a steep mountain slope. Thus, taking a wave-off means doing it early or never again.
"Pictures were taken": The Naval Photographic Service eventually assembled an excellent movie of the various facets of the Skijump Operation, including the final abandonment of the oceanographic aircraft deep in the Polar Basin.

*The International Indian Ocean Expedition and its
Scientific Aftermath
— Dr. Wilbert M. Chapman[11]*

The *I*nternational *I*ndian *O*cean *E*xpedition (IIOE) was not an expedition in any ordinary sense of that term.[12] It was an oceanographic "happening" in the sense of "Woodstock" and other recent manifestations of hippydom in the United States. Oceanographers went to the Indian Ocean and did their "thing" in a mildly orgiastic way. Those were the days of "oceanography for fun". Science was rampant in the world. Laymen had a mystic feeling that anything labeled "science" was likely to be a pretty good thing, and perhaps, even useful. Oceanography was a far-out science even by normal standards, and the Indian Ocean was particularly far away and romantic. The bill came to at least $60 million; it is likely that if all the odds and ends and continuations could be tracked down, it might come to twice that.

There were rational voices heard. Georg Wüst, the German, tried to get station patterns laid out in space and time so that rational results could be expected in the end, but we were laughed out of the Councils. Colin Ramage took the bit in his teeth and, by strenuous personal effort, succeeded in getting some rhyme and reason injected into the meteorological end of the thing. I complained bitterly in whatever forum I could reach that whereas the promoters of IIOE were using as a principal money-raising excuse the need of the region to increase its yield of fish, (yet) none of the organizers knew the slightest thing about the fish business, and nothing they were planning was likely to affect fish (catch) in the region much within their lifetimes. I got even less attention than Wüst.

I had dreamed for some years of getting meteorologists, oceanographers, marine biologists, and fishery scientists crowded together so tightly that it was obviously mutually beneficial to work together jointly and could not escape doing so. The fragile, complex, and everchanging weather, surface circulation, and biological productivity of the Arabian Sea are so apparently hooked together that even the most purblind scientist in one of these fields can't help but see the relationship between what he is doing and what the scientists of other

11. Remarks consolidated from letters of 14 April 1969 and 31 January 1970 to John Marr of the United Nations Food and Agricultural Organization and to an associated distribution list of 75 oceanic specialists and bureaucrats. At the time, "Wib" was serving on a score of committees, panels, and working groups, traveling over 200,000 miles a year, and never at home. He tried to slow down by taking only work assignments that provided first-class air transportation (he was a large man), but death still came at age 57 during 1972.

12. A chartlet of IIOE cruise tracks is carried as Fig. 6.12 (page 227).

persuasions are doing. There is no part of the world I know where great biological productivity is so closely tuned to mixed layer circulation and the mixed layer circulation so at the whim of movement in the lower atmosphere. And I assume the action in the latter is governed by what is going on in the troposphere, and that, in turn, is governed by the flow of energy into it from the ocean. Here it is most obvious that the year-to-year changes in energy flows in that part of the (temporal) spectrum between perhaps three to fifteen years is more important in its biological consequences than are diurnal, monthly, or seasonal changes. We need to get the descriptive oceanographers to thinking on the geographic scale that Jakob Bjerknes does, which would be a revelation to them, and a boon to us fishery types down the trail. I have said before, and will say again, that Bjerknes is probably the most effective fishery oceanographer at work in the world today without even trying; he would smile his wry smile to hear it because that is not even what he is trying to do. But as we have proceeded with our internal company prediction work on fishery production here and there about the world, the economic effects on risk investment, management, and growth of large-scale fishery development in the fairly short term, it is evident that climatic variations loom larger all of the time.[13]

In the two years from mid-1961 to mid-1963, I made a series of reconnaissance trips throughout the (IIOE) region which touched most countries between Cape Town and Bali. I concocted a scheme to concentrate on (the Arabian Sea) part of the Expedition and see if a little sense could be made of it from the fishery viewpoint. This was circulated to SCOR, FAO, UNDP, and IOC.[14] The only spark it struck was at the newly organized IOC where (Warren) Wooster was just putting things in order. He put together a scheme to be financed by UNDP through IOC to do these things. It fell flat. IOC was science; UNDP was organized to promote economic development, not science. FAO was the chosen instrument for UNDP schemes related to fishery development; FAO did not want to tackle the job and did not want IOC horning into the "fish racket". Wooster's scheme, therefore, laid a big, fat egg. We persisted in fishery development planning in the Arabian Sea region without any notable effect. The revolutionary fervor that was already wracking the Aden region came to full flower

---

13. *Author's note*: At the time of writing, Dr. Chapman was Director of Marine Resources, Ralston Purina Company, and based in San Diego, California.

14. The abbreviations indicate: SCOR: *S*cientific *C*ommittee on *O*ceanic *R*esearch (an international group of oceanographic academicians); FAO: *F*ood and *A*gricultural *O*rganization of the United Nations based in Rome, Italy; UNDP: *U*nited *N*ations *D*evelopment *P*rogram which has its own funding; and IOC: *I*nternational *O*ceanographic *C*ommission of the United Nations based in Paris, France.

during 1967 . . . amid a certain amount of shooting and roughing up that suspended (UNDP) activity directed towards fishery development there.

All of this rumpus looked, in 1967, as if it might blow away in a year or two and, when it did, it would be possible to begin fishery development where we had left off. The field work phase of IIOE had been completed. The oceanographers had been so strongly attracted to the Arabian Sea that their ship-tracks intertwined and overlapped like a web built by a drunken spider. There could not help being a yield from this immense effort that would be informative to fishery development efforts in the region if the useful parts could be distilled from the dross. We concocted a scheme to do this and at the same time get some use out of the comatose Aden Fishery Project. With the consent of all hands, FAO contracted with the Institute of Marine Resources, University of California with Dr. Warren S. Wooster as Principal Investigator. The result was the volume by Wooster, Schaefer and Robinson, *Atlas of the Fishery Oceanography of the Arabian Sea*.[15] Doing jobs of analysis and synthesis of this sort is cheap if done completely outside the region where scientists competent to the task have all the computer facilities and other implements of their trade conveniently at hand, while it cannot be done in the region at all, or hardly at all. This whole effort cost something less than $15,000 and was accomplished to the printed and distribution stage in only a few months after the contract was let.

The Wooster–Schaefer–Robinson atlas was a quick, dirty job done on the cheap, if competently, to get such data as were then available out immediately in useful form for fishery development planning purposes. It served its purpose exceedingly well. Prior to its publication, personal observations, odds and ends of data, suspicions, and hunches had given a few observers the belief that the unexploited living resources of the Arabian Sea were substantial, but the evidence was almost hearsay and there were sceptics. Subsequent to its publication, there was a broad data base that sceptics and all (others) could examine, indicating that the Western Arabian Sea, where substantially no commercial fisheries existed, was so rich in primary biological productivity as to warrant comparison with the Peru Current and the Benguela Current areas which supported the two most prolific fisheries in the world. It (also) provided a solid basis for suggesting to the countries of the Indian Ocean Fishery Commission and to UNDP that the cost of a second-stage inquiry was warranted. Thus arose the

---

[15.] Dr. Milner B. Schaefer was an international fishery expert, while Margaret Robinson of Scripps was an outstanding compiler of oceanographic data.

Indian Ocean Fishery Survey and Development Program. It also illustrated that special contracts designed to dig specialized information for special fishery purposes out of the treasure-holds of the IIOE are both practical, economically cheap, and effective. The lesson should be applied in the egg and larvae field, in further detail within the physical oceanography field, and inquired into in the meteorological field.

The whole Arabian Sea, and particularly several of its parts, seems to be biologically fragile in a way, and on a scale not experienced elsewhere. Examples are given by the massive fish kills observed far off shore in 1957–1958 where it is difficult to understand not only the fish kill, but why there were so many fish there to be killed to begin with, and the smaller fish kills observed from time to time off northern Somalia and the Muscat coast, as well as the substantial fish kills observed on almost an annual basis along the Kerala and Mysore coasts of India. There is so much uncertainty, unexplained variability, and fluctuations in yield from present fisheries, etc., that it is a brave man or government who would invest much money in the development of these resources until some of these uncertainties are "riddled out", and that part of the risk reduced.

While the UNDP was not, and is not, set up to support science, when a reasonable case is made that practical development results are likely to be stimulated to flow by the doing of a piece of quite scientific science, UNDP is quite prepared to bend its criteria in a practical manner so as to get the job done. This is particularly important to you from here on in. If, in any instance, you can couch your proposal in a manner that useful developmental results can reasonably be expected to flow from what scientists would call science, UNDP will go along. The trouble is more commonly with the scientists than with UNDP. The scientists want to do science and feel that they will be demeaned in their trade if practical results flow therefrom. Wooster and Schaefer are not that kind of scientist. They are quite pleased to get the job done competently, under whatever terminology will please the funding agency, and let the chips fall where they may. There are other competent scientists who will do likewise.

Hank Stommel (has) told me that Nimbus III produced extraordinarily useful data for that region, but you will have to ask him what that means, and whether anybody is working those data up into any useful form.[16] Paul Laviolette at NAVOCEANO has actually

16. *Author's note*: Next to Walter Munk, Henry Stommel of Harvard University became probably the most brilliant of the post-war generation of physical oceanographers within the United States. His first major work of note was mathematically explaining why "jet streams" in the ocean, such as the Gulf Stream and the Kuroshio Current off Japan, flanked the east side of continents in the Northern Hemisphere.

published a "first shot" at sea-surface temperatures in the region both from ships of opportunity and from Nimbus III which you should get. In any event, you must learn about ocean variability in your domain. You don't have to be as fancy about it as do oceanographers and meteorologists. Reasonably precise surface temperatures and barometric pressures from several points in your region taken over a long enough period of time will probably provide you more of what you need to know in the long run than several other sorts of more sophisticated and expensive exercises. To get these series you have to start at some point in time, and there is no time like now. I have been trying to talk others into doing this for some time to no avail, and I think a "docks-of-opportunity" program globally in the tropics and sub-tropics would be a hell of a lot more useful for IOC to undertake than a lot of the other more fancy stuff they spin their wheels on. But I have been a voice crying in the wilderness. As usual, if fish folks want things *done*, they've got to do them themselves. The oceanographers will (then) trail along at their leisure and fancy things up.

APPENDIX A

Significant Events in the Evolution of Geophysical Service, Inc. and its Offshoot, Texas Instruments, Inc.[1]

1930 GSI is first independent contractor to offer reflection seismograph prospecting service; places 11 crews in field by end of year. John C. Karcher and Eugene McDermott President and Vice President respectively.

1931 First foreign operation begins in Mexico. Price of crude oil drops to 10 cents per barrel. Many GSI employees, including Cecil H. Green, laid off for a year.

1932 First work in western Canada, as well as first work in Louisiana marsh and swamp land. GSI rotary drilling rig developed.

1934 Newark, New Jersey, laboratory and machine shop, with six employees under the supervision of J. Erik Jonsson, moves to Dallas, Texas. Jonsson then made office manager, personnel director, purchasing agent, and Secretary of the Corporation. Three crews by boat to start first reflection seismograph work ever done in Venezuela.

1937 Seismic parties sent by boat to Saudi Arabia, Java, Sumatra, and Ecuador.

1939 Open-water work in Persian Gulf proves effective. Parties to Panama and India. Because of Karcher's increasing interest in oil production generating a conflict of interest, Coronado Corporation formed with Karcher as President. GSI becomes Coronado subsidiary with McDermott President.

1941 On 6 December (one day before Japanese attack on Pearl Harbor), McDermott, Jonsson, Green, and H. Bates Peacock acquire GSI for $300,000. Green and Peacock become Vice Presidents and Jonsson Secretary-Treasurer. Everette DeGolyer

---

1. Compiled by Robert B. Rice from various sources of information, including GSI's in-house journal, *Grapevine*.

and Karcher sell remaining portion of Coronado to Standard Oil of India for $7 million.

1945 End of hostilities and all military contract work ceases. Fifteen parties in field by late December. Lt. Patrick E. Haggerty, USNR, and Robert W. Olson hired from Navy's Bureau of Aeronautics.

1946 *L*aboratory and *M*anufacturing (L&M) Division set up to supply electronic equipment to the armed services and industrial users. GSI has 554 employees and achieves a net income of $123,000 from revenues of $2.25 million.

1948 Peacock becomes President, while McDermott moves up to Chairman of the Board and devotes full time to research on exploration problems.

1949 Conversion of war surplus "weasels" permit surveying across Canadian muskeg. Sales reached $5.8 million, generating net income of $263,000. 792 employees on board.

1950 Major corporate reorganization. Rapidly increasing military sales, mainly due to Korean War, calls for making L&M Group a separate corporate entity.

1951 In January, L&M Group incorporated as General Instruments with McDermott Chairman of the Board, Jonsson President, Haggerty Executive Vice President, and Olson Chief Engineer. During December, General Instruments renamed Texas Instruments, with GSI as its fully owned subsidiary wherein Peacock is Chairman of the Board, Green is President, Fred J. Agnich Executive Vice President of Operations, Robert C. Dunlap, Jr. Vice President, and Kenneth E. Burg Vice President (Special Problems and Engineering). First annual student cooperative plan inaugurated by Green and Professor R. R. Shrock, Head, Department of Geology and Geophysics, MIT.

1952 Fifty-three crews in the field (26 in the U.S. and 27 outside). Edgar J. Stulken, Chief Seismologist, establishes correspondence training program.

1953 TI acquires stock of all GSI subsidiaries and purchases Engineering Supply Company, Industrial Electronic Supply Company, *H*ouston *T*echnical *L*aboratories (HTL), and Geomarine. Geomarine's M/V SONIC provides one-ship seismic shooting in open waters for first time. TI produces first germanium transistors in commercial quantities. Merger with Intercontinental Rubber Company creates enough stockholders to have TI listed on the New York Stock Exchange. Peacock resigns to do consulting and Jonsson becomes GSI's Chairman of the Board while also serving as TI's President.

1954 Seismic instrument manufacture shifted to HTL under Olson. Seismic operations improved by using 0.5-kg seismometers, the 7000-Amplifier Recording System, and the MagneDISC. Mark K. Smith hired as Research Seismologist from MIT's Geophysical Analysis Group and continues research under Burg on digital computer data-processing methods. Gravity Department formed with Frederick E. Romberg (formerly with LaCoste and Romberg) as Manager and Richard A. Geyer as Chief Geophysicist. HTL begins marketing the famous Worden® gravity meter acquired by purchasing HTL. TI announces first commercial availability of silicon transistors.

1955 Over 60 seismic and gravity parties in the field, with some 1,000 employees located throughout the world. Green moves up to GSI's Chairman of the Board and Agnich becomes President.

1956 Dr. Milo M. Backus, a participant in the Student Cooperative Plan, hired as Research Seismologist from MIT and begins work on methods for overcoming water-layer-induced reverberations. Robert J. Graebner, who began work as a GSI computer during 1949 and moved to research in 1953, continues work on other improved data-processing methods. First seisMAC analog data-processing center installed in Dallas.

1958 Because of rapidly declining U.S. seismic exploration market, oil-exploration activities now comprise only about 20 percent of TI's total company business as compared to 85 percent in 1948.

1959 Dunlap becomes GSI's President, while Agnich becomes Chairman of the Board and also assumes role as a TI Director and Vice President, Geosciences and Instrumentation Division. Haggerty now TI's President and Jonsson Board Chairman. First experimental digital field data taken and processed. Dr. John Gerrard serves on Berkner Panel for Seismic Improvement, U.S. Department of State.

1961 Series 9000 DFS® (Digital Field System) prototype completed. Edward O. Vetter, Industrial Products Group, obtains financial and technical help from Mobil and Texaco for constructing and field testing digital seismic equipment. Industrial Products under J. Fred Bucy, Jr. also delivers first *Texas Instruments Automatic Computer* (TIAC) to GSI, thereby providing the

®indicates registered trademarks of Texas Instruments Inc.

first computer specifically designed for processing seismic data. GSI/TI become major Vela Uniform program contractor with Edward Kinsley, Richard Arnett, Hugh Racketts, Lawrence Strickland, Milo Backus, and John Burg particularly active in this program.

1962 First digital seismic recording crews fielded for Texaco. TI's Geosciences Division (formerly the Geosciences and Instrumentation Division) has name change to Science Services Division and all GSI companies merge into it.

1963 GSI begins use of nondynamite energy sources on land. Many GSI field crews now equipped with digital recording. First transistorized Series 10,000 Digital Field System completed.

1964 Dunlap announces existence of digital seismic technology to the press during June, by which time TIAC systems had been delivered to four processing centers in the U.S. and to one in Calgary, Alberta. Kenneth Burg presents "Total Seismic Exploration Systems and the Role of Digital Technology" to 18 groups in the U.S. and Canada as an SEG Distinguished Lecturer, while E. J. Toomey, Manager of Client Services, does the same via seminars for clients' geophysical staffs in Dallas and several European cities.

1965 GSI now operating over 50 company-owned and leased seismic survey vessels. Green retires as a TI Vice President and made "Honorary Admiral of the GSI Navy".

1966 TI invents the miniature electronic calculator.

1967 TIAC 870 Processing System introduced, including a cathode-ray tube on-line display. GSI pioneers in use of air-guns for marine surveying. TI acquires Geophoto Services. Haggerty becomes TI's Chairman of the Board and Mark Shepherd, Jr. President.

1968 Sixth generation seismic software (the "300 Package") introduced and is the first to perform automatic computer picking of seismic reflections and three-dimensional integration and map contouring.

1969 Vetter becomes GSI's President.

1970 TI invents the single chip microprocessor.

1971 TI invents the single chip microcomputer.

1972 TI unveils its *Advanced Scientific Computer* (ASC) as being a "super-computer with several times the capacity of the most powerful computer now installed in the world". Combining the ASC with the new "700 Package" of seismic software provides the most advanced seismic data-processing capability in existence, including the capability to perform practical three-dimensional data processing.

1973 Graebner assumes GSI Presidency.

1974 TIMAP microcomputer systems introduced for terminal, display and stand-alone seismic data-processing applications.

1975 GSI introduces practical three-dimensional land and marine seismic surveying capability to aid in better prospect evaluation and to reduce unnecessary drilling in reservoir development. This capability utilizes a proprietary technique for determining accurately the location of various points along the hydrophone cable in order to position correctly subsurface reflection points. Associated processing capabilities using the ASC then permit construction of vertical cross-sections along any desired hypothetical traverse line or of any desired series of horizontal time slices which can be viewed successively in the form of a movie. Grant A. Dove becomes GSI's President.

1976 Haggerty retires, Shepherd becomes TI's Chairman of the Board, and Bucy becomes President.

1977 Dolan K. McDaniel becomes GSI's President.

1980 GSI introduces very advanced processing scheme to remove effects of complex source wavelets from land or marine seismic data to produce more accurate resolution of subsurface features. TI's assets are now valued at $1.32 billion, and the total number of employees stands at 89,875 on 31 December. Haggerty dies, age 66, on 1 October after kidney surgery.

1981 Despite continued strong geophysical and government electronic operations, TI's earnings drop 32 percent during the first quarter because of severe price cuts for metal oxide semiconductors and weak demand for such consumer products as calculators and digital watches. Capital spending reduced to $341 million from 1980's record $542 million. Research and development expenditures increased, however, by 16 percent to $219 million, while annual earnings came to only $108.5 million.

References

The Prologue

BOLT, B. A. (1976) *Nuclear Explosions and Earthquakes — The Parted Veil.* W. H. Freeman & Company.

BYERLY, P. (1942) *Seismology.* Prentice-Hall.

DAMPIER, W. C. (1942) *History of Science.* Cambridge University Press.

DEACON, M. (1971) *Scientists and the Sea, 1650—1900. A Study of Marine Science.* Academic Press.

DEACON, M. (1978) *Oceanography — Concepts and History.* Dawson, Hutchinson & Ross.

DOBRIN, M. (1976) *Introduction to Geophysical Prospecting.* McGraw-Hill (3rd edition).

GREENBERG, D. S. (1967) *The Politics of Pure Science.* New American Library.

JAKOSKY, J. J. (1950) *Exploration Geophysics.* Times Mirror Press (2nd edition).

JASTRAM, R. (1981) *The Golden Constant.* John Wiley & Sons.

KELLY, S. F. (Editor)(1940) *Geophysics, 1940. Trans. Amer. Inst. Min. Met. Eng.* 138, 1–489.

MACELWANE, J. B. (1950) *Jesuit Seismological Association, 1925—1950.* Central Station, St. Louis University, St. Louis, Missouri.

PETTY, O. S. (1976) *Seismic Reflections.* (Privately printed by Geosource, Inc., Houston, Texas.)

RAITT, H. and B. MOULTON (1967) *Scripps Institution of Oceanography: First Fifty Years.* Ward Ritchie Press.

RICHTER, C. F. (1958) *Elementary Seismology.* W. H. Freeman & Company.

SCHLEE, S. (1973) *The Edge of an Unfamiliar World — A History of Oceanography.* E. P. Dutton & Company.

SHERIFF, R. E. (1980) *Geophysical Exploration and Interpretation.* International Human Resources Development Corporation, Boston, Massachusetts.

SHOR, E. N. (1978) *Scripps Institution of Oceanography: Probing the Oceans 1936 to 1976.* Tofua Press.

SULLIVAN, W. (1974) *Continents in Motion — the New Earth Debate.* McGraw-Hill Book Company.

SVERDRUP, H. U., M. W. JOHNSON, and R. H. FLEMING (1942) *The Oceans: Their Physics, Chemistry and General Biology.* Prentice-Hall.

SWEET, G. E. (1978) *The History of Geophysical Prospecting* (Volumes 1 and 2). Neville Spearman (3rd edition).

U.S. NATIONAL ACADEMY OF SCIENCES (1970) *Polar Research: A Survey.*

U.S. NATIONAL ACADEMY OF SCIENCES (1976) *Directions for Naval Oceanography.*

U.S. NATIONAL ACADEMY OF SCIENCES (1977) *Trends and Opportunities in Seismology.*

U.S. NATIONAL ACADEMY OF ENGINEERING (1972) *Toward Fulfillment of a National Ocean Commitment.* National Academy of Sciences, Washington, D.C.

WENK, E., Jr. (1972) *The Politics of the Ocean.* University of Washington Press.

Chapter 1

BRUSH, S. G. (1980) Discovery of the Earth's core. *Amer. J. Phys.* **48**, 705—21.

BURSTYN, H. L. and S. B. SCHLEE (1979) The study of ocean currents in America before 1930. (Chapter 13 of *Proc., Bicentennial Conf. on History of Geology,* University Press of New England.)

EWING, M. and J. L. WORZEL (1948) Long range sound transmission. *Memoir 27*, Geol. Soc. Amer.

FAY, H. J. W. (1963) *Submarine Signal Log.* Submarine Signal Division, Raytheon Company, Portsmouth, Rhode Island.

HERBERT-GUSTAR, L. K. and P. A. NOTT (1980) *John Milne: Father of Modern Seismology.* Paul Norbury Publications, Ltd.

LYMAN, J. (1948) The centennial of pressure-pattern navigation, U.S. *Naval Inst. Proc.* **74**, 309—14.

STANTON, W. (1975) *The Great United States Exploring Expedition of 1838—1942.* University of California (Berkeley) Press.

Chapter 2

ALLAUD, L. A. and M. H. MARTIN (1977) *Schlumberger, the History of a Technique.* John Wiley & Sons.

ALLEN, W. E. (1929) Letter of 25 July 1929 to Dr. R. G. Sproul, Vice President, University of California.

ASQUITH, G. B. (1980) *Log Analysis by Mini-Computer.* PennWell Books.

BARTON, D. C. (1938) *Tech. Pub. 950.* Amer. Inst. Min. Met. Eng., Geology Section, 5, 8.

BATES, R. L. (1941) Review of Internal Constitution of the Earth by B. Gutenberg *et al. Bull. Amer. Assoc. Petr. Geol.* **25**, 172—74.

BIGELOW, H. B. (1931) *Oceanography: Its Scope, Problems and Economic Importance.* U.S. National Academy of Sciences.

BILLINGS, M. H. (1940) Geophysical interpretations — report of Houston Geological Society Study Group. *Bull. Amer. Assoc. Petr. Geol.* **24**, 372—73.

BLAU, L. W. (1941) Review of Exploration Geophysics by J. J. Jakosky. *Bull. Amer. Assoc. Petrol. Geol.* **25**, 170—71.

BOLT, B. A. (1979) Perry Byerly — A Memorial. *Bull. Seism. Soc. Amer.* **69**, 928—45.

BULLARD, E. C. and T. F. GASKELL (1941) Submarine seismic investigations. *Proc. Roy. Soc. London* (A), **177**, 476—99.

ECKHARDT, E. A. (1940) A brief history of the gravity method of prospecting for oil. *Geophysics*, **5**, 231—42.

GUTENBERG, B., (Editor) (1939) *Internal Constitution of the Earth.* McGraw-Hill.

GUYOD, H. and L. E. SHANE (1969) *Geophysical Well Logging* (Volume 1). H. Guyod, Houston, Texas.

HEILAND, C. A. (1940) *Geophysical Exploration.* Prentice-Hall.

HILCHIE, D. W. (1979) *Applied Openhole Log Interpretation.* Institutes for Engineering Development, Tulsa, Oklahoma.

HILCHIE, D. W. (1979) *Old Electrical Well Log Interpretation.* Institutes for Engineering Development.

ISELIN, C. O'D. (1959) Environmental Factors Influencing the Performance of Naval Weapon Systems — An Introduction. Lectures sponsored by the Woods Hole Oceanographic Institution, Woods Hole, Massachusetts.

JAKOSKY, J. J. (1940) *Exploration Geophysics.* Times Mirror Press.

JEFFREYS, H. (1924) *The Earth, Its Origin, History and Physical Constitution.* Cambridge Press. (Currently in 6th edition, 1976.)

JOHNSON, C. H. (1938) Locating and detailing fault formations by means of the Geo-Sonograph. *Geophysics,* **3**, 273—94.

JOHNSON, H. M. (1962) A history of well logging. *Geophysics,* **27**, 507—27.

KNOPOFF, L., R. E. HOLZER, and C. F. KENNEL (1979) Memorial to Louis B. Slichter. *Bull. Seism. Soc. Amer.* **69**, 655—57.
LEET, L. D. (1938) *Practical Seismology and Seismic Prospecting.*
MACELWANE, J. B. (1940) Geophysical education in a Department of Geophysics. *Geophysics*, **5**, 80—90.
NETTLETON, L. L. (1940) *Geophysical Prospecting for Oil.* McGraw-Hill.
PIRSON, S. J. (1978) *Geologic Well Log Analysis.* Gulf Publishing Company (2nd edition).
RAITT, H. and B. MOULTON (1967) *Scripps Institution of Oceanography — First Fifty Years.* Ward Ritchie Press.
REVELLE, R. and W. MUNK (1948) Harald Ulrik Sverdrup — an appreciation. *J. Mar. Res.* **7**, 127.
ROSAIRE, E. E. (1940) A perspective of exploration for petroleum. *Geophysics*, **5**, 259—71.
SARGENT, P. (1979) *The Sea Acorn.* Printed privately by Peter Sargent, San Diego, California.
SCHLEE, S. (1973) *The Edge of an Unfamiliar World — A History of Oceanography.* E. P. Dutton & Company.
SCHLEE, S. (1978) *On Almost Any Wind — the Story of the Oceanographic Research Vessel ATLANTIS.* Cornell Univ. Press.
SHERIFF, R. E. and L. P. GELDART (1981) *Exploration Seismology.* Cambridge University Press.
SLOTNICK, M. M. and R. GEYER (1959) *Lessons in Seismic Computing.* Soc. Expl. Geophysicists.
SVERDRUP, H. U., M. W. JOHNSON, and R. FLEMING (1942) *The Oceans.* Prentice-Hall.
SWEET, G. E. (1978) *The History of Geophysical Prospecting* (Volumes 1 and 2). Neville Spearman (3rd edition).
U.S. NATIONAL ACADEMY OF SCIENCES (1930) *Final Report, Committee on Oceanography.* (Mimeographed version only.)
VAUGHAN, T. W. (1937) *International Aspects of Oceanography.* U.S. National Academy of Sciences.

Chapter 3

BATES, C. C. (1949) Utilization of wave forecasting in the invasions of Normandy, Burma and Japan. *Ann. New York Acad. Sci.* **51**, 545—69.
BENNETT, G. (1975) *Naval Battles of World War II.* David McKay.
BORN, W. T. (1941) The future of geophysics. *Geophysics*, **6**, 213—20.
DEACON, G. E. R. (1946) Ocean waves and swell. *Occasional Papers, Challenger Society*, **1**, 1—12.
DEGOLYER, E. (1942) Notes on the present status of problem of exploration. *Bull. Amer. Assoc. Petrol. Geol.* **26**, 1214—20.
EISENHOWER, D. (1948) *Crusade in Europe.* Doubleday & Company.
FIELD, R. M. (1941) Geophysics and world affairs: a plea for geoscience. *Trans. Amer. Geophys. Union*, **22**, 225—34.
GILMORE, M. H. (1947) Tracking ocean storms with the seismograph. *Bull. Amer. Meteor. Soc.* **28**, 73—86.
GOLDSTONE, F. (1943) Maintaining an adequate level of geophysical exploration. *Geophysics*, **8**, 237—43.
GORDON, A. R., Jr. (1979) Personal communication.
GUTENBERG, B. (1946) Interpretation of records obtained from the New Mexico atomic bomb test, 16 July 1945. *Bull. Seis. Soc. Amer.* **36**, 327—29.
JAKOSKY, J. J. (1947) Whither exploration? *Geophysics*, **12**, 361—68.
JOHNSON, E. A. and D. A. KATCHER (1973) *Mines Against Japan.* U.S. Naval Ordnance Laboratory; also Government Printing Office.
MACELWANE, J. B. (1951) Practical application of microseisms to forecasting. *Compendium of Meteorology*, pp. 1312—16. American Meteorological Society.
SVERDRUP, H. U. and W. H. MUNK (1944) *Wind Waves and Swell; Principles in Forecasting* (U.S. Navy Hydrographic Office Misc. 11275).

TATE, J. T. (1946) A survey of subsurface warfare in World War II. *Summary Technical Report of Division 6,* National Defense Research Committee, Office of Scientific Research and Development.

VAJK, R. (1949) Geophysical developments in Europe during the war. *Geophysics,* 14, 101−108.

WILSON, R. E. (1941) Petroleum and the war. *Bull. Amer. Assoc. Petrol. Geol.* 25, 1264−82.

WYCKOFF, R. D. (1948) The Gulf airborne magnetometer. *Geophysics,* 13, 182−208.

Chapter 4

DOBRIN, M. (1976) *Introduction to Geophysical Prospecting.* McGraw-Hill (3rd edition).

ECKHARDT, E. A. (1948) Geophysical activity in the oil industry in the United States in 1947. *Geophysics,* 13, 529−34.

HOLZMAN, B. G. and A. A. CUMBERLEDGE (1946) Weather and the atomic bomb tests at Bikini. *Bull. Amer. Meteor. Soc.* 27, 435−37.

JAKOSKY, J. J. (1950) *Exploration Geophysics.* Times Mirror Press (2nd edition).

JOINT TASK FORCE ONE (1946) *Operation Crossroads − the Official Pictorial Record.* William H. Wise & Company.

KNUDSEN, V. O., A. C. REDFIELD, R. REVELLE, and R. R. SHROCK (1950) Education and training for oceanographers. *Science,* 111, 700−703.

PAUL, C. K. and A. C. MASCARENHAS (1981) Remote sensing in development. *Science,* 214, 139−45.

REFORD, M. S. and J. SUMNER (1964) Review of aeromagnetics. *Geophysics,* 29, 482−516.

RICE, R. B. (1953) New seismic computing method fast and efficient. *World Oil,* 137 (2), 93−98, 104.

RICE, R. B. (1955) Additional notes on the resolved-time computing method. *Geophysics,* 20, 104−22.

VON ARX, W. S. (1962) *An Introduction to Physical Oceanography.* Addison−Wesley Publishing Company.

WEEKS, L. G. (1949) Highlights of 1948 developments in foreign petroleum fields. *Bull. Amer. Assoc. Petrol. Geol.* 33, 1029−1124.

WYCKOFF, R. D. (1944) Geophysics looks forward. *Geophysics,* 9, 287−98.

Chapter 5

AMERICAN GEOLOGICAL INSTITUTE (1951) Policy statement regarding status of geological and geophysical manpower.

BASCOM, W. (1961) *A Hole in the Bottom of the Sea.* Doubleday.

BASCOM, W. (1981) Personal communication.

BATES, C. C. (1958) Current status of sea ice reconnaissance and forecasting for the American Arctic. *Polar Atmosphere Symposium,* AGARDOGRAPH, 29 (I), 285−322. Pergamon Press.

BOLT, B. A. (1976) *Nuclear Explosions and Earthquakes − the Parted Veil.* W. H. Freeman & Company.

CHAPMAN, S. (1959) *IGY − Year of Discovery.* University of Michigan Press.

CRAWFORD, J. M., W. DOTY, and M. R. LEE (1960) Continuous signal seismograph. *Geophysics,* 25, 95−105.

DOBRIN, M. B. and R. G. VAN NOSTRAND (1956) Review of current developments in exploration geophysics. *Geophysics,* 21, 142−55.

DUNBAR, M. (1954) The pattern of ice distribution in Canadian arctic seas. *Trans. Roy. Soc. Canada,* 48, III, 9−18.

ECKHARDT, E. A. (1951) Geophysical activity in 1950. *Geophysics,* 16, 391−400.

ECKHARDT, E. A. (1952) Geophysical activity in 1951. *Geophysics,* 17, 441−51.

FLETCHER, J. O. and L. S. KOENIG (1951) *Floating Ice Islands.* Spec. Rept. 5, 58th Strategic Reconnaissance Squadron (Weather), Eilesen Air Force Base, Alaska.

GASKELL, T. F. (1954) Seismic refraction work by HMS *Challenger* in the deep oceans. *Proc. Royal Soc.* (A), 222, 356—61.

GASKELL, T. F., J. SWALLOW, and G. S. RITCHIE (1953) Further notes on the greatest oceanic sounding and the topography of the Marianas Trench. *Deep Sea Research,* 1, 60—63.

GOULD, L. M. (1981) The last terrestrial frontier. Response upon receiving the Cosmos Club (Washington, D.C.) Award, 14 May 1981.

GREEN, C. H. (1953) A cooperative plan in student education. *Geophysics,* 18, 525—29.

GUTENBERG, B. and C. F. RICHTER (1954) *Seismicity of the Earth.* Princeton University Press (2nd edition).

HALL, H. E. (1945) Use of pressure pattern flying over the North Atlantic. *Bull. Amer. Meteor. Soc.* 26, 37.

HASKELL, N. A. (1959) *The Detection of Nuclear Explosions by Seismic Means.* Geophysics Research Directorate Technical Note 60-632.

HUBBERT, M. K. (1955) Discussion of paper by G. P. Woollard entitled, "An educational program in geophysics". *Geophysics,* 20, 681—82.

JAFFE, G., W. WITTMAN, and C. C. BATES (1954) Radioactive hailstones in the District of Columbia area, 26 May 1953. *Bull. Amer. Meteor. Soc.* 35, 245—49.

JAMES, R. W. (1957) *Application of Wave Forecasts to Marine Navigation.* Special Publication No. 1, U.S. Navy Hydrographic Office.

KEAN, C. H. and F. N. TULLOS (1948) Acoustic impedance logging (abstract). *Oil and Gas J.* 47, 95.

KILLOUGH, J. R. (1953) Petroleum exploration on our public lands. *Geophysics,* 18, 201—11.

KOMONS, N. A. (1966) *Science and the Air Force (A History of the Air Force Office of Scientific Research).* Office of Aerospace Research, Arlington, Virginia.

LEE, O. S. and L. S. SIMPSON (1954) *A Practical Method of Predicting Sea Ice Formation and Growth.* Tech. Rept. 4, U.S. Navy Hydrographic Office.

LIBBY, W. F. (1958) Radioactive fallout. Speech presented at Swiss Academy of Medical Sciences' "Radioactive Fallout Symposium", Lausanne, Switzerland, 27 March 1958.

LILL, G. G. (1971) Oceanographic activities in the Geophysics Branch, Office of Naval Research, 1950—1959. *Proc. Royal Soc. Edinburgh* (B), 72, 253—61.

LINEHAN, D. (1956) A seismic problem in St. Peter's Basilica, Vatican City, Rome, Italy. *Geophysical Case Histories,* 2, 615—22.

LONGWELL, C. R. (1955) In support of the American Geological Institute. *Geophysics,* 20, 683—87.

LOPER, G. B. and PITTMAN, R. R. (1954) Seismic recording on magnetic tape. *Geophysics,* 19, 104—15.

LUDWIG, W. J. and R. E. HOUTZ (1979) Determining crustal structures at sea from airgun-sonobuoy measurements. *Yearbook 1979,* Lamont-Doherty Geological Observatory of Columbia University.

LYONS, P. L. (1951) A seismic reflection quality map of the United States. *Geophysics,* 16, 506—10.

LYONS, P. L. (1956) Crossroads of geophysics. *Geophysics,* 21, 1—11.

MACELWANE, J. B. (1954) Annual survey of geophysical education, 1953—1954. *Geophysics,* 19, 549—64.

MARTELL, E. A. (1959) *Global Fallout and its Variability.* Air Force Cambridge Research Center. *Tech. Rept.* 59-268, Bedford, Massachusetts.

MAYNE, W. H. (1956) U.S. Patent 2,732,906. (Abstracted in *Geophysics,* 21, 856.)

MAYNE, W. H. (1962) Common reflection point horizontal data stacking techniques. *Geophysics,* 27, 927—38.

PETERSON, R. (1960) Investigation of Proposed Use of Large Seismometer Arrays in the U.S.S.R. for the Detection of Underground Explosions. (Testimony presented to Joint Committee on Atomic Energy, U.S. Congress on 21 April 1960, Part I, pp. 318—335.)

RICE, R. B. (1962) Inverse convolution filters. *Geophysics,* 27, 4—18.

RICHARDSON, P. L., J. F. PRICE, W. B. OWENS, W. J. SCHMITZ, H. T. ROSSBY, A. M.

BRADLEY, J. R. VALDES, and D. C. WEBB (1981) North Atlantic Subtropical Gyre: SOFAR floats tracked by moored listening stations. *Science*, 213, 435—37.

ROBINSON, E. (1967) Predictive decomposition of time series with application to seismic exploration. *Geophysics*, 32, 418—84.

ROLL, H. U. (1979) *A Focus for Ocean Research — Intergovernmental Oceanographic Commission's History, Functions, Achievements.* Int. Ocean. Comm. *Tech. Series*, 20, UNESCO.

SHROCK, R. R. (1966) *A Cooperative Plan in Geophysical Education — The GSI Student Cooperative Plan, 1951—1965.* Geophysical Service, Inc., Dallas, Texas.

SHROCK, R. R. (1982) *A History of the Geology Department, Massachusetts Institute of Technology* (in press).

SILVERMAN, D. (1967) The digital processing of seismic data. *Geophysics*, 32, 988—1002.

STERNE, W. P. (1954) New seismic tool. *Oil and Gas J.* 52 (2), 106—107.

SULLIVAN, W. (1961) *Assault on the Unknown — the International Geophysical Year.* McGraw-Hill.

SUMMERS, G. C. and R. A. BRODING (1952) Continuous velocity logging. *Geophysics*, 17, 598—614.

SWALLOW, J. (1955) A neutral-buoyancy float for measuring deep ocean currents. *Deep Sea Research*, 3, 74—81.

TELLER, E. and D. GRIGGS (1956) *Deep Underground Test Shots.* Univ. Calif. Lawrence Radiation Laboratory (Livermore) Report 4659.

U.S. CONGRESS JOINT COMMITTEE ON ATOMIC ENERGY (1959) *Fallout from Nuclear Weapons Tests.* Hearings of 5—8 May 1959 before Sub-Committee on Radiation (pp. 1—2618) (four volumes).

U.S. NATIONAL ACADEMY OF SCIENCES (1959) *Oceanography 1960 to 1970.* Report of Committee on Oceanography.

U.S. STATE DEPARTMENT (1959) *The Need for Fundamental Research in Seismology* (A Report of the Panel on Seismic Improvement, Dr. L. V. Berkner, Chairman).

U.S. STATE DEPARTMENT (1961) *Geneva Conference on the Discontinuance of Nuclear Weapons Tests — History and Analysis of Negotiations.* Department of State Publication 7258 (Disarmament Series 4).

VOGEL, C. B. (1952) A seismic velocity logging method. *Geophysics*, 17, 586—97.

VOSS, E. H. (1963) *Nuclear Ambush — the Test-Ban Trap.* Henry Regnery Company.

WALSH, D. (1980) Personal communication of 31 January 1980.

WEATHERBY, B. B. (1959) An appraisal of the present state of geophysical activity. *Oil and Gas J.* 57 (32), 48—49.

WILSON, J. T. (1961) *IGY — the Year of the New Moons.* Knopf.

WOOLLARD, G. P. (1955) An educational program in geophysics. *Geophysics*, 20, 671—80.

WOOLLARD, G. P. and V. M. GODLEY (1980) *The New Gravity System: Changes in International Gravity Base Values and Anomaly Values.* Hawaii Institute of Geophysics Report 80-1. University of Hawaii.

Chapter 6

AMERICAN ASSOCIATION OF PETROLEUM GEOLOGISTS (1928) *Theory of Continental Drift — A Symposium on the Origin and Movement of Land Masses.* Amer. Assoc. Petroleum Geologists.

BARK, E. V. D. and O. THOMAS (1980) Ekofisk: first of the giant oil fields in western Europe. *Memoir 30, Amer. Assoc. Petrol. Geologists*, pp. 195—224.

BELDERSON, R. H., N. H. KENYON, A. H. STRIDE, and A. R. STUBBS (1972) *Sonographs of the Sea Floor.* Elsevier North-Holland.

BELOUSSOV, V. V. (1963) Upper mantle and its influence on the development of the earth's crust (Upper Mantle Project). Intl. Union Geod. and Geophys. *Circular 2025* of 21/8/63.

BENIOFF, H. (1949) Seismic evidence for the fault origin of oceanic deeps. *Bull. Geol. Soc. Amer.* 60, 1837—56.

BENKERT, W. M. (1965) Statement concerning "Oceanographic Research Vessels Exemption", HR 3419 and HR 7320. Serial No. 89-8 Print for Committee of Merchant Marine and Fisheries, U.S. House of Representatives. (Hearings of 4—5 May 1965.)

BEN-AVRAHAM, Z. (1981) The movement of continents. *Amer. Sci.* 69, 291—99.

BEN-AVRAHAM, Z., A. NUR, D. JONES, and A. COX (1981) Continental accretion: from oceanic plateaus to allochthonous terranes. *Science,* 213, 47—54.

BOLT, B. (1976) *Nuclear Explosions and Earthquakes — The Parted Veil.* W. H. Freeman & Company.

BULLARD, E., J. E. EVERETT, and A. G. SMITH (1965) The fit of the continents around the Atlantic. *Philos. Trans. Royal Soc. London,* Series A, 258, 41—51.

CARPENTER, E. W. (1965) An historical review of seismometer array development. *Proc. Inst. Electrical and Electronic Engineers,* 53, 1816—20.

CLEGG, J. A., M. ALMOND, and P. H. S. STUBBS (1954) The remnant magnetism of some sedimentary rocks in Britain. *Philosophical Magazine* 45, 583—98.

COX, A., R. R. DOELL, and G. B. DALYRUMPLE (1964) Reversals in the earth's magnetic field. *Science,* 144, 1537—43.

DAHLMAN, O. and H. ISRAELSON (1977) *Monitoring Underground Nuclear Explosions.* Elsevier North-Holland.

DICKSON, G. O., W. C. PITMAN III, and J. R. HEIRTZLER (1968) Magnetic anomalies in the South Atlantic and ocean floor spreading. *J. Geophys. Res.* 73, 2087—99.

DIETZ, R. S. (1961) Continent and ocean basin evolution by spreading of the seafloor. *Nature,* 190, 854—57.

DOBRIN, M. (1962) Exploration geophysics: today and tomorrow. *Geophysics,* 27, 109—10.

DRUCKER, P. (1969) *The Age of Discontinuity.* Harper & Row.

GRAHAM, J. W. (1949) The stability and significance of magnetism in sedimentary rocks. *J. Geophys. Res.* 54, 131—67.

HARRIS, G. (1962) How Livermore Lab survived the test ban. *Fortune,* 65, 127 and 236—41.

HEEZEN, B. C. (1960) The rift in the ocean floor. *Sci. Amer.* 203, 98—100.

HEIRTZLER, J. R., X. LE PICHON, and J. G. BARON (1966) Magnetic anomalies over the Reykjanes Ridge. *Deep Sea Research,* 13, 427—43.

HERRIN, E. and J. TAGGART (1962) Regional variations in P_n velocity and their effect on the location of epicenters. *Bull. Seismological Soc. Amer.* 52, 1037—1046.

HESS, H. H. (1962) History of ocean basins. *Petrologic Studies, Geol. Soc. Amer.,* pp. 599—620.

HORNIG, D. (1965) The atmosphere and the nation's future. *Bull. Amer. Meteor. Soc.* 46, 438—42.

ISACKS, B., J. OLIVER, and L. R. SYKES (1968) Seismology and the new global tectonics. *J. Geophys. Res.* 73, 2565—77.

IYER, H. M., L. C. PAKISER, D. J. STUART, and D. H. WARREN (1969) Project EARLY RISE: seismic probing of the upper mantle. *J. Geophys. Res.* 74, 4409—41.

JAMISON, H. C., L. D. BROCKETT, and R. A. McINTOSH (1980) Prudhoe Bay — a 10-year perspective. *Memoir 30, Amer. Assoc. Petrol. Geologists,* pp. 289—314.

JEFFREYS, H. (1924) *The Earth, Its Origin, History and Physical Constitution.* Cambridge Press. (Currently in 6th edition, 1976.)

JOHNSON, J. D. and L. DENNIS FARMER (1971) Use of side-looking air-borne radar for sea ice identification. *J. Geophys. Res.* 76, 2138—55.

KELLY, E. J. (1966) *LASA On-line Detection, Location and Signal-to-Noise Enhancement, Tech. Note 1966-36,* Lincoln Laboratory (Lexington, Massachusetts).

KISSLINGER, C. (1971) Science and public policy — lessons from Vela Uniform. *EOS,* 52, 939 and 967.

LATTER, A. L., R. E. LeLEVIER, E. A. MARTINELLI, and W. G. McMILLAN (1959) *A Method of Concealing Underground Nuclear Explosions. Report 348,* RAND Corporation (Santa Monica, California).

LAUGHTON, A. S. (1981) The first decade of GLORIA. *EOS,* 62,

LEVIN, F., J. F. BAYHI, W. A. GRIEVES, R. J. WATSON and G. M. WEBSTER (1966) Developments in exploration geophysics, 1962—1965. *Geophysics* 31, 320—25.

LEWIS, L. (1967) *One of Our H-Bombs Is Missing.* McGraw-Hill.

LYONS, P. L. (1981) Letter dated 2 August 1981.

MANCHEE, E. and W. D. COOPER (1968) *Operation and Maintenance of the Yellowknife Seismological Array. Seismological Series 1968-2*, Dominion Observatory, Ottawa.

MANSFIELD, R. H. and J. F. EVERNDEN (1966) Long-range seismic data from the Lake Superior seismic experiment, 1963—1964. *Geophys. Monograph* 10, 249—69. American Geophysical Union.

MASON, R. G. (1958) A magnetic survey off the west coast of the United States between latitudes $32°$ and $36°N$, longitude $121°$ and $128°W$. *Geophys. J.* 1, 320—29.

MAXWELL, A. E., R. P. VON HERZEN, K. J. HSÜ, J. E. ANDREWS, T. SAITO, S. F. PERCIVAL, Jr., E. D. MILOW, and R. E. BOYCE (1970) Deep sea drilling in the South Atlantic. *Science*, 168, 1047—59.

MENARD, H. W. (1958) Development of median elevations in the ocean basins. *Bull. Geol. Soc. Amer.* 69, 1179—86.

MOORES, E. M. (1981) Ancient suture zones within continents. *Science*, 213, 41—46.

MORLEY, L. W. and A. LAROCHELLE (1964) Paleomagnetism as a means of dating geological events. *Geochronology in Canada.* Spec. Pub. No. 8, Royal Soc. Canada.

PETERSON, J., H. M. BUTLER, L. G. HOLCOMB, and C. R. HUTT (1976) The seismic research observatory. *Bull. Seismo. Soc. Amer.* 66, 2049—68.

PETERSON, J. and N. A. ORSINI (1976) Seismic research observatories — upgrading the worldwide seismic data network. *EOS*, 57, 548—556.

POMEROY, P. W., G. HADE, J. SAVINO, and R. CHANDER (1969) Preliminary results from high-gain wide-band long-period electromagnetic seismograph systems. *J. Geophys. Res.* 74, 3295—98.

REVELLE, R. and A. E. MAXWELL (1952) Heat flow through the floor of the eastern North Pacific Ocean. *Nature*, 170, 199—200.

RICE, R. B. (1962) Inverse convolution filters. *Geophysics*, 27, 4—18.

ROCARD, Y. (1962) *Le Signal du Sourcier.* Dunod, Paris.

ROCKWELL, D. (1967) The digital computer's role in the enhancement and interpretation of North Sea seismic data. *Geophysics*, 32, 259—81.

RODEAN, H. C. (1971) *Cavity Decoupling of Nuclear Explosions. Report 51097*, Lawrence Radiation Laboratory (Livermore, California).

ROMNEY, C. and C. BEYER (1960) U.S. offers aid to seismographic stations participating in global net using standardized equipment. *Science*, 131, 1877.

ROMNEY, C., B. G. BROOKS, R. H. MANSFIELD, D. S. CARDER, J. N. JORDAN, and D. W. GORDON (1962) Travel times and amplitudes of principal body phases recorded from GNOME. *Bull. Seismological Soc. Amer.* 52, 1057—74.

ROTHÉ, J. P. (1954) La zone séismique médiane Indo-Atlantique. *Proc. Royal. Soc. London*, Series A, 222, 387—397.

RUNCORN, S. K. (1955) Rock magnetism — geophysical aspects. *Advances in Physics*, 4, 244—91.

SABINS, F. F. Jr., R. BLOM, and C. ELACHI (1980) SEASAT radar image of San Andreas Fault, California. *Bull. Amer. Assoc. Petrol. Geol.* 64, 619—28.

SAVIT, C. H. (1960) Preliminary report: a stratigraphic seismogram. *Geophysics*, 25, 312—21.

SHOR, G. C. (1976) Interview of 3 July 1976 with R. A. Calvert (filed Texas A&M University Oral History Collection).

SILVERMAN, D. (1967) The digital processing of seismic data. *Geophysics*, 32, 988—1002.

SPECHT, R. N., A. E. BROWN, J. H. CARLISLE, and C. H. SELMAN (1981) Prudhoe Bay Field — a geophysical case history. *Abstracts, Technical Program, Soc. Expl. Geophys. Annual Meeting*, page 73.

SPRINGER, D., M. DENNY, J. HEALY, and W. MICKEY (1968) The STERLING experiment: decoupling of seismic waves by a shot-generated cavity. *J. Geophys. Res.* 73, 5995—6011.

STEINHART, J. S. and T. J. SMITH (1966) Preface to *The Earth Beneath the Continents*, *Geophys. Monograph 10*, American Geophysical Union.

SUNDSTROM, O. J. (1970) The Thule affair. *USAF Nuclear Safety, 65 (Part 2),* 5–22.
SZULC, T. (1967) *The Bombs of Palomares.* Viking Press.
TAYLOR, D. (1974) Understanding bright spot. *Ocean Industry,* 9.
TELLER, E. (1979) *Energy from Heaven and Earth.* W. H. Freeman.
U.S. COMMISSION ON MARINE SCIENCE, ENGINEERING AND RESOURCES (1969) *Our Nation and the Sea: A Plan for National Action* (4 volumes). U.S. Government Printing Office.
U.S. CONGRESS (1971) Status of Current Technology to Identify Seismic Events as Natural or Man Made. Hearings of 27–28 October 1971 before Subcommittee on Research, Development, and Radiation, Joint Committee on Atomic Energy.
U.S. COUNCIL ON MARINE RESOURCES AND ENGINEERING DEVELOPMENT (1967) *United States Activities in Spacecraft Oceanography.* Government Printing Office.
U.S. NAVY (1967) *Lessons and Implications for the Navy (Aircraft Salvage Operation, Mediterranean),* CNO Technical Advisory Group (Chaired by RADM L. V. Swanson, USN), Department of the Navy.
VAN MELLE, F. A. (1964) Editorial note concerning Vela Uniform special issue, *Geophysics,* 29, 299–300.
VAN NOSTRAND, R. and W. HELTERBRAN (1964) *A Comparative Study of the SHOAL Event.* United ElectroDynamics Seismic Data Laboratory Report 109 prepared for the Air Force Technical Applications Center, Alexandria, Virginia.
VINE, F. J. and D. H. MATTHEWS (1963) Magnetic anomalies over ocean ridges. *Nature,* 199, 947–49.
VINE, F. J. and J. T. WILSON (1965) Magnetic anomalies over a young oceanic ridge off Vancouver Island. *Science,* 150, 485–89.
WARME, J. E., R. G. DOUGLAS, and E. L. WINTERER (Editors)(1981) *The Deep Sea Drilling Project; A Decade of Progress. Spec. Pub. 32,* Soc. Econ. Paleon. and Mineralogists.
WEGENER, A. (1929) *The Origin of Continents and Oceans* (4th edition). (English translation, 1966, Methuen.)
WENK, E., Jr. (1972) *The Politics of the Ocean.* University of Washington Press.
WILSON, J. T. (1963) Evidence from islands on the spreading of ocean floors. *Nature,* 198, 536–38.
WILSON, J. T. (1965) A new class of faults and their bearing on continental drift. *Nature,* 207, 343–47.
WILSON, J. T. (1968) Static or mobile earth: the current scientific revolution. *Proc. Amer. Phil. Soc.* 112, 309–20.
WISEMAN, J. D. H. and R. B. S. SEWELL (1937) The floor of the Arabian Sea. *Geol. Magazine,* 74, 219–30.

Chapter 7

BARRY, K. (1982) Geophysics is alive and well. *Geophysics,* 47 (in press).
BRYSON, R. A. and C. PADOCH (1981) On the climates of history. *Climate and History* (Chap. 1). Princeton University Press.
COOK, R. J., J. C. BARRON, R. I. PAPENDICK, and G. J. WILLIAMS III (1981) Impact on agriculture of the Mount St. Helens' eruptions. *Science,* 211, 16–22.
CRANDELL, D. R., D. R. MULLINEAUX, and M. RUBIN (1975) Mount St. Helens Volcano, recent and future behavior. *Science,* 187, 438–441.
DOMENICO, N. (1974) Effect of water saturation on seismic reflectivity of sand reservoirs encased in shale, *Geophysics,* 39, 759–69.
DONN, W. L. and N. B. BALACHANDRAN (1981) Mount St. Helens' eruption of 18 May 1980: air waves and explosive yield. *Science,* 213, 539–41.
EOS (1980) Australian microfossils may be oldest known organisms. *EOS,* 61, 578.
FLOWERS, B. S. (1976) An overview of exploration geophysics — recent breakthroughs in geophysics and recognition of challenging new problems. *Bull. Amer. Assoc. Petrol. Geol.* 60, 3.

FOUGERE, P. F. and C. W. TSACOYEANES (1980) AFGL magnetometer observations of Mount St. Helens' eruption. *EOS*, 61, 1209.

HOOPER, P. F. (1979) The Administrative Policy Process for Science: *A Case Study of Organizational—Environmental Dynamics*, WHOI-79-69, Woods Hole Oceanographic Institution.

HUBBERT, M. K. (1950) Energy from fossil fuels. *Proc. Centennial Celebration*, Amer. Assoc. Adv. Science, pp. 171—77.

LAMB, H. H. (1977) *Climatic History and the Future*. Barnes & Noble.

LINDSETH, R. (1979) Synthetic sonic logs — a process for stratigraphic interpretation. *Geophysics*, 44, 3—26.

LOWMAN, P. (1981) *A Global Tectonic Activity Map with Orbital Photography Supplement*. *NASA Tech. Memo*. 82073, Goddard Space Flight Center.

LOWMAN, P. and H. V. FREY (1979) *A Geophysical Atlas for Interpretation of Satellite Derived Data*. *NASA Tech. Memo*. 79722, Goddard Space Flight Center.

MASON, B. J. (1976) Towards the understanding and prediction of climatic variations. *Qtrly J. Royal Met. Soc*. 102, 473—98.

McKELVEY, V. E. (1980) Seabed minerals and the law of the sea. *Science*, 209, 464—72.

NAMIAS, J. (1980) Some concomitant regional anomalies associated with hemispherically averaged temperature variations. *J. Geophys. Res*. 85, 1585—1590.

PRESS, F. (1981) Science and technology in the White House, 1977 to 1980. *Science*, 211, 139—45 and 249—56.

PRESS, F. and R. SIEVER (1974) *Earth*. W. H. Freeman & Company.

RICE, C. J. and D. K. WATSON (1981) Satellite observations of Mt. St. Helens. *EOS*, 62, 577.

RICE, R. B. (1976) Today's geophysical dilemmas — and challenges. *Proc. Southwestern Legal Foundation, Exploration and Economics of the Petroleum Industry* (Volume 14).

RICE, R. B., S. J. ALLEN, O. J. GANT, Jr., R. N. HODGSON, D. E. LARSON, J. P. LINDSEY, J. R. PATCH, T. R. LaFEHR, G. R. PICKETT, W. A. SCHNEIDER, J. E. WHITE, and J. C. ROBERTS (1981) Developments in exploration geophysics, 1975—1980. *Geophysics*, 46, 1088—99.

RIFKIN, J. and T. HOWARD (1980) *Entropy: A New World View*. Viking Press.

RITSEMA, A. R. (1980) Observations of St. Helens' eruption. *EOS*, 61, 1201.

ROBINSON, R. and H. M. IYER (1981) Delineation of a low-velocity body under the Roosevelt Hot Springs geothermal area, Utah, using teleseismic P-wave data. *Geophysics*, 46, 1456—66.

SAVIT, C. H. (1960) Preliminary report concerning a stratigraphic seismogram. *Geophysics*, 25, 312—21.

SAVIT, C. H. (1974) Bright spot in the energy picture. *Ocean Industry*, 9.

SAVIT, C. H. (1980) Geophysics yesterday, today and tomorrow. (Revision from talk originally given to 25th Meeting, Assoc. Reciprocal Assistance among Latin American State Oil Companies, Santa Cruz, Bolivia, 27 July 1976.)

SIEMS, L. and F. HEFER (1967) A discussion on seismic binary gain switching amplifiers. *Geophys. Prosp*. 15, 23—34.

TARANIK, J. V. (1981) Global geophysical investigations with satellite-acquired data. Abstract within technical program of 51st Annual Meeting of Society of Exploration Geophysicists.

TAYLOR, S. R. (1975) *Lunar Science — a Post Apollo View*. Pergamon Press.

U.S. COUNCIL ON ENVIRONMENTAL QUALITY (1980) *Environmental Quality — 1980*. U.S. Government Printing Office.

U.S. COUNCIL ON ENVIRONMENTAL QUALITY (1981) *Global Energy Futures and the Carbon Dioxide Problem*. U.S. Government Printing Office.

U.S. NATIONAL ACADEMY OF SCIENCE (COMMITTEE ON OCEANOGRAPHY) (1969) *An Oceanic Quest — The International Decade of Ocean Exploration*.

U.S. NATIONAL ACADEMY OF SCIENCE (1979) *The Continuing Quest — Large-scale Ocean Science for the Future*.

U.S. OFFICE OF TECHNOLOGY ASSESSMENT (1980) *Ocean margin drilling — a technical memorandum*. U.S. Government Printing Office.

VAIL, P. R., J. HARDENBOL, R. M. MITCHUM, Jr., S. THOMPSON, III, R. G. TODD, J. B. SANGREE, J. M. WIDMIER, J. N. BUBB, and W. G. HATLELID (1977) Seismic stratigraphy and global changes in sea level. *Amer. Assoc. Petrol. Geologists, Memoir* 26, 49—212.
WARD, S. H., W. R. SILL, D. S. CHAPMAN, J. R. BOWMAN, W. T. PARRY, K. L. COOK, F. H. BROWN, W. P. NASH, R. B. SMITH, and J. A. WHELAN (1978) A summary of the geology, geochemistry and geophysics of the Roosevelt Hot Springs thermal area, Utah. *Geophysics*, 43, 1515—42.
WENK, E., Jr. (1972) *The Politics of the Ocean.* University of Washington Press.
WHITHAM, K. (1972) *Bibliography of DEMR papers relevant to seismological verification problems.* Mimeographed report. Department of Energy, Mines and Resources, Canada.

Chapter 8

CARLILE, R. E. (1981) Forecasting worldwide seismic activity, 1980—1984. *World Oil,* 192 (3), 36—56.
DARDEN, Mrs. D. B. (1980) Reminiscences. *Geophysics,* 45, 1720—23.
DOBRIN, M. B., A. L. INGALLS, and J. A. LONG (1965) Velocity and frequency filtering of seismic data using laser light. *Geophysics,* 30, 1144—78.
GRAEBNER, R. J., G. STEEL, and C. B. WASON (1980) Evolution of seismic technology into the 1980's. *A.P.E.A. J.* 20 (Pt. 1), 110—20.
HAGGERTY, P. E. (1965) *Management Philosophies and Practices of Texas Instruments, Inc.* Texas Instruments, Dallas, Texas.
HALBOUTY, M. T. (1981) New frontiers where future world oil and gas will be found. Presentation to Ann. Mtg, Soc. Expl. Geophysicists.
HAMMER, S. (1975) The energy crisis and geophysics — a challenge. *Oil and Gas J.* 73, Feb. 3, 140, 145—46.
HAMMER, S. and R. ANZOLEAGA (1975) Exploring for stratigraphic traps with gravity gradients. *Geophysics,* 40, 256—68.
HEINRICHS, W. E., Jr. and R. E. THURMOND (1956) A case-history of the geophysical discovery of the Pima Mine, Pima County, Arizona. *Geophysical Case Histories,* Vol. 2, 600—14, Soc. Expl. Geophys.
LaCOSTE, J. B. L. (1934) A new type long period vertical seismograph. *Physics,* 5, 178.
LAYAT, C., A. C. CLEMENT, and G. POMMIER (1961) Some technical aspects of refraction seismic prospecting in the Sahara. *Geophysics,* 26, 437—46.
LOEWENTHAL, D., L. LU, R. ROBERSON, and J. SHERWOOD (1976) The wave-equation applied to migration. *Geophys. Prosp.* 24, 380—99.
McCOLLUM, E. V. (1941) Water prospecting with the gravity meter. *World Petroleum,* June, p. 74.
McCOLLUM, E. V. and A. BROWN (1943) Use of the gravity-meter in establishment of gravity bench marks. *Geophysics,* 8, 379—90.
MANSFIELD, R. H. (1980) Personal communication.
MARR, J. D. (1971) Seismic stratigraphic exploration, Parts I, II and III. *Geophysics,* 36, 311—29, 533—53, 676—89.
MEISSNER, R. and M. A. HEGAZY (1981) The ratio of the PP- to the SS-reflection coefficient as a possible future method to estimate oil and gas reservoirs. *Geophys. Prosp.* 29, 533—40.
NATIONAL PETROLEUM COUNCIL (1979) *Materials and Manpower Requirements for U.S. Oil and Gas Exploration and Production — 1979—1990.* Chapter two: geological and geophysical services, Washington, D.C.
NEIDELL, N. S. and M. TURHAN TANER (1971) Semblance and other coherency measures for multichannel data. *Geophysics,* 36, 482—97.
PEMBERTON, R. H. (1962) Airborne electromagnetics in review. *Geophysics,* 27, 691—713.
PETERSON, R. A., W. R. FILLIPPONE, and F. B. COKER (1955) The synthesis of seismograms from well log data. *Geophysics,* 20, 516—38.
PETTY, O. SCOTT (1976) *Seismic Reflections,* Geosource Inc., Houston, Texas.

PIEUCHOT, M. and H. RICHARD (1958) Some technical aspects of reflection seismic prospecting in the Sahara. *Geophysics*, 23, 557–73.

REFORD, M. S. and J. SUMNER (1964) Review article: aeromagnetics. *Geophysics*, 29, 482–516.

RICE, R. B. (1976) A bond of achievement. *Geophysical Prospecting*, 24, 418–22.

RICE, R. B., S. J. ALLEN, O. J. GANT, Jr., R. N. HODGSON, D. E. LARSON, J. P. LINDSEY, J. R. PATCH, T. R. LaFEHR, G. R. PICKETT, WM. A. SCHNEIDER, J. E. WHITE, and J. C. ROBERTS (1981) Developments in exploration geophysics, 1975–1980. *Geophysics*, 46, 1088–1099.

SAVIT, C. H. and L. E. SIEMS (1977) A 500-channel streamer system, 9th Annual Offshore Technology Conference, Houston, Texas, 2–5 May 1977.

SCHNEIDER, W. A. and M. M. BACKUS (1968) Dynamic correlation analysis. *Geophysics*, 33, 105–26.

SENTI, R. J. (1981) Geophysical activity in 1980. *Geophysics*, 46, 1316–33.

SHERIFF, R. E. (1980) Advertisements in *Geophysics* – a history, paper presented at the 50th Annual International Meeting of the SEG, Houston, Texas, 16–20 November 1980.

SWEET, G. E. (1966, 1969) *The History of Geophysical Prospecting*, Vols. 1 and 2. Science Press, Los Angeles, Calif.

TANER, M. TURHAN and F. KOEHLER (1969) Velocity spectra – digital computer derivation and applications of velocity functions. *Geophysics*, 34, 859–81.

TANER, M. TURHAN, F. KOEHLER and K. A. ALHILALI (1974) Estimation and correction of near-surface time anomalies. *Geophysics*, 39, 441–63.

TANER, M. T., F. KOEHLER, and R. E. SHERIFF (1979) Complex seismic trace analysis. *Geophysics*, 44, 1041–63.

Chapter 9

BROECKER, W. (1979) Response upon receiving the Maurice Ewing Medal. *EOS*, 60, 629.

WERTENBAKER, W. (1974) *The Floor of the Sea, Maurice Ewing and the Search to Understand the Earth.* Little, Brown & Company.

Name Index

Subject Index

About the Authors

Charles C. Bates was born in northern Illinois. After graduating *cum laude* from De Pauw University (B.S., geology), he joined a seismic field party of the Carter Oil Company. Upon being drafted into the army prior to World War II, he rose to the rank of Captain, U.S. Army Air Force, while simultaneously earning a master's degree in physics—meteorology from the University of California at Los Angeles. After the war, he eventually obtained the first doctoral degree awarded in oceanography by Texas A&M University (1953), received the President's Award of the American Association of Petroleum Geologists during the same year for research pertaining to delta formation, and became involved in many applications of the geophysical sciences. Positions held in this regard included those of Environmental Systems Coordinator, Office of the Deputy Chief of Naval Operations (Development) (1957–1960); Chief, Vela Uniform Branch, Nuclear Test Detection Office, Office of the Secretary of Defense (1960–1964); Scientific and Technical Director, U.S. Naval Oceanographic Office (1965–1968), and Science Advisor to the Commandant, U.S. Coast Guard (1968–1979). He also served as Vice President of the Society of Exploration Geophysicists (1964–1965) and was named an Honorary Member of that society in 1981.

Thomas F. Gaskell was born in Lancashire and obtained first-class Honours in Physics, Chemistry, and Mineralogy when he graduated from Cambridge University in 1937. He received his Ph.D. at the same school during 1940 after specializing in terrestrial and marine seismic investigations. During the initial phase of World War II, he worked with Dr. (later Sir) Edward Bullard in the solution of mine warfare problems. Then, during 1944 and 1945, he served as a specialist in beach intelligence and bombardment with Headquarters, Combined Operations in London and with the Southeast Asia Command in Kandy, Sri Lanka. Between 1946 and 1949, he was Chief Petroleum Physicist in the oilfields of Iran for Anglo-Iranian Petroleum, Ltd., and followed this by serving as Chief Scientist of the global oceanographic expedition aboard HMS *Challenger* between 1950 and 1952. After returning to what had become British Petroleum, Ltd., Gaskell worked in research and then became Scientific Adviser to the Information Department. While in the latter role, he frequently appeared on radio and television relative to topics involving the earth sciences. In addition to editing the *Geophysical Journal* of the Royal Astronomical Society and *The Earth's Mantle*, he has written several popular science books, including *Under The Deep Oceans, The Adventure of North Sea Oil* (with B. Cooper), *Using the Oceans, Physics of the Earth, The Gulf Stream,* and *World Climate* (with Martin Morris).

Robert B. Rice was born in a major petroleum center, Bartlesville, Oklahoma, and has been closely associated with the petroleum industry throughout his lifetime. After earning a bachelor's degree in physics and mathematics *cum laude* from the College of Wooster in 1941, he simultaneously did graduate work in mathematics and taught in the same field at Ohio State University during the war years. In 1946, he became a research geophysicist with the Geophysical Research Department of Phillips Petroleum in Bartlesville, where he made significant advances in seismic data processing. After joining the Denver Research Center of the Marathon Oil Company in 1956, four years later he became Manager, Physics and Mathematics Department and was responsible for research and technical service work on geophysical data processing and interpretation. He became an independent geophysical consultant in 1978. During 1961–1963, Rice served as Editor of the journal, *Geophysics*, and in 1975–1976, he was President of the Society of Exploration Geophysicists.